AF179552

Elements of Mammalian Fertilization

Volume I
Basic Concepts

Editor

Paul M. Wassarman

Department of Cell and Developmental Biology
Roche Institute of Molecular Biology
Roche Research Center
Nutley, New Jersey

CRC Press
Boca Raton Ann Arbor Boston

Library of Congress Cataloging-in-Publication Data

Elements of mammalian fertilization/editor, Paul M. Wassarman.
p. cm.
Includes bibliographical references and index.
Contents: v. 1. Basic concepts -- v. 2. Practical applications.
ISBN 0-8493-6386-1. -- ISBN 0-8493-6387-X
1. Fertilization (Biology) 2. Fertilization in vitro.
I. Wassarman, Paul M.
QP273.E44 1991
599′.0166--dc20 90-2299
CIP

This book represents information obtained from authentic and highly regarded sources. Reprinted material is quoted with permission, and sources are indicated. A wide variety of references are listed. Every reasonable effort has been made to give reliable data and information, but the author and the publisher cannot assume responsibility for the validity of all materials or for the consequences of their use.

All rights reserved. This book, or any parts thereof, may not be reproduced in any form without written consent from the publisher.

Direct all inquiries to CRC Press, Inc., 2000 Corporate Blvd., N.W., Boca Raton, Florida 33431.

© 1991 by CRC Press, Inc.

International Standard Book Number 0-8493-6386-1 (Vol. I)
International Standard Book Number 0-8493-6387-X (Vol. II)

Library of Congress Card Number 90-2299
Printed in the United States

INTRODUCTION

Fertilization of eggs by sperm, the means by which sexual reproduction takes place in nearly all multicellular organisms, is fundamental to maintenance of life. Consequently, fertilization traditionally has been of great interest to biologists and fertilization research has a long and rich history. From theoretical considerations of its evolutionary origins to experimental analysis of its molecular basis, today the subject of fertilization continues to pique the curiosity of scientists and remains a major area for both basic and applied research.

Mammalian fertilization has been studied in the laboratory for more than a century. However, only in the past 30 years or so have the many steps that lead to the union of male and female mammalian gametes been examined in some detail. This detailed examination essentially began in the mid-to-late 1950s with the introduction of experimental protocols that supported mammalian fertilization *in vitro*. Today, *in vitro* fertilization has become almost commonplace and, as frequently reported in the popular press, has dramatically affected both agriculture and medicine. In a world in which test tube babies and genetically engineered domestic animals have become realities, research on fertilization has taken on a new sense of importance, relevance, and excitement. In some quarters, such research has even been viewed with alarm. Certainly, fertilization research is an area of investigation that already has profoundly affected our lives and promises to continue to do so in the future.

The term fertilization itself refers specifically to the fusion of male and female gametes to form a zygote. However, the overall "pathway" to fertilization encompasses a host of events that precede gamete fusion. In this context, *Elements of Mammalian Fertilization* considers a wide range of critical events that ultimately lead to formation of a zygote and, wherever possible, these events are described at the cellular and molecular level (Volume I, *Basic Concepts*). *Elements of Mammalian Fertilization* also considers some practical consequences of our expanding knowledge of mammalian fertilization, with particular emphasis placed on human *in vitro* fertilization (Volume II, *Practical Applications*).

Volume I of this treatise begins at the beginning of the fertilization pathway with development of eggs and sperm (gametogenesis) and progresses to the point at which gametes are rendered capable of producing a viable zygote (Chapters 1 to 5). This includes the final "maturation stages" for both eggs ("meiotic maturation"; Chapter 4) and sperm ("capacitation"; Chapter 5). A description of specific macromolecules involved in gamete adhesion (Chapter 6) is then followed by discussions of two exocytotic events, the acrosome reaction of sperm (Chapter 7) and cortical reaction of eggs (Chapter 8). A discussion of the characteristics and possible roles of a sperm enzyme, acrosin (Chapter 9), precedes the final two chapters that deal with morphological and ultrastructural aspects of the fertilization process (Chapters 10 and 11). Overall, this volume provides an up-to-date assessment of the status of basic research on some major events of the mammalian fertilization pathway.

Volume II of this treatise begins with a discussion of the origins and history of human *in vitro* fertilization (Chapter 1) and is followed by three chapters that discuss some of the technical considerations and diverse factors that often determine the success or failure of human *in vitro* fertilization (Chapters 2, 3, and 4). The latter themes are continued in a chapter dealing with the application of *in vitro* fertilization to domestic animals (Chapter 5). The final two chapters consider the application of gamete biochemistry and molecular biology toward development of novel methods of contraception (Chapters 6 and 7). Overall, this volume highlights areas of current excitement in *in vitro* fertilization technology and provides insight as to some future directions for application of basic research in mammalian fertilization.

The process of mammalian fertilization includes events that represent some of the major themes currently under investigation in cellular biology (e.g., cell differentiation, cell-cell interactions, signal transduction, and membrane fusion), and advances in mammalian fertilization research have had extremely important practical consequences in both medicine and agriculture. Thus, *Elements of Mammalian Fertilization* was designed with a relatively diverse audience in mind. Hopefully, it will prove to be of interest and use to both graduate and medical students, as well as to research scientists and clinicians.

Finally, I am delighted to extend my thanks to all of the authors for their fine contributions, as well as for their cooperation and patience during preparation of this treatise. I also extend a very special thanks to my assistant, Mrs. Alice O'Connor, who brought her considerable organizational skills and doggedness to bear on a sometimes frustrating and chaotic endeavor. Without her assistance and advice, my job would have been considerably more difficult and the final product less useful.

THE EDITOR

Paul M. Wassarman, Ph.D., is the Chairman of Department of Cell and Developmental Biology at the Roche Institute of Molecular Biology, Roche Research Center, Nutley, New Jersey, and Adjunct Professor of Cell Biology, New York University School of Medicine, New York City, New York.

Dr. Wassarman received B.S. (1961) and M.S. (1964) degrees from the University of Massachusetts, Amherst, Massachusetts, and a Ph.D. (1968) in Biochemistry from Brandeis University, Waltham, Massachusetts. He was a postdoctoral fellow (1967 to 1969) at the Medical Research Council, Laboratory of Molecular Biology, Cambridge, England, prior to joining the staff of the Department of Biological Sciences, Purdue University, West Lafayette, Indiana (1969 to 1971). From 1972 to 1985, Dr. Wassarman was a member of the staff of the Department of Biological Chemistry, Harvard Medical School, Boston, Massachusetts. He moved to the Roche Institute of Molecular Biology in 1985 and assumed his present position as Chairman of the Department of Cell and Developmental Biology in 1986.

Dr. Wassarman has been the recipient of research fellowships from the Helen Hay Whitney Foundation (1967 to 1969), Rockefeller Foundation (1972 to 1974), and National Science Foundation (1971). His research at Harvard Medical School was supported in part by the National Institutes of Health, National Science Foundation, American Cancer Society, and Rockefeller Foundation. Dr. Wassarman has served in an advisory capacity to the National Institutes of Health, National Science Foundation, and various other governmental agencies. He is currently a member of several professional societies, serves on the editorial boards of several professional journals, and is editor of two research annuals that address topics in developmental biology. During the past 15 years or so, Dr. Wassarman's laboratory has published extensively on cellular and molecular aspects of mammalian gametogenesis, fertilization, and early embryogenesis.

CONTRIBUTORS

Donner F. Babcock, Ph.D
Research Associate Professor
Department of Biochemistry
University of Washington
Seattle, Washington

Jeffrey D. Bleil, Ph.D
Assistant Member
Molecular Biology Department
Research Institute of Scripps Clinic
La Jolla, California

Judith Cebra-Thomas, Ph.D.
Research Staff Member
Department of Molecular Biology
Princeton University
Princeton, New Jersey

Erma Z. Drobnis, Ph.D.
Assistant Research Zoologist
Department of Zoology
University of California Davis
Davis, California

Thomas Ducibella, Ph.D.
Associate Professor
Department of Obstetrics and Gynecology
Tufts University School of Medicine
Boston, Massachusetts

E. M. Eddy, Ph.D.
Head, Gamete Biology Section
Laboratory of Reproductive and Developmental Toxicology
National Institute of Environmental Sciences, National Institutes of Health
Research Triangle Park, North Carolina

John J. Eppig, Ph.D.
Senior Staff Scientist
The Jackson Laboratory
Bar Harbor, Maine

Harvey M. Florman, Ph.D.
Staff Scientist
The Worcester Foundation for Experimental Biology
Shrewsbury, Massachusetts

George L. Gerton, Ph.D.
Associate Professor of Obstetrics and Gynecology and Anatomy
Department of Obstetrics and Gynecology
Division of Reproductive Biology
University of Pennsylvania Medical Center
Philadelphia, Pennsylvania

David F. Katz, Ph.D.
Professor
Department of Obstetrics and Gynecology
University of California Davis
Davis, California

Gregory S. Kopf, Ph.D.
Associate Professor of Obstetrics and Gynecology and Biology
Department of Obstetrics and Gynecology
Division of Reproductive Biology
University of Pennsylvania Medical Center
Philadelphia, Pennsylvania

Deborah A. O'Brien, Ph.D.
Research Associate Professor
Laboratories for Reproductive Biology
Departments of Pediatrics and Cell Biology and Anatomy
University of North Carolina at Chapel Hill
Chapel Hill, North Carolina

David M. Phillips, Ph.D.
Senior Scientist
The Population Council
New York, New York

Richard M. Schultz, Ph.D.
Professor
Department of Biology
University of Pennsylvania
Philadelphia, Pennsylvania

Lee M. Silver, Ph.D.
Associate Professor
Department of Molecular Biology
Princeton University
Princeton, New Jersey

Umbert A. Urch, Ph.D.
Assistant Research Biochemist
Department of Biochemistry and Biophysics
University of California
Davis, California

Paul M. Wassarman, Ph.D.
Chairman
Department of Cell and Developmental Biology
Roche Institute of Molecular Biology
Roche Research Center
Nutley, New Jersey

Jeffrey E. Welch, Ph.D.
Staff Fellow
Laboratory of Reproductive and Developmental Toxicology
National Institute of Environmental Sciences, National Institutes of Health
Research Triangle Park, North Carolina

TABLE OF CONTENTS

Volume I

TABLE OF CONTENTS

Volume II

Chapter 1

MAMMALIAN SPERM DEVELOPMENT *IN VIVO* AND *IN VITRO*

E. M. Eddy, D. A. O'Brien, and J. E. Welch

TABLE OF CONTENTS

I. THE PROCESS OF SPERMATOGENESIS

The highly specialized spermatozoon is the end result of the unique developmental process of spermatogenesis. The novel structural and functional features of the spermatozoon[1] give it the ability to carry out fertilization, thereby delivering the paternal DNA to the egg and triggering initiation of development. Spermatogenesis occurs in three distinct phases, a mitotic phase, a meiotic phase, and a postmeiotic phase. During the mitotic phase, spermatogonial stem cells give rise to proliferating spermatogonia. These cells undergo a series of doublings which result in a substantial expansion in the number of germ cells that will eventually give rise to spermatozoa. In the meiotic phase, spermatocytes accomplish the unique events of chromosomal synapsis and genetic recombination, and the two reductional divisions of meiosis. As a result of the reductional divisions, spermatozoa contain a single copy of each gene, while the result of recombination is that a particular gene may have come from either parent. During the postmeiotic phase, the spermatid is dramatically remodeled into a spermatozoon. The acrosome and the flagellum are formed and previously existing structures are reshaped and redistributed to produce the male gamete (Figure 1).

The process of spermatogenesis in mammals takes place within the seminiferous epithelium lining the seminiferous tubules. Using the rat as an example, about 30 seminiferous tubules make up the testis and each tubule is between 200 to 250 μm in diameter. The tubules are surrounded by a peritubular wall, in most species consisting of two or three layers of peritubular myoid cells enclosed by collagen and other extracellular matrix materials.[2]

The seminiferous epithelium is composed solely of spermatogenic cells and Sertoli cells. The Sertoli cells envelop and support the germinal cells, providing the framework and environment for their development in the seminiferous epithelium.[3] Each Sertoli cell stretches across the full height of the epithelium, with its basal surface sitting against the peritubular wall of the seminiferous tubule and its apical surface facing the lumen. Specialized tight junctions between Sertoli cells separate the seminiferous epithelium into two compartments, the basal and adluminal compartments.[4] The basal compartment contains spermatogonia and early spermatocytes, while the adluminal compartment contains later-stage spermatocytes and spermatids. Germ cells in the basal compartment sit between the peritubular wall peripherally and the Sertoli cells centrally, while germ cells in the adluminal compartment are completely surrounded by Sertoli cells until the spermatid flagellum forms and extends into the lumen of the seminiferous tubule. Germ cells in the adluminal compartment are isolated from the immune system and differentiate in a fluid environment of unusual ionic and molecular composition.[5]

Although spermatogenesis encompasses three phases of germ cell development—from spermatogonia to spermatocyte to spermatid—these phases do not occur independently of one another. Spermatogenic cells in one particular developmental phase are always found in association with four or five cohorts of spermatogenic cells in specific phases of development. These constant cell associations have been classified into 12 different stages in the mouse.[6] The stages have precise kinetics and altogether make up one cycle of spermatogenesis. The duration of spermatogenesis, the time that it takes for development to proceed from a spermatogonial stem cell to the end of spermatozoon formation, is 34.6 days in the mouse. The duration of a cycle, the time required for the 12 stages of spermatogenesis to occur in a particular segment of a seminiferous tubule, is 8.6 days in the mouse.[7] In the human, there are six cell associations which correspond to stages of spermatogenesis. However, the patterns of differentiation are not synchronized and there is neither a regular association of stages in a tubule cross section nor a seminiferous cycle in humans like that found in rodents.[8]

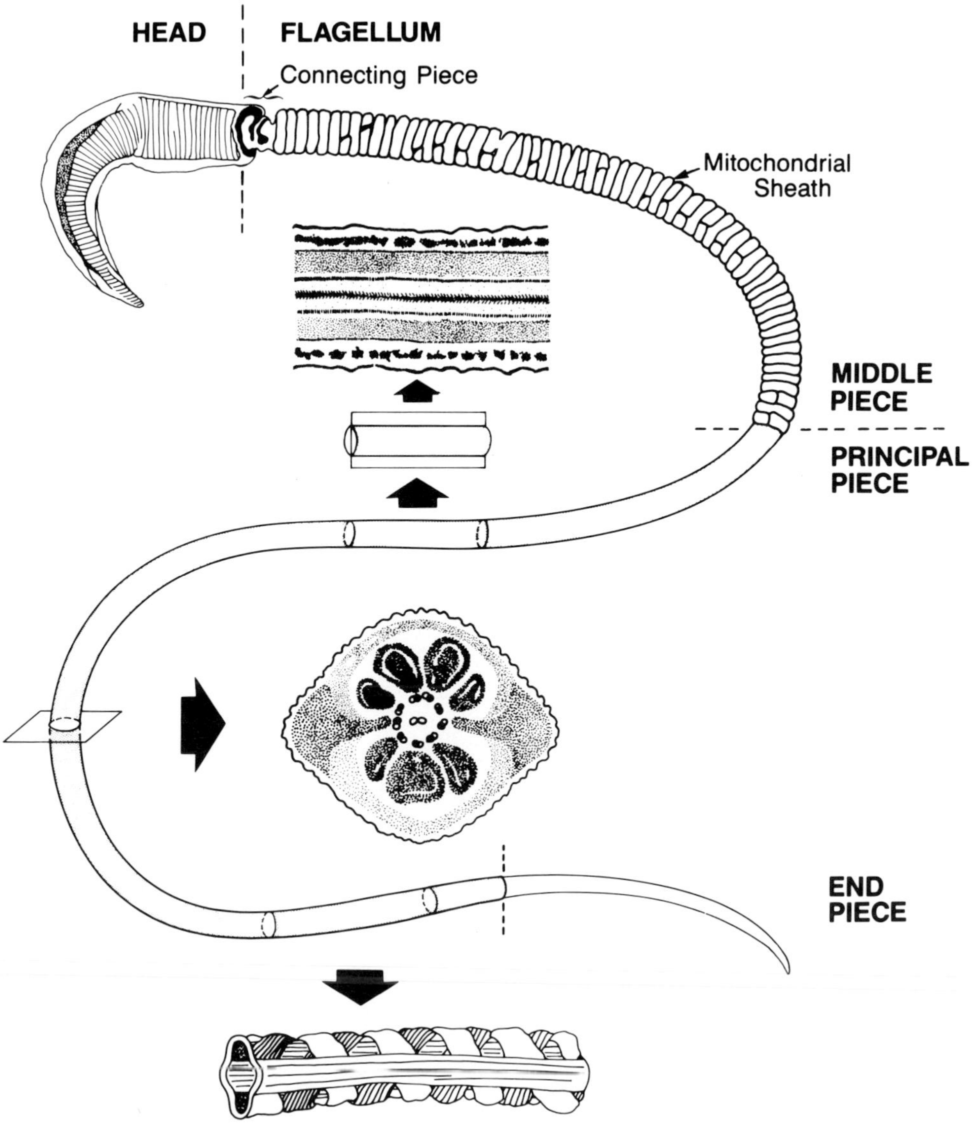

FIGURE 1. The structure of a mammalian spermatozoon. The sperm head contains the nucleus and acrosome, while the flagellum contains the mitochondria, axoneme, outer dense fibers, and fibrous sheath. The mitochondria are contained in the middle piece segment, immediately underlying the plasma membrane, and provide the source of energy for flagellar motion. The axoneme lies at the center of the flagellum and extends from the connecting piece to the distal tip. The outer dense fibers immediately surround the axoneme and extend from the connecting piece into the end piece. The fibrous sheath underlies the plasma membrane and surrounds the outer dense fibers in the principal piece. (From Eddy, E. M., *The Physiology of Reproduction,* Knobil, E. and Neill, J., Eds., Raven Press, New York, 1988, 27. With permission.)

II. MITOTIC PHASE OF SPERMATOGENESIS

The spermatogonia are the earliest members of the germ cell lineage in adult males. The number and pattern of divisions that spermatogonia undergo vary between species, but some divisions renew the stem cell population, while others produce cells which enter the proliferative pathway leading to later stages of spermatogenesis and the eventual formation of spermatozoa. The divisions by spermatogonial stem cells are infrequent, while mitoses by the proliferating spermatogonia occur more frequently and in coordination with the spermatogenic cycle. There is controversy over whether a class of long-cycling stem cells exists with a higher probability for self-renewal, or if the stem cells are a homogeneous population, the proliferative activity and developmental fate of which vary during the cycle of the seminiferous epithelium.[9]

A. THE DEVELOPMENT OF SPERMATOGONIA

The proliferating spermatogonia are classified as type A, Intermediate, and type B, based on their stage of progression through spermatogonial development. In the rat and mouse there are four generations of type A cells (A_1, A_2, A_3, and A_4), with the division of the last of these producing Intermediate cells. These cells divide once to produce type B cells, which also divide once (the last mitotic division of spermatogenesis) to produce spermatocytes.[6,10] The mitotic divisions throughout the period of spermatogonial proliferation occur in synchrony for each group of progeny of an original type A_1 cell. However, division of the cytoplasm (cytokinesis) is incomplete during spermatogonial proliferation, resulting in a clone of 64 preleptotene spermatocytes connected by cytoplasmic bridges.[11] Germ cells remain connected by patent cytoplasmic bridges throughout spermatogenesis until sperm formation is completed.

The regulation of spermatogonial proliferation is not well understood. Studies of the thymidine-labeling index of rat and Chinese hamster spermatogonia indicate that stimulation of proliferative activity occurs in specific stages of spermatogenesis.[12,13] Because Sertoli cells are known to produce growth factors,[14] it has been suggested that they may be the source of a stimulatory factor.[9] It has also been shown that proliferation of spermatogonia can be partly inhibited by testicular extracts. This inhibitory activity appears to be tissue specific, but not species specific. It is present in extracts of testes containing Intermediate and type B spermatogonia and not type A spermatogonia, but is absent from extracts of testes lacking all spermatogonia.[15] The molecular nature of these stimulatory and inhibitory factors and their mechanisms of action have not been characterized. However, spermatogonial proliferation may be controlled by regulatory factors for mitosis like those identified recently in somatic cells. Although the factors were originally described in frog eggs, yeast, and sea urchin embryos, they now have been detected in mammalian cells. Such factors include the maturation-promoting factor (MPF) protein complex, which has protein kinase activity and induces mitosis,[16] the *cdc2* protein, another protein kinase essential for the execution of mitosis,[17,18] and the cyclins, a small family of proteins that participate in mitotic induction.[19] A recent model proposes that the *cdc2* protein kinase is an essential component of MPF and that the production of cyclins leads to the activation of the *cdc2* protein and generation of MPF activity.[20] It has been suggested that the product of a unique male-specific cyclin mRNA present in *Drosophila* testes maintains stem cell proliferation in gametogenesis or mediates the transition between mitosis and meiosis.[21]

B. GENE EXPRESSION IN SPERMATOGONIA

Spermatogonia contain histones and high-mobility-group (HMG) proteins characteristic of germinal cells. Histones, the basic nuclear proteins responsible for local structure of DNA within the chromatin, have been studied most thoroughly in the rat where several germ cell variants have been identified. Three of these (H1a, H2A.X, and TH3) are synthesized at high levels in spermatogonia and may have roles in stabilizing DNA during spermatogonial proliferation.[22] Histone variant TH3 is found only in spermatogenic cells, while the other two histones are present at low levels in somatic cells.[22] Histones H1a, H2A.X, and TH3 are also present in the mouse testis[23-25] and are assumed to have the same developmental pattern of expression as in the rat.

HMG proteins are nonhistone chromosomal proteins involved in maintaining higher order chromatin structure. Rat spermatogonia and early spermatocytes synthesize three electrophoretic variants of HMG2, while somatic cells contain only the slowest migrating band.[26,27]

The presence of high levels of faster migrating bands of HMG2 is common to the testes of most mammalian species.[27] However, it has not been shown that there is a germ cell-specific HMG2. Since no amino acid differences are observed between HMG2 from rat testis and somatic cells, the faster migrating bands might arise by posttranslational modifications of the slow forms.[28] It will be necessary to clone the genes for the HMG proteins to clarify this issue.

Other nuclear proteins associated with cell proliferation are the products of certain proto-oncogenes. One study demonstrated that *c-myc* mRNA and protein are expressed in a cell cycle-dependent manner in mouse spermatogonia, but not in spermatocytes or spermatids. It was suggested that *c-myc* plays a role in mitotic cell divisions of both somatic and germinal cells, but is not involved in meiotic cell divisions.[29] A more recent study demonstrated that the *c-raf* proto-oncogene and three members of the *c-ras* proto-oncogene family are expressed in spermatogonia as well as later in spermatogenesis.[30] However, the nuclear proto-oncogenes *c-fos* and *c-jun,* as well as *c-myc,* are expressed at high levels in type B spermatogonia. Transcripts for *c-myc* and *c-fos* are also present at high levels in preleptotene spermatocytes, but not in later stages, while the *c-jun* transcript level is low in preleptotene spermatocytes. The *c-fos* and *c-jun* proto-oncogenes have been identified recently as components of the transcriptional regulatory factor AP-1.[31] It was suggested that these proto-oncogenes may act as transcriptional regulatory factors in type B spermatogonia to modulate changes in gene expression and to alter programs of gene expression at the spermatogonial-to-spermatocyte developmental transition.[30]

Although the roles of such nuclear proteins in spermatogonia have not been defined, their expression marks the first phase of the developmental program of spermatogenesis and demonstrates that the mitotic phase is a unique period of gene expression in the history of the germ cell lineage. Another observation suggests that spermatogonia have unusual structural or functional nuclear characteristics compared to other germ cells. Spermatogonia apparently contain appreciably more Z-DNA than do other rat spermatogenic cells and it has been suggested that the Z-conformation is mediated by chromosomal proteins.[32]

Spermatogonia are difficult to isolate in large quantities and have been studied less frequently with biochemical and molecular biology methods than other spermatogenic cells. However, it appears that not only the expression of nuclear proteins is developmentally regulated in spermatogenesis. A 2.7-kb β-actin mRNA is appreciably more abundant in spermatogonia than in later stage spermatogenic cells,[33] suggesting that this cytoskeletal protein isoform has an important role in spermatogonial development or function.

III. MEIOTIC PHASE OF SPERMATOGENESIS

A. THE PROCESS OF MEIOSIS

The development of spermatocytes begins with the preleptotene stage, the stage which immediately precedes the initiation of meiosis. During this stage, DNA replication occurs in a unique prolonged S phase,[34] resulting in the formation of a pair of daughter chromatids for each chromosome. Because of this, the nucleus of each primary spermatocyte contains the diploid number of chromosomes, but twice the diploid amount of DNA. After DNA replication ceases, chromosome condensation is initiated, signaling the beginning of the leptotene stage of primary spermatocyte development and the unique process of meiosis. During this period of spermatogenesis in the rat[35] and presumably in other species, spermatocytes begin moving between the Sertoli cells from the basal compartment to the adluminal compartment of the seminiferous epithelium. The Sertoli cell junctions separating these two compartments part in front of the migrating germ cells and reform behind them, maintaining the blood-testis barrier.

In addition to the leptotene stage, the long prophase of the first meiotic division in primary spermatocytes includes zygotene, pachytene, diplotene, and diakinesis stages. During these stages, the unique synaptonemal complex structures form, the chromosomes condense and undergo pairing or synapsis, and crossing-over and exchange of chromosomal material between homologous chromosomes occur.[36] The nuclear membrane subsequently disperses, a spindle apparatus forms, the paired homologous chromosomes are separated one to each pole, and two daughter cells are formed. The first meiotic division produces the secondary spermatocytes, which contain the haploid number of chromosomes, but the diploid amount of DNA. These cells soon proceed through the second meiotic division without undergoing DNA replication, forming the spermatids that are haploid both in number of chromosomes and amount of DNA. No further DNA replication or cell divisions occur in spermatogenesis. However, germ cells remain connected by cytoplasmic bridges throughout the meiotic phase of spermatogenesis.

The synaptonemal complex is a prominent and unique nuclear structure of meiotic cells. It serves a key role in chromosome pairing by forming the axis of the chromosomal bivalents in meiotic prophase. Biochemical studies of the synaptonemal complex recently became possible when methods were developed for isolating this structure. Synaptonemal complexes from rat spermatocytes analyzed by gel electrophoresis were found to contain major protein components of 67 to 60 and 57 to 55 kDa, and minor proteins of 90, 35, 33, 28, and 26 kDa were found.[37] The 67- to 60-kDa proteins co-migrated with lamin proteins associated with nuclear pore complexes and the nuclear envelope, suggesting that the synaptonemal complexes are related to these nuclear components. Another study identified a 110-kDa phosphoprotein constituent of nuclear matrix/synaptonemal complex structures isolated from rat pachytene spermatocytes.[38] This protein is not present in the nuclear matrix of liver cells and is most prominent at the pachytene stage of spermatogenesis.

Antibodies offer a promising approach for dissecting the synaptonemal complex and characterizing its constituent proteins. Human autoimmune sera often react with various nuclear structures, including synaptonemal complexes.[39] In addition, two monoclonal antibodies have been found which label the central region of the synaptonemal complex and the attachment plaques at the ends.[39] Immunohistochemical and Western blotting studies suggest that at least one of the antibodies may bind to a synaptonemal complex protein that shares an epitope with actin, vimentin, and other proteins, but that is distinct from these proteins.[39] Another monoclonal antibody reacts specifically with a 30- to 33-kDa protein of isolated rat synaptonemal complex.[40,41] The protein is localized to the cores of chromosomes during pairing in the zygotene stage, the lateral elements of the synaptonemal complex of pachytene stage spermatocytes, and along the length of the cores when the chromosomes separate in the diplotene stage.[42] The protein was suggested to have a role in chromatin binding, and/or maintaining structural integrity of the synaptonemal core. A different monoclonal antibody reacts specifically with the region between the lateral elements of the synaptonemal complex during pairing of homologous chromosomes.[42] These observations suggest that the synaptonemal complex is composed of multiple proteins and that some are restricted to particular regions of the structure.

B. REGULATION OF MEIOSIS

The initiation of meiosis is controlled in a sex-specific manner, with meiosis occurring during embryogenesis in the female and upon puberty in the male. Meiosis-inducing substances are present in media in which ovaries or testes with germ cells undergoing meiosis have been cultured,[43] and meiosis-preventing substances are produced by fetal testes in which meiosis is not occurring.[44] These two substances have not been well characterized, but are small molecules that are not species specific.[45] Other factors appear to be involved in the control of meiosis in oocytes, including some also implicated in the control of mitosis. These include MPF, which acts as a histone H1 kinase,[18] and the *c-mos* proto-oncogene product, which may trigger initiation of meiotic division.[46-48] However, the possible roles of MPF and *c-mos* in regulating meiosis in spermatogenic cells apparently have not been studied.

Other studies have focused on processes affecting the DNA during chromosome synapsis and recombination in meiotic prophase. Such studies have identified specific DNA sequences involved in the recombination process,[49] and recombination-related factors expressed in the zygotene and pachytene stages.[50] These factors include an endonuclease that introduces site-selective nicks, a U-protein that unwinds duplex DNA from nicked sites, and an R-protein that facilitates homologous recombination of single-stranded DNA.[50] However, these results came primarily from studies on meiosis in the lily and comparable studies have been limited in mammals.

DNA polymerase β mRNA and enzyme activity in mice are most abundant in the testis and early pachytene spermatocytes.[51] DNA polymerase β is a 40-kDa polypeptide that appears to be the main enzyme used for DNA repair in meiotic and postmeiotic germ cells of male mice.[51,52] In other studies, recA-like proteins (which catalyze recombination in *Escherichia coli*) have been purified from extracts of mouse somatic and meiotic cells and shown to have the capacity to catalyze homologous recombination.[53] The recA-like protein purified from meiotic cells (m-rec) differs from that of somatic cells (s-rec). The m-rec from mouse spermatocytes is a 45-kDa protein with highest activity at 32°C, while the 75-kDa s-rec has highest activity at 37°C. The different temperature optima are of physiological significance; the testis normally is maintained 4 to 5°C below body temperature and spermatogenesis is disrupted when the testis is heated to 37°C.[54,55] The peak m-rec activity occurs in the late zygotene through early pachytene stages of meiosis and is reported to be at least ten times greater during this period than during the premeiotic S phase or following termination of meiosis.[53] However, these findings have not yet been independently confirmed.

In addition to chromosomal recombination during meiosis, chromatin structure is altered to a less compact form.[56,57] The chromosomal basic proteins, which include multiple variants of histones that are either enriched or found exclusively in germ cells, may have important roles in these structural and functional alterations of the nucleus. During meiosis, there is a progressive replacement of somatic-type histones with germ cell variants.[22,58,59] The three histone variants first synthesized in spermatogonia in the rat show different levels of synthesis in spermatocytes. H1a synthesis continues at a lower rate in spermatocytes, H2A.X synthesis is unchanged in early spermatocytes and decreased in later stages, while TH3 is apparently not synthesized in spermatocytes.[22,58]

Three additional testis-specific histone variants (H1t, TH2A, and TH2B) are synthesized during the meiotic phase of spermatogenesis in the rat. The maximum H1t synthesis occurs in late pachytene spermatocytes, while the maximum synthesis of TH2A and TH2B occurs in early spermatocytes.[58] Similar testis-specific H1 and H2B histones are synthesized during meiosis in the mouse.[59] In addition, the HMG2 protein variants which first appear in spermatogonia reach maximal levels of synthesis during the mid- to late-pachytene stages in the rat.[26,28] Although the histone H4 found in germ cells appears to be identical to the somatic H4 by biochemical criteria, it has been found that a spermatogenic cell-specific histone H4 gene is expressed at high levels in pachytene spermatocytes, while the somatic H4 gene is not expressed in germ cells.[60] The predicted amino acid sequences of the two gene products appear to be identical, but the genes are regulated quite differently.

Posttranslational modifications of nuclear proteins may also contribute to alterations in the structure and function of the meiotic nucleus. Recent studies have shown that the specific activity and apparent level of the 115-kDa poly(ADP)-ribose synthetase enzyme is significantly higher in the testis than in a variety of somatic tissues.[61] Furthermore, the meiotic histone H1t and three nonhistone proteins have been identified as major acceptor proteins for poly(ADP)-ribose in the rat.[62] In both studies it was suggested that ADP-ribosylation, which appears to be involved in modulating DNA repair mechanisms and changes in chromatin structure, may play a role in meiotic recombination. Inhibition of ADP-ribosylation has been shown to increase X-ray-induced chromosomal damage in mouse testis.[63] It has been reported that histone H1t and a 20-kDa protein are selectively enriched in poly(ADP)-ribosylated chromatin domains of rat pachytene spermatocytes and suggested that these domains represent highly enriched DNA-repair regions.[64]

C. GENE EXPRESSION DURING MEIOSIS

Many spermatogenic cell-specific proteins are variants or isoforms of known somatic cell proteins. It may be that these unique proteins are needed to meet the special metabolic, morphologic, developmental, or genetic requirements of gametogenesis in the male. For example, the special metabolic requirements of spermatogenic cells may have been accommodated by gene duplication and subsequent mutation to produce enzymes particularly suited to this environment or to replace the products of cognate genes inactivated during spermatogenesis.[65] At least 15 isozymes with testis-specific substrate specificities, cofactor requirements, or electrophoretic mobilities have been reported.[66] Several of these isozymes have been characterized at the protein and gene level, and found to be products of spermatogenic cell-specific genes. Two particularly well-studied examples are phosphoglycerate kinase and lactate dehydrogenase.

Phosphoglycerate kinase catalyzes the conversion of 1,3-diphosphoglycerate to 3-phosphoglycerate and ATP during glycolysis. The spermatogenic cell-specific isozyme has been designated PGK-2 and its expression is regulated both by transcriptional and translational control mechanisms. PGK-2 mRNA synthesis is initiated in prophase spermatocytes and continues in spermatids,[67] but PGK-2 protein is detected only in spermatids.[68] Cloning of the somatic (PGK-1) and germ cell (PGK-2) genes revealed that PGK-1 and PGK-2 proteins have only a small number of amino acid differences. However, much less sequence similarity was seen in the 5′ and 3′ untranslated regions of the two mRNAs, and the PGK-2 gene was found to lack introns and to possess a possible remnant of a poly(A) tail.[65,69] This inferred that the PGK-2 gene arose by the retro-insertion of a PGK-1 cDNA into the genome and subsequent accumulation of mutations in the nucleic acid sequence. PGK-1 is located on the X chromosome and inactivation of this chromosome during spermatogenesis would deprive spermatogenic cells of an enzyme essential in the glycolytic pathway. Since the PGK-2 gene is located on an autosome, its expression during spermatogenesis apparently substitutes for the loss of PGK-1 gene expression and enzyme activity.[65,69]

Spermatogenic cells also contain a unique isozyme of lactate dehydrogenase which catalyzes the interconversion of lactate and pyruvate.[70-72] There are three genes for this tetrameric enzyme, but the LDH-C gene (also referred to as LDH-X) is expressed only in spermatogenic cells.[73] The LDH-C_4 isozyme may be present as early as preleptotene spermatocyte development in mice,[74] but the synthesis of LDH-C mRNA continues after meiosis and translatable mRNA is still present in elongating spermatids.[75,76] The mouse[77,78] and human[79,80] cDNA sequences for LDH-C have been determined; they have only 73% homology with each other and similar degrees of homology with somatic LDH cDNA sequences.[79] Compared to somatic forms, the LDH-C_4 isozyme has a broader substrate specificity, a higher affinity for lactate, is more sensitive to pyruvate inhibition, and less sensitive to lactate inhibition.[66,81] These properties would favor ATP production by germ cells in the presence of high levels of lactate, a preferred metabolic substrate for spermatids in particular.[81]

Germ cell differentiation antigens first appearing on the cell surface during the pachytene stage of meiosis or subsequent haploid stages have been identified with a variety of polyclonal and monoclonal antibodies.[82-84] Antigens appearing during meiosis that have been characterized biochemically include small molecular weight glycoproteins such as the RSA-1 rabbit sperm autoantigen[83] and significantly larger lactosaminoglycans with distinct carbohydrate epitopes.[85] An unusual glycolipid, sulfogalactosylglycerolipid, and its associated binding protein also appear on the germ cell surface during meiotic prophase.[86,87] The functions of these glycosylated surface molecules have not been defined, although several of them persist on the spermatozoon where they may play a role in sperm-egg interaction.[83]

IV. THE POSTMEIOTIC PHASE OF SPERMATOGENESIS

During the postmeiotic phase of spermatogenesis, haploid spermatids are converted into spermatozoa by the morphogenetic processes of spermiogenesis. As spermatids differentiate, the acrosome forms (see Chapter 8), the nucleus condenses, the flagellum develops, and the cell elongates and sheds much of its cytoplasm. The size and shape of the acrosome and nucleus have been used to precisely classify spermiogenesis into 16 stages in the mouse[6] and 19 stages in the rat.[19] There are two general populations of spermatids present during this time, the round spermatids and the elongating spermatids. Acrosome development and flagellar formation begin in round spermatids, while in elongating spermatids the nucleus condenses, the cell becomes highly polarized, and the acrosome and flagellar formation continues. Spermiogenesis lasts about 2 weeks in mice and 3 weeks in rats, with round spermatids being present during the first half and elongating spermatids occupying the remainder of the process. Once formed, spermatozoa are released from the surrounding Sertoli cells into the fluid-filled lumen of the seminiferous tubule by the process of spermiation.

Since meiosis results in the segregation of a single copy of each chromosome (and therefore a single copy of each gene) into each spermatid, a particular allele will be present in only half of the spermatids in a heterozygous animal. However, haploid expression of such an allele does not ensure that its product will be confined to half of the cells. Because spermatids in a clone arising from a single spermatogonium remain connected by cytoplasmic bridges, small products can pass from one cell to another. This was shown in hemizygous transgenic mice (mice carrying the human growth hormone transgene on one chromosome) in which the product of the transgene expressed during spermiogenesis was seen in >90% of spermatids, although 50% of the spermatids lacked the transgene.[88] *In situ* hybridization showed that the transgene mRNA was present in all spermatids, strongly suggesting that these transcripts had passed through cytoplasmic bridges. Mice heterozygous for certain *t*-haplotypes appear to be an exception to the phenotypic equivalence of spermatids. Although half the sperm contain the wild-type allele in these mice, the *t* mutation can be transmitted to over 90% of progeny, in defiance of Mendelian predictions.[89] Although the basis for such transmission ratio distortion is unknown, it would presumably require confining both the mRNA and the product of the *t* allele to individual spermatids.

The structures which form during development of the spermatozoon require production of new proteins. Unique nuclear proteins have been identified that are probably associated with the compaction of DNA in spermatids, and it is apparent that other germ cell-specific proteins are required to form the cytoskeleton of the head and the flagellum, and to accomplish remodeling of the spermatid. Although proteins synthesized during the postmeiotic phase of spermatogenesis have been identified, little is yet known about the mechanisms regulating the synthesis and assembly of most of these. However, as more proteins involved in spermiogenesis are identified and their genes characterized, it is likely that extrinsic and intrinsic factors responsible for expression of these genes will become known.

A. CONDENSATION OF THE HAPLOID NUCLEUS

As the haploid nucleus condenses during the transformation of round into elongating spermatids, the nucleosomal organization of chromatin is replaced by smooth chromatin fibrils which associate in parallel arrays.[90-92] In addition, there is progressive inactivation of the genome, although the synthesis and accumulation of messenger RNAs continue during the early stages of spermiogenesis.

The structural transformation of the spermatid nucleus correlates temporally with changes in the chromosomal basic proteins. Studies in several species have shown that the histones are removed and replaced with small, basic transition proteins (TP) during midspermiogenesis, the period when nuclear condensation begins and the chromatin becomes organized into smooth fibrils.[59,93,94] However, the TP are the major DNA binding proteins present during only a relatively short period of spermiogenesis (steps 13 to 15 in the rat).

Mechanisms for the removal of histones during spermiogenesis are not fully understood, although posttranslational modifications appear to be involved in this process. Histones of spermatids are hyperacetylated, both in mammalian[57] and nonmammalian species.[95] Hyperacetylation of histones, particularly histone H4, has been correlated with a transient decrease in chromatin thermal stability, perhaps related to chromatin decondensation and nucleosome disassembly.[57] In addition, ubiquitinated histones have been identified in spermatogenic cells of the rooster[96] and trout.[97] The proportion of ubiquitinated H2A increases during rooster spermiogenesis and it has been proposed that these modified histones play a role in haploid chromatin transitions.[96] Furthermore, mammalian testes are particularly rich in ubiquitin[98] and this 8.5-kDa protein is synthesized by pachytene spermatocytes and round spermatids isolated from rats.[26] Further studies are needed to determine if histones become ubiquitinated during mammalian spermatogenesis and if this modification is related to their subsequent removal.

Multiple forms of the small, basic chromosomal TP have been identified in rat, mouse, and ram.[99-101] In addition, a specific antiserum to rat TP1 was shown by immunoblotting to react with a similar protein in bull, rabbit, human, boar, and hamster testes.[102] Biochemical studies suggest as many as four TP may occur, but cDNA clones have been isolated only for rat TP1 and TP2[103,104] and mouse TP1.[105] The carboxyl terminal domain was found to be responsible for the basic charge of rat TP2.[106] *In vitro* studies indicated that TP have a high affinity for single-stranded nucleic acids and may behave like a DNA-melting protein to destabilize double-stranded DNA.[107] It has been suggested that TP1 plays a role in the completion of chromatin condensation in elongating spermatids.[103,108]

The TP are, in turn, replaced in elongating spermatids by protamines, small arginine- and cysteine-rich basic proteins that persist as the predominant chromosomal proteins of mammalian spermatozoa.[109] A single protamine (P1) has been identified in most mammalian species, although multiple forms are present in mouse and human.[109,110] However, the cDNAs for mouse protamines 1 (mP1) and 2 (mP2) hybridize with DNA sequences in a variety of mammals.[111] Although approximately equal levels of P1 mRNA are present in hamster, rat, and mouse testes, P2 mRNA is 2- to 5-fold lower in hamster and 50-fold lower in rat than in mouse.[112] While protamines from different mammals have divergent amino acid sequences, these proteins generally have a highly basic central domain of arginine clusters, with less basic domains on either end.[113] The DNA-protamine complex is apparently stabilized during epididymal maturation by the formation of disulfide bridges between protamine molecules.[114-116] Because of this, the sperm nucleus is extremely resistant to disruption by physical or chemical methods.[59]

B. CYTOSKELETAL PROTEINS AND FORMATION OF THE SPERM HEAD

Much of the sparse cytoplasm of the spermatozoon head is occupied by cytoskeletal structures. Although it has not been established that cytoskeletal components are responsible for determining and maintaining the shape and organization of the head of the spermatozoon, it is likely that they have major roles in these processes.

The sperm head in most mammals is spatulate in shape, with the nucleus and acrosome flattened in the plane of the anterior-posterior axis of the sperm (Figure 2). The nucleus and acrosome are usually symmetrical structures in these species. However, the head of rodent sperm is falciform in shape and an asymmetric acrosome overlies the convex margin of the hook-shaped nucleus.[1]

In testicular spermatozoa of rabbits, filamentous actin accumulates in the subacrosomal

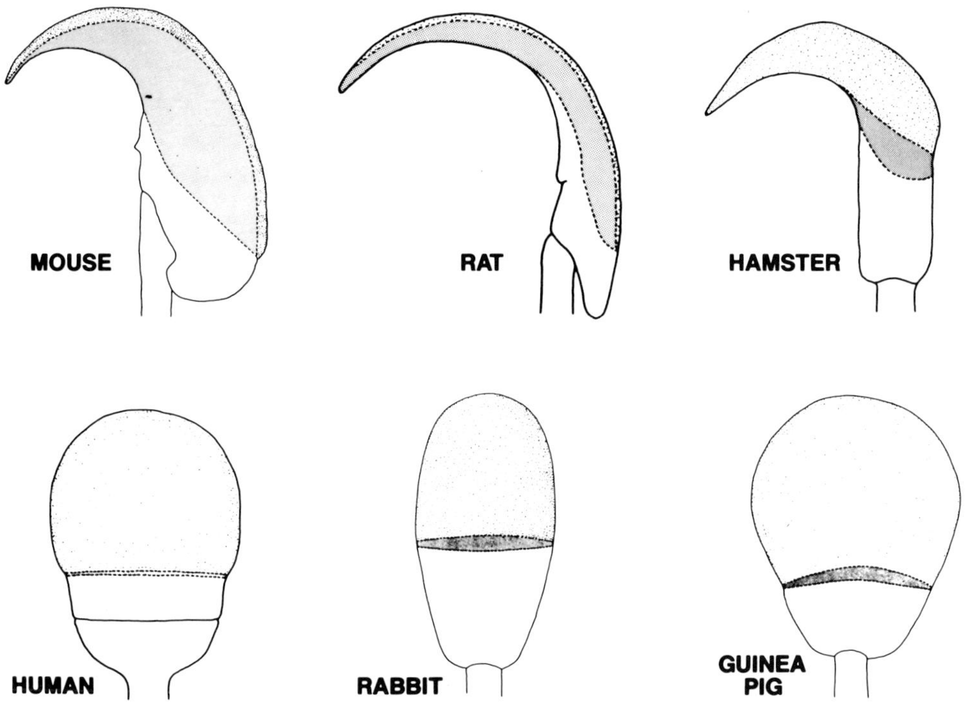

FIGURE 2. The heads of mammalian spermatozoa. The heads of mouse, rat, and hamster sperm are asymmetric, being falciform in shape and flattened laterally. The acrosome overlies the convex margin of the hook-shaped nucleus. The sperm heads of most other mammals, including human, rabbit, and guinea pig, are spatulate in shape and the nucleus is flattened symmetrically in the plane of the anterior-posterior axis of the sperm. The shaded area is the equatorial segment of the acrosome. The acrosome is similar to the nucleus in shape in most species, however the anterior part of the guinea pig acrosome curves out of the flattened plane of the nucleus, making the head spoon-shaped. (From Eddy, E. M., *The Physiology of Reproduction,* Knobil, E. and Neill, J., Eds., Raven Press, New York, 1988, 27. With permission.)

space and in a band associated with the developing acrosome.[117] The band lengthens coincident with elongation of the nucleus and becomes continuous with actin in the postacrosomal region of the head of mature sperm. Actin has also been identified in the postacrosomal region of epididymal sperm of other species,[118,119] including rat and hamster sperm with falciform heads.[120,121] The location and timing of appearance of actin suggest that it may be involved in shaping the acrosome and elongating the sperm head.[117] A 1.5-kb actin mRNA first detected during spermatogenesis in round spermatids has been found to be a smooth muscle gamma-actin.[122-123] The postmeiotic appearance of this actin isoform might compensate for down regulation of other actin isoforms or provide a unique actin required for processes such as acrosome shaping or sequestering excess cytoplasm into residual bodies.[123]

There are two major cytoskeletal structures composed of microtubules in the spermatid. One of those, the manchette, is a transient feature, while the other, the axoneme, forms the core of the flagellum and persists in the spermatozoon. The manchette is a sheath of microtubules that appears at the beginning of spermatid elongation. The microtubules attach to a dense fibrillar band on the internal surface of the plasma membrane surrounding the nucleus and the band migrates from the anterior to the posterior end of the nucleus during spermatid elongation.[124] The band later becomes the posterior ring demarking the junction between sperm head and flagellum, while the microtubules of the manchette disappear.[124] Microtubules are known to serve important roles in determining cell shape, and it has been suggested that the manchette may be responsible for the elongation and shaping of the sperm nucleus and head.[125]

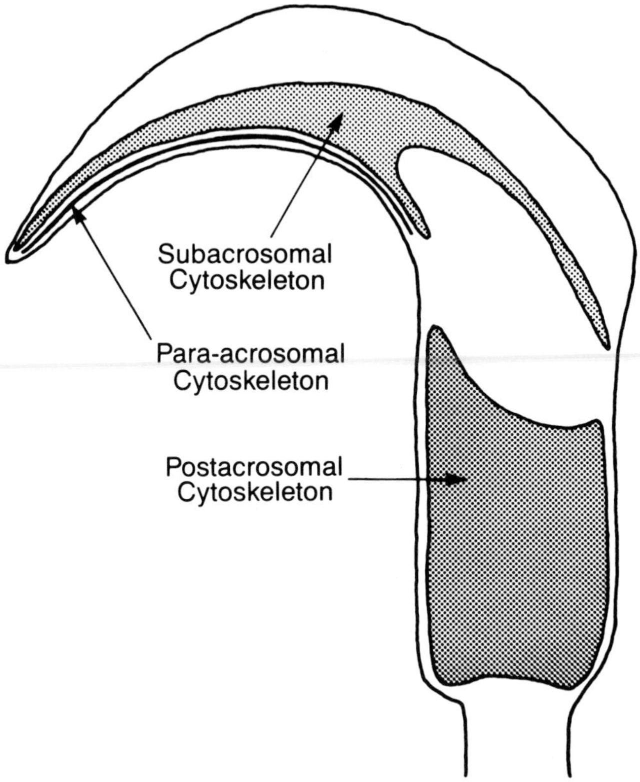

FIGURE 3. The cytoskeleton of the head of a mammalian spermatozoon. The subacrosomal cytoskeleton (perforatorium) of a hamster spermatozoon lies between the acrosome and the anterior part of the nucleus and extends to the anterior tip of the falciform-shaped head. The postacrosomal cytoskeleton surrounds the posterior part of the nucleus and lies between the nuclear envelope and plasma membrane. The para-acrosomal cytoskeleton (filamentous cytoskeletal complex) is present between the plasma membrane and acrosome on the concave aspect of the hamster sperm head. (From Eddy, E. M., *The Physiology of Reproduction,* Knobil, E. and Neill, J., Eds., Raven Press, New York, 1988, 27. With permission.)

Microtubules contain closely related α- and β-tubulin isoforms and at least one new variant of each appears in spermatids.[126] A cDNA for a spermatid α-tubulin, 70% homologous with other vertebrate tubulins, has been characterized.[127] This tubulin variant could give germ cell microtubules different structural or functional characteristics. However, it has not yet been determined if this tubulin variant is present in the microtubules of the manchette or axoneme.

Other cytoskeletal structures have been identified in the head of the spermatozoon (Figure 3). These include the subacrosomal cytoskeleton (between the nucleus and the acrosome), the postacrosomal cytoskeleton (between the plasma membrane and the postacrosomal portion of the nucleus), and the para-acrosomal cytoskeleton (between the plasma membrane and convex surface of the acrosome of falciform sperm).[1]

The subacrosomal and postacrosomal cytoskeletons together have been referred to as the perinuclear theca.[128] The isolated perinuclear theca contains a diverse population of proteins from 80 to 8 kDa, with four major proteins of 25, 16.5, 15, and 13 kDa.[129] It retains the shape of the rat sperm nucleus after isolation, suggesting that it may be an extrinsic determinant of nuclear shape.[129] The subacrosomal cytoskeleton is a minor component in spermatozoa with spatulate heads,[114,130] but is a more extensive feature in falciform spermatozoa, where it is referred to as the perforatorium. The putative precursor to the subacrosomal cytoskeleton is visible in early elongating spermatids as a dense layer between the acrosomal granule and the nucleus which progressively extends posteriorly over the nucleus as the sperm head elongates.[128,131] The perforatorium does not form until the last step of spermiogenesis in the rat, when it condenses from whispy cytoplasmic material in the subacrosomal space.[128] The perforatorium contains extensive disulfide bonds and is highly insoluble.[114] The rat perforatorium has been isolated and initial studies found it to consist mainly of a 13-kDa protein and to be rich in cysteine.[132,133] A more recent study identified six prominent polypeptides of 43, 35, 33, 16, 13.4, and 13 kDa in the rat perforatorium, with a 16-kDa polypeptide being the major component.[134] An antiserum to rat perforatorium also reacts with the para-acrosomal cytoskeleton and flagellar cytoskeletal components, but not with the postacrosomal cytoskeleton. It was suggested that these structures share a 16-kDa protein and contain immunologically similar proteins.[134]

The postacrosomal cytoskeleton (also referred to as postacrosomal sheath, postnuclear sheath, postacrosomal dense lamina, postnuclear cap, postnuclear body, and postacrosomal calyx)[1] is attached anteriorly to the subacrosomal cytoskeleton and terminates at the posterior ring caudally. It begins to form coincident with nuclear elongation and development of the manchette as an accumulation of dense material just caudal to the posterior margin of the acrosome.[128,135] Assembly of the postacrosomal cytoskeleton occurs as the manchette moves caudally to form the posterior ring.[136] It has been isolated from bull and rat spermatozoa[137] and two kinds of basic proteins are found to be major constituents. A 60-kDa protein, termed calicin, is present only in the postacrosomal cytoskeleton, and a group of multiple-band polypeptides (MBP) of 74 to 56 kDa are present in both subacrosomal and postacrosomal cytoskeleton.[137] The polypeptides of the MBP are related to each other, but unrelated to calicin by antibody binding and peptide mapping characteristics. Although the antibody to calicin reacts with a 60-kDa protein from sperm of several other mammals, the protein does not always appear to be localized to the postacrosomal cytoskeleton.[138]

The para-acrosomal cytoskeleton (filamentous cytoskeletal complex) has been described as an organized array of cytoplasmic filaments associated with specific regions of the ventral surface of the hamster spermatozoon acrosome.[139] It is a tripartite structure, consisting of a cone at the anterior tip and a bifurcated sheet on the convex surface of the head, lying between the acrosome and the plasma membrane. Although filaments and striations have been described along the ventral surface of the acrosome in other species,[140,141] the para-acrosomal cytoskeleton has not yet been identified in other species or characterized biochemically.

C. CYTOSKELETAL PROTEINS AND FORMATION OF THE FLAGELLUM

The long, slender flagellum is divided into middle piece, principal piece, and end piece segments and is composed mainly of cytoskeletal structures. The middle piece occupies the initial one fourth or so of the flagellum. It contains the mitochondria, which are wrapped in a tight helix around the underlying cytoskeletal structures and are the source of ATP necessary for propelling the sperm.[1] The principal piece makes up most of the remainder of the flagellum (Figure 1). The cytoskeletal structures of the flagellum are the axoneme, the outer dense fibers, and the fibrous sheath. While the axoneme provides the motile force for flagellar activity, the latter two structures stiffen the flagellum and probably serve a passive role in flagellar motion. The outer dense fibers are suggested to influence the form of the beat by determining the elastic properties of the flagellum,[142] while the fibrous sheath may modulate the plane of the beat.[143] Men whose sperm are defective in these cytoskeletal structures are often infertile[144] and males of several strains of mutant mice that are sterile when homozygous have flagellar defects.[145]

The axoneme extends from the base of the head to the distal tip of the flagellum. It consists of two central microtubules surrounded by nine outer microtubule doublets (Figure 4), the same "9 + 2" organization present in cilia.[146] The axoneme forms in the round

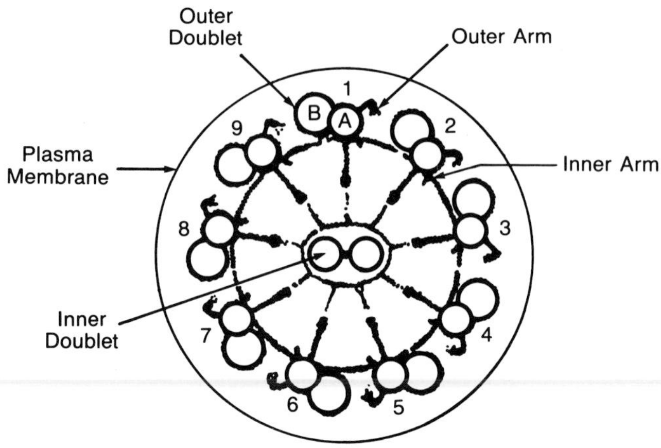

FIGURE 4. The axoneme of the sperm flagellum. The axoneme consists of an inner doublet of microtubules surrounded by nine pairs of outer doublets of microtubules. The arms extending from the outer doublets contain dynein proteins with ATPase activity. (From Eddy, E. M., *The Physiology of Reproduction,* Knobil, E. and Neill, J., Eds., Raven Press, New York, 1988, 27. With permission.)

spermatid, before other flagellar structures appear, from a centriole lying against the nucleus opposite to the forming acrosome.[143] Extending from the outer microtubule doublet are two arms which contain dynein, composed of two or three >300-kDa proteins with ATPase activity and several other proteins of lower molecular weight.[147,148] The ATPase activity generates the forces that cause sliding between adjacent outer doublets to produce flagellar movement.[149]

The outer dense fibers are nine rods which immediately surround the axoneme and extend most of the length of the flagellum (Figures 1 and 5). They attach to the head through the connecting piece and taper from the head to the posterior end of the principal piece. Similar structures are present in some nonmammalian species,[150] but their relationship in composition and function to outer dense fibers in mammalian sperm has not been determined. The outer microtubule doublets of the axoneme appear to be templates for formation of the outer dense fibers.[151-153] In the rat, small fibers are seen in the most proximal part of the flagellum of step 8 spermatids and these extend into the principal piece by step 14.[154] The fibers increase in length and diameter during step 15, enlarge and take their nearly final shape in step 16, and continue slight growth during steps 17 to 19.[154] Radioautographic studies with labeled proline and cysteine indicated that the major period of outer dense fiber protein synthesis is during step 15.[154]

The outer dense fibers in the rat are reported to contain about six major proteins of 87 to 84, 80, 32 to 30, 26 to 25, 20 to 18, and 14 to 11 kDa.[155-158] These proteins are phosphorylated,[151] rich in cysteine,[156,157,159] and relatively insoluble because of abundant disulfide cross-links.[114,160] The proteins differ in amino acid composition, amino-terminal amino acid, net charge, and degree of phosphorylation.[156,157,159,161] This suggests that the outer dense fibers are not composed of a family of closely related proteins and that the smaller proteins are not processed forms of the larger proteins. In addition, the higher molecular weight proteins are rich in α-helical configuration[159] and the lower molecular weight proteins are the major zinc-binding components in spermatozoa.[155]

The fibrous sheath lies in the principal piece segment of the flagellum, between the plasma membrane and the outer dense fibers. It is a tapered cylinder formed by two longitudinal columns connected by circumferential ribs. Fibrous sheath assembly in the rat is reported to proceed throughout much of spermiogenesis and in a distal to proximal gradient

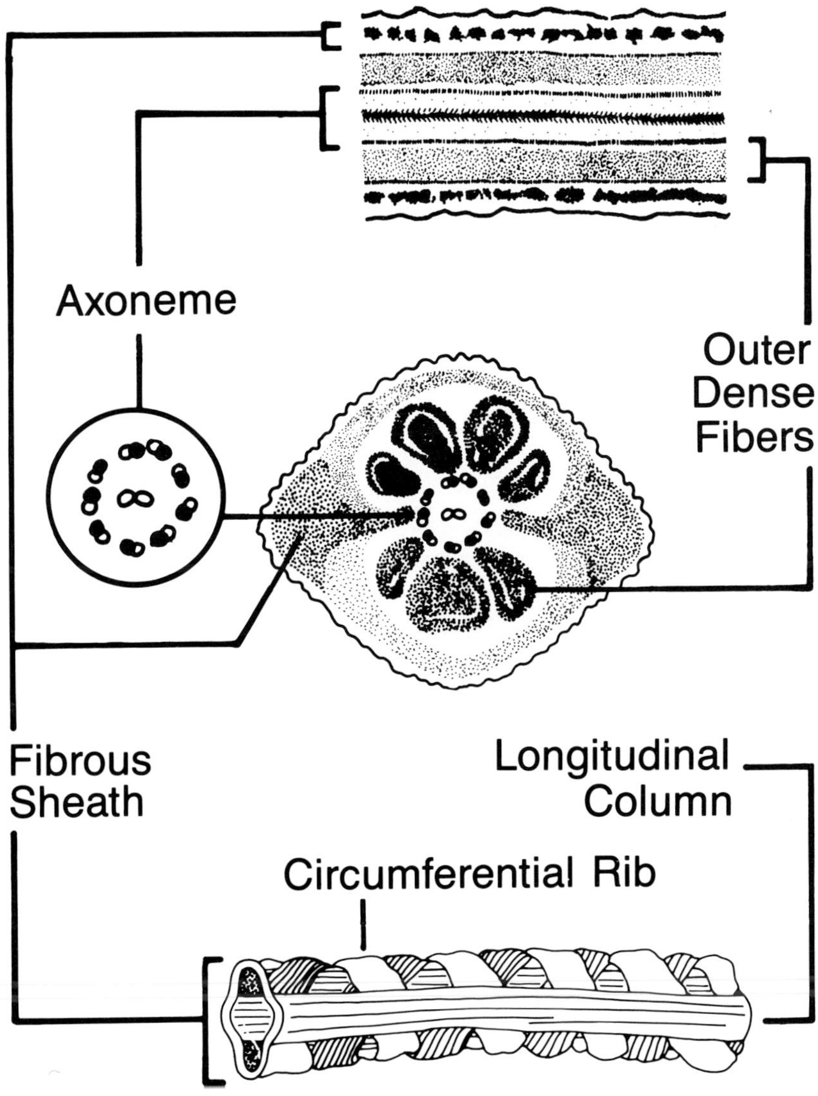

FIGURE 5. The cytoskeleton of the flagellum of a mammalian spermatozoon. The axoneme immediately underlies the plasma membrane in the end piece, but is surrounded by the outer dense fibers in the remainder of the flagellum, as well as mitochondria in the middle piece, and fibrous sheath in the principal piece of the flagellum. Nine outer dense fibers are present in the proximal part of the flagellum, but two usually terminate in the first part of the principal piece and their place is taken by inward extensions of the fibrous sheath. The fibrous sheath consists of two longitudinal columns connected by ribs that spiral obliquely around the axoneme. The axoneme provides the motile force for the flagellum, while the outer dense fibers and fibrous sheath probably stiffen the flagellum and shape the form and plane of the flagellar beat. (From Eddy, E. M., *The Physiology of Reproduction,* Knobil, E. and Neill, J., Eds., Raven Press, New York, 1988, 27. With permission.)

along the flagellum.[162] The longitudinal columns first appear in the rat as thin rods in the distal end of the flagellum in step 2 and increase in length during steps 2 to 10. The ribs begin to form in the distal part of the flagellum during steps 12 to 14, and additional ribs form in the proximal flagellum during steps 15 to 17.[162] Radioautographic studies indicated that fibrous sheath proteins are synthesized throughout the 15-day period of fibrous sheath formation.[162] However, a fibrous sheath protein in the mouse was first detected with a monoclonal antibody in the cytoplasm of step 14, in the principal piece in step 15, and in the proximal part of the fibrous sheath in step 16 spermatids.[163] This suggests that not all fibrous sheath proteins are synthesized throughout spermiogenesis or are incorporated into the fibrous sheath from distal to proximal regions.

The fibrous sheath in the rat is composed of three major proteins of approximately 80 to 75, 27 to 24, and 14 to 11 kDa, with the largest protein being phosphorylated and the most abundant.[132,158,164] Peptide map analysis indicated the two larger proteins are unrelated to proteins of similar size in the outer dense fibers, but that the small protein may be shared by the two structures.[164] The reactions of antisera to outer dense fiber and fibrous sheath proteins also indicated that the smallest protein might be present in both structures and that antigenic epitopes are shared by several other proteins of the outer dense fibers and the fibrous sheath.[158] A monoclonal antibody has been used to identify a 67 to 65-kDa fibrous sheath protein in mouse, rat, and hamster sperm.[165] This protein is first seen by immunohistochemistry in step 13 spermatids,[165] but is detected by immunoblotting in extracts of round spermatids.[166]

V. *IN VITRO* STUDIES OF SPERMATOGENIC CELLS

The study of mechanisms regulating germ cell differentiation has been hampered by the complex organization of the testis. However, techniques have been developed for isolating germ cells at distinct stages of development and for maintaining these cells in short-term cultures. These procedures have been standardized in several laboratories and have facilitated studies of spermatogenic cell metabolism, biochemical characterization of spermatogenic cells at different stages of development, and the mechanisms regulating synthesis of lineage- and stage-specific gene products.

Isolation of germ cells involves sequential enzymatic and mechanical treatments of the testis to remove the interstitial cells, to dissociate the seminiferous tubules, and then to disrupt the intercellular bridges to produce single-cell suspensions of germ cells. Because of the substantial changes in cell size during spermatogenesis, distinct cell populations can be isolated from these suspensions by velocity sedimentation either at unit gravity or by centrifugal elutriation.[167-170] Differences in germ cell density have also allowed equilibrium density gradient centrifugation to be used as a second purification step.[93,171,172] With these procedures, mid- to late-pachytene spermatocytes, round spermatids, and condensing spermatids have been isolated from adult animals of several species. In mouse and rat, spermatogonia and spermatocytes at earlier stages of meiosis also have been isolated from prepuberal animals.[168,172] Using these procedures, germ cells representing most of the major stages of development have been isolated with sufficient yields and purities to allow biochemical analysis.

One of the uses of isolated germ cells has been to determine the energy requirements of spermatogenic cells at different stages of development. Such studies have shown that glucose is not an adequate metabolic substrate for pachytene spermatocytes or round spermatids from the rat. Lactate is the preferred metabolic substrate of round spermatids, while lactate or pyruvate will maintain ATP levels of pachytene spermatocytes.[70,72,173,174] These findings, in turn, have allowed culture conditions to be defined (medium supplemented with lactate, pyruvate, and fetal bovine serum) that allow mouse pachytene spermatocytes and round spermatids to be maintained for up to 48 h with viabilities exceeding 85%.[175,176] Rat germ cells have been maintained *in vitro* for 20 to 24 h in similar medium containing albumin instead of serum.[177] Pachytene spermatocytes and round spermatids have been used for *in vitro* studies of RNA synthesis[177] and protein synthesis and glycosylation.[175,176,178,179] In addition, mouse round spermatids cultured under these conditions are capable of reforming flagella removed during the isolation process.[175] Although serum is toxic to prepachytene spermatocytes isolated from juvenile mice, metabolic labeling for up to 6 h has been possible in serum-free medium.[180]

Isolated germ cells have been used to characterize several spermatogenic cell-specific molecules and to determine when they are present during germ cell development. Examples from this laboratory include cell surface lactosaminoglycans,[85] acrosomal glycoconjugates,[181,182] a structural protein of the fibrous sheath,[165] and unusual members of the hsp70 heat shock protein family.[183-185] Specific uses of isolated spermatogenic cells in these studies were to verify that these molecules are constituents of germ cells and not somatic cells, to determine their subcellular localization, to define the time of their appearance during spermatogenesis, and to identify and characterize the presence of particular proteins during different stages of germ cell development by one- and two-dimensional immunoblotting with antibodies specific to these proteins.

Isolated spermatogenic cells also have been used to monitor the synthesis of specific proteins at different stages of germ cell development. For example, an hsc71 heat shock protein variant that appears in condensing spermatids was shown to arise by *de novo* synthesis rather than modification of a preexisting protein,[185] and P70, a germ cell-specific protein apparently related to hsp70 heat shock protein, was found to be synthesized predominantly during the pachytene stage of meiosis.[183,184]

Another use of isolated spermatogenic cells has been to identify specific mRNA transcripts synthesized during restricted periods of spermatogenesis. Several studies have used DNA probes on isolated spermatogenic cells to identify lineage- and stage-specific transcripts.[109,110,186] The analysis of mRNAs and cDNAs from isolated spermatogenic cells has begun to give new insight into the mechanisms involved in regulating spermatogenic cell differentiation (discussed below).

Although various studies show that short-term cultures with isolated spermatogenic cells are quite useful, more could be learned if germ cell differentiation could be supported *in vitro*. Longer survival times and limited differentiation have been observed when germ cells are cocultured with Sertoli cells, either as small segments of seminiferous tubules[187] or as cell aggregates.[188-190] These approaches have been used to monitor the effects of reproductive toxicants on spermatogenic cells.[191,192] However, biochemical analyses of cell-specific events or of mechanisms for cell interaction are complicated by the diversity of cells present in these preparations.

A promising alternative approach for some questions is the coculture of distinct stages of isolated germ cells with Sertoli cells or with Sertoli cell-conditioned medium. With this approach it has been shown that germ cells, particularly pachytene spermatocytes, adhere specifically to Sertoli cells[193,194] and surface glycoproteins of pachytene spermatocytes that appear to mediate this interaction have been identified.[195] In addition to direct cell contact, secretory products may mediate germ cell-Sertoli cell interactions. *In vitro* studies suggest that lactate and pyruvate secreted by Sertoli cells may provide a source of energy substrates required by germ cells.[173] However, other factors are clearly involved in the complex interactions between these cell types. Protein factor(s) in Sertoli cell-conditioned medium, but not pyruvate and lactate, will maintain viabilities and ATP levels of pachytene spermatocytes and round spermatids cultured for 2 days in serum-free medium.[196,197] Furthermore, Sertoli cells secrete growth factors that may be important modulators of spermatogenesis.[14,198] The multifunctional receptor for insulin-like growth factor II and mannose 6-phosphate has recently been found on the surfaces of pachytene spermatocytes and round spermatids.[199] Information gained from such studies on cell interactions in the seminiferous epithelium should lead to refinements in culture conditions and media for isolated germ cells, as well as improving our understanding of the regulation of spermatogenesis.

VI. GENE REGULATION IN SPERMATOGENESIS

Three peaks of RNA synthesis occur during spermatogenesis which correspond to the three main phases of male germ cell development.[110] The first occurs in spermatogonia and may largely represent the general synthetic activity associated with the considerable proliferation and growth of these cells. In addition, the high level of expression of certain oncogenes,[29,30] histones,[22] and a β-actin[33] specific to this phase of spermatogenesis has been discussed previously. The second peak of RNA synthesis occurs in the midpachytene spermatocyte stage of meiosis, in the absence of active cell proliferation, but during a period of cell enlargement and the expression of several germ cell-specific genes. Some of these genes have been considered already, including PGK-2,[65,69] LDH-C,[77,79,80,96] and certain histones.[26,28,58-60] Additional spermatogenic cell-specific products expressed during meiosis that have been identified include an enolase,[200] P70,[183,184] and hsp70-related transcripts.[201,202]

The third peak of RNA synthesis occurs during spermiogenesis. While earlier reports indicated little or no production of mRNA in spermatids, more recent studies correcting for the decrease in volume from the two meiotic divisions and employing high resolution autoradiography have shown that this is not the case.[110] The ratio of total RNA/DNA remains constant for pachytene spermatocytes and round spermatids, but the concentration of poly(A) + mRNA in round spermatids is double that in pachytene spermatocytes[19] and transcription continues in spermatids until nuclear condensation begins.[109] Like the transition proteins[104,105,203] and protamines[204] discussed earlier, calspermin appears to be another unique gene product of postmeiotic cells.[205]

In addition to gene products shown to be spermatogenic cell-specific and for which the stage of expression is known, other testis-specific variants have been reported recently which remain to be fully characterized. These include a calmodulin-dependent phosphodiesterase,[206] an alkaline phosphatase,[207] and cytochrome c_t.[208] It remains to be determined if the products of these genes are present in germ cells or somatic cells, and if present in germ cells, when they are expressed during spermatogenesis.

Another category of unique gene products occurring in spermatogenic cells and the testis are transcripts which differ in size from those seen in somatic cells. Germ cell-specific or testis-specific transcripts have been reported for angiotensin converting enzyme,[209] type I[210] and type II[211] regulatory subunits for cAMP-dependent protein kinase, ATP-dependent RNA helicase,[212] phosphoribosyl phosphate synthetase,[213] glyceraldehyde 3-phosphate dehydrogenase,[214,215] cAMP phosphodiesterase,[216] catalytic subunit of phosphorylase kinase,[217] proenkephalin,[218,219] proopiomelanocortin,[220] cholecystokinin,[221,222] and several protooncogenes.[186,223,224] Although some of the mRNAs might represent products of genes uniquely expressed in the testis, they could also be transcripts from more widely expressed genes that differ in size due to use of alternative start sites or polyadenylation signals, differential splicing, or other processing events.

Translational control appears to be an important mechanism for regulating gene expression in spermatogenesis. During this process, translationally inactive mRNA is sequestered in a nonpolysomal or ribonucleoprotein compartment until it is recruited into the polysomal compartment for translation into protein.[110] The components involved both in the sequestration and recruitment are unknown, but mRNA isolated from either compartment translates efficiently *in vitro*.[225] There appear to be at least two categories of translational control in germ cells, one in which transcript size is unaltered prior to translation, and another in which a decrease in the length of 3′ poly(A) tracts is associated with translation.

The PGK-2 transcript appears in the cytoplasm during meiotic prophase and increases substantially in amount after meiosis.[226] It is found mainly in the nonpolysomal mRNA fraction in spermatocytes and in the polysomal fraction in spermatids.[67] Since the PGK-2 protein is not detected until after meiosis,[68] at least some of the mRNA is probably stored in the cytoplasm and not translated until several days later. The mRNA for PGK-2 is the same size in mouse spermatocytes and spermatids[65] and appears not to be modified during storage or prior to translation.

The transcripts for transition proteins and protamines are synthesized in round spermatids, but the proteins do not appear until several days later in elongating spermatids.[69,104,105,108,203] The mRNAs for mouse TP1, TP2, mP1, and mP2, and an unidentified mRNA (HEM1050) are not associated with polysomes in round spermatids or early elongating spermatids, but about 30 to 40% of each is associated with polysomes in elongating spermatids.[227] The translationally repressed forms of the transition protein and protamine nonpolysomal mRNAs are homogeneous in size, whereas the polysomal forms are shorter and more heterogenous due to shortening of their poly(A) tracts. The poly(A) shortening is thought to be a consequence of, rather than a prerequisite for, translation.[227] Although the majority of the HEM1050 mRNA shows no obvious poly(A) shortening, it has been suggested that the relationship between translational activity and poly(A) size may be typical of mRNAs expressed in round spermatids and translated in elongating spermatids.[227]

LDH-C transcripts also undergo a change in size during spermatogenesis, but in reverse of the situation for transition proteins and protamines, there is an increase in the length of the poly(A) tails in spermatids compared to those in spermatocytes.[228] However, LDH-C mRNA and active enzyme are both synthesized in spermatocytes and spermatids.[74,75] Although differences in LDH-C transcript size do not appear to affect translation or to be involved in translational control, it has been suggested that the longer poly(A) tracts of the LDH-C mRNA in postmeiotic cells could stabilize the mRNA for continued synthesis of the enzyme in elongating spermatids.[228]

The pattern of expression of the testis-specific histones presents an interesting contrast to that of somatic histones. The rat H1t[229] and TH2B[230] sequences display features in common with many somatic histones, including a 3′ palindromic sequence (inverted sequence repeat) which usually specifies the 3′ end of DNA replication-dependent somatic cell histone mRNA. Evidence for both H1t and TH2B indicates that their mRNAs are nonpolyadenylated, as are most somatic histone messages.[231] However, these testis-specific histone transcripts appear well after DNA synthesis is complete, in contrast to their somatic counterparts.[229,230] It was previously thought that only a few poly(A)-containing histone messages were expressed independent of DNA replication. These differences in expression of germ cell and somatic cell histones make it doubtful that sequences previously thought to regulate replication-dependent histones have this role. More recently a germ cell-specific polyadenylated H2B transcript has been identified in mouse spermatids.[232] While lacking the 3′ palindromic sequence, the predicted protein is >95% identical with somatic H2B and the germ cell mRNA appears to be actively translated.

Transcriptional regulation of lineage- and stage-specific gene expression in spermatogenic cells is an attractive possibility, but common motifs suggestive of regulatory sequences have not yet been identified among the genes known to be expressed uniquely during spermatogenesis. However, a 2.4-kb DNA sequence containing the mouse protamine 1 (mP1) coding sequence and 880 nucleotides of the 5′ flanking noncoding sequence was sufficient to achieve spermatid-specific expression of this sequence in transgenic mice.[233] Further studies in transgenic mice using a mP1-human growth hormone fusion gene showed that all of the *cis*-acting sequence elements required for spermatid-specific transcription of mP1 lie within the 0.46-kb mP1 5′ sequence.[234] Subsequent studies found that the mP1-human growth hormone fusion gene product was first detected in elongating spermatids at about stage 12.[235] However, when the mP1 3′-untranslated region was replaced with the human growth hormone 3′-untranslated region, the fusion gene product was detected in round spermatids at about stage 3 to 4.[235] These studies showed that 156 nucleotides of mP1 3′-untranslated region are sufficient to confer mP1-type translational regulation on the transgene in spermatids.[235]

Using such approaches in future studies, it should be possible to identify additional sequence elements responsible for the proper tissue and temporal expression of spermatogenic cell genes and to begin to identify transcriptional regulatory factors for these genes. These elegant *in vivo* systems, used in combination with *in vitro* studies, will be valuable for identifying and characterizing intrinsic and extrinsic mechanisms regulating expression of genes important to mammalian sperm development.

REFERENCES

1. **Eddy, E. M.,** The spermatozoon, in *The Physiology of Reproduction,* Knobil, E. and Neill, J., Eds., Raven Press, New York, 1988, 27.
2. **De Kretser, D. M. and Kerr, J. B.,** The cytology of the testis, in *The Physiology of Reproduction,* Knobil, E. and Neill, J., Eds., Raven Press, New York, 1988, 837.
3. **Bardin, C. W., Cheng, C. Y., Musto, N. A., and Gunsalus, G. L.,** The Sertoli cell, in *The Physiology of Reproduction,* Knobil, E. and Neill, J., Eds., Raven Press, New York, 1988, 933.
4. **Dym, M. and Fawcett, D. W.,** The blood-testis barrier in the rat and the physiological compartmentation of the seminiferous epithelium, *Biol. Reprod.,* 3, 308, 1970.
5. **Fritz, I. B.,** Sites of action of androgens and follicle-stimulating hormone on cells of the seminiferous tubule, in *Biochemical Actions of Hormones,* Vol. 5, Litwak, G., Ed., Academic Press, New York, 1978, 249.
6. **Oakberg, E. F.,** A description of spermiogenesis in the mouse and its use in the analysis of the cycle in the seminiferous epithelium and germ cell renewal, *Am. J. Anat.,* 99, 391, 1956.
7. **Oakberg, E. F.,** Duration of spermatogenesis in the mouse and timing of stages of the cycle of the seminiferous epithelium, *Am. J. Anat.,* 99, 507, 1956.
8. **Schulze, W. and Rehder, U.,** Organization and morphogenesis of the human seminiferous epithelium, *Cell Tissue Res.,* 237, 395, 1984.
9. **De Rooij, D. G.,** Regulation of the proliferation of spermatogonial stem cells, *J. Cell Sci.,* Suppl. 10, 181, 1989.
10. **Leblond, C. P. and Clermont, Y.,** Definition of the stages of the cycle of the seminiferous epithelium in the rat, *Ann. N.Y. Acad. Sci.,* 55, 548, 1952.
11. **Moens, P. B. and Go, V. L. W.,** Intercellular bridges and division patterns of rat spermatogonia, *Z. Zellforsch.,* 127, 201, 1972.
12. **Huckins, C.,** The spermatogonial stem cell population in adult rats. I. Their morphology, proliferation and maturation, *Anat. Rec.,* 169, 533, 1971.
13. **Lok, D. and De Rooij, D. G.,** Spermatogonial multiplication of the Chinese hamster. I. Cell cycle properties and synchronization of differentiating spermatogonia, *Cell Tissue Kinet.,* 16, 7, 1983.
14. **Bellvé, A. R. and Zheng, W.,** Growth factors as autocrine and paracrine modulators of male gonadal functions, *J. Reprod. Fertil.,* 85, 771, 1989.
15. **De Rooij, D. G., Lok, D., and Weenk, D.,** Feedback regulation of the proliferation of the undifferentiated spermatogonia in the Chinese hamster by the differentiating spermatogonia, *Cell Tissue Kinet.,* 18, 71, 1985.
16. **Lohka, M. J., Hayes, M. K., and Maller, J. L.,** Purification of maturation-promoting factor, an intracellular regulator of early mitotic events, *Proc. Natl. Acad. Sci. U.S.A.,* 85, 3009, 1988.
17. **Labbe, J. C., Lee, M. G., Nurse, P., Picard, A., and Doree, M.,** Activation at M-phase of a protein kinase encoded by a starfish homologue of the cell control gene cdc2+, *Nature,* 335, 251, 1988.
18. **Arion, D., Meijer, L., Brizuela, L., and Beach, D.,** cdc2 is a component of the M-phase specific histone H1 kinase: evidence for identity with MPF, *Cell,* 55, 371, 1988.
19. **Evans, T., Rosenthal, E. T., Youngblom, J., Distel, D., and Hunt, T.,** Cyclin: a protein specified by maternal mRNA in sea urchin eggs that is destroyed at each cleavage division, *Cell,* 33, 389, 1983.
20. **Dunphy, W. G. and Newport, J. W.,** Unraveling of mitotic control mechanisms, *Cell,* 55, 925, 1988.
21. **Whitfield, W. G. F., Gonzalez, C., Sanchez-Herrero, E., and Glover, D. M.,** Transcripts of one of two *Drosophila* cyclin genes become localized in pole cells during embryogenesis, *Nature,* 338, 337, 1989.
22. **Meistrich, M. L. and Brock, W. A.,** Proteins of the meiotic cell nucleus, in *Meiosis,* Moens, P. B., Ed., Academic Press, Orlando, FL, 1987, 333.
23. **Bhatnagar, Y. M. and Bellvé, A. R.,** Two-dimensional electrophoretic analysis of major histone species and their variants from somatic and germ-line tissue, *Anal. Biochem.,* 86, 754, 1978.
24. **Lennox, R. W. and Cohen, L. H.,** The alterations in H1 histone complement during mouse spermatogenesis and their significance for H2 subtype function, *Dev. Biol.,* 163, 80, 1984.

25. **Zweidler, A.,** Core histone variants in the mouse: primary structure and differential expression, in *Histone Genes: Structure, Organization and Regulation,* Stein, G. S., Stein, J. L., and Marzluff, W. F., Eds., John Wiley & Sons, New York, 1984, 339.
26. **Bucci, L. R., Brock, W. A., Goldknopf, I. L., and Meistrich, M. L.,** Characterization of high mobility group protein levels during spermatogenesis in the rat, *J. Biol. Chem.,* 259, 8840, 1984.
27. **Bucci, L. R., Brock, W. A., and Meistrich, M. L.,** Heterogeneity of high-mobility-group protein 2. Enrichment of a rapidly migrating form in the testis, *Biochem. J.,* 229, 233, 1985.
28. **Bucci, L. R., Brock, W. A., Johnson, T. S., and Meistrich, M. L.,** Isolation and biochemical characterization of enriched populations of spermatogonia and early primary spermatocytes from rat testes, *Biol. Reprod.,* 34, 195, 1986.
29. **Koji, T., Izumi, S., Tanno, M., Moriuchi, T., and Nakane, P. K.,** Localization in situ of *c-myc* mRNA and *c-myc* protein in adult mouse testis, *Histochem. J.,* 20, 551, 1988.
30. **Wolfes, H., Kogawa, K., Millette, C. F., and Cooper, G. M.,** Specific expression of nuclear proto-oncogenes before entry into meiotic prophase of spermatogenesis, *Science,* 245, 740, 1989.
31. **Curran, T. and Franza, B. R., Jr.,** Fos and Jun: the AP-1 connection, *Cell,* 55, 395, 1988.
32. **Morhenegg, G., Celio, M. R., Malfoy, B., Leng, M., and Kuenzle, C. C.,** Z-DNA immunoreactivity in rat tissues, *Nature,* 303, 540, 1983.
33. **Slaughter, G. R., Needleman, D. S., and Means, A. R.,** Developmental regulation of calmodulin, actin, and tubulin RNAs during rat testis differentiation, *Biol. Reprod.,* 37, 1259, 1987.
34. **Stern, H. and Hotta, Y.,** The biochemistry of meiosis, in *Meiosis,* Moens, P. B., Ed., Academic Press, Orlando, FL, 1987, 303.
35. **Russell, L. D.,** Movement of spermatocytes from the basal to the adluminal compartment of the rat testis, *Am. J. Anat.,* 148, 313, 1977.
36. **Moens, P. B.,** Introduction to meiosis, in *Meiosis,* Moens, P. B., Ed., Academic Press, Orlando, FL, 1987, 1.
37. **Heyting, C., Dietrich, A. J. J., Redeker, E. J., and Vink, A. C. G.,** Structure and composition of synaptonemal complexes, isolated from rat spermatocytes, *Eur. J. Cell Biol.,* 36, 307, 1985.
38. **Behal, A., Prakash, K., and Rao, M. R. S.,** Identification of a meiotic prophase-specific nuclear matrix protein in the rat, *J. Biol. Chem.,* 262, 10898, 1988.
39. **Dresser, M. E.,** The synaptonemal complex and meiosis; an immunohistochemical approach, in *Meiosis,* Moens, P. B., Ed., Academic Press, Orlando, FL, 1987, 245.
40. **Heyting, C., Dettmers, R. J., Dietrich, A. J. J., Redeker, E. J. W., and Vink, A. C. G.,** Two major components of synaptonemal complexes are specific for meiotic prophase nuclei, *Chromosoma,* 96, 325, 1988.
41. **Heyting, C., Moens, P. B., van Raamsdonk, W., Dietrich, A. J. J., Vink, A. C. G., and Redeker, E. J. W.,** Identification of two major components of the lateral elements of synaptonemal complexes of the rat, *Eur. J. Cell Biol.,* 43, 148, 1987.
42. **Moens, P. B., Heyting, C., Dietrich, A. J. J., van Raamsdonk, W., and Chen, Q.,** Synaptonemal complex antigen location and conservation, *J. Cell Biol.,* 105, 93, 1987.
43. **Byskov, A. G. and Saxen, L.,** Induction of meiosis in foetal mouse testis in vitro, *Dev. Biol.,* 52, 193, 1976.
44. **Byskov, A. G.,** Regulation of initiation of meiosis in fetal gonads, *J. Androl.,* 2(Suppl.), 29, 1978.
45. **Byskov, A. G. and Hoyer, P. E.,** Embryology of mammalian gonads and ducts, in *The Physiology of Reproduction,* Knobil, E. and Neill, J., Eds., Raven Press, New York, 1988, 265.
46. **Sagata, N., Daar, I., Oskarsson, M., Showalter, S. D., and Vande Woude, G. F.,** The product of the *mos* proto-oncogene as a candidate "initiator" for oocyte maturation, *Science,* 245, 643, 1989.
47. **Sagata, N., Oskarsson, M., Copeland, T., Brumbaugh, J., and Vande Woude, G. F.,** Function of *c-mos* proto-oncogene product in meiotic maturation of *Xenopus* oocytes, *Nature,* 335, 519, 1988.
48. **Paules, R. S., Buccione, R., Moschel, R. C., Vande Woude, G. F., and Eppig, J. J.,** Mouse *Mos* protooncogene product is present and functions during oogenesis, *Proc. Natl. Acad. Sci. U.S.A.,* 86, 5395, 1989.
49. **Hotta, Y. and Stern, H.,** The organization of DNA segments undergoing repair synthesis during pachytene, *Chromosoma,* 89, 127, 1984.
50. **Hotta, Y. and Stern, H.,** The effect of dephosphorylation on the properties of a helix-destabilizing protein from meiotic cells and its partial reversal by a protein kinase, *Eur. J. Biochem.,* 95, 31, 1979.
51. **Hirose, F., Hotta, Y., Yamaguchi, M., and Matsukage, A.,** Difference in the expression level of DNA polymerase β among mouse tissues: high expression in the pachytene spermatocyte, *Exp. Cell Res.,* 181, 169, 1989.
52. **Orlando, P., Geremia, R., Frusciante, C., Tedeschi, B., and Grippo, P.,** DNA repair synthesis in mouse spermatogenesis involves DNA polymerase β activity, *Cell Differ.,* 23, 221, 1988.
53. **Hotta, Y., Tabata, S., Bouchard, R. A., Pinon, R., and Stern, H.,** General recombination mechanisms in extracts of meiotic cells, *Chromosoma,* 93, 140, 1985.

54. **Chowdhury, A. K. and Steinberger, E.,** A quantitative study of the effect of heat on germinal epithelium of rat testes, *Am. J. Anat.*, 115, 509, 1964.
55. **Chowdhury, A. K. and Steinberger, E.,** Early changes in the germinal epithelium of rat testis following exposure to heat, *J. Reprod. Fertil.*, 22, 205, 1970.
56. **Rao, B. J., Brahmachari, S. K., and Rao, M. R. S.,** Structural organization of the meiotic prophase chromatin in the rat testis, *J. Biol. Chem.*, 258, 13478, 1983.
57. **Grimes, S. R. and Smart, P. G.,** Changes in the structural organization of chromatin during spermatogenesis in the rat, *Biochim. Biophys. Acta*, 824, 128, 1985.
58. **Meistrich, M. L., Bucci, L. R., Trostle-Weige, P. K., and Brock, W. A.,** Histone variants in rat spermatogonia and primary spermatocytes, *Dev. Biol.*, 112, 230, 1985.
59. **Grimes, S. R., Jr.,** Nuclear proteins in spermatogenesis, *Comp. Biochem. Physiol. B*, 83, 495, 1986.
60. **Wolfe, S. A., Anderson, J. V., Grimes, S. R., Stein, G. S., and Stein, J. S.,** Comparison of the structural organization and expression of germinal and somatic rat histone H4 genes, *Biochim. Biophys. Acta*, 1007, 140, 1989.
61. **Concha, I. I., Figuoeroa, J., Concha, M. I., and Burzio, L. O.,** Intracellular distribution of poly(ADP-ribose) synthetase in rat spermatogenic cells, *Exp. Cell Res.*, 180, 353, 1989.
62. **Quesada, P., Farina, B., and Jones, R.,** Poly(ADP-ribosylation) of nuclear proteins in rat testis correlates with active spermatogenesis, *Biochim. Biophys. Acta*, 1007, 167, 1989.
63. **Zwanenburg, T. S. B. and van Buul, P. P. W.,** Inhibition of ADP-ribosylation increases X-ray-induced chromosomal damage in mouse testis and bone-marrow cells in vivo, *Mutat. Res.*, 175, 33, 1986.
64. **Satyanarayana, K. and Rao, M. R.,** Characterization of poly(ADP-ribosyl)ated domains of rat pachytene chromatin, *Biochem. J.*, 261, 775, 1989.
65. **McCarrey, J. R. and Thomas, K.,** Human testis-specific PGK gene lacks introns and possesses characteristics of a processed gene, *Nature*, 326, 501, 1987.
66. **Goldberg, E.,** Isozymes in testes and spermatozoa, in *Isozymes*, Ratazzi, M. C., Scandalios, J. G., and Whitt, G. S., Eds., Alan R. Liss, New York, 1977, 80.
67. **Gold, B., Fujimoto, H., Kramer, J. M., Erickson, R. P., and Hecht, N. B.,** Haploid accumulation and translational control of phosphoglycerate kinase-2 messenger RNA during mouse spermatogenesis, *Dev. Biol.*, 98, 392, 1983.
68. **Kramer, J. M.,** Immunofluorescent localization of PGK-1 and PGK-2 isozymes within specific cells of the mouse testis, *Dev. Biol.*, 87, 30, 1981.
69. **Boer, P. H., Adra, C. N., Lau, Y.-F., and McBurney, M. W.,** The testis-specific phosphoglycerate kinase gene pgk-2 is a recruited retroposon, *Mol. Cell. Biol.*, 7, 3107, 1987.
70. **Grootegoed, J. A., Jansen, R., and van der Molen, H. J.,** The role of glucose, pyruvate and lactate in ATP production by spermatocytes and spermatids, *Biochim. Biophys. Acta*, 767, 248, 1984.
71. **Nakamura, M., Okinaga, S., and Arai, K.,** Metabolism of pachytene primary spermatocytes from rat testes: pyruvate maintenance of adenosine triphosphate level, *Biol. Reprod.*, 30, 1187, 1984.
72. **Nakamura, M., Okinaga, S., and Arai, K.,** Metabolism of round spermatids: evidence that lactate is preferred substrate, *Am. J. Physiol.*, 247, E234, 1984.
73. **Markert, C. L., Shakelee, J. B., and Whitt, G. S.,** Evolution of a gene, *Science*, 189, 102, 1975.
74. **Li, S. S.-L., O'Brien, D. A., Hou, E. W., Versola, J., Rockett, D. L., and Eddy, E. M.,** Differential activity and synthesis of lactate dehydrogenase isozymes A (muscle), B (heart), and C (testis) in mouse spermatogenic cells, *Biol. Reprod.*, 40, 173, 1989.
75. **Wieben, E. D.,** Regulation of the synthesis of lactate dehydrogenase-X during spermatogenesis in the mouse, *J. Cell Biol.*, 88, 492, 1981.
76. **Fujimoto, H., Erickson, R. P., Quinto, M., and Rosenberg, M. P.,** Post-meiotic transcription in mouse testes detected with spermatid cDNA clones, *Biosci. Rep.*, 4, 1037, 1984.
77. **Sakai, I., Sharief, F. S., and Li, S. S.-L.,** Molecular cloning and nucleotide sequence of the cDNA for sperm-specific lactate dehydrogenase-C from mouse, *Biochem. J.*, 242, 619, 1987.
78. **Wu, K. C., Chan, K., Lee, C.-Y. G., and Lau, Y.-F.,** Molecular isolation and sequence determination of the cDNA for the mouse sperm-specific lactate dehydrogenase-X gene, *Biochem. Biophys. Res. Commun.*, 146, 946, 1987.
79. **Millan, J. L., Driscoll, C. E., LeVan, K. M., and Goldberg, E.,** Epitopes of human testis-specific lactate dehydrogenase deduced from a cDNA sequence, *Proc. Natl. Acad. Sci. U.S.A.*, 84, 5311, 1987.
80. **Takano, T. and Li, S. S.-L.,** Human testicular lactate dehydrogenase-C gene is interrupted by six introns at positions homologous to those of LDH-A (muscle) and LDH-B (heart) genes, *Biochem. Biophys. Res. Commun.*, 159, 579, 1989.
81. **Mita, M. and Hall, P.,** Metabolism of round spermatids from rats: lactate as the preferred substrate, *Biol. Reprod.*, 26, 445, 1982.
82. **Bellvé, A. R. and O'Brien, D. A.,** The mammalian spermatozoon: structural and temporal assembly, in *Mechanism and Control of Animal Fertilization*, Hartman, J. F., Ed., Academic Press, Orlando, FL, 1983, 55.

83. **O'Rand, M. G.,** Differentiation of mammalian sperm antigens, in *Biology of Fertilization,* Vol. 2, Metz, C. B. and Monroy, A., Eds., Academic Press, Orlando, FL, 1985, 103.
84. **Eddy, E. M., Vernon, R. B., Muller, C. H., Hahnel, A. C., and Fenderson, B. A.,** Immunodissection of sperm surface modifications during epididymal maturation, *Am. J. Anat.,* 174, 225, 1985.
85. **Fenderson, B. A., O'Brien, D. A., Millette, C. F., and Eddy, E. M.,** Stage-specific expression of three cell surface carbohydrate antigens during murine spermatogenesis detected with monoclonal antibodies, *Dev. Biol.,* 103, 117, 1984.
86. **Law, H., Itkonnen, O., and Lingwood, C. A.,** Sulfogalactolipid binding protein SLIP 1: a conserved function for a conserved protein, *J. Cell. Physiol.,* 137, 462, 1988.
87. **Lingwood, C. A.,** Developmental regulation of galactoglycerolipid and galactosphingolipid sulphation during mammalian spermatogenesis, *Biochem. J.,* 231, 393, 1985.
88. **Braun, R. E., Behringer, R. R., Peschon, J. J., Brinster, R. L., and Palmiter, R. D.,** Genetically haploid spermatids are phenotypically diploid, *Nature,* 337, 373, 1989.
89. **Silver, L. M.,** Mouse t haplotypes, *Annu. Rev. Genet.,* 19, 179, 1985.
90. **Kierszenbaum, A. L. and Tres, L. L.,** Structural and transcriptional features of the mouse spermatid genome, *J. Cell Biol.,* 65, 258, 1975.
91. **Kierszenbaum, A. L. and Tres, L. L.,** RNA transcription and chromatin structure during meiotic and postmeiotic stages of spermatogenesis, *Fed. Proc.,* 37, 2512, 1978.
92. **Loir, M., Bouvier, D., Fornells, M., Lanneau, M., and Subirana, J. A.,** Interactions of nuclear proteins with DNA during sperm differentiation in the ram, *Chromosoma,* 92, 304, 1985.
93. **Meistrich, M. L., Longtin, J., Brock, W. A., Grimes, S. R., and Mace, M. L.,** Purification of rat spermatogenic cells and preliminary biochemical analysis of these cells, *Biol. Reprod.,* 25, 1065, 1981.
94. **Kistler, W. S., Heidaran, M. A., Cole, K. D., Kandala, J. C., and Showman, R. M.,** Genes for chromosomal proteins expressed before and after meiosis, *Ann. N.Y. Acad. Sci.,* 513, 102, 1987.
95. **Oliva, R., Mezquita, J., Mezquita, C., and Dixon, G. H.,** Haploid expression of the rooster protamine mRNA in the postmeiotic stages of spermatogenesis, *Dev. Biol.,* 125, 332, 1988.
96. **Agell, N., Chiva, M., and Mezquita, C.,** Changes in nuclear content of protein conjugate histone H2A-ubiquitin during rooster spermatogenesis, *FEBS Lett.,* 155, 209, 1983.
97. **Nickel, B. E., Roth, S. Y., Cook, R. G., Allis, C. D., and Davie, J. R.,** Changes in the histone H2A.Z and polyubiquitinated histone species in developing trout testis, *Biochemistry,* 26, 4417, 1987.
98. **Loir, M., Caraty, A., Lanneau, M., Menezo, Y., Muh, J. P., and Sautiere, P.,** Purification and characterization of ubiquitin from mammalian testis, *FEBS Lett.,* 169, 199, 1984.
99. **Meistrich, M. L., Brock, W. A., Grimes, S. R., Platz, R. D., and Hnilica, L. S.,** Nuclear protein transitions during spermatogenesis, *Fed. Proc.,* 37, 2522, 1978.
100. **Balhorn, R., Weston, S., Thomas, C., and Wyrobek, A. J.,** DNA packaging in mouse spermatids. Synthesis of protamine variants and four transition proteins, *Exp. Cell Res.,* 150, 298, 1984.
101. **Dupressoir, T., Sautiere, P., Lanneau, M., and Loir, M.,** Isolation and characterization of the ram spermatidal nuclear proteins P1, 2, and T, *Exp. Cell Res.,* 161, 63, 1985.
102. **Heidaran, M. A., Showman, R. M., and Kistler, W. S.,** A cytochemical study of the transcriptional and translational regulation of nuclear transition protein 1 (TP1), a major chromosomal protein of mammalian spermatids, *J. Cell Biol.,* 106, 1427, 1988.
103. **Heidaran, M. A. and Kistler, W. S.,** Isolation of a cDNA clone for transition protein 1 (TP1), a major chromosomal protein of mammalian spermatids, *Gene,* 54, 281, 1987.
104. **Luerssen, H., Maier, W. M., Hoyer-Fender, S., and Engel, W.,** The nucleotide sequence of rat transition protein 2 (TP2) cDNA, *Nucleic Acids Res.,* 17, 3585, 1989.
105. **Kleene, K. C., Borzorgzadah, A., Flynne, J. F., Yelick, P. C., and Hecht, N. B.,** Nucleotide sequence of a cDNA clone encoding mouse transition protein 1, *Biochim. Biophys. Acta,* 950, 215, 1988.
106. **Cole, K. D. and Kistler, W. S.,** Nuclear transition protein 2 (TP2) of mammalian spermatids has a very basic carboxyl terminal domain, *Biochem. Biophys. Res. Commun.,* 147, 437, 1987.
107. **Singh, J. and Satyanatayana Rao, M.,** Interaction of rat testis protein, TP, with nucleic acids in vitro: fluorescence quenching, UV absorption and thermal denaturation studies, *J. Biol. Chem.,* 262, 734, 1987.
108. **Heidaran, M. A. and Kistler, W. S.,** Transcriptional and translational control of the message for transition protein 1, a major chromosomal protein of mammalian spermatids, *J. Biol. Chem.,* 262, 13309, 1987.
109. **Hecht, N. B.,** Regulation of gene expression during mammalian spermatogenesis, in *Experimental Approaches to Mammalian Embryonic Development,* Rossant, J. and Pedersen, R., Eds., Cambridge University Press, New York, 1986, 151.
110. **Hecht, N. B.,** Post-meiotic gene expression during spermatogenesis, in *Meiotic Inhibition: Molecular Control of Meiosis,* Vol. 267, Haseltine, F. P. and First, N. L., Eds., Alan R. Liss, New York, 1988, 291.
111. **Johnson, P. A., Yelick, P. C., Liem, H., and Hecht, N. B.,** Differential distribution of the P1 and P2 protamine gene sequences in eutherian and marsupial mammals and a monotreme, *Gamete Res.,* 19, 169, 1988.

112. **Bower, P. A., Yelick, P. C., and Hecht, N. B.,** Both P1 and P2 protamine genes are expressed in mouse, hamster, and rat, *Biol. Reprod.*, 37, 479, 1987.
113. **Balhorn, R.,** A model for the structure of chromatin in mammalian sperm, *J. Cell Biol.*, 93, 298, 1982.
114. **Calvin, H. I. and Bedford, J. M.,** Formation of disulphide bonds in the nucleus and accessory structures of mammalian spermatozoa during maturation in the epididymis, *J. Reprod. Fert.*, Suppl. 13, 65, 1971.
115. **Pellicciari, C., Hosokawa, Y., Fukuda, M., and Manfredi Romanini, M. G.,** Cytofluorometric study of nuclear sulphydryl and disulfide groups during sperm maturation in the mouse, *J. Reprod. Fertil.*, 68, 371, 1983.
116. **Shalgi, R., Seligman, J., and Kosower, N. S.,** Dynamics of the thiol status of rat spermatozoa during maturation: analysis with the fluorescent labeling agent monobromobimane, *Biol. Reprod.*, 40, 1037, 1989.
117. **Welch, J. E. and O'Rand, M. G.,** Identification and distribution of actin in spermatogenic cells and spermatozoa of the rabbit, *Dev. Biol.*, 109, 411, 1985.
118. **Marsi, B. A., Russell, L. D., and Vogl, A. W.,** Distribution of actin in spermatids and adjacent Sertoli cell regions in the rat, *Anat. Rec.*, 218, 20, 1989.
119. **Flaherty, S. P., Winfrey, V. P., and Olson, G. E.,** Localization of actin in human, bull, rabbit, and hamster sperm by immunoelectron microscopy, *Anat. Rec.*, 221, 599, 1988.
120. **Aumüller, G. and Seitz, J.,** Immunocytochemical localization of actin and tubulin in rat testis and spermatozoa, *Histochemistry*, 89, 261, 1988.
121. **Fouquet, J.-P., Kann, M.-L., and Dadoune, J.-P.,** Immunogold distribution of actin during spermiogenesis in the rat, hamster, monkey, and human, *Anat. Rec.*, 223, 35, 1989.
122. **Waters, S. H., Distel, R. J., and Hecht, N. B.,** Mouse testes contain two size classes of actin mRNA that are differentially expressed during spermatogenesis, *Mol. Cell. Biol.*, 5, 1649, 1985.
123. **Kim, E., Waters, S. H., Hake, L. E., and Hecht, N. B.,** Identification and developmental expression of a smooth-muscle γ-actin in postmeiotic male germ cells of mice, *Mol. Cell. Biol.*, 9, 1875, 1989.
124. **Fawcett, D. W., Anderson, W., and Phillips, D. M.,** Morphogenetic factors influencing the shape of the sperm head, *Dev. Biol.*, 26, 220, 1971.
125. **McIntosh, J. R. and Porter, K. R.,** Microtubules in the spermatids of the domestic fowl, *J. Cell Biol.*, 35, 153, 1967.
126. **Hecht, N. B., Kleene, K. C., Distel, R. J., and Silver, L. M.,** The differential expression of the actins and tubulins during spermatogenesis in the mouse, *Exp. Cell Res.*, 153, 275, 1984.
127. **Distel, R. J., Kleene, K. C., and Hecht, N. B.,** Haploid expression of a mouse testis alpha-tubulin gene, *Science*, 224, 68, 1984.
128. **Lalli, M. and Clermont, Y.,** Structural changes of the head components of the rat spermatid during late spermiogenesis, *Am. J. Anat.*, 160, 419, 1981.
129. **Bellvé, A. R.,** Biogenesis of the mammalian spermatozoon, in *Prospects for Sexing Mammalian Sperm*, Amann, R. P. and Seidel, G. E., Jr., Eds., Colorado Association University Press, Boulder, 1982, 69.
130. **Huang, T. T. F. and Yanagimachi, R.,** Inner acrosomal membrane of mammalian spermatozoa: its properties and possible functions in fertilization, *Am. J. Anat.*, 174, 249, 1985.
131. **Courtens, J. L., Courot, M., and Flechon, J. E.,** The perinuclear substance of boar, bull, ram and rabbit spermatozoa, *J. Ultrastruct. Res.*, 57, 54, 1976.
132. **Olson, G. E., Hamilton, D. W., and Fawcett, D. W.,** Isolation and characterization of the perforatorium of rat spermatozoa, *J. Reprod. Fertil.*, 47, 293, 1976.
133. **Olson, G. E.,** Isolation of the fibrous sheath and perforatorium of rat spermatozoa, in *The Spermatozoon*, Fawcett, D. W. and Bedford, J. M., Eds., Urban & Schwarzenberg, Baltimore, 1979, 395.
134. **Oko, R. and Clermont, Y.,** Isolation, structure and protein composition of the perforatorium of rat spermatozoa, *Biol. Reprod.*, 39, 673, 1988.
135. **Maxwell, W. L.,** The acrosomal zonule, *Tissue Cell*, 14, 283, 1982.
136. **Czaker, R.,** Morphogenesis and cytochemistry of the postacrosomal dense lamina during mouse spermiogenesis, *J. Ultrastruct. Res.*, 90, 26, 1985.
137. **Longo, F. J., Krohne, G., and Franke, W. W.,** Basic proteins of the perinuclear theca of mammalian spermatozoa and spermatids: a novel class of cytoskeletal elements, *J. Cell Biol.*, 105, 1105, 1987.
138. **Paranko, J., Longo, F., Potts, J., Krohne, G., and Franke, W. W.,** Widespread occurrence of calicin, a basic cytoskeletal protein of sperm cells, in diverse mammalian species, *Differentiation*, 38, 21, 1988.
139. **Olson, G. E. and Winfrey, V. P.,** Substructure of a cytoskeletal complex associated with the hamster sperm acrosome, *J. Ultrastruct. Res.*, 92, 167, 1985.
140. **Koehler, J. K.,** Observations on the fine structure of vole spermatozoa with particular reference to cytoskeletal elements in the mature sperm head, *Gamete Res.*, 1, 247, 1978.
141. **Phillips, D. M.,** Cell surface structure of rodent sperm heads, *J. Exp. Zool.*, 191, 1, 1975.
142. **Phillips, D. M.,** Comparative analysis of mammalian sperm motility, *J. Cell Biol.*, 53, 561, 1972.
143. **Fawcett, D. W.,** The mammalian spermatozoon, *Dev. Biol.*, 44, 394, 1975.

144. **Escalier, D. and Serres, C.,** Aberrant distribution of the peri-axonemal structures in the human spermatozoon: possible role of the axoneme in the spatial organization of the flagellar organization of the flagellar components, *Biol. Cell,* 53, 239, 1985.
145. **Handel, M. A.,** Genetic control of spermatogenesis in mice, in *Spermatogenesis: Genetic Aspects,* Henning, W., Ed., Springer-Verlag, Berlin, 1987, 1.
146. **Olson, G. E. and Linck, R. W.,** Observations on the structural components of flagellar axonemes and central pair microtubules from rat sperm, *J. Ultrastruct. Res.,* 61, 21, 1977.
147. **Gibbons, I. R. and Rowe, A. J.,** Dynein: a protein with ATPase activity from cilia, *Science,* 149, 424, 1965.
148. **McConnell, J. D., Stone, D. K., Johnson, L., and Wilson, J. D.,** Partial purification and characterization of dynein adenosine triphosphatase from bovine sperm, *Biol. Reprod.,* 37, 385, 1987.
149. **Johnson, K. A., Porter, M. E., and Shimizu, T.,** Mechanism of force production for microtubule-dependent movements, *J. Cell Biol.,* 99, 1325, 1984.
150. **Baccetti, B. and Afzelius, B. A.,** *The Biology of the Sperm Cell,* Vol. 10, Wolsky, A., Ed., S. Karger, Basel, 1976.
151. **Fawcett, D. W. and Phillips, D. M.,** The fine structure and development of the neck region of the mammalian spermatozoon, *Anat. Rec.,* 165, 153, 1969.
152. **De Kretser, D. M.,** Ultrastructural features of human spermiogenesis, *Z. Zellforsch.,* 98, 229, 1969.
153. **Sapsford, C. S., Rae, C. A., and Cleland, K. W.,** Ultrastructural studies on the development and form of the principal piece sheath of the Bandicoot spermatozoon, *Aust. J. Zool,* 8, 21, 1970.
154. **Irons, M. J. and Clermont, Y.,** Formation of the outer dense fibers during spermiogenesis in the rat, *Anat. Rec.,* 202, 463, 1982.
155. **Calvin, H. I.,** Electrophoretic evidence for the identity of the major zinc-binding polypeptides in the rat sperm tail, *Biol. Reprod.,* 21, 873, 1979.
156. **Olson, G. E. and Sammons, D. W.,** Structural chemistry of outer dense fibers of rat sperm, *Biol. Reprod.,* 22, 319, 1980.
157. **Vera, J. C., Brito, M., Zuvic, T., and Burzio, L. O.,** Polypeptide composition of rat sperm outer dense fibers, *J. Biol. Chem.,* 259, 5970, 1984.
158. **Oko, R.,** Comparative analysis of proteins from the fibrous sheath and outer dense fibers of rat spermatozoa, *Biol. Reprod.,* 39, 168, 1988.
159. **Baccetti, B., Pallini, V., and Burrini, A. G.,** The accessory fibers of the sperm tail. III. High-sulfur and low-sulfur components in mammals and cephalopods, *J. Ultrastruct. Res.,* 57, 289, 1976.
160. **Bedford, J. M. and Calvin, H. I.,** Changes in -S-S- linked structures of the sperm tail during epididymal maturation, with comparative observations in sub-mammalian species, *J. Exp. Zool.,* 187, 181, 1974.
161. **Bradley, F. M., Meth, B. M., and Bellvé, A. R.,** Structural proteins of the mouse spermatozoan tail: an electrophoretic analysis, *Biol. Reprod.,* 24, 691, 1981.
162. **Irons, M. J. and Clermont, Y.,** Kinetics of fibrous sheath formation in the rat spermatid, *Am. J. Anat.,* 165, 121, 1982.
163. **Sakai, Y., Koyama, Y.-I., Fujimoto, H., Nakamoto, T., and Yamashina, S.,** Immunocytochemical study on fibrous sheath formation in mouse spermiogenesis using a monoclonal antibody, *Anat. Rec.,* 215, 119, 1986.
164. **Brito, M., Figueroa, J., Maldonado, E. U., Vera, J. C., and Burzio, L. O.,** The major component of the rat sperm fibrous sheath is a phosphoprotein, *Gamete Res.,* 22, 205, 1989.
165. **Fenderson, B. A., Toshimori, K., Muller, C. H., Lane, T. F., and Eddy, E. M.,** Identification of a protein in the fibrous sheath of the sperm flagellum, *Biol. Reprod.,* 38, 345, 1988.
166. **Eddy, E. M., O'Brien, D. A., Rockett, D. L., and Chapman, L. M.,** Appearance of flagellar cytoskeletal antigens during spermatogenesis, *J. Cell Biol.,* 103 (Abstr.), 82a, 1986.
167. **Romrell, L. J., Bellvé, A. R., and Fawcett, D. W.,** Separation of mouse spermatogenic cells by sedimentation velocity. A morphological characterization, *Dev. Biol.,* 49, 119, 1976.
168. **Meistrich, M. L.,** Separation of spermatogenic cells and nuclei from rodent testes, *Methods Cell Biol.,* 15, 15, 1977.
169. **Bellvé, A. R., Cavicchia, J. C., Millette, C. F., O'Brien, D. A., Bhatnagar, Y. M., and Dym, M.,** Spermatogenic cells of the prepubertal mouse: isolation and morphological characterization, *J. Cell Biol.,* 74, 68, 1977.
170. **Bellvé, A. R., Millette, C. F., Bhatnagar, Y. M., and O'Brien, D. A.,** Dissociation of the mouse testis and characterization of isolated spermatogenic cells, *J. Histochem. Cytochem.,* 25, 480, 1977.
171. **Stern, L., Gold, B., and Hecht, N. B.,** Gene expression during mammalian spermatogenesis. I. Evidence for stage-specific synthesis of polypeptides in vivo, *Biol. Reprod.,* 28, 483, 1983.
172. **Slaughter, G. R., Meistrich, M. L., and Means, A. R.,** Expression of RNAs for calmodulin, actins, and tubulins in rat testis cells, *Biol. Reprod.,* 40, 395, 1989.
173. **Grootegoed, J. A., Jansen, R., and van der Molen, H. J.,** Effect of glucose on ATP dephosphorylation in rat spermatids, *J. Reprod. Fertil.,* 77, 99, 1986.

174. **Nakamura, M., Okinaga, S., and Arai, K.,** Studies of metabolism of round spermatids: glucose as unfavorable substrate, *Biol. Reprod.*, 35, 927, 1986.
175. **Gerton, G. L. and Millette, C. F.,** Generation of flagella by cultured mouse spermatids, *J. Cell Biol.*, 98, 619, 1984.
176. **Gerton, G. L. and Millette, C. F.,** Stage-specific synthesis and fucosylation of plasma membrane proteins by mouse pachytene spermatocytes and round spermatids in culture, *Biol. Reprod.*, 35, 1025, 1986.
177. **Grootegoed, J. A., Grolle-Hey, A. H., Rommerts, F. F. G., and van der Molen, H. J.,** Ribonucleic acid synthesis in vitro in primary spermatocytes isolated from rat testes, *Biochem. J.*, 168, 23, 1977.
178. **Grootegoed, J. A., Kruger-Sewnarain, B. C., Jutte, N. H. P. M., Rommerts, F. F. G., and van der Molen, H. J.,** Fucosylation of glycoproteins in rat spermatocytes and spermatids, *Gamete Res.*, 5, 303, 1982.
179. **Jutte, N. H. P. M., Jansen, R., Grootegoed, J. A., Rommerts, F. F. G., and van der Molen, H. J.,** Protein synthesis by isolated pachytene spermatocytes in the absence of Sertoli cells, *J. Exp. Zool.*, 233, 285, 1985.
180. **O'Brien, D. A.,** Stage-specific protein synthesis by isolated spermatogenic cells throughout meiosis and early spermiogenesis in the mouse, *Biol. Reprod.*, 37, 147, 1987.
181. **O'Brien, D. A., Gerton, G. L., and Eddy, E. M.,** Acrosomal constituents identified with a monoclonal antibody are modified during late spermiogenesis in the mouse, *Biol. Reprod.*, 38, 955, 1988.
182. **Gerton, G. L., O'Brien, D. A., and Eddy, E. M.,** Antigens recognized by monoclonal antibody to mouse acrosomal components differ in guinea pig spermatogenic cells and sperm, *Biol. Reprod.*, 39, 431, 1988.
183. **Allen, R. L., O'Brien, D. A., Jones, C. C., Rockett, D. L., and Eddy, E. M.,** Expression of heat shock proteins by isolated mouse spermatogenic cells, *Mol. Cell. Biol.*, 8, 3260, 1988.
184. **Allen, R. L., O'Brien, D. A., and Eddy, E. M.,** A novel hsp70-like protein (P70) is present in mouse spermatogenic cells, *Mol. Cell. Biol.*, 8, 828, 1988.
185. **Maekawa, M., O'Brien, D. A., Allen, R. L., and Eddy, E. M.,** Heat-shock congnate protein (hsc71) and related proteins in mouse spermatogenic cells, *Biol. Reprod.*, 40, 843, 1989.
186. **Willison, K. and Ashworth, A.,** Mammalian spermatogenic gene expression, *Trends Genet.*, 3, 351, 1987.
187. **Parvinen, M.,** Regulation of the seminiferous epithelium, *Endocr. Rev.*, 3, 404, 1982.
188. **Tres, L. L. and Kierszenbaum, A. L.,** Viability of rat spermatogenic cells in vitro is facilitated by their coculture with Sertoli cells in serum-free hormone-supplemented medium, *Proc. Natl. Acad. Sci. U.S.A.*, 80, 3377, 1983.
189. **Hadley, M. A., Byers, S. W., Suarez-Quian, C. A., Kleinman, H. K., and Dym, M.,** Extracellular matrix regulates Sertoli cell differentiation, testicular cord formation, and germ cell development, *J. Cell Biol.*, 101, 1511, 1985.
190. **Kierszenbaum, A. L. and Tres, L. L.,** An automated perifusion system for the study of rat spermatogenesis in vitro, *Ann. N.Y. Acad. Sci.*, 513, 146, 1987.
191. **Parvinen, M., Lahdetie, J., and Parvinen, L. M.,** Toxic and mutagenic influences on spermatogenesis, *Arch. Toxicol.*, Suppl. 7, 128, 1984.
192. **Gray, T. J. B.,** Applications of in vitro systems in male reproductive toxicology, in *Physiology and Toxicology of Male Reproduction*, Lamb, J. C. and Foster, P. M. D., Eds., Academic Press, New York, 1988, 225.
193. **D'Agostino, A., Stefanini, M., and Geremia, R.,** Study of the interation between germ cells and Sertoli cells in vitro, *Exp. Cell Res.*, 150, 430, 1984.
194. **Enders, G. C. and Millette, C. G.,** Pachytene spermatocyte and round spermatid binding to Sertoli cells in vitro, *J. Cell Sci.*, 90, 105, 1988.
195. **D'Agostino, A. and Stefanini, M.,** A rat spermatocyte surface protein is involved in adhesion of pachytene spermatocytes to Sertoli cells in vitro, *Mol. Cell Biol.*, 7, 1250, 1987.
196. **O'Brien, D. A., Frøysa, A., Rockett, D. L., Hansson, V., and Eddy, E. M.,** Retinoids decrease viability and ATP levels of round spermatids, *Ann. N.Y. Acad. Sci.*, 513 (Abstr.), 477, 1987.
197. **O'Brien, D. A., Frøysa, A., and Rockett, D. L.,** Sertoli cell-conditioned medium increases viability and ATP levels of pachytene spermatocytes and round spermatids in vitro, *J. Cell Biol.*, 103 (Abstr.), 485a, 1986.
198. **Griswold, M. D.,** Protein secretions of Sertoli cells, *Int. Rev. Cytol.*, 110, 133, 1988.
199. **O'Brien, D. A., Gabel, C. A., Rockett, D. L., and Eddy, E. M.,** Receptor-mediated endocytosis and differential synthesis of mannose 6-phosphate receptors in isolated spermatogenic and Sertoli cells, *Endocrinology*, 125, 2973, 1989.
200. **Edwards, Y. H. and Grootegoed, J. A.,** A sperm-specific enolase, *J. Reprod. Fertil.*, 68, 305, 1983.
201. **Zakeri, Z. F., Ponzetto, C., and Wolgemuth, D. J.,** Translational regulation of the novel haploid-specific transcripts for the *c-abl* proto-oncogene and a member of the 70 kDa heat-shock protein gene family in the male germ line, *Dev. Biol.*, 125, 417, 1988.

202. **Krawczyk, Z., Mali, P., and Parvinen, M.,** Expression of a testis-specific hsp70 gene-related RNA in defined stages of rat seminiferous epithelium, *J. Cell Biol.*, 107, 1317, 1988.
203. **Heidaran, M. A., Kozak, C. A., and Kistler, W. S.,** Nucleotide sequence of the Stp-1 gene coding for rat spermatid nuclear transition protein 1 (TP1): homology with protamine P1 and assignment of the mouse Stp-1 gene to chromosome 1, *Gene*, 75, 39, 1989.
204. **Hecht, N. B.,** Mammalian protamines and their expression, in *Histones and Other Basic Nuclear Proteins*, Hnilica, L. S. and Stein, G. S., Eds., CRC Press, Boca Raton, FL, 1989, 347.
205. **Ono, T., Slaughter, G. R., Cook, R. G., and Means, A. R.,** Molecular cloning, sequence and distribution of rat calspermin, a high affinity calmodulin-binding protein, *J. Biol. Chem.*, 264, 2081, 1989.
206. **Rossi, P., Giorgi, M., Geremia, R., and Kincaid, R. L.,** Testis-specific calmodulin-dependent phosphodiesterase. A distinct high affinity cAMP isoenzyme immunologically related to brain calmodulin-dependent cGMP phosphodiesterase, *J. Biol. Chem.*, 263, 15521, 1988.
207. **Millan, J. L. and Manes, T.,** Seminoma-derived Nagao isozyme is encoded by a germ-cell alkaline phosphatase gene, *Proc. Natl. Acad. Sci. U.S.A.*, 85, 3024, 1988.
208. **Virbasius, J. V. and Scarpulla, R. C.,** Structure and expression of rodent genes encoding the testis-specific cytochrome c. Differences in gene structure and evolution between somatic and testicular variants, *J. Biol. Chem.*, 263, 6791, 1988.
209. **Roy, S. N., Kusari, J., Soffer, R. L., Lai, C. Y., and Sen, G. C.,** Isolation of cDNA clones of rabbit angiotensin converting enzyme: identification of two distinct mRNAs for the pulmonary and the testicular isozymes, *Biochem. Biophys. Res. Commun.*, 155, 678, 1988.
210. **Sandberg, M., Tasken, K., Øyen, O., Hansson, V., and Jahnsen, T.,** Molecular cloning, cDNA structure and deduced amino acid sequence for a type I regulatory subunit of cAMP-dependent protein kinase from human testis, *Biochem. Biophys. Res. Commun.*, 149, 939, 1987.
211. **Øyen, O., Scott, J. D., Cadd, G. G., McKnight, G. S., Krebs, E. G., Hansson, V., and Johnsen, T.,** A unique mRNA species for a regulatory subunit of cAMP-dependent protein kinase is specifically induced in haploid germ cells, *FEBS Lett.*, 229, 391, 1988.
212. **Leroy, P., Alzari, P., Sassoon, D., Wolgemuth, D., and Fellous, M.,** The protein encoded by a murine male germ cell-specific transcript is a putative ATP-dependent RNA helicase, *Cell*, 57, 549, 1989.
213. **Taira, M., Iizasa, T., Yamada, K., Shimada, H., and Tatibana, M.,** Tissue-differential expression of two distinct genes for phosphoribosyl pyrophosphate synthetase and existence of the testis-specific transcript, *Biochim. Biophys. Acta*, 1007, 203, 1989.
214. **Fort, P., Marty, L., Piechaczyk, M., El Sabrouty, S., Dani, C., Jeanteur, P., and Blanchard, J. M.,** Various rat adult tissues express only one major mRNA species from the glyceraldehyde-3-phosphate dehydrogenase multigenic family, *Nucleic Acids Res.*, 13, 1431, 1985.
215. **Krawetz, S. A., Connor, W., Cannon, P. D., and Dixon, G. H.,** A vector-primer-cloner-sequencer plasmid for the construction of cDNA libraries: evidence for a rat glyceraldehyde-3-phosphate dehydrogenase-like mRNA and a ferritin mRNA within testis, *DNA*, 5, 427, 1986.
216. **Swinnen, J. V., Joseph, D. R., and Conti, M.,** Molecular cloning of rat homologues of the *Drosophila melanogaster* dunce cAMP phosphodiesterase: evidence for a family of genes, *Proc. Natl. Acad. Sci. U.S.A.*, 86, 5325, 1989.
217. **Hanks, S. K.,** Messenger ribonucleic acid encoding an apparent isoform of phosphorylase kinase catalytic subunit is abundant in the adult testis, *Mol. Endocrinol.*, 3, 110, 1989.
218. **Kilpatrick, D. L. and Millette, C. F.,** Expression of proenkephalin messenger RNA by mouse spermatogenic cells, *Proc. Natl. Acad. Sci. U.S.A.*, 83, 5015, 1986.
219. **Yoshikawa, K. and Aizawa, T.,** Enkephalin precursor gene expression in postmeiotic germ cells, *Biochem. Biophys. Res. Commun.*, 151, 664, 1988.
220. **Gizang-Ginsberg, E. and Wolgemuth, D. J.,** Expression of the proopiomelanocortin gene is developmentally regulated and affected by germ cells in the male mouse reproductive system, *Proc. Natl. Acad. Sci. U.S.A.*, 84, 1600, 1987.
221. **Persson, H., Ericsson, A., Schalling, M., Rehfeld, J. F., and Hokfelt, T.,** Detection of cholecystokinin in spermatogenic cells, *Acta Physiol. Scand.*, 134, 565, 1988.
222. **Persson, H., Rehfeld, J. F., Ericsson, A., Schalling, M., Pelto-Huikko, M., and Hokfelt, T.,** Transient expression of the cholecystokinin gene in male germ cells and accumulation of the peptide in the acrosomal granule: possible role of cholecystokinin in fertilization, *Proc. Natl. Acad. Sci. U.S.A.*, 86, 6166, 1989.
223. **Propst, F., Rosenberg, M. P., and Vande Woude, G. F.,** Proto-oncogene expression in germ cell development, *Trends Genet.*, 4, 183, 1988.
224. **Cooper, G. M., Coldman, D. S., Kiessling, A. A., and Millette, C. F.,** The expression of oncogenes in germ cells, in *Meiotic Inhibitors: Molecular Control of Meiosis*, Vol. 267, Haseltine, F. P. and First, N. L., Eds., Alan R. Liss, New York, 1988, 275.
225. **Gold, B. and Hecht, N. B.,** Differential compartmentalization of messenger ribonucleic acid in murine testis, *Biochemistry*, 20, 4871, 1981.

226. **Erickson, R., Kramer, J., Rittenhouse, J., and Salkeld, A.,** Quantitation of mRNAs during mouse spermatogenesis: protamine-like histone and PGK-2 mRNAs increase after meiosis, *Proc. Natl. Acad. Sci. U.S.A.,* 77, 6086, 1980.
227. **Kleene, K. C.,** Poly(A) shortening accompanies the activation of translation of five mRNAs during spermiogenesis in the mouse, *Development,* 106, 367, 1989.
228. **Fujimoto, H., Erickson, R. P., and Tone, S.,** Changes in polyadenylation of lactate dehydrogenase-X mRNA during spermatogenesis in mice, *Mol. Reprod. Dev.,* 1, 27, 1988.
229. **Cole, K. D., Kandala, J. C., and Kisler, W. S.,** Isolation of the gene for the testis-specific H1 histone variant H1t, *J. Biol. Chem.,* 261, 7178, 1986.
230. **Kim, Y.-J., Hwang, I., Tres, L. L., Kierszenbaum, A. L., and Chae, C.-B.,** Molecular cloning and differential expression of somatic and testis-specific H2B histone genes during rat spermatogenesis, *Dev. Biol.,* 124, 23, 1987.
231. **Birnstiel, M. L., Busslinger, M., and Strub, K.,** Transcriptional termination and 3′ processing: the end is in site, *Cell,* 41, 349, 1985.
232. **Moss, S. B., Challoner, P. B., and Groudine, M.,** Expression of a novel histone 2B during mouse spermiogenesis, *Dev. Biol.,* 133, 83, 1989.
233. **Peschon, J. J., Behringer, R. R., Brinster, R. L., and Palmiter, R. D.,** Spermatid-specific expression of protamine 1 in transgenic mice, *Proc. Natl. Acad. Sci. U.S.A.,* 84, 5316, 1987.
234. **Peschon, J. J., Behringer, R. R., Palmiter, R. D., and Brinster, R. L.,** Expression of mouse protamine 1 genes in transgenic mice, *Ann. N.Y. Acad. Sci.,* 564, 186, 1989.
235. **Braun, R. E., Peschon, J. J., Behringer, R. R., Brinster, R. L., and Palmiter, R. D.,** Protamine 3′-untranslated sequences regulate temporal translational control and subcellular localization of growth hormone in spermatids of transgenic mice, *Genes Dev.,* 3, 793, 1989.

Chapter 2

GENETIC ASPECTS OF SPERM DIFFERENTIATION AND FUNCTION IN THE MOUSE

Judith A. Cebra-Thomas and Lee M. Silver

TABLE OF CONTENTS

I. INTRODUCTION

Throughout the history of biology, a major quest has been to understand the process of development in complex organisms. For many contemporary developmental biologists, the distant unspoken goal has been to describe, at a molecular level, the mechanisms involved in the generation of an adult animal from a single-celled fertilized egg. To even think about achieving this goal in mammals, it has long been realized that compartmentalization of the problem is essential. For this reason, all present-day workers in the field concentrate their studies on particular stages of development in the early embryo or on particular tissues or cell types during later developmental periods.

The choice of compartment, or differentiative system, is based on a number of criteria including the ease with which it is possible to (1) recognize and (2) isolate sufficient quantities of cell types from different stages of differentiation, as well as (3) the availability of additional tools for analysis and dissection. Spermatogenesis presents a unique opportunity to study cell commitment and differentiation because all of the cell types involved are constitutively present in the testis of the adult male. Furthermore, the process of sperm differentiation has been divided into a large number of morphologically distinct phases, and protocols have been developed for the efficient separation of germ cells at different stages. Finally, a large set of independent mutations and polymorphisms has been identified in laboratory and wild mice with well-defined effects at particular stages of differentiation. The purpose of this review is two-fold: first, to present a unified summary of the various "spermatogenic mutations" that are known, and second, to summarize current information on genes and products known to be expressed in a male germ cell-specific manner.

II. BRIEF DESCRIPTION OF SPERMATOGENESIS

A detailed description of the morphologic stages of mammalian spermatogenesis has been developed based on the appearance of the nucleus and the developing acrosome.[1-3] In addition, there are several excellent reviews describing the current understanding of the biochemistry, molecular biology, and endocrinology of mammalian spermatogenesis.[4-10]

Accordingly, we will only provide here a brief description of the salient features associated with the major stages of spermatogenesis to serve as a framework to facilitate discussion of mutations that affect this process. (See Figure 1.) The mouse presents an ideal organism in which to study spermatogenesis because of the availability of genetically identical animals, and a wealth of genetic information (coupled with high fecundity and a short generation time). Therefore, we will concentrate primarily on a description of spermatogenesis in the mouse, supplementing our discussion with results obtained by study of the rat.

A. THE SEMINIFEROUS EPITHELIUM

The adult testis consists primarily of a bundle of seminiferous tubules. Each tubule is composed of a ring of Sertoli cells linked by tight junctions which are responsible for forming the blood-testis barrier.[11,12] The Sertoli cells support the growth and differentiation of the germ cells, which comprise the major cell type in the tubule. The blood-testis barrier isolates the developing germ cells in the central (adluminal) compartment from serum components and from the immune system. The tubule is surrounded by a basement membrane bounded by peritubular myoid cells, with Leydig cells clustered around the blood vessels in the interstices between the tubules.

B. INITIATION OF MEIOSIS

The type A spermatogonia are the most primitive germ cell type in the adult testis. They serve as a self-renewing stem cell population, and are localized around the periphery of the tubules, along the basement membrane (outside the blood-testis barrier). They undergo periodic waves of proliferation.[13,14] A certain percentage of the population divides and differentiates to form histologically distinct Intermediate, and then type B spermatogonia. Commitment to differentiation is linked to incomplete cytokinesis, so that each developing clone of spermatogenic cells is linked by cytoplasmic bridges.[15] The extent to which molecules can cross these bridges is still unknown, but they may play a role in synchronizing the development of the population. The cytoplasmic bridges appear to be closed during cell division by flattened membranous sacs that form perpendicular to the axis of the bridge.[15,16] However, during the interphase and throughout most of spermatogenesis, the bridges appear open and are traversed by the endoplasmic reticulum (ER).[15,17] This suggests that proteins that contain a signal peptide, and are capable of entering the ER and the Golgi network, may be transported between the meiotic partners in the developing syncytium. Intercellular transport has been demonstrated using transgenic mouse strains producing human growth hormone under the control of the haploid-specific protamine 1 promoter. Human growth hormone protein is detectable in all of the cells, although only 50% carry the transgene.[18,19] It is not yet clear if proteins that are not compartmentalized in the ER, or other macromolecules such as mRNA, are capable of moving through the bridges.

After a series of mitotic divisions, the spermatogonia enter meiotic prophase I and become 1^0 spermatocytes. The earliest 1^0 spermatocytes (preleptotene) are still in the basal compartment. During leptotene, the Sertoli cells form a new set of tight junctions below the spermatocytes and transfer them into the adluminal compartment, where they are isolated from the immune system and from serum components. Thereafter, as the germ cells develop, they progress toward the central lumen of the tubule, while remaining associated with cytoplasmic processes of the Sertoli cells.[11] The 1^0 spermatocytes proceed through zygotene to a prolonged pachytene stage. The cells remain in the pachytene stage for approximately 1 week,[20] during which there is a large wave of poly(A)$^+$RNA synthesis.[5,21,22] After a brief diplotene/diakinesis stage, the spermatocytes undergo the two meiotic divisions to produce first 2^0 spermatocytes, and then haploid spermatids.

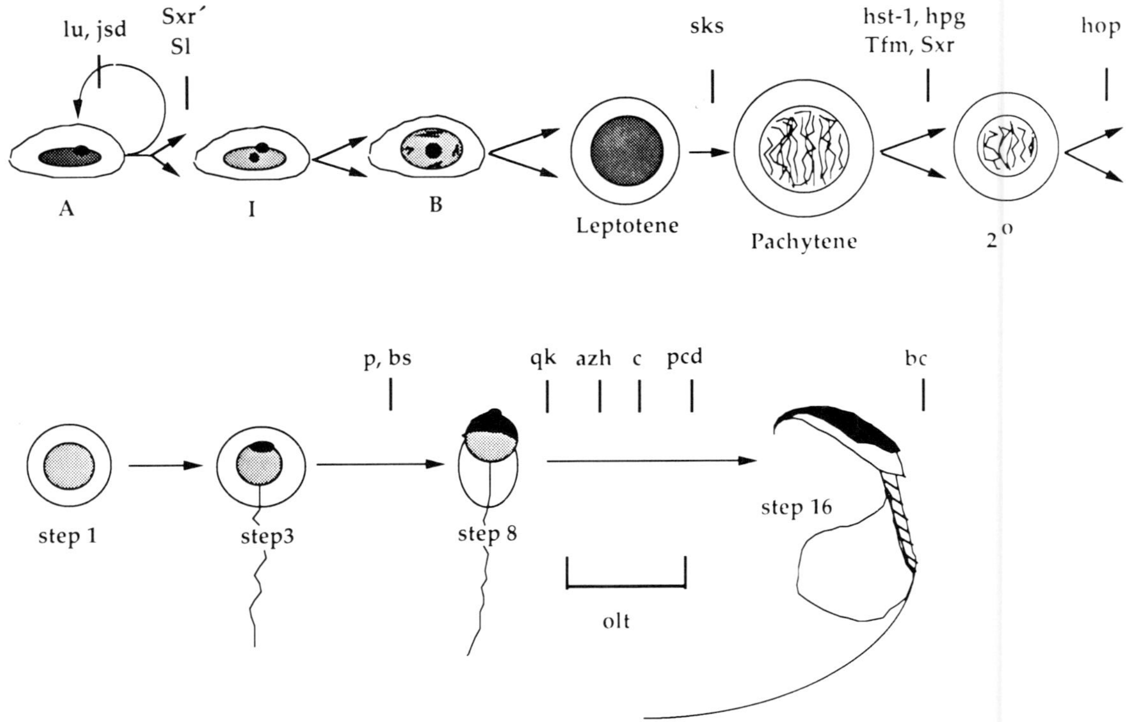

FIGURE 1. Mutations that affect spermatogenesis in the mouse. A schematic representation of the major stages of spermatogenesis, and the approximate stages at which male-infertility mutations appear to affect that process (see text). A, I, and B = type A, Intermediate, and type B spermatogonia. 2° = secondary spermatocytes. Steps 1, 3, 8, and 16 refer to steps of spermiogenesis as defined by Oakberg.[3]

C. SPERMIOGENESIS

1. Development of the Nucleus and the Acrosome

Spermiogenesis, the series of morphological transitions undergone by the haploid spermatid during the formation of the mature spermatozoon, has been divided into 16 steps in the mouse based on changes in the nucleus and the developing acrosome visualized by periodic acid-Schiff staining.[3] The newly formed spermatid (step 1) is a small round cell with the weakly staining Golgi apparatus associated with the nucleus. Darkly staining proacrosomal granules are formed by the Golgi (step 2), fuse together (step 3), and flatten to form a cap over the nucleus (step 4). The cap gradually enlarges to cover approximately one third of the nucleus (steps 5 to 7). At step 8, the spermatids realign toward the basement membrane of the tubule. At the same time, the cytoplasm is displaced toward the opposite pole of the cell and the nucleus becomes ovoid. The nucleus elongates and flattens while the acrosome covers its dorsal edge (steps 9 to 11).[21] The chromatin loses the typical nucleosome structure and becomes more fibrous, and condensation begins in the anterior region.[23] The nucleus becomes tapered toward the anterior region (step 12) and the bulk of the acrosome moves caudally to give the appearance of a hook shape (step 13). At step 14, the acrosome appears as a crescent with a sharp point. The nucleus narrows and the acrosome appears less heavily stained by step 15, and at step 16 the process is finished with the production of a mature spermatozoon. The spermatozoa are released into the lumen of the tubules, leaving behind the majority of their cytoplasmic contents as residual bodies. The entire process of spermiogenesis lasts approximately 2 weeks in the mouse.[20]

2. Nuclear Protein Transitions

Chromatin condensation is accompanied by the displacement of the histones from the chromatin by transitional proteins (TP) 1 and 2 late in step 12.[24-28] At this time, the chromatin becomes transcriptionally inactive.[21,22,29] The TPs are subsequently replaced by protamines 1 and 2 during steps 14 to 16.[24-26] The transitional proteins and protamines are all transcribed in round spermatids, and the mRNAs are stored as translationally inactive forms.[30-40] The translatable mRNA isolated from polysomes contains a shorter poly(A) tract than that in the nonpolysomal fraction.[32,34,36-38] Sequences sufficient to confer translational regulation on a heterologous mRNA are located in the 3′-untranslated region of mouse protamine 1 mRNA.[19] These testis-specific proteins are responsible for the highly compact and stable nature of sperm chromatin.

3. Manchette Formation

Just prior to the onset of chromatin condensation, a sheath of parallel microtubules known as the manchette forms around the caudal pole of the nucleus and extends back to encircle the flagellum.[41] At step 12, the manchette encloses the posterior region of the nucleus and appears to constrain it.[42,43] As the spermatid nucleus elongates, the manchette extends along the ventral surface of the nucleus, and covers the entire surface that is not associated with the acrosome.[42] Later, it is displaced backward to surround what will become the middle piece of the mature spermatozoon. Treatment with vinblastin, an inhibitor of microtubule function, disrupts normal nuclear elongation.[42] The manchette microtubules may play a role in the shaping of the elongated nucleus or in the movement of the cytoplasm laterally during elongation.[41,43]

4. Development of the Tail

A number of other organelles are also involved in the production of the highly specialized spermatozoon;[4,8] the most striking of these is the flagellum. After meiosis, the two centrioles migrate to the distal pole of the round spermatid. The tail axoneme then grows as a typical 9 + 2 microtubule doublet flagellum from the distal centriole.[44] The pair of centrioles and the base of the flagellum subsequently migrate inward and become associated with the caudal pole of the nucleus.[44] The flagellum of the mature spermatozoon is composed of a number of specialized cytoskeletal structures.[4] In the middle piece, the microtubule axoneme is surrounded by nine outer dense fibers, which in turn are surrounded by the mitochondrial sheath.[4] In the principle piece, two longitudinal columns replace two of the outer dense fibers, and are connected by numerous transverse ribs.[4] The longitudinal columns grow along doublets 3 and 8 slowly beginning early in spermiogenesis, starting from the distal end of the principle piece.[45] Beginning just prior to elongation (in step 8), nine very fine fibers develop at the proximal ends of the microtubule doublets, and gradually grow along the doublets.[46] Late in spermiogenesis, the outer dense fibers suddenly increase in diameter, and the ribs of the fibrous sheath form, due to the deposition of proteins that have been synthesized earlier.[30,45-47] During midspermiogenesis, the ER forms a close network of cisternae surrounding the flagellum, and presumably providing the site for local synthesis of the tail proteins.[17] The outer dense fibers and fibrous sheath contain both unique and shared proteins.[48] At least two of the outer dense fiber proteins appear to bind zinc through protein sulfhydryl groups.[49]

5. Structural Proteins

Coincident with the morphologic changes of spermiogenesis and the formation of novel organelles such as the flagellum and the manchette, variants of structural proteins have been detected. Two-dimensional gel analysis of testicular proteins during different stages of spermatogenesis have shown a change in the relative levels of expression of γ- and β-actin, and the expression of novel isoforms of α- and β-tubulin in round and elongating spermatids.[50] Villasante and co-workers have isolated cDNA clones for four different α-tubulin genes expressed in the testis.[51] A pair of closely related genes, Mα3 and Mα7, encode testis-specific transcripts of ~1.7 kb that differ from each other in only silent nucleotide positions. The proteins specified by Mα3/Mα7 differ at multiple amino acid positions from the somatic α-tubulins, especially in the carboxy-terminal regions.[51] Mα6 is expressed at low levels in all tissues, but the testis contains an additional transcript of ~2.4 kb. The fourth gene, Mα4, also produces different somatic (1.7 kb) and testicular (2.4 kb) transcripts; these contain different 5′-ends generated by alternative splicing. However, the testis-specific form of Mα4 does not contain the initiator ATG codon, and may not be translated.[51] A testis-specific α-tubulin has been identified in the chicken that, like Mα4, terminates prior to the usual carboxy-terminal tyrosine which is involved in the tyrosination/detyrosination cycle of microtubule polymerization/disassembly.[52] A cDNA clone (pRDαTT1) representing a fifth testis-expressed α-tubulin identified by a unique 3′-untranslated region has been isolated by Distel and co-workers.[53] Testis-specific transcripts of 2.1 and 1.6 kb are detectable in round and elongating spermatids.[53] These encode a highly divergent α-tubulin protein that differs at 30% of the 212 amino acids from both the somatic α-tubulins and Mα3/Mα7.[54] In addition, αTT1 extends for 16 amino acids beyond the usual termination point, and does not contain a carboxy-terminal tyrosine. Antisera raised against a unique epitope of αTT1 stains the manchette and the meiotic spindle fibers in testis sections, but does not react with brain sections.[54] Expression of αTT1 and Mα3/Mα7 therefore corresponds with the formation of several unique microtubule systems during spermatogenesis.

The testis contains two different size changes of actin mRNA, a 2.1-kb band which is roughly the same size as that found in somatic tissues, and a testis-specific 1.5-kb band.[55] The 1.5-kb transcript is first detected at 22 days after birth and appears to be restricted to the haploid stages of spermatogenesis.[35,56] It is detectable in round and elongating spermatids, but not in pachytene spermatocytes.[35,55,56] The 2.1-kb band contains both β- and γ-actin mRNA, but the 1.5-kb transcript does not hybridize with 3′-untranslated region probes specific for α-, β-, or γ-actin.[55] It may represent either a novel actin gene product or an alternatively spliced variant of one of the somatic actins. Its restriction to the haploid stage

TABLE 1
Cell Associations during the Cycle of the Seminiferous Epithelium

Stage	Spermatogonia			Spermatocyte		Spermatid (step)		Duration (h)	
I	A				Pachytene	1	13	22.2	
II	*	In			*	2	14	18.1	
III	*	*			*	3	14	8.7	
IV	*	*	B		*	4	15	18.6	
V	*		*		*	5	15	11.3	
VI	*		*	Preleptotene	*	6	15	18.1	
VII	*			*	*	7	16	20.6	
VIII	*			Leptotene	*	8	16	20.8	Release
IX	*			*	*	9		15.2	
X	*			Zygotene	*	10		11.3	
XI	*			*	Diplotene	11		21.4	
XII	*			Pachytene	Meiosis	12		20.8	

Note: In = Intermediate-type spermatogonia. Cell types persisting through more than one cycle are indicated by *.

Data from Oakberg.[3,20]

cells suggests that it may also play a role in the formation of the specialized organelles formed during spermiogenesis.

6. The Chromatoid Body

The function of the chromatoid body, a dense mass of fine, filamentous material found associated with the nucleus, is also intriguing. It first appears in the cytoplasm of primary spermatocytes.[57] In early round spermatids, the chromatoid body is associated with the Golgi complex and the developing acrosomal granule, and gradually migrates back toward the caudal pole.[57,58] Filamentous material extends to the neighboring nuclear pores, which are concentrated in the vicinity of the chromatoid body.[57] By midspermiogenesis, it encircles the flagellum at the caudal pole of the nucleus.[57,59] Histochemical staining suggests that the chromatoid body contains RNA and basic proteins.[60,61] The structure of the chromatoid body is disrupted by treatment of round spermatids with actinomycin D, and it can be labeled with [^{3}H]-uridine after a 14-h chase in spermatids from steps 1 to 8.[62-64] This suggests the exciting possibility that the chromatoid body may be involved in the storage of mRNA synthesized in pachytene spermatocytes or early round spermatids for translation during elongation.

D. THE CYCLE OF THE SEMINIFEROUS EPITHELIUM

In mammals, spermatogenesis proceeds in waves along a stretch of seminiferous epithelium. The timing of both the initiation of the waves (every 8.6 days in the mouse) and the time required to complete differentiation (34.5 days) is constant for a particular species.[20] Therefore, particular cell types are always found in association with each other.[1-3,65] In the mouse, the cycle of the seminiferous epithelium can be divided into 12 stages based on these associations (see Table 1).[3,20] The timing of the onset of the first meiotic cycle is also constant for a given species or strain, and thus prepuberal animals of a given age can reliably be shown to have reached a specific stage of spermatogenesis.[66] The first cells enter meiotic prophase on day 8 of life. Cells in midpachytene are detectable by days 15 to 17, and round spermatids are detectable by days 22 to 24.[66] In addition, cell separation techniques have been developed that allow the isolation of enriched populations of cells at many of the developmental stages.[67-69] These features can be exploited to identify genes that are expressed at either the mRNA or the protein level in particular cell types, or at particular times during the spermatogenic cycle.

Spermiogenesis in the rat can be even more finely subdivided than in the mouse (into 19 steps), and the cycle of the seminiferous epithelium can be divided into 14 stages.[1,65] Stages I to VIII (spermatozoa release) are roughly equivalent, and stages IX to XII (from the mouse) have been expanded. The murine stage X is roughly equivalent to stages X and XI in the rat, and murine stage XII is roughly equivalent to stages XIII and XIV.

III. GENETIC EFFECTS ON SPERM DIFFERENTIATION

A. GENES THAT AFFECT SPERMATOGENESIS

As we indicated above, a major advantage in the choice of mouse spermatogenesis as a model system for developmental studies has been the identification, characterization, and availability of a large number of mutations at different loci with clearly defined effects on sperm differentiation. In this section of our review, we provide a brief description of the best-characterized of these mutations. In nearly all cases, ''sperm mutations'' were identified as such through their deleterious effects on fertility—usually in males only, but occasionally both males and females are affected. In most cases, the description of the phenotype is purely morphologic, and the biochemical basis for the defect is unknown. However, with recent improvements in ''reverse genetics'' techniques that allow one to go from a chromosomal locus to the cloning of a gene, these mutations could provide a starting point for dissecting spermatogenesis at the molecular level.

Although many of these defects are believed to be germ cell-autonomous, this has been rigorously demonstrated for only a few mutations. The best method of determining whether a breakdown in spermatogenesis is due to an intrinsic defect of the germ cells (cell-autonomous) or to a defect in the supportive testicular environment is to analyze chimeras. If progeny of the mutant genotype are produced, the defect is not cell-autonomous. Two interesting mutations (*Steel* and *testicular feminization*) have to be shown to affect the somatic elements of the testicular environment, and not the germ cells. Unfortunately, as many male chimeras are sterile, or produce offspring of only one parental type even when both are fertile;[70] a significant number of offspring must be produced to be sure that a mutation is germ cell autonomous. Nonetheless, if the chimeric testis contains arrested germ cells with defects characteristic of a particular mutation, it suggests that the mutation is not corrected by the wild-type testicular environment.

1. Luxoid (*lu*), Chromosome 9

Heterozygotes and homozygotes exhibit a number of skeletal abnormalities including preaxial polydactyly of the feet. Females are fertile, but *lu/lu* males are sterile.[71] The spermatogonia differentiate through first a normal, and then a delayed cycle of spermatogenesis during the first 6 weeks. After 6 weeks, there is a complete lack of spermatogonia (D. Elkins and P. F. Forsthoefel, cited by Johnson and Hunt).[72]

2. Juvenile Spermatogonial Depletion (*jsd*), Chromosome 1

With a phenotype reminiscent of *lu/lu* (but without the accompanying skeletal defects), *jsd/jsd* males experience one wave of spermatogenesis at puberty, but fail to continue mitotic proliferation of type A spermotogonia. After 8 weeks of age, the tubules are composed primarily of Sertoli cells, with only occasional spermatogonia. Serum testosterone levels are normal, but follicle-stimulating hormone (FSH) levels are elevated, and the authors suggest that the defect may lie in the environment provided by the Sertoli cells.[73] Sertoli cells produce inhibin, which negatively regulates FSH secretion.[74] This prediction could be tested by the production of chimeras; if the *jsd/jsd* germ cells can develop in the context of +/+ testicular elements, then the defect must lie in the developmental environment. Purified Sertoli cells can now be cultured *in vitro*, and can reform tubules if embedded in an extracellular matrix gel.[75] Under these conditions, the Sertoli cells are able to support limited differentiation of associated spermatogonia.[75] It is possible that this system could be used to investigate whether or not *jsd/jsd* Sertoli cells could support the differentiation of +/+ spermatogonia up to meiosis, and if *jsd/jsd* spermatogonia can differentiate in a +/+ environment. The phenotypes expressed by *lu* and *jsd* suggest that the initial round of spermatogenesis may be under a different regulatory control mechanism than the subsequent rounds.

3. Steel (*Sl*), Chromosome 10

Mice homozygous for the *Steel* mutation (*Sl/Sl*) die *in utero* of severe macrocytic anemia. Compound Sl/Sl^d animals are viable, but anemic and sterile.[76,77] Both male and female sterility is caused by a failure of primordial germ cells to persist and proliferate in the gonadal ridge during embryogenesis, leading to a complete absence of germ cells in the gonads.[78] The defect is not germ cell-autonomous; $Sl/Sl^d \leftrightarrow +/+$ male chimeras can breed from their Sl/Sl^d component.[79] However, the testes of the chimeras contain both tubules differentiating normally, and ones which contain only Sertoli cells and type A spermatogonia. Differentiated and nondifferentiated segments alternate in a single tubule; the size of the segments corresponds to 150 to 180 Sertoli cells.[79] The authors propose that the segments represent clones of Sertoli cells, and that the Sl/Sl^d Sertoli cells are unable to support the differentiation of type A spermatogonia.

4. Skeletal Fusions with Sterility (*sks*), Chromosome 4

Homozygous *sks/sks* mice are characterized by fusions of ribs and vertebrae; both males and females are sterile.[80] No normal spermatids are seen in *sks/sks* males. Most frequently, the developing spermatocytes arrest at the pachytene stage; occasionally they undergo one meiotic division. Some aberrantly large spermatids are observed that may have initiated differentiation without completing both meiotic divisions. Abnormal and degenerating germ and Sertoli cells were observed sloughed into the lumen of the tubules. It is an interesting coincidence that both *sks* and proto-oncogene *Mos* map to chromosome 4, and both appear to affect meiosis.[81-83]

5. Hybrid Sterility (*hst-1*), Chromosome 17

The *hst-1* locus controls the fertility of F_1 males produced by mating wild *Mus musculus* males and females of certain lab strains (B10, A, BALB/c, DBA/1, and AKR). Spermatogenesis arrests between 1^0 and 2^0 spermatocytes, and a high frequency of cells with two to six univalents and X-Y dissociation was observed at the first meiotic division.[84] A high frequency of premature X-Y dissociation was also observed in both fertile[85] and oligospermic[86] male hybrids between wild Japanese mice (*M. musculus molossinus*) and certain laboratory strains (BALB/c and B10.BR,[85] and random-bred Swiss).[86] In another system, failure of normal pairing between the X and Y chromosomes, or between autosomes due to translocation heterozygosity can result in sterility due to meiotic arrest.[87-90] However, all of the species and subspecies involved in *hst-1* associated sterility are karyotypically identical. Therefore, in these cases, biochemical factors could be responsible for the disruption of normal chromosome pairing. Alternately, a recent study has shown a reduced level of testosterone secretion from testis $hst\text{-}1^s/hst\text{-}1^{ws}$ sterile mice, and suggested that Leydig cell function may be impaired.[91] A defect in Leydig cell function could be responsible for the spermatogenic arrest, or the two could be separate defects. It is interesting to note that when only one sex is affected by hybrid sterility, or by translocation-induced sterility, it is the heterogametic (e.g., XY) sex that is affected, as first observed by Haldane.[90,92]

A second hybrid-sterility locus, *hst-2*, that controls the fertility of males produced by backcrossing BALB/c X *M. spretus* F_1 females to BALB/c males has been identified, but has not been characterized further.[93]

6. Hypogonadal (*hpg*)

Both male and female *hpg/hpg* animals fail to show any postnatal gonadal development, and are consequently sterile.[94] In the testis, spermatogenesis arrests at the diplotene stage immediately prior to the meiotic divisions.[94] The primary defect appears to be a deficiency in the production of the hypothalamic gonadotropin-releasing hormone (GnRH), leading to a reduction in the circulating levels of both luteinizing hormone (LH) and FSH.[94] Both testes and ovaries develop normally when transplanted into normal recipients, and *hpg/hpg* males injected with testosterone after birth and implanted with a capsule releasing GnRH are fertile.[95,96] Finally, molecular studies have shown that there is a deletion in the gene for GnRH in *hpg* mice, and that the defect can be corrected in *hpg/hpg* mice transgenic for the wild-type GnRH gene.[97]

7. Testicular Feminization (*Tfm*), X Chromosome

Genotypic males carrying the X-linked mutation *Tfm* ($X^{Tfm}Y$) possess testes, but their epididymides, vasa deferentia, and seminal vesicles are vestigial, and they outwardly appear as females.[98] The germ cells arrest as primary spermatocytes, and *Tfm* mice fail to respond to testosterone.[98] The lack of responsiveness to testosterone is due to a deficiency in the cytosolic androgen-receptor protein.[99,100] Two fertile XY $\leftrightarrow$ $X^{Tfm}Y$ chimeras have been produced, and both produce offspring from the $X^{Tfm}Y$ component, demonstrating that the *Tfm* defect is not germ cell-autonomous.[101] It is likely that the *Tfm* defect affects Sertoli cells; Sertoli cells in $X^{Tfm}Y$ animals fail to form the tight junctions that make up the blood-testis barrier.[96]

8. Hop-Sterile (*hop*), Chromosome 6

hop and *hpy* (*hydrocephalic-polydactyl*) have been shown to be allelic.[102] Homozygous animals exhibit a hopping gait, preaxial polydactyly of all feet, hydrocephaly, and poor growth.[72,103] Homozygous males are sterile. The second meiotic division is frequently incomplete, resulting in 20% multinucleate spermatids. The early steps of spermiogenesis are apparently normal, and the acrosome develops normally until step 7. However, the centrioles fail to migrate to the distal pole of the cells and the flagellum is either absent or abnormal. Abnormalities have also been observed in the cilia of the tracheal epithelium of homozygous animals.[104] Nuclear elongation commences normally, but many heads are abnormally shaped with incursions of Sertoli cell processes.[72,105] The latter is a common feature of degenerating spermatids and may not reflect a specific defect caused by *hop*. The observed defects in the flagella could be due to a defect in one of the structural proteins involved (e.g., tubulin or one of the microtubule-associated proteins).

9. Pink-Eyed Dilution (*p*), Chromosome 7

Four radiation-induced alleles covering the *p* locus have been obtained that cause male sterility. Abnormalities in developing spermatids appear soon after meiosis in p^{6H}, p^{25H}, p^{s} (*pink sterile*), and p^{bs} (*pink-black-eyed sterile*).[106-108] During spermiogenesis steps 1 to 7 the Golgi appears disorganized and produces multiple proacrosomal granules. Chromosome condensation is often delayed or incomplete, and the resulting spermatozoa have deformed heads. There is no evidence for complementation between the alleles to alleviate the sperm defects.[108] Sperm from p^{6H}/p^{6H} and p^{25H}/p^{25H} males have been reported to bind reduced levels of colloidal iron hydroxide, implying a reduction in negative surface charges, possibly on neuraminidase-resistant gangliosides.[109] Endocrine defects have also been reported for many of the *p* alleles;[110-112] however, these are not consistently associated with sterility. The four male-sterile alleles do not always share the same endocrine defects. For example, p^{25H}, but not p^{bs}, contains degenerating nerve axons in the pituitary pars nervosa,[112] while p^{11H}, which is not male-sterile, also appears to have endocrine defects.[112] Preliminary studies have indicated that the defect in p^{6H}/p^{6H} germ cells is not rescued in chimeras, which suggests that it is germ cell-autonomous.[113]

10. Blind-Sterile (*bs*), Chromosome 2

Homozygous *bs/bs* animals are blind due to cataracts. Females are fertile, but males are sterile.[114] Sterile males appear to have two defects. First, they have significant germ cell depletion, which may be caused by an insufficient number of gonocytes populating the developing testis.[115] In addition, there is a specific defect in spermiogenesis. Disorganized proacrosomal granules are observed during the early haploid stages (up to step 3), but they fail to form an acrosomal cap. The sperm that are produced have grossly abnormal heads and normal tails. Multinucleate clusters are also observed.[116] There is a reduction in the amount of protamine 1 mRNA, but only proportional to the reduction in the numbers of elongating spermatids relative to wild-type.[34]

11. Quaking (*qk*), Chromosome 17

The original spontaneous, recessive *qk* mutation is characterized by seizures and quaking resulting from a severe deficiency of myelin in the central and peripheral nervous system.[117] Homozygous males are sterile, with azoospermia. Spermiogenesis is apparently normal until the onset of nuclear elongation (step 10). The chromatin condenses on schedule, but irregularities and vacuoles form, and Sertoli cell processes indent into the spermatid. The flagellum forms normally at first, but is lost at step 9. In addition, the mitochondria appear poorly organized.[118] The expression of protamine 1 mRNA appears normal in *qk/qk* males; both the 580-nuc. and the partially deadenylated, translatable 450-nuc. species are observed.[34] Although the defects are not apparent until haploid stages, *qk/*; + males can transmit both chromosomes. However, the dysfunction of *qk* germ cells is not rescued in chimeras. In one study, nine male chimeras of the genotype XY; *qk/qk*↔ XY; +/+ were formed. Seven were fertile and contained 15 to 50% *qk/qk* cells, but none transmitted the *qk* component. All contained abnormal spermatids with the characteristic *qk/qk* defects.[113]

Four new alleles at the *qk* locus have been induced by ethylnitrosourea mutagenesis,[119-121] which primarily induced point mutations. All four fail to complement the myelination defect of *qk*, but do complement the spermatogenesis defect. In addition, unlike the classical *qk* mutation, all four cause homozygous embryo lethality. This suggests that the original *qk* mutation may be complex, and that the genes involved in male fertility and neurologic function may be distinct.

12. Abnormal Spermatozoon Head (*azh*)

azh is an autosomal-recessive mutation that causes the formation of abnormally shaped sperm heads.[122] Up to 40% of the sperm heads separate from the flagella. Homozygous *azh/azh* males are fertile,[122] but their fertility may be lower than normal. Under *in vitro* fertilization conditions, significantly fewer *azh/azh* sperm bind to ova than do control sperm.[123] The spermatids develop normally until step 10, after which the packaging of the chromatin in the head is morphologically aberrant. The manchette appears to constrict prematurely, and to split and detach from the nucleus.[43,124] As the manchette initially forms normally in steps 7 and 8, the abnormalities in *azh/azh* spermatids may be due to a defect in a factor that regulates the changes in manchette shape during nuclear elongation.

13. Albino (*c*), Chromosome 7

Two different radiation-induced mutations that cover the *albino* locus can complement for lethality, c^{3H}/c^{6H}, but the resulting males are sterile.[125] The sperm produced are nonmotile and abnormal, with misshapen heads and multiple tails.[126] The spermatids are histologically normal through step 12, after which the heads become vacuolated, with abnormally condensed chromatin.[127] Cattanach's insertion, Is(7:X)Ct, in which a region of approximately

21 cM from chromosome 7 including the C locus is translocated to the X chromosome, can rescue both the albino-deletion-associated lethality and c^{3H}/c^{6H} male sterility.[128] This result is somewhat surprising because of the functional inactivation of the X chromosome during spermatogenesis.[29] There are several possible explanations: the autosomal insertion may not be completely inactivated, the essential gene product may be transcribed early in meiosis, or the defect may be due to the somatic cells in the testis.

14. Purkinje Cell Degeneration (*pcd*), Chromosome 13

Homozygous *pcd/pcd* animals exhibit rapid degeneration of nearly all of their Purkinje cells beginning at 15 to 28 days of life, and a slower degeneration of the photoreceptor cells of the retina and the mitral cells of the olfactory bulb.[129] In chimeras, the action of *pcd* is cell-autonomous to the Purkinje cells.[130] *pcd/pcd* males have low numbers (>5% of wild-type) of nonmotile sperm; 93% of these have abnormal heads and/or tails in both the testis and epididymis. The early stages of spermatogenesis are normal. Acrosome formation, the initiation of nuclear condensation, and manchette formation occur on schedule. At later stages, heads appear abnormal, with incursions of Sertoli cell processes. In addition, there are defects in the assembly of outer dense fibers in the tail with extraneous or ectopic fibers formed.[131] Recently, an insertional mutant in *pcd* generated in forming a transgenic mouse line has been obtained.[132] The new *pcd* allele exhibits both the neurologic and male-fertility defects, suggesting that a single, pleiotropic locus is responsible for both phenotypes.

15. Oligotriche (*olt*)

Homozygous *olt/olt* individuals are characterized by sparse fur; females are fertile, but males are sterile. Spermatogenesis proceeds normally until the haploid spermatid stages, and then becomes abortive. No mature spermatozoa are found in the seminiferous tubules or epididymis.[133]

16. Wobbler (*wr*)

Homozygous animals exhibit symptoms of motor neuron disease consisting of progressive weakness, followed by paralysis in the forelimbs, a wobbly gait, and neuromuscular atrophy. Both males and females are sterile.[134] *wr/wr* males contain grossly normal, but nonmotile sperm in the vas deferens. The flagella are missing between one and four outer doublets. A higher percentage of defective sperm are observed in the vas deferens (70%) than in the testis (5%), suggesting that these outer doublets are unstable and dissociate or degenerate.[135] As in the case of *hop*, the phenotype exhibited by *wr/wr* males suggests that they may have a defect in a structural component of the flagella.

17. Bouncy (*bc*)

Homozygous *bc/bc* males are infertile, but spermatogenesis, sperm production, levels of serum testosterone, and mating behavior are apparently all normal.[136] However, after mating only 1% of the recovered ova have been fertilized. In addition, there is a significant decrease both in the percentage of ova fertilized *in vitro* and in the number of sperm bound per egg.[136] An 18-kDa protein normally synthesized in the epididymis is missing in *bc/bc* males.[136] The *bc* mutation thus appears to directly affect the process of sperm-egg interaction, either through a defect in a specific sperm membrane protein, or through a breakdown in sperm maturation during transit in the epididymis.

B. EFFECTS OF THE Y CHROMOSOME ON SPERMATOGENESIS

The Y chromosome plays a central role in primary sex determination in the male, and contains the putative testis-determining gene (*Tdy*).[137-140] XX↔XY and XO↔XY chimeras usually develop as fertile males.[141,142] However, both the differentiating germ cells and the Sertoli cells in the chimeric testes are exclusively XY.[142-144] This suggests that, in addition to *Tdy*, the Y chromosome carries one or more genes that are required for spermatogenesis and that act in a germ cell-autonomous manner. Evidence for the existence of several genes required during spermatogenesis has been obtained from phenotypically sex-reversed animals that carry varying amounts of Y chromosome DNA (Y DNA) translocated to an X chromosome.

1. Sex-Reversed Phenotypic Males

Three different examples of sex reversal leading to phenotypic males are known. All involve the translocation of a portion of the Y chromosome to the X chromosome. All three produce testes, but the extent to which spermatogenesis progresses is directly correlated with the extent of Y DNA present. XX germ cells are unable to survive in the adult testicular environment, but XO cells can be supported to varying degrees. Normal XO germ cells can contribute to spermatogonia in chimeric XO↔XY testes, but fail to differentiate further.[142] X^YO males, which contain the majority of the Y chromosome, produce motile sperm and are fertile.[139] $X^{Sxr}O$ animals are also phenotypic males, but spermatogenesis breaks down at the meiotic divisions, and the spermatids produced are abnormal.[96,145-147] The *Sxr* phenotype is caused by the translocation of approximately 5 × 10^6 bp of Y DNA to the X chromosome.[140,148-151] A variant of *Sxr* (*Sxr'*) arose through a partial deletion within the Y DNA, and has lost the expression of the male-specific antigen H-Y.[150-152] Spermatogenesis is severely affected in $X^{Sxr'}O$ mice, and few cells differentiate beyond type A spermatogonia.[153] The spermatogenic block in $X^{Sxr'}O$ germ cells is similar to that seen with XO germ cells in an XO↔XY chimera, and the defect first becomes evident during spermatogonial proliferation soon after birth.[153-155] Taken together, these data suggest that there may be several genes on the Y chromosome that are required for spermatogenesis, and at least one appears to function early in the differentiation of spermatogonia.

2. Y Chromosome Effects on Sperm Morphology

The percentage of spermatozoa with morphologically abnormal head shapes is very high (16%) in males of the KE strain of mice.[156,157] In addition, the fertilization rate after normal matings is reduced compared with other strains, and more than 20% of the ova are not fertilized.[156,157] Some of the factors involved have been localized to the Y chromosome using reciprocal backcrosses between the KE strain and the CBA strain, which has a high fertilization rate and a low percentage (5.9%) of abnormal sperm.[156,157] CBA spermatozoa also have a competitive advantage over KE spermatozoa, but not KE.CBA-Y congenic spermatozoa, when the two types are mixed and used in heterospermic artificial insemination experiments.[158]

C. EFFECTS OF *t* HAPLOTYPES ON SPERMATOGENESIS AND SPERM FUNCTION

t haplotypes are variant forms of mouse chromosome 17 that are present as polymorphisms in wild populations at levels of up to 30% (a more detailed description is presented in several recent reviews).[159-162] *t* haplotypes are distinguished from the wild-type form of chromosome 17 by the presence of four large relative inversions which suppress recombination over a 15- to 20-cM region of the proximal third of chromosome 17, including the mouse major histocompatibility complex.[163-168] This region of chromosome 17 contains the *t* complex, a set of linked genes affecting sperm function and embryonic development. Using *in situ* hybridization, the *t* complex has been estimated to comprise approximately 1.2% of the mouse genome.[169] Although most *t* haplotypes carry recessive embryonic lethality mutations, they are maintained in wild mouse populations by their transmission from heterozygous +/*t* males at levels far greater than 50% (up to 99%). This non-Mendelian transmission ratio distortion (TRD) was first described in 1936 by Chesley and Dunn.[170] In contrast, males that are balanced-lethal heterozygotes for two different *t*-lethal mutations (t^x/t^y) are sterile.[171,172]

1. The Physiology of Transmission Ratio Distortion

TRD could potentially result either from an inherent superiority of *t*-sperm or through a functional inactivation of +-sperm.[172,173] Two lines of data support the latter proposal. First, four male chimeras ($+/+ \leftrightarrow +/t$) have been produced which do not show TRD with respect to the chimeric +/+ partner, but continue to exhibit TRD with respect to the +-meiotic partner.[174,175] Second, after artificial insemination with a mixture of equal numbers of motile +/+ and $+/t^{w32}$ sperm, the t^{w32}- chromosome was transmitted to only 22% of the progeny, instead of the 95% obtained following artificial insemination with $+/t^{w32}$ sperm alone.[176] This again suggests that the *t*-sperm are not inherently superior, but rather their wild-type meiotic partners are dysfunctional.

The physiologic basis for TRD is not yet understood, but it is likely to prove to be a complex phenomenon. Spermatogenesis and meiosis proceed normally in +/*t* males.[172,177] Sperm populations from +/*t* males do not contain a significantly higher number of abnormal sperm than those from +/+ males, nor do they contain sperm with unique abnormalities.[178-180] DNA analysis has demonstrated that equal numbers of sperm bearing the wild-type form of chromosome 17 (+-sperm) and bearing the *t* haplotype form (*t*-sperm) are produced, ejaculated, and maintained in the female reproductive tract (uterus) for at least 90 min after coitus.[181] However, sperm populations from +/*t* males have elevated levels of galactosyltransferase activity[182,183] and contain a higher percentage of sperm exhibiting nonprogressive motility or "hyperactivation"[184] than +/+-sperm populations. Hyperactivation requires the presence of Ca^{2+} and occurs earlier in +/*t*-sperm than in +/+-sperm populations.[185,186]

Recently, we examined the status of the acrosome before and after *in vitro* capacitation in populations of sperm from congenic +/+ and +/*t* males.[187] Sperm require an intact acrosome in order to bind to the zona pellucida of the egg.[188-190] We found that a significant proportion of sperm from +/*t* mice spontaneously undergo a premature acrosome reaction during *in vitro* capacitation. There was no change in the proportion of acrosome-reacted sperm in the +/+-population during the incubation period. Since the +/+ and +/*t* animals were congenic, the observed difference must be due to the presence of the *t* haplotype. These data strongly suggest that the presence of the *t* complex during spermatogenesis has induced the resulting +-sperm to undergo a premature acrosome reaction, rendering them dysfunctional. We have examined two independent *t* haplotypes that are transmitted at very high ratios (<95%), and both exhibit this phenomenon. In contrast, *t* haplotypes that exhibit a moderate transmission ratio do not exhibit this effect.[186] Thus, it is likely that TRD is the result of a number of additive effects, and that the induction of a premature acrosome reaction in the +-sperm is one of these effects.

2. *t* Complex-Associated Male Sterility

Male mice heterozygous for two *t* haplotypes carrying different lethal alleles (t^x/t^y) or homozygous for semilethal *t* haplotypes (e.g., t^{w2}/t^{w2}) are sterile.[171,172] In addition, males that carry one complete and one partial *t* haplotype (see below) or certain combinations of partial *t* haplotypes are also sterile or show reduced fertility.[178,191] The extent of spermatogenic impairment may depend on the extent of the *t* haplotype form of *t* complex DNA (*t*-DNA), the particular *t* haplotypes involved, and the genetic background of the mice. Although some studies have shown oligospermia[192] or spermatogenic breakdown,[193] in most cases near-normal levels[178,194,195] of morphologically normal sperm are formed by *t*/*t* males.[196,197] Sperm from t^x/t^y mice exhibit reduced motility and few reach the oviduct after natural mating.[172,178,194,195,198] In addition, t^x/t^y sperm are unable to fertilize even if injected into the ovarian bursa after ovulation[199] or introduced *in vitro* to zona-free eggs.[200] Sperm from t^x/t^y males are able to disperse the cumulus cells around the egg, indicating that they do not have a defect in hyaluronidase activity. After coculture with eggs, all t^x/t^y sperm have undergone the acrosome reaction, but there is no evidence of membrane fusion.[200]

3. The Genetics of Transmission Ratio Distortion and Sterility

A major tool employed in the dissection of the myriad genes located within the *t* complex is a panel of partial *t* haplotypes that have been derived through rare recombination events. Each partial *t* haplotype contains only a portion of *t*-DNA and expresses a subset of *t*-associated effects.[201] Until the recent molecular analysis of the *t* complex region, the precise extent of *t*-DNA in different partial *t* haplotypes was not known.[202] In addition, differences in both the individual *t* haplotypes and the genetic backgrounds employed have made it difficult to compare results between different laboratories. However, recently, it has been possible to assemble the complex data and develop a genetic model for TRD.

Lyon has proposed that TRD is caused by the interaction of two types of genes and/or their products.[203,204] She suggested that three or more *t*-alleles at a series of *t complex distorter (Tcd)* loci could act *in cis* or *in trans* to promote the transmission of the chromosome bearing the *t*-allele at the *t complex responder (Tcr)* locus. If the Tcr^t allele is absent, or present on both chromosomes, no TRD occurs. Molecular analysis of the extent of *t*-DNA present in a panel of partial *t* haplotypes provided strong support for this proposal, and has allowed the mapping of the *Tcr* locus and each of the three *Tcd* loci to genomic intervals defined by a series of DNA markers.[202] In further studies, we have identified and mapped two additional *Tcd* loci and provided evidence that the Tcd^t loci can act additively to increase the transmission ratio.[205,206] The genetics of TRD indicate that individual *Tcd* loci could function either before or after meiosis; however, postmeiotic gene expression would have to be accompanied by free diffusion of the products among both + and *t* haploid cells. In contrast, the genetics indicate that the *Tcr* locus must function postmeiotically, and that the *Tcr* products must be sequestered within the haploid cells in which they are expressed.

The production of *t*-complex-associated male sterility also appears to require the additive influences of a series of factors, and they are genetically inseparable from the *Tcd* loci.[191,203,206,207] The presence of either a Tcr^t or a Tcr^+ allele does not seem to be required for sterility.[191] However, in the absence of at least one copy of Tcr^t, even one set of *Tcd* loci can impair fertility.[191,207]

4. Candidate Genes for Loci Involved in Transmission Ratio Distortion

The *Tcr* locus plays the central role in the TRD phenotype; in the absence of a Tcr^t allele, no distortion of transmission ratio is observed. In fact, the *Tcr* locus represents the only known mammalian gene to exert a haploid-specific effect. We have used reverse genetics to clone a candidate gene for the *Tcr* locus.[202,208,209] The center of the *t* complex contains a family of highly homologous transcription units that are expressed specifically in meiotic and postmeiotic male germ cells, and one of these genes maps to the genetically defined *Tcr* region.[210] We have cloned and sequenced all three wild-type and *t* haplotype members of this gene family[210] (unpublished data). They potentially could encode a series of highly conserved polypeptides with a predicted molecular weight of 48 kDa. The predicted amino acid sequence does not show any strong homologies with previously reported sequences. However, an interesting and unusual feature of the predicted sequence is the presence of a 9-amino acid motif that is repeated 20 times in tandem in the C terminus of the coding region. The consensus motif is Glu-Thr-Val-(Phe/Tyr)-Pro-(Asp/Asn)-Gly-Thr-Val. We have recently identified a unique, alternatively spliced variant of the Tcr^t candidate that is expressed only in haploid cells, and which generates a novel carboxy terminus (unpublished observations). We are currently analyzing the expression and localization of the products of these genes during spermatogenesis.

Candidate genes have also been identified for two of the *Tcd* loci,[211,212] and polypeptides have been identified by high-resolution, two-dimensional gel electrophoresis that can be localized to intervals corresponding to two of the remaining *Tcd* loci.[202,213,214] The gene for the *t* complex polypeptide-1 (TCP1) polypeptide, which maps to the region of *Tcd-4*, has been cloned.[211] TCP1 is expressed in all cell types examined, but it is very abundant in both pachytene spermatocytes and round spermatids.[215-217] Recently, TCP1 has been shown to be associated with the cytoplasmic aspect of the membranes of the *trans*-Golgi network.[217] Thus, it may be involved in the transport of proteins through the Golgi network, and in the development of the acrosome. A candidate for a second *Tcd* locus (*Tcd-3*) has also been cloned.[212] Transcripts corresponding to this clone, 117c3, are first detectable in primary spermatocytes, and continue to be synthesized in spermatids. 117c3 could potentially encode a 21-kDa polypeptide. TCP4, 5, 8, and 9 map to the region defining *Tcd-1*, and TCP3, 6, and 7 map to the region defining *Tcd*-2.[202,214] TCP3 and 7 are not detectable in somatic tissues, and are expressed at the highest levels during the haploid stages of spermatogenesis.[215] A number of other testis-expressed genes also map to the *t* complex region of chromosome 17, including the *Pim-1* proto-oncogene,[218,219] *tctex-1*,[220] *Hst-1*,[84] and *qk*.[221] In addition, the structural gene for *pgk-2*,[222-224] and two testis-expressed genes, *tcte-1* (formerly D17Si11)[225] (unpublished observations) and *Tpx-1*,[226,227] are located just outside the *t* complex. It is intriguing that so many testis-expressed genes are located in this region, but this may simply reflect the fact that few regions of the mouse genome have been as extensively analyzed.

IV. MOLECULAR CANDIDATES FOR GENES INVOLVED IN THE REGULATION OF SPERM CELL DIFFERENTIATION

The results of two-dimensional gel studies indicate that the complexity of gene expression in male germ cells is not significantly different from that observed in other tissues of the adult mouse.[214,228] Furthermore, with standard Northern blot surveys of newly obtained mammalian gene clones, a large number of new testis-expressed transcripts have been identified.[229,230] However, it is obvious that the vast majority of these testis-expressed products will not be involved in the regulation of differentiation. We describe below a number of the best candidates obtained to date that could play a regulatory role in the process of spermatogenesis based on their known functions (or the functions of their homologues) in other tissues.

A. PROTO-ONCOGENES

1. Nuclear Proteins

Several nuclear proto-oncogenes have been detected in the testis and are restricted to the early stages of differentiation. The highest levels of c-*myc* and c-*jun* transcripts are found in type B spermatogonia and early preleptotene spermatocytes; *c-fos* is even more restricted, with maximal levels only in type B spermatogonia. Some expression is also detectable in type A spermatogonia and through the pachytene-stage spermatocytes.[219,231,232] The $p62^{fos}$ protein is detectable only in type B spermatogonia.[232] Transcripts for *mdm-1*, the gene for a 77-kDa nuclear protein that may be involved in transformation, are also expressed in the testis.[233] Smaller transcripts that are generated by alternative splicing and polyadenylation site usage, and may encode truncated proteins, are more common in testis than in somatic tissues.[233] *c-myc, c-fos,* and *c-jun* have all been implicated in transcriptional regulation;[234-236] their absence from later spermatogenic stages suggests that transcription in these cells may be under different controls than in somatic cells.

2. c-*mos*

The product of the c-*mos* proto-oncogene has been implicated in the regulation of meiosis. c-*mos* transcripts are expressed at much higher levels in testis and ovary than in other adult tissue, and both tissues express a unique RNA species.[237] The 1.4-kb ovarian and 1.7-kb testicular transcripts are produced from different promoters.[238] c-*mos* transcripts are not detectable in the testis of mutant mice arrested prior to meiosis.[56,239] Low levels of transcripts have been detected in purified pachytene spermatocytes in some studies, but the highest expression is found in round spermatids.[56,240] c-*mos* transcripts are not detectable in residual bodies, which suggests that the gene acts earlier, during meiosis or spermiogenesis.[56,240] The presence of c-*mos* transcripts in ovaries is solely due to a very high level of expression in oocytes. The levels of c-*mos* RNA increase as the oocytes grow, and the concentration of RNA is ~100X higher in oocytes than in round spermatids.[56,240] c-*mos* has been shown to be involved in the resumption of meiosis and oocyte maturation in both *Xenopus* and mice.[83,241-243] The c-*mos* product persists through oocyte maturation, which suggests that it may also be required at a later stage. The function of the c-*mos* product in spermatogenesis is still unknown, and very little c-*mos* expression is detectable before meiosis. c-*mos* may provide a second function, common to both spermatids and oocytes during the postmeiotic stages of differentiation.

3. *int-1*

The proto-oncogene *int-1* is expressed only in the developing nervous system of mid-gestation embryos and the adult testis.[244-247] In testis, *int-1* expression is detectable at approximately day 26, and cell separation experiments have demonstrated *int-1* transcripts in round spermatids and residual bodies, but not in pachytene spermatocytes.[247] *int-1* has recently been demonstrated to be the homologue of the *Drosophila* segment polarity gene *wingless,* which is thought to be a diffusible extracellular inductive factor.[248]

4. c-*abl*

Low levels of ~6- and ~7-kb transcripts from the c-*abl* gene, a tyrosine-kinase that is the cellular homologue of the Abelson murine leukemia virus-transforming gene, are expressed in most somatic tissues. The highest level of expression is seen in testis, which also contains an additional 4.7-kb testis-specific transcript.[249] This 4.7-kb transcript is not seen in RNA isolated from immature testis, or in testis from sterile males that fail to produce haploid spermatids.[239,247,250] High levels of the testis-specific transcript are seen in isolated round spermatids and in the cytoplasmic fragment/residual body fraction representing late spermatids.[250] Utilization of different promoters and alternative splicing pathways generate at least four somatic forms of c-*abl* transcripts with different 5′ exons that could result in the generation of distinct protein N termini.[251,252] The 4.7-kb testis-specific transcript arises from the use of an unusual polyadenylation signal that removes 1.2 kb of 3′-untranslated region from that 6-kb somatic transcript, but does not affect the coding region.[253,254] The 4.7-kb transcript is most abundant in elongating spermatids, is very stable, and is found in both the polysomal and nonpolysomal fractions.[254,255] This suggests the possibility that c-*abl* expression is under translational control, and that the novel 3′-untranslated region may be involved in this control, or in RNA stability. P150$^{c\text{-}abl}$ protein is detectable in the testis, and the highest levels are found in late stage elongating spermatids and residual bodies.[256] The protein is indistinguishable from that seen in somatic tissues, but a unique 74-kDa phosphoprotein is labeled and coprecipitated from haploid spermatids.[256]

B. HOMEOBOX-CONTAINING GENES

A number of murine genes identified by homology to the *Drosophila* homeobox domain (*HOX* genes) are expressed during embryogenesis and also in the testis of the adult.[230,257-259] The expression of the *HOX-1.4* gene has been characterized in detail. *HOX-1.4* transcripts are detectable in midgestation embryos and in adult testis, but not in other somatic tissues in adult animals.[260-262] During embryogenesis, the highest levels of *HOX-1.4* expression is found in the developing spinal cord.[263] Cell separation and *in situ* hybridization experiments have shown that *HOX-1.4* expression is first detectable in primary spermatocytes, and continues through the round and elongating spermatid stages.[261,263] However, *HOX-1.4* transcripts are not detectable in the testis of *hpg/hpg*, $X^{Tfm}Y$, or X:autosome translocation sterile male mice which arrest during meiotic prophase, suggesting that it may only be expressed immediately before meiosis.[239] The *HOX-1.4* transcript in pachytene spermatocytes (1.35 kb) is slightly smaller than that in spermatids (1.45 kb), and much smaller than the transcripts expressed during embryogenesis (1.7 and 2.4 kb).[262,263] The differences appear to arise in the 5′-untranslated regions through the use of alternative promoters or splicing patterns.[262] The homeobox domain of *HOX-1.4* is highly homologous to the homeobox region of the *Antennapedia* gene of *Drosophila,* and thus it provides a good candidate for a regulatory gene expressed during spermatogenesis.[260]

C. POTENTIAL TESTIS-SPECIFIC RNA-BINDING PROTEIN

The D1Pas1 locus defined by the cDNA probe PL10 maps to murine chromosome 1. PL10 hybridizes to two transcripts of 3.7 and 3.3 kb that are expressed at high levels in the testis; in addition, PL10 detects a weak band of 4.6 kb in somatic tissues.[264,265] The testis-specific transcripts are first detectable at day 16, and *in situ* hybridization and cell separation experiments demonstrate expression beginning at the pachytene stage and persisting through the round and elongating spermatid stages.[265] No expression was detected in oocytes.[265] The predicted protein product of the D1Pas1 locus has extensive homology to a set of proteins proposed to belong to the newly defined ATP-dependent helicase family.[266] This family includes the murine protein synthesis initiation factors eIF-4AI and eIF-4AII.[267] The D1Pas1 product is most closely related to *vasa,* a *Drosophila* maternal-effect gene whose product is localized in the polar granules of the oocyte, and is required for the formation of the germ line in the embryo.[268,269] The product of the D1Pas1 locus could provide a novel translation factor for testis-specific expression, or alternately, it could be involved in translational control by binding to and sequestering mRNA.

V. FINAL THOUGHTS

We have attempted in this review to present a survey of mouse genes characterized to date that could play a role in male germ cell differentiation. As is clear from the organization of this review, two very different approaches have been used to identify such genes. The first approach, followed over the last half century, has been the identification of mutations and other chromosomal alterations that result in male sterility. The second approach, possible only in the last decade, has been the identification of cloned genes expressed in a unique manner during germ cell differentiation that are known, from other studies, to represent good candidates for regulatory elements. With very few exceptions, the two lists of genes do not overlap. Although classical genetics and molecular biology each provide powerful tools for the analysis of germ cell differentiation, it is when these approaches are combined that we can expect a rapid increase in our understanding of this aspect of developmental biology as well as others. The present challenge to workers in this field is to bridge the gap from both directions—with the use of reverse genetics to clone mutant loci, and with the use of transgenic mice and embryonic stem cells to produce mutations at cloned loci identified as molecular candidates for regulators of germ cell differentiation in mammals.

ACKNOWLEDGMENTS

The authors would like to thank our colleagues L. C. Snyder, J.-Y. Tsai, and S. Pilder for critical review and discussion of this manuscript during its preparation. Work in the laboratory of L.M.S. is supported by NIH grants (RO1HD20275, PO1CA41086, and HD24383). J.C.T. is supported by a postdoctoral fellowship from the American Cancer Society.

REFERENCES

1. **Leblond, C. P. and Clermont, Y.,** Spermiogenesis of rat, mouse, hamster and guinea pig as revealed by the periodic acid-fuchsin sulfurous acid technique, *Am. J. Anat.,* 90, 167, 1952.
2. **Clermont, Y. and Perey, B.,** The stages of the cycle of the seminiferous epithelium of the rat: practical definitions in PA-Schiff-hematoxylin and hematoxylin-eosin stained sections, *Rev. Can. Biol.,* 16, 451, 1957.
3. **Oakberg, E. F.,** A description of spermiogenesis in the mouse and its use in the analysis of the cycle of the seminiferous epithelium and germ cell renewal, *Am. J. Anat.,* 99, 391, 1956.
4. **Fawcett, D. W.,** The mammalian spermatozoon, *Dev. Biol.,* 44, 394, 1975.
5. **Monesi, V., Geremia, R., D'Agostino, A, and Boitani, C.,** Biochemistry of male germ cell differentiation in mammals: RNA synthesis in meiotic and postmeiotic cells, *Curr. Top. Dev. Biol.,* 12, 11, 1978.
6. **Bellvé, A. R.,** The molecular biology of mammalian spermatogenesis, in *Oxford Reviews of Reproductive Biology,* Vol. I, Finn, C. A., Ed., Clarendon Press, Oxford, 1979, 159.
7. **Hecht, N. B.,** Regulation of gene expression during mammalian spermatogenesis, in *Experimental Approaches to Mammalian Embryonic Development,* Rossant, J. and Pedersen, R. A., Eds., Cambridge University Press, Cambridge, 1986, 151.
8. **Guraya, S. S.,** *Biology of Spermatogenesis and Spermatozoa in Mammals,* Springer-Verlag, Berlin, 1987.
9. **Stefanini, M., Conti, M., Geremia, R., and Ziparo, E.,** Regulatory mechanisms of mammalian spermatogenesis, in *Biology of Fertilization,* Vol. II, Metz, C. B. and Monroy, A., Eds., Academic Press, New York, 1985, 59.
10. **Sharpe, R. M., Kerr, J. B., Cooper, I., and Bartlett, J. M. S.,** Paracrine control mechanisms in the testis, in *Molecular and Cellular Endocrinology of the Testis,* Stefanini, M., Conti, M., Geremia, R., and Ziparo, E., Eds., Elsevier, Amsterdam, 1986, 15.
11. **Dym, M. and Fawcett, D. W.,** The blood-testis barrier in the rat and the physiological compartmentation of the seminiferous epithelium, *Biol. Reprod.,* 3, 308, 1970.
12. **Russell, L. D. and Peterson, R. N.,** Sertoli cell junctions: morphological and functional correlates, *Int. Rev. Cytol.,* 94, 177, 1985.
13. **Clermont, Y.,** Kinetics of spermatogenesis in mammals: seminiferous epithelium cycle and spermatogonial renewal, *Physiol. Rev.,* 52, 198, 1972.
14. **DeRooij, D. G.,** Regulation of the proliferation of spermatogonial stem cells, in *Stem Cells,* Lord, B. I. and Dexter, T. M., Eds., Company of Biologists, Cambridge, 1988, 181.
15. **Dym, M. and Fawcett, D. W.,** Further observations on the numbers of spermatogonia, spermatocytes, and spermatids connected by intercellular bridges in the mammalian testis, *Biol. Reprod.,* 4, 195, 1971.
16. **Fawcett, D. W.,** Observations on cell differentiation and organelle continuity in spermatogenesis, in *Proc. Int. Symp. the Genetics of the Spermatozoon,* Beatty, R. A. and Gluecksohn-Waelsch, S., Eds., University of Edinburgh, Edinburgh, 1972, 37.
17. **Clermont, Y. and Rambourg, A.,** Evolution of the endoplasmic reticulum during rat spermiogenesis, *Am. J. Anat.,* 151, 191, 1978.
18. **Braun, R. E., Behringer, R. R., Peschon, J. J., Brinster, R. L., and Palmiter, R. D.,** Genetically haploid spermatids are phenotypically diploid, *Nature (London),* 337, 373, 1989.
19. **Braun, R. E., Peschon, J. J., Behringer, R. R., Brinster, R. L., and Palmiter, R. D.,** Protamine 3'-untranslated sequences regulate temporal translational control and subcellular localization of growth hormone in spermatids of transgenic mice, *Genes Dev.,* 3, 793, 1989.
20. **Oakberg, E. F.,** Duration of spermatogenesis in the mouse and timing of stages of the cycle of the seminiferous epithelium, *Am. J. Anat.,* 99, 507, 1956.
21. **Kierszenbaum, A. L. and Tres, L. L.,** Structural and transcriptional features of the mouse spermatid genome, *J. Cell Biol.,* 65, 258, 1975.
22. **Kierszenbaum, A. L. and Tres, L. L.,** RNA transcription and chromatin structure during meiotic and postmeiotic stages of spermatogenesis, *Fed. Proc.,* 37, 2512, 1978.
23. **Dooher, G. B. and Bennett, D.,** Fine structural observations on the development of the sperm head in the mouse, *Am. J. Anat.,* 136, 339, 1973.
24. **Grimes, S. R., Jr., Meistrich, M. L., Platz, R. D., and Hnilica, L. S.,** Nuclear protein transitions in rat testis spermatids, *Exp. Cell Res.,* 110, 31, 1977.
25. **Meistrich, M. L., Brock, W. A., Grimes, S. R., Platz, R. D., and Hnilica, L. S.,** Nuclear protein transitions during spermatogenesis, *Fed. Proc.,* 37, 2522, 1978.
26. **Mayer, J. F., Jr., Chang, T. S. K., and Zirkin, B. R.,** Spermatogenesis in the mouse. II. Amino acid incorporation into basic nucleoproteins of mouse spermatids and spermatozoa, *Biol. Reprod.,* 25, 1041, 1981.
27. **Kistler, W. S. and Williams-Ashman, H. G.,** On three vartieties of specific basic proteins associated with mammalian spermatogenesis, in *Hormonal Regulation of Spermatogenesis,* French, F. S., Hansson, V., Ritzen, E. M., and Nayfeh, S. N., Eds., Plenum Press, New York, 1975, 423.

28. **Kistler, W. S., Noyes, C., Hsu, R., and Heinrickson, R. L.,** The amino acid sequence of a testis-specific basic protein that is associated with spermatogenesis, *J. Biol. Chem.*, 250, 1847, 1975.
29. **Monesi, V.,** Synthetic activities during spermatogenesis in the mouse, *Exp. Cell Res.*, 39, 197, 1965.
30. **O'Brien, D. A. and Bellvé, A. R.,** Protein constituents of the mouse spermatozoon. II. Temporal synthesis during spermatogenesis, *Dev. Biol.*, 75, 405, 1980.
31. **Kleene, K. C., Distel, R. J., and Hecht, N. B.,** cDNA clones encoding cytoplasmic poly(A)+ RNAs which first appear at detectable levels in haploid phases of spermatogenesis in the mouse, *Dev. Biol.*, 98, 455, 1983.
32. **Kleene, K.C., Distel, R. J., and Hecht, N. B.,** Translational regulation and deadenylation of a protamine mRNA during spermiogenesis in the mouse, *Dev. Biol.*, 105, 71, 1984.
33. **Kleene, K. C., Distel, R. J., and Hecht, N. B.,** Nucleotide sequence of a cDNA clone encoding mouse protamine 1, *Biochemistry*, 24, 719, 1985.
34. **Hecht, N. B., Bower, P. A., Kleene, K. C., and Distel, R. J.,** Size changes of protamine 1 mRNA provide a molecular marker to monitor spermatogenesis in wild-type and mutant mice, *Differentiation*, 29, 189, 1985.
35. **Hecht, N. B., Bower, P. A., Waters, S. H., Yelick, P. C., and Distel, R. J.,** Evidence for haploid expression of mouse testicular genes, *Exp. Cell Res.*, 164, 183, 1986.
36. **Heidaran, M. A. and Kistler, W. S.,** Transcriptional and translational control of the message for transition protein 1, a major chromosomal protein of mammalian spermatids, *J. Biol. Chem.*, 262, 13309, 1987.
37. **Kleene, K. C. and Flynn, J. F.,** Characterization of a cDNA clone encoding a basic protein, TP2, involved in chromatin condensation during spermatogenesis in the mouse, *J. Biol. Chem.*, 262, 17272, 1987.
38. **Yelick, P. C., Balhorn, R., Johnson, P. A., Corzett, M., Mazrimas, J. A., Kleene, K. C., and Hecht, N. B.,** Mouse protamine 2 is synthesized as a precursor whereas mouse protamine 1 is not, *Mol. Cell. Biol.*, 7, 2173, 1987.
39. **Mali, P., Sandberg, M., Vuorio, E., Yelick, P. C., and Hecht, N. B.,** Localization of protamine 1 mRNA in different stages of the cycle of the rat seminiferous epithelium, *J. Cell Biol.*, 107, 407, 1988.
40. **Kleene, K. C.,** Poly(A) shortening accompanies the activation of translation of five mRNAs during spermatogenesis in the mouse, *Development*, 106, 367, 1989.
41. **Fawcett, D. W., Anderson, W. A., and Phillips, D. M.,** Morphogenetic factors influencing the shape of the sperm head, *Dev. Biol.*, 26, 220, 1971.
42. **Cherry, L. M. and Hsu, T. C.,** Antitubulin immunofluorescence studies of spermatogenesis in the mouse, *Chromosoma*, 90, 265, 1984.
43. **Meistrich, M. L., Cole, A., Cherry, L. M., and Trostle-Weige, P. K.,** Nuclear and manchette development during spermiogenesis of normal and *azh* mutant mice, *Ann. N.Y. Acad. Sci.*, 513, 304, 1987.
44. **Fawcett, D. W. and Phillips, D. M.,** The fine structure and development of the neck region of the mammalian spermatozoon, *Anat. Rec.*, 165, 153, 1969.
45. **Irons, M. J. and Clermont, Y.,** Kinetics of fibrous sheath formation in the rat spermatid, *Am. J. Anat.*, 165, 121, 1982.
46. **Irons, M. J. and Clermont, Y.,** Formation of the outer dense fibers during spermiogenesis in the rat, *Anat. Rec.*, 202, 463, 1982.
47. **Fawcett, D. W. and Phillips, D. M.,** Recent observations on the ultrastructure and development of the mammalian spermatozoon, in *Comparative Spermatology*, Baccetti, B., Ed., Academic Press, New York, 1970, 13.
48. **Oko, R.,** Comparative analysis of proteins from the fibrous sheath and outer dense fibers of rat spermatozoa, *Biol. Reprod.*, 39, 169, 1988.
49. **Calvin, H. I.,** Electrophoretic evidence for the identity of the major zinc-binding polypeptides in the rat sperm tail, *Biol. Reprod.*, 21, 873, 1979.
50. **Hecht, N. B., Kleene, K. C., Distel, R. J., and Silver, L. M.,** The differential expression of the actins and tubulins during spermatogenesis in the mouse, *Exp. Cell Res.*, 153, 275, 1984.
51. **Villasante, A., Wang, D., Dobner, P., Dolph, P., Lewis, S. A., and Cowan, N. J.,** Six mouse α-tubulin mRNAs encode five distinct isotypes: testis-specific expression of two sister genes, *Mol. Cell. Biol.*, 6, 2409, 1986.
52. **Pratt, L. F., Okamura, S., and Cleveland, D. W.,** A divergent testis-specific α-tubulin isotype that does not contain a coded C-terminal tyrosine, *Mol. Cell. Biol.*, 7, 552, 1987.
53. **Distel, R. J., Kleene, K. C., and Hecht, N. B.,** Haploid expression of a mouse testis α-tubulin gene, *Science*, 224, 68, 1984.
54. **Hecht, N. B., Distel, R. J., Yelick, P. C., Tanhauser, S. M., Driscoll, C. E., Goldberg, E., and Tung, K. S. K.,** Localization of a highly divergent mammalian testicular α tubulin that is not detectable in brain, *Mol. Cell. Biol.*, 8, 996, 1988.
55. **Waters, S. H., Distel, R. J., and Hecht, N. B.,** Mouse testes contain two size classes of actin mRNA that are differentially expressed during spermatogenesis, *Mol. Cell. Biol.*, 5, 1649, 1985.
56. **Goldman, D. S., Kiessling, A. A., Millette, C. F., and Cooper, G. M.,** Expression of *c-mos* RNA in germ cells of male and female mice, *Proc. Natl. Acad. Sci. U.S.A.*, 84, 4509, 1987.

57. **Fawcett, D. W., Eddy, E. M., and Phillips, D. M.,** Observations on the fine structure and relationships of the chromatoid body in mammalian spermatogenesis, *Biol. Reprod.*, 2, 129, 1970.
58. **Parvinen, M. and Parvinen, L.-M.,** Active movements of the chromatoid body. A possible transport mechanism for haploid gene products, *J. Cell Biol.*, 80, 621, 1979.
59. **Leblond, C. P. and Clermont, Y.,** Definition of the stages of the cycle of the seminiferous epithelium in the rat, *Ann. N.Y. Acad. Sci.*, 55, 548, 1952.
60. **Daoust, R. and Clermont, Y.,** Distribution of nucleic acids in germ cells during the cycle of the seminiferous epithelium in the rat, *Am. J. Anat.*, 96, 255, 1955.
61. **Sud, B.,** Morphological and histochemical studies of the chromatoid body and related elements in spermiogenesis of the rat, *Q. J. Microsc. Sci.*, 102, 495, 1961.
62. **Söderström, K.-O. and Parvinen, M.,** Incorporation of [^{3}H]uridine by the chromatoid body during rat spermatogenesis, *J. Cell Biol.*, 70, 239, 1976.
63. **Söderström, K.-O.,** Effect of actinomycin D on the structure of the chromatoid body in the rat spermatids, *Cell Tissue Res.*, 184, 411, 1977.
64. **Parvinen, L.-M., Jokelainen, P. T., and Parvinen, M.,** Chromatoid body and haploid gene activity: actinomycin D induced morphological alterations, *Hereditas*, 88, 75, 1978.
65. **Perey, B., Clermont, Y., and Leblond, C. P.,** The wave of the seminiferous epithelium in the rat, *Am. J. Anat.*, 108, 47, 1961.
66. **Nebel, B.R., Amarose, A. P., and Hackett, E. M.,** Calendar of gametogenic development in the prepuberal male mouse, *Science*, 134, 832, 1961.
67. **Meistrich, M. L., Bruce, W. R., and Clermont, Y.,** Cellular composition of fractions of mouse testis cells following velocity sedimentation separation, *Exp. Cell Res.*, 79, 213, 1973.
68. **Romrell, L. J., Bellvé, A. R., and Fawcett, D. W.,** Separation of mouse spermatogenic cells by sedimentation velocity, *Dev. Biol.*, 49, 119, 1976.
69. **Wolgemuth, D. J., Gizang-Ginsberg, E., Engelmyer, E., Gavin, B. J., and Ponzetto, C.,** Separation of mouse testis cells on a Celsep apparatus and their usefulness as a source of high molecular weight DNA or RNA, *Gamete Res.*, 12, 1, 1985.
70. **McLaren, A.,** Germ cell lineages, in *Chimeras in Developmental Biology*, LeDouarin, N. and McLaren, A., Eds., Academic Press, London, 1984, 111.
71. **Forsthoefel, P. E.,** The skeletal effects of the luxoid gene in the mouse, including its interaction with the luxate gene, *J. Morphol.*, 102, 247, 1958.
72. **Johnson, D. R. and Hunt, D. M.,** Hop-sterile, a mutant gene affecting sperm tail development in the mouse, *J. Embryol. Exp. Morphol.*, 25, 223, 1971.
73. **Beamer, W. G., Cunliffe-Beamer, T. L., Shultz, K. L., Langley, S. H., and Roderick, T. H.,** Juvenile spermatogonial depletion (*jsd*): a genetic defect of germ cell proliferation of male mice, *Biol. Reprod.*, 38, 899, 1988.
74. **Steinberger, A. and Steinberger, E.,** Secretion of an FSH-inhibiting factor by cultured Sertoli cells, *Endocrinology*, 99, 918, 1976.
75. **Hadley, M. A., Byers, S. W., Suarez-Quian, C. A., Kleinman, H. K., and Dym, M.,** Extracellular matrix regulates Sertoli cell differentiation, testicular cord formation, and germ cell development in vitro, *J. Cell Biol.*, 101, 1511, 1985.
76. **Bennett, D.,** Developmental analysis of a mutation with pleiotropic effects in the mouse, *J. Morphol.*, 98, 199, 1956.
77. **Green, M. C.,** Catalog of mutant genes and polymorphic loci, in *Genetic Variants and Strains of the Laboratory Mouse*, Green, M. C., Ed., Gustav Fischer Verlag, Stuttgart, 1981, 8.
78. **McCoshen, J. A. and McCallion, D. J.,** A study of primordial germ cells during their migratory phase in steel mutant mice, *Experientia*, 31, 589, 1975.
79. **Nakayama, H., Kuroda, H., Onoue, H., Nishimune, Y., Matsumoto, K., Nagano, T., Suzuki, F., and Kitamura, Y.,** Studies of $Sl/Sl^d \leftrightarrow +/+$ mouse aggregation chimeras. II. Effect of the steel locus on spermatogenesis, *Development*, 102, 117, 1988.
80. **Handel, M. A., Lane, P. W., Schroeder, A. C., and Davisson, M. T.,** New mutation causing sterility in the mouse, *Gamete Res.*, 21, 409, 1988.
81. **Swan, D., Oskarsson, M., Keithley, D., Ruddle, F. H., D'Eustachio, P., and Vande Woude, G. F.,** Chromosomal localization of the Maloney sarcoma virus mouse cellular (*c-Mos*) sequence, *J. Virol.*, 44, 752, 1982.
82. **Threadgill, D. W. and Womack, J. E.,** Regional localization of mouse *Abl* and *Mos* proto-oncogenes by *in situ* hybridization, *Genomics*, 3, 82, 1988.
83. **Paules, R. S., Buccione, R., Moschel, R. C., Vande Woude, G. F., and Eppig, J. J.,** Mouse *Mos* protooncogene product is present and functions during oogenesis, *Proc. Natl. Acad. Sci. U.S.A.*, 86, 5395, 1989.
84. **Forejt, J. and Iványi, P.,** Genetic studies on male sterility of hybrids between laboratory and wild mice (*Mus musculus* L.), *Genet. Res.*, 24, 189, 1975.

85. **Imai, H. T., Matsuda, Y., Shiroishi, T., and Moriwaki, J.,** High frequency of X-Y chromosome dissociation in primary spermatocytes of F_1 hybrids between Japanese wild mice (*Mus musculus molossinus*) and inbred laboratory mice, *Cytogenet. Cell Genet.,* 29, 166, 1981.
86. **de Boer, P. and Nijhoff, J. H.,** Incomplete sex chromosome pairing in oligospermic male hybrids of *Mus musculus* and *M. musculus molossinus* in relation to the source of the Y chromosome and the presence or absence of a reciprocal translocation, *J. Reprod. Fertil,* 62, 235, 1981.
87. **Lyon, M. F. and Meredith, R.,** Autosomal translocations causing male sterility and viable aneuploidy in the mouse, *Cytogenetics,* 5, 335, 1966.
88. **Forejt, J. and Gregorová, S.,** Meiotic studies of translocations causing male sterility in the mouse. I. Autosomal reciprocal translocations, *Cytogenet. Cell Genet.,* 19, 159, 1977.
89. **Forejt, J.,** X-Y involvement in male sterility caused by autosome translocations—a hypothesis, in *Genetic Control of Gamete Production and Function,* Crosignani, P.G. and Rubin, B. L., Eds., Academic Press, London, 1982, 135.
90. **Handel, M. A.,** Genetic control of spermatogenesis in mice, in *Spermatogenesis: Genetic Aspects,* Hennig, W., Eds., Springer-Verlag, Berlin, 1987, 1.
91. **Chubb, C. and Nolan C.,** Mouse hybrid sterility and testicular function, *Biol. Reprod.,* 36, 1343, 1987.
92. **Haldane, J. B. S.,** Sex ratio and unisexual sterility in hybrid animals, *J. Genet.,* 12, 101, 1922.
93. **Bonhomme, F., Guenet, J.-L., and Catalan, J.,** Présence d'un facteur de stérilité mâle, *Hst-2*, segrégant dans les croisements interspécifiques *M. musculus* L. × *M. Spretus* Lastaste et lié à *Mod-1* et *Mpi-1* sur le chromosome 9, *C. R. Acad. Sci. paris, Ser. III,* 294, 691, 1982.
94. **Cattanach, B. M., Iddon, C. A., Charlton, H. M., Chiappa, S. A., and Fink, G.,** Gonadtrophin-releasing hormone deficiency in a mutant mouse with hypogonadism, *Nature (London),* 269, 338, 1977.
95. **Bamber, S., Iddon, C. A., Charlton, H. M., and Ward, B. J.,** Transplantation of the gonads of hypogonadal (*hpg*) mice, *J. Reprod. Fertil.,* 58, 249, 1980.
96. **Lyon, M. F., Cattanach, B. M., and Charlton, H. M.,** Genes affecting sex determination in mammals, in *Mechanisms of Sex Differentiation in Animals and Man,* Austin, C. R. and Edwards, R. G., Eds., Academic Press, London, 1981, 329.
97. **Mason, A. J., Pitts, S. L., Nikolics, K., Szonyi, E., Wilcox, J. N., Seeburg, P. H., and Stewart, T. A.,** Gonadal development and gametogenesis in the hypogonadal mouse are restored by gene transfer, *Ann. N.Y. Acad. Sci.,* 513, 16, 1987.
98. **Lyon, M. F. and Hawkes, S. G.,** X-linked gene for testicular feminization in the mouse, *Nature (London),* 227, 1217, 1970.
99. **Attardi, B. and Ohno, S.,** Cytosol androgen receptor from kidney of normal and testicular feminized (*Tfm*) mice, *Cell,* 2, 205, 1974.
100. **Gehring, U. and Tomkins, G. M.,** Characterization of a hormone receptor defect in the androgen-insensitivity mutant, *Cell,* 3, 59, 1974.
101. **Lyon, M. F., Glenister, P. H., and Lanoreux, M. L.,** Normal spermatozoa from androgen-resistant germ cells of chimaeric mice and the role of androgen in spermatogenesis, *Nature (London),* 258, 620, 1975.
102. **Handel, M. A., Park, C., and Sotomayor, R.,** *Mouse News Lett.,* 72, 124, 1985.
103. **Hollander, W. F.,** Hydrocephalic-polydactyl, a recessive and pleiotropic mutant in the mouse, and its location in Chromosome 6, *Iowa State J. Res.,* 51, 13, 1976.
104. **Bryan, J. H. D. and Chandler, D. B.,** Tracheal ciliary defects in mice homozygous for a recessive pleiotropic mutation hydrocephalic-polydactyl, *J. Cell Biol.,* 79, 281, 1978.
105. **Bryan, J. H. D.,** Spermatogenesis revisited. IV. Abnormal spermiogenesis in mice homozygous for another male-sterility-inducing mutation, *hpy* (hydrocephalic-polydactyl), *Cell Tissue Res.,* 180, 187, 1977.
106. **Hunt, D. M. and Johnson, D. M.,** Abnormal spermiogenesis in two pink-eyed sterile mutants in the mouse, *J. Embryol. Exp. Morphol.,* 26, 111, 1971.
107. **Bryan, J. H. D.,** Spermatogenesis revisited. III. The course of spermatogenesis in a male-sterile pink-eyed mutant type in the mouse, *Cell Tissue Res.,* 180, 173, 1977.
108. **Wolfe, H. G., Erickson, R. P., and Schmidt, L. C.,** Effects on sperm morphology by alleles at the pink-eyed dilution locus in mice, *Genetics,* 85, 303, 1977.
109. **Hash, D. C. and Wolfe, H. G.,** Pink-eyed dilution alleles affect negative surface charges of mouse spermatozoa, *Dev. Genet.,* 1, 61, 1979.
110. **Wolfe, H. G.,** Genetic influence on gonadotropic activity in mice, *Biol. Reprod.,* 4, 161, 1971.
111. **Melvold, R. W.,** The effects of mutant *p*-alleles on the reproductive system in mice, *Genet. Res.,* 23, 319, 1974.
112. **Johnson, D.R. and Hunt, D. M.,** Endocrinological findings in sterile pink-eyed mice, *J. Reprod. Fertil.,* 42, 51, 1975.
113. **Handel, M. A., Washburn, L. L., Rosenberg, M. P., and Eicher, E. M.,** Male sterility caused by p^{6H} and qk mutations is not corrected in chimeric mice, *J. Exp. Zool.,* 243, 81, 1987.
114. **Varnum, D. S.,** Blind-sterile: a new mutation on Chromosome 2 of the house mouse, *J. Hered.,* 74, 206, 1983.

115. **Sotomayor, R. E., Clark, J., and Handel, M. A.,** Developmental expression of the blind-sterile (*bs*) mutation in mouse testes, *Biol. Reprod.*, 34(Suppl. 1), 60, 1986.
116. **Sotomayor, R. E. and Handel, M. A.,** Failure of acrosome assembly in a male sterile mouse mutant, *Biol. Reprod.*, 34, 171, 1986.
117. **Sidman, R. L., Dickie, M. M., and Appel, S. H.,** Mutant mice (quaking and jimpy) with deficient myelination in the central nervous system, *Science*, 144, 309, 1964.
118. **Bennett, W. I., Gall, A. M., Southard, J. L., and Sidman, R. L.,** Abnormal spermiogenesis in quaking, a myelin-deficient mutant mouse, *Biol. Reprod.*, 5, 30, 1971.
119. **Justice, M. J. and Bode, V. C.,** Induction of new mutations in a mouse *t*-haplotype using ethylnitrosourea mutagenesis, *Genet. Res.*, 47, 187, 1986.
120. **Justice, M. J. and Bode, V. C.,** Three ENU-induced alleles of the murine quaking locus are recessive embryonic lethal mutations, *Genet. Res.*, 51, 95, 1988.
121. **Shedlovsky, A., King, T. R., and Dove, W. F.,** Saturation germ line mutagenesis of the murine *t* region including a lethal allele at the quaking locus, *Proc. Natl. Acad. Sci. U.S.A.*, 85, 180, 1988.
122. **Hugenholtz, A. P., Eng, W. S., and Bruce, W. R.,** *Mouse News Lett.*, 71, 34, 184.
123. **Kot, M. C. and Handel, M. A.,** Binding of morphologically abnormal sperm to mouse egg zonae pellucida in vitro, *Gamete Res.*, 18, 57, 1987.
124. **Cole, A., Meistrich, M. L., Cherry, L. M., and Trostle-Weige, P. K.,** Nuclear and manchette development in spermatids of normal and *azh/azh* mutant mice, *Biol. Reprod.*, 38, 385, 1988.
125. **Gluecksohn-Waelsch, S., Schiffman, M. B., Thorndike, J., and Cori, C. F.,** Complementation studies of lethal alleles in the mouse causing deficiencies of glucose-6-phosphatase, tyrosine aminotransferase, and serine dehydratase, *Proc. Natl. Acad. Sci. U.S.A.*, 71, 825, 1974.
126. **Lewis, S. E., Turchin, H. A., and Gluecksohn-Waelsch, S.,** Partial complementation of lethal alleles at the albino locus in the mouse resulting in sterility, *Genetics*, 80, s52, 1975.
127. **Lewis, S. E., Turchin, H. A., and Wojtowicz, T. E.,** Fertility studies of complementing genotypes at the albino locus of the mouse, *J. Reprod. Fertil.*, 53, 197, 1978.
128. **Erickson, R. P.,** Cattanach's translocation [Is(7:X)Ct] corrects male sterility due to homozygosity for chromosome 7 deletions, *Genet. Res.*, 43, 35, 1984.
129. **Mullen, R. J., Eicher, E. M., and Sidman, R. L.,** Purkinje cell degeneration, a new neurological mutation in the mouse, *Proc. Natl. Acad. Sci. U.S.A.*, 73, 208, 1976.
130. **Mullen, R. J.,** Site of *pcd* gene action and Purkinje cell mosaicism in cerebella of chimaeric mice, *Nature (London)*, 270, 245, 1977.
131. **Handel, M. A. and Dawson, M.,** Effects on spermiogenesis in the mouse of a male sterile neurological mutation, Purkinje cell degeneration, *Gamete Res.*, 4, 185, 1981.
132. **Krulewski, T. F., Neumann, P. E., and Gordon, J. W.,** Insertional mutation in a transgenic mouse allelic with Purkinje cell degeneration, *Proc. Natl. Acad. Sci. U.S.A.*, 86, 3709, 1989.
133. **Moutier, R.,** New mutation causing sterility restricted to the male in rats and mice, in *The Laboratory Animal in the Study of Reproduction*, Ankikatizides, T., Ericksen, S., and Spiegel, A., Eds., Gustav Fischer, Stuttgart, 1976, 115.
134. **Duchen, L. W., Strich, S. J., and Falconer, D. S.,** An hereditary motor neuron disease with progressive denervation of muscle in the mouse: the mutant "Wobbler", *J. Neurol. Neurosurg. Psychiatry*, 31, 535, 1968.
135. **Leestma, J. E. and Sepsenwol, S.,** Sperm tail axoneme alterations in the Wobbler mouse, *J. Reprod. Fertil.*, 58, 267, 1980.
136. **Holland, M. K. and Orgebin-Crist, M.-C.,** Epididymal protein synthesis and secretion in strains of mice bearing single gene mutations which affect fertility, *Biol. Reprod.*, 38, 497, 1988.
137. **Welshons, W. J. and Russell, L. B.,** The Y-chromosome as the bearer of male determining factors in the mouse, *Proc. Natl. Acad. Sci. U.S.A.*, 45, 560, 1959.
138. **Ohno, S.,** Major regulatory genes for mammalian sexual development, *Cell*, 7, 315, 1976.
139. **Eicher, E. M. and Washburn, L. L.,** Genetic control of primary sex determination in mice, *Annu. Rev. Genet.*, 20, 327, 1986.
140. **Goodfellow, P. N. and Darling, S. M.,** Genetics of sex determination in man and mouse, *Development*, 102, 251, 1988.
141. **McLaren, A.,** Chimeras and sexual differentiation, in *Chimeras in Developmental Biology*, Le Douarin, N. and McLaren, A., Eds., Academic Press, London, 1984, 381.
142. **Levy, E.R. and Burgoyne, P. S.,** The fate of XO germ cells in the testes of XO/XY and XO/XY/XYY mouse mosaics: evidence for a spermatogenesis gene on the mouse Y chromosome, *Cytogenet. Cell Genet.*, 42, 208, 1986.
143. **Singh, L., Matsukuma, S., and Jones, K. W.,** Testis development in a mouse with 10% of XY cells, *Dev. Biol.*, 122, 287, 1987.

144. **Burgoyne, P. S., Buehr, M., Koopman, P., Rossant, J., and McLaren, A.,** Cell-autonomous action of the testis-determining gene: Sertoli cells are exclusively XY in XY↔XY chimaeric mouse testes, *Development,* 102, 443, 1988.
145. **Cattanach, B. M., Pollard, C. E., and Hawkes, S. G.,** Sex-reversed mice: XX and XO males, *Cytogenetics,* 10, 318, 1971.
146. **Levy, E. R. and Burgoyne, P. S.,** Diploid spermatids: a manifestation of spermatogenic impairment in XO*Sxr* and T31H/+ male mice, *Cytogenet. Cell Genet.,* 42, 159, 1986.
147. **Kot, M. C. and Handel, M. A.,** Spermatogenic kinetics and differentiation in XO, *Sxr* mice, *Biol. Reprod.,* 36 (Suppl. 1), 81, 1987.
148. **Singh, L. and Jones, K. W.,** Sex reversal in the mouse (Mus musculus) is caused by a recurrent nonreciprocal crossover involving the X and an aberrant Y chromoosome, *Cell,* 28, 205, 1982.
149. **Evans, E. P., Burtenshaw, M. D., and Cattanach, B. M.,** Meiotic crossing-over between the X and Y chromosomes of male mice carrying the sex-reversing (*Sxr*) factor, *Nature (London),* 300, 443, 1982.
150. **McLaren, A., Simpson, E., Epplen, J. T., Studer, R., Koopman, P., Evans, E. P., and Burgoyne, P. S.,** Location of the genes controlling H-Y antigen expression and testis determination on the mouse Y chromosome, *Proc. Natl. Acad. Sci. U.S.A.,* 85, 6442, 1988.
151. **Roberts, C., Weith, A., Passage, E., Michot, J. L., Mattei, M. G., and Bishop, C. E.,** Molecular and cytogenetic evidence for the location of *Tdy* and *Hya* on the mouse Y chromosome, *Proc. Natl. Acad. Sci. U.S.A.,* 85, 6446, 1988.
152. **McLaren, A., Simpson, E., Tomonari, K., Chandler, P., and Hogg, H.,** Male sexual differentiation in mice lacking H-Y antigen, *Nature (London),* 312, 552, 1984.
153. **Burgoyne, P. S., Levy, E. R., and McLaren, A.,** Spermatogenic failure in male mice lacking H-Y antigen, *Nature (London),* 320, 170, 1986.
154. **Burgoyne, P. S.,** The role of the mammalian Y chromosome in spermatogenesis, in *The Mammalian Y Chromosome: Molecular Search for the Sex-Determining Factor,* Goodfellow, P. N., Craig, I. W., Smith, J. C., and Wolfe, J., Eds., The Company of Biologists, Cambridge, 1987, 133.
155. **Sutcliffe, M. J. and Burgoyne, P. S.,** The spermatogenic block in H-Y negative XOSxr^b mice, *Genet. Res.,* 53, 228, 1989.
156. **Krzanowska, H.,** Factor responsible for spermatozoan abnormality located on the Y chromosome, *Genet. Res.,* 13, 17, 1969.
157. **Krzanowska, H.,** Influence of Y chromosome on fertility in mice, in *Proc. Int. Symp. the Genetics of the Spermatozoon,* Beatty, R. A. and Gluecksohn-Waelsch, S., Eds., University of Edinburgh, Edinburgh, 1972, 370.
158. **Krzanowska, H.,** Interstrain competition amongst mouse spermatozoa inseminated in various proportions, as affected by the genotype of the Y chromosome, *J. Reprod. Fertil.,* 77, 265, 1986.
159. **Silver, L. M.,** Mouse *t* haplotypes, *Annu. Rev. Genet.,* 19, 179, 1985.
160. **Frischauf, A.-M.,** The *T/t* complex of the mouse, *Trends Genet.,* 1, 100, 1985.
161. **Klein, J.,** *Natural History of the Major Histocompatibility Complex,* John Wiley & Sons, New York, 1986.
162. **Silver, L. M.,** Mouse *t* haplotypes: a tale of tails and a misunderstood selfish chromosome, *Curr. Top. Microbiol. Immunol.,* 137, 64, 1988.
163. **Silver, L. M. and Artzt, K.,** Recombination suppression of mouse *t*-haplotypes is due to chromatin mismatching, *Nature (London),* 290, 68, 1981.
164. **Shin, H.-S., Flaherty, L., Artzt, K., Bennett, D., and Ravetch, J.,** Inversion in the *H-2* complex of *t*-haplotypes in mice, *Nature (London),* 306, 380, 1983.
165. **Shin, H.-S., Bennett, D., and Artzt, K.,** Gene mapping within the *T/t* complex of the mouse. IV. The inverted *MHC* is intermingled with several *t*-lethal genes, *Cell,* 39, 573, 1984.
166. **Herrmann, B., Búcan, M., Mains, P. E., Frischauf, A.-M., Silver, L. M., and Lehrach, H.,** Genetic analysis of the proximal portion of the mouse *t* complex: evidence for a second inversion within *t* haplotypes, *Cell,* 44, 469, 1986.
167. **Herrmann, B. G., Barlow, D. P., and Lehrach, H.,** A large inverted duplication allows homologous recombination between chromosomes heterozygous for the proximal *t* complex inversion, *Cell,* 48, 813, 1987.
168. **Hammer, M. F., Schimenti, J., and Silver, L. M.,** The evolution of mouse chromosome 17 and the origin of *t* complex inversions, *Proc. Natl. Acad. Sci. U.S.A.,* 86, 3261, 1989.
169. **Lyon, M. F., Zenthon, J., Evans, E. P., Burtenshaw, M. D., and Willison, K. R.,** Extent of the mouse *t* complex and its inversions shown by in situ hybridization, *Immunogenetics,* 27, 375, 1988.
170. **Chesley, P. and Dunn, L. C.,** The inheritance of taillessness (anury) in the house mouse, *Genetics,* 21, 525, 1936.
171. **Dunn, L. C.,** A third lethal in the *T* (brachy) series in the house mouse, *Proc. Natl. Acad. Sci. U.S.A.,* 23, 474, 1937.
172. **Bryson, V.,** Spermatogenesis and fertility in Mus musculus as affected by factors at the T locus, *J. Morphol.,* 74, 131, 1944.

173. **Gluecksohn-Waelsch, S.,** Models for mechanisms of segregation distortion, in *Proc. Int. Symp. the Genetics of the Spermatozoon,* Beatty, R. A. and Gluecksohn-Waelsch, S., Eds., University of Edinburgh, Edinburgh, 1972, 306.
174. **Papaioannou, V. E., Artzt, K., Dooher, G. B., and Bennett, D.,** Effects of chimaerism for two *T/t* complex mutations on sperm function, *Gamete Res.,* 2, 147, 1979.
175. **Seitz, A. W. and Bennett, D.,** Transmission distortion of *t*-haplotypes is due to interactions between meiotic partners, *Nature (London),* 313, 143, 1985.
176. **Olds-Clarke, P. and Peitz, B.,** Fertility of sperm from *t*/+ mice: evidence that +-bearing sperm are dysfunctional, *Genet. Res.,* 47, 49, 1985.
177. **Hammerberg, C. and Klein, J.,** Evidence for postmeiotic effect of *t* factors causing segregation distortion in the mouse, *Nature (London),* 253, 137, 1975.
178. **Braden, A. W. H. and Gluecksohn-Waelsch, S.,** Further studies of the effect of the T locus in the house mouse on male fertility, *J. Exp. Zool.,* 138, 431, 1958.
179. **Hillman, N. and Nadijcka, M.,** A comparative study of spermiogenesis in wild type and *T:t*-bearing mice, *J. Embryol. Exp. Morphol.,* 44, 243, 1978.
180. **Hillman, N. and Nadijcka, M.,** A study of spermatozoan defects in wild-type and *T:t*-bearing mice, *J. Embryol. Exp. Morphol.,* 44, 263, 1978.
181. **Silver, L. and Olds-Clarke, P.,** Transmission ratio distortion of mouse *t* haplotypes is not a consequence of wild type sperm degeneration, *Dev. Biol.,* 105, 250, 1984.
182. **Shur, B. D.,** Galactosyltransferase activities on mouse sperm bearing multiple t^{lethal} and t^{viable} haplotypes of the *T/t*-complex, *Genet. Res.,* 38, 225, 1981.
183. **Scully, N. F. and Shur, B. D.,** Stage-specific increase in cell surface galactosyltransferase activity during spermatogenesis in mice bearing *t* alleles, *Dev. Biol.,* 125, 195, 1988.
184. **Olds-Clarke, P.,** Nonprogressive sperm motility is characteristic of most complete *t* haplotypes in the mouse, *Genet. Res.,* 42, 151, 1983.
185. **Olds-Clarke, P.,** The nonprogressive motility of sperm populations from mice with a t^{w32} haplotype, *J. Androl.,* 4, 136, 1983.
186. **Olds-Clarke, P.,** Sperm from t^{w32}/+ mice: capacitation is normal, but hyperactivation is premature and non-hyperactivated sperm are slow, *Dev. Biol.,* 131, 475, 1989.
187. **Brown, J., Cebra-Thomas, J. A., Bleil, J. D., Wassarman, P. M., and Silver, L. M.,** A premature acrosome reaction is programmed by mouse *t* haplotypes during sperm differentiation and could play a role in transmission ratio distortion, *Development,* 106, 769, 1989.
188. **Saling, P. M., Sowinski, J., and Storey, B. T.,** An ultrastructural study of the epididymal mouse spermatozoa binding to zona pellucida *in vitro:* sequential relationship to the acrosome reaction, *J. Exp. Zool.,* 209, 229, 1979.
189. **Saling, P. M. and Storey, B. T.,** Mouse gamete interactions during fertilization in vitro. Chlortetracycline as a fluorescent probe for the mouse sperm acrosome reaction, *J. Cell Biol.,* 83, 544, 1979.
190. **Florman, H. M. and Storey, B. T.,** Mouse gamete interactions: the zona pellucida is the site of the acrosome reaction leading to fertilization *in vitro, Dev. Biol.,* 91, 121, 1982.
191. **Lyon, M. F.,** Male sterility of the mouse *t*-complex is due to homozygosity of the distorter genes, *Cell,* 44, 357, 1986.
192. **Tucker, M. J.,** Explanation of sterility in t^xt^y male mice, *Nature (London),* 288, 367, 1980.
193. **Dooher, G. B. and Bennett, D.,** Abnormal microtubular systems in mouse spermatids associated with a mutant gene at the T-locus, *J. Embryol. Exp. Morphol.,* 32, 749, 1974.
194. **Bennett, D. and Dunn, L. C.,** Studies of effects of *t*-alleles in the house mouse on spermatozoa. I. Male sterility effects, *J. Reprod. Fertil.,* 13, 421, 1967.
195. **Olds, P. J.,** Effect of the T locus on sperm distribution in the house mouse, *Biol. Reprod.,* 2, 91, 1970.
196. **Hillman, N. and Nadijcka, M.,** Sterility in mutant (t^{Lx}/t^{Ly}) male mice. I. A morphological study of spermiogenesis, *J. Embryol. Exp. Morphol.,* 59, 27, 1980.
197. **Nadijcka, M. and Hillman, N.,** Sterility in mutant (t^{Lx}/t^{Ly}) male mice. II. A morphological study of spermatozoa, *J. Embryol. Exp. Morphol.,* 59, 39, 1980.
198. **Olds-Clarke, P.,** Motility characteristics of sperm from the uterus and oviducts of female mice after mating to congenic males differeing in sperm transport and fertility, *Biol. Reprod.,* 34, 453, 1986.
199. **Olds, P. J.,** Effect of the T locus on fertilization in the house mouse, *J. Exp. Zool.,* 177, 417, 1971.
200. **McGrath, J. and Hillman, N.,** Sterility in mutant (t^{Lx}/t^{Ly}) male mice. III. *In vitro* fertilization, *J. Embryol. Exp. Morphol.,* 59, 49, 1980.
201. **Lyon, M. F. and Phillips, R. J. S.,** Crossing-over in mice heterozygous for *t*-alleles, *Heredity,* 13, 23, 1959.
202. **Fox, H. S., Martin, G. R., Lyon, M. F., Herrmann, B., Frischauf, A.-M., Lehrach, H. and Silver, L. M.,** Molecular probes define different regions of the mouse *t* complex, *Cell,* 40, 63, 1985.
203. **Lyon, M. F. and Mason, I.,** Information on the nature of *t*-haplotypes from the interaction of mutant haplotypes in male fertility and segregation ratio, *Genet. Res.,* 29, 255, 1977.

204. **Lyon, M. F.,** Transmission ratio distortion in mouse *t*-haplotypes is due to multiple distorter genes acting on a responder locus, *Cell,* 37, 621, 1984.
205. **Silver, L. M. and Remis, D.,** Five of the nine genetically defined regions of most *t* haplotypes are involved in transmission ratio distortion, *Genet. Res.,* 49, 51, 1987.
206. **Silver, L. M.,** Gene dosage effects on transmission ratio distortion and fertility in mice that carry *t* haplotypes, *Genet. Res.,* 54, 221, 1989.
207. **Lyon, M. F.,** Distorter genes of the mouse *t*-complex impair male fertility when heterozygous, *Genet. Res.,* 49, 57, 1987.
208. **Röhme, D., Fox, H., Herrmann, B., Frischauf, A.-M., Edström, J.-E., Mains, P., Silver, L. M., and Lehrach, H.,** Molecular clones of the mouse *t* complex derived from microdissected metaphase chromosomes, *Cell,* 36, 783, 1984.
209. **Schimenti, J., Vold, L., Socolow, D., and Silver, L. M.,** An unstable family of large DNA elements in the center of the mouse *t* complex, *J. Mol. Biol.,* 194, 583, 1987.
210. **Schimenti, J., Cebra-Thomas, J. A., Decker, C., Islam, S., Pilder, S. H., and Silver, L. M.,** A candidate gene family for the mouse *t complex responder* (*Tcr*) locus responsible for haploid effects on sperm function, *Cell,* 55, 71, 1988.
211. **Willison, K. R., Dudley, K., and Potter, J.,** Molecular cloning and sequence analysis of a haploid expressed gene encoding *t* complex polypeptide-1, *Cell,* 44, 727, 1986.
212. **Rappold, G. A., Stubbs, L., Labeit, S., Crkvenjakov, R. B., and Lehrach, H.,** Identification of a testis-specific gene from the mouse *t*-complex next to a CpG-rich island, *EMBO J.,* 6, 1975, 1987.
213. **Silver, L. M., Artzt, K., and Bennett, D.,** A major testicular cell protein specified by a mouse *T/t* complex gene, *Cell,* 17, 275, 1979.
214. **Silver, L. M., Uman, J., Danska, J., and Garrels, J. I.,** A diversified set of testicular cell proteins specified by genes within the mouse *t* complex, *Cell,* 35, 35, 1983.
215. **Silver, L. M., Kleene, K. C., Distal, R. J., and Hecht, N. B.,** Synthesis of mouse *t* complex proteins during haploid stages of spermatogenesis, *Dev. Biol.,* 119, 605, 1987.
216. **Dudley, K., Potter, J., Lyon, M. F., and Willison, K. R.,** Analysis of male sterile mutations in the mouse using haploid stage expressed cDNA probes, *Nucl. Acids Res.,* 12, 4281, 1984.
217. **Willison, K., Lewis, V., Zuckerman, K. S., Cordell, J., Dean, C., Miller, K., Lyon, M. F., and Marsh, M.,** The *t* complex polypeptide 1 (TCP-1) is associated with the cytoplasmic aspect of golgi membranes, *Cell,* 57, 621, 1989.
218. **Nadeau, J. H. and Phillips, S. J.,** The putative oncogene *Pim-1* in the mouse: its linkage and variation among *t* haplotypes, *Genetics,* 117, 533, 1987.
219. **Sorrentino, V., McKinney, M. D., Giorgi, M., Geremia, R., and Fleissner, E.,** Expression of cellular protooncogenes in the male germ line: a distinctive 2.4-kilobase *pim-1* transcript is expressed in haploid postmeiotic cells, *Proc. Natl. Acad. Sci. U.S.A.,* 85, 2191, 1988.
220. **Lader, E., Ha, H.-S., O'Neill, M., Artzt, K., and Bennett, D.,** *tctex-1*: a candidate gene family for a mouse *t* complex sterility locus, *Cell,* 58, 969, 1989.
221. **Dickie, M. M.,** *Mouse News Lett.,* 40, 29, 1969.
222. **VandeBerg, J. L., Cooper, D. W., and Close, P. J.,** Mammalian testis phosphoglycerate kinase, *Nature New Biol.,* 243, 48, 1973.
223. **Eicher, E. M., Cherry, M., and Flaherty, L.,** Autosomal phosphoglycerate kinase linked to mouse major histocompatibility complex, *Mol. Gen. Genet.,* 158, 225, 1978.
224. **Randolph, N. S. and VandeBerg, J. L.,** Linkage disequilibrium between the *T/t* complex and *Pgk-2* in wild mice, *J. Exp. Zool.,* 217, 455, 1981.
225. **Sarvetnick, N., Tsai, J.-Y., Fox, H., Pilder, S. H., and Silver, L. M.,** A mouse chromosome 17 gene encodes a testis-specific transcript with unusual properties, *Immunogenetics,* 30, 34, 1989.
226. **Kasahara, M., Figueroa, F., and Klein, J.,** Random cloning of genes from mouse chromosome 17, *Proc. Natl. Acad. Sci. U.S.A.,* 84, 3325, 1987.
227. **Kasahara, M., Passmore, H. C., and Klein, J.,** A testis-specific gene *Tpx-1* maps between *Pgk-2* and *Mep-1* on mouse chromosome 17, *Immunogenetics,* 29, 61, 1989.
228. **O'Brien, D. A. and Bellvé, A. R.,** Protein constituents of the mouse spermatozoon. I. An electrophoretic characterization, *Dev. Biol.,* 75, 386, 1980.
229. **Willison, K. and Ashworth, A.,** Mammalian spermatogenic gene expression, *Trends Genet.,* 3, 351, 1987.
230. **Wolgemuth, D. J., Gizang-Ginsberg, E., Ponzetto, C., and Zakeri, Z. F.,** Genetic control of germ cell function: developmentally regulated gene expression during gametogenesis, in *The Molecular Biology of Fertilization,* Schatten, H. and Schatten, G., Eds., Academic Press, San Diego, 1989, 235.
231. **Stewart, T.A., Bellvé, A. R., and Leder, P.,** Transcription and promoter usage of the *myc* gene in normal somatic and spermatogenic cells, *Science,* 226, 707, 1984.
232. **Wolfes, H., Kogawa, K., Millette, C. F., and Cooper, G. M.,** Specific expression of nuclear proto-oncogenes before entry into meiotic prophase of spermatogenesis, *Science,* 245, 740, 1989.

233. **Snyder, L.C., Trusko, S. P., Freeman, N., Eshleman, J. R., Fakharzadeh, S. S., and George, D. L.,** A gene amplified in a transformed mouse cell line undergoes complex transcriptional processing and encodes a nuclear protein, *J. Biol. Chem.*, 263, 17150, 1988.
234. **Bohmann, D., Bos, T. J., Adman, A., Nishimura, T., Vogt, P. K., and Tijian, R.,** Human proto-oncogene *c-jun* encodes a DNA binding protein with structural and functional properties of transcription factor AP-1, *Science,* 238, 1386, 1987.
235. **Lech, K., Anderson, K., and Brent, R.,** DNA-bound fos proteins activate transcription in yeast, *Cell,* 52, 179, 1988.
236. **Sassone-Corsi, P., Ransone, L. J., Lamph, W. W., and Vermu, I. M.,** Direct interaction between fos and jun nuclear oncoproteins: role of the 'leucine zipper' domain, *Nature (London),* 336, 692, 1988.
237. **Propst, F. and Vande Woude, G. F.,** Expression of *c-mos* proto-oncogene transcripts in mouse tissues, *Nature (London),* 315, 516, 1985.
238. **Propst, F., Rosenberg, M. P., Iyer, A., Kaul, K., and Vande Woude, G. F.,** *c-mos* proto-oncogene RNA transcripts in mouse tissues: structural features, developmental regulation, and localization in specific cell types, *Mol. Cell. Biol.*, 7, 1629, 1987.
239. **Propst, F., Rosenberg, M. P., Oskarsson, M. K., Russell, L. B., Nguyen-Huu, M. C., Nadeau, J., Jenkins, N. A., Copeland, N. G., and Vande Woude, G. F.,** Genetic analysis and developmental regulation of testis-specific RNA expression of *Mos, Abl,* actin and *Hox-1.4, Oncogene,* 2, 227, 1988.
240. **Mutter, G. L. and Wolgemuth, D. J.,** Distinct developmental patterns of *c-mos* expression in female and male mouse germ cells, *Proc. Natl. Acad. Sci. U.S.A.*, 84, 5301, 1987.
241. **Sagata, N., Oskarsson, M., Copeland, T., Brumbaugh, J., and Vande Woude, G. F.,** Function of *c-mos* proto-oncogene product in meiotic maturation in *Xenopus* oocytes, *Nature (London),* 335, 519, 1988.
242. **Sagata, N., Daar, I., Oskarsson, M., Showalter, S.D., and Vande Woude, G. F.,** The product of the *mos* proto-oncogene as a candidate "initiator" for oocyte maturation, *Science,* 245, 643, 1989.
243. **Freeman, R. S., Pickham, K. M., Kanki, J. P., Lee, B. A., Pena, S. V., and Donoghue, D. J.,** *Xenopus* homologue of the *mos* protooncogene transforms mammalian fibroblasts and induces maturation of *Xenopus* oocytes, *Proc. Natl. Acad. Sci. U.S.A.*, 86, 5805, 1989.
244. **Nusse, R. and Varmus, H. E.,** Many tumors induced by the mouse mammary tumor virus contain a provirus integrated in the same region of the host genome, *Cell,* 31, 99, 1982.
245. **Jakobovits, A., Shackleford, G. M., Varmus, H. E., and Martin, G. R.,** Two proto-oncogenes implicated in mammary carcinogenesis, *int-1* and *int-2*, are indpendently regulated during mouse development, *Proc. Natl. Acad. Sci. U.S.A.*, 83, 7806, 1986.
246. **Wilkinson, D. G., Bailes, J. A., and McMahon, A. P.,** Expression of the proto-oncogene *int-1* is restricted to specific neural cells in the developing mouse embryo, *Cell,* 50, 79, 1987.
247. **Shackleford, G. M. and Varmus, H. E.,** Expression of the proto-oncogene *int-1* is restricted to postmeiotic male germ cells and the neural tube of mid-gestational embryos, *Cell,* 50, 89, 1987.
248. **Rijsewijk, F., Schuermann, M., Wagenaar, E., Parren, P., Wiegel, D., and Nusse, R.,** The Drosophila homologue of the mouse mammary oncogene *int-1* is identical to the segment polarity gene *wingless, Cell,* 50, 649, 1987.
249. **Muller, R., Slamon, D. J., Tremblay, J. M., Cline, M. J., and Verma, I. M.,** Differential expression of cellular oncogenes during pre- and post-natal development of the mouse, *Nature (London),* 299, 640, 1982.
250. **Ponzetto, C. and Wolgemuth, D. J.,** Haploid expression of a unique *c-abl* transcript in the mouse male germ line, *Mol. Cell. Biol.*, 5, 1791, 1985.
251. **Ben-Neriah, Y., Bernards, A., Paskind, M., Daley, G. Q., and Baltimore, D.,** Alternative 5′ exons in *c-abl* mRNA, *Cell,* 44, 577, 1986.
252. **Shtivelman, E., Lifshitz, B., Gale, R. P., Roe, B. A., and Canaani, E.,** Alternative splicing of RNAs transcribed from the human *abl* gene and from the *bcr-abl* fused gene, *Cell,* 47, 277, 1986.
253. **Duggal, R. N., Zakeri, Z. F., Ponzetto, C., and Wolgemuth, D. J.,** Differential expression of the c-*abl* proto-oncogene and the homeobox-containing gene Hox 1.4 during mouse spermatogenesis, *Ann. N.Y. Acad. Sci.*, 513, 112, 1987.
254. **Meijer, D., Hermans, A., von Lindern, M., van Agthoven, T., de Klein, A., Mackenbach, P., Grootegoed, A., Talarico, D., Della Valle, G., and Grosveld, G.,** Molecular characterization of the testis specific *c-abl* mRNA in the mouse, *EMBO J.*, 6, 4041, 1987.
255. **Zakeri, Z. F., Ponzetto, C., and Wolgemuth, D. J.,** Translational regulation of the novel haploid-specific transcripts for the c-*abl* proto-oncogene and a member of the 70 kDa heat-shock protein gene family in the male germ line, *Dev. Biol.*, 125, 417, 1988.
256. **Ponzetto, C., Wadewitz, A. G., Pendergast, A. M., Witte, O. N., and Wogemuth, D. J.,** $P150^{c\text{-}abl}$ is detected in mouse male germ cells by an *in vitro* kinase assay and is associated with stage-specific phosphoproteins in haploid cells, *Oncogene,* 4, 685, 1989.
257. **Colberg-Poley, A. M., Voss, S. D., Chowdhury, K., Stewart, C. L., Wagner, E. F., and Gruss, P.,** Clustered homeoboxes are differentially expressed during murine development, *Cell,* 43, 39, 1985.

258. **Odenwald, W. F., Taylor, C. F., Palmer-Hill, F. J., Friedrich, V., Jr., Tani, M., and Lazzarini, R. A.,** Expression of a homeo domain protein in noncontact-inhibited cultured cells and postmitotic neurons, *Genes Dev.*, 1, 482, 1987.
259. **Dollé, P. and Duboule, D.,** Two gene members of the murine Hox-5 complex show regional and cell-type specific expression in developing limbs and gonads, *EMBO J.*, 8, 1507, 1989.
260. **Wolgemuth, D. J., Engelmyer, E., Duggal, R. N., Gizang-Ginsberg, E., Mutter, G. L., Ponzetto, C., Viviano, C., and Zakeri, Z. F.,** Isolation of a mouse cDNA coding for a developmentally regulated, testis-specific transcript containing homeo box homology, *EMBO J.*, 5, 1229, 1986.
261. **Rubin, M. R., Toth, L. E., Patel, M. D., D'Eustachio, P. D., and Nguyen-Huu, M. C.,** A mouse homeo box gene is expressed in spermatocytes and embryos, *Science*, 233, 663, 1986.
262. **Duboule, D., Baron, A., Mähl, P., and Galliot, B.,** A new homeo-box is present in overlapping cosmid clones which define the mouse *HOX*-1 locus, *EMBO J.*, 5, 1973, 1986.
263. **Wolgemuth, D. J., Viviano, C. M., Gizang-Ginsberg, E., Frohman, M. A., Joyner, A. L., and Martin, G. R.,** Differential expression of the mouse homeobox-containing gene *Hox-1.4* during male germ cell differentiation and embryonic development, *Proc. Natl. Acad. Sci. U.S.A.*, 84, 5813, 1987.
264. **Leroy, P., Seboun, E., Mattei, M. G., Fellous, M., and Bishop, C. E.,** Testis-specific transcripts detected by a human Y-DNA-derived probe, in *The Mammalian Y Chromosome: Molecular Search for the Sex-Determining Factor,* Goodfellow, P. N., Craig, I. W., Smith, J. C., and Wolfe, J., Eds., The Company of Biologists, Cambridge, 1987, 177.
265. **Leroy, P., Alzari, P., Sassoon, D., Wolgemuth, D., and Fellous, M.,** The protein encoded by a murine male germ cell-specific transcript is a putaive ATP-dependent RNA helicase, *Cell,* 57, 549, 1989.
266. **Linder, P., Lasko, P. F., Leroy, P., Nielsen, P. J., Nishi, K., Schnier, J., and Slonimski, P. P.,** Birth of the D-E-A-D box, *Nature (London),* 337, 212, 1989.
267. **Nielsen, P. J. and Trachsel, H.,** The mouse protein synthesis initiation factor 4A gene family includes two related functional genes which are differentially expressed, *EMBO J.*, 7, 2097, 1988.
268. **Lasko, P. F. and Ashburner, M.,** The product of the *Drosophila* gene *vasa* is very similar to eukaryotic initiation factor-4A, *Nature (London),* 335, 611, 1988.
269. **Hay, B., Jan, L. Y., and Jan, Y. N.,** A protein component of Drosophila polar granules is encoded by *vasa* and has extensive sequence similarity to ATP-dependent helicases, *Cell,* 55, 577, 1988.

Chapter 3

MAMMALIAN OOCYTE DEVELOPMENT *IN VIVO* AND *IN VITRO*

John J. Eppig

TABLE OF CONTENTS

I. INTRODUCTION

Oogonia are mitotic cells of the germ line and are located in the primitive ovary. In response to an unknown signal, the oogonia enter meiosis and become oocytes. The progression of meiosis in oocytes, however, is arrested at prophase I. Meiotic prophase I is divided into substages: leptotene, zygotene, pachytene, diplotene. Chromosome condensation occurs during leptotene and the homologous chromosomes pair during the zygotene stage, but the actual exchange of chromosomal segments that results in recombination does not occur until the pachytene stage. The prolonged diplotene stage in oocytes is often referred to as the dictyotene stage. Oocytes normally remain arrested at the diplotene stage until they have completed their growth phase and the resumption of meiosis is induced by the preovulatory surge of gonadotropins. Resumption of meiosis also occurs in atretic (degenerating) follicles containing oocytes that have completed, or nearly completed, their growth. Oocytes can, therefore, be arrested at the diplotene stage for periods ranging from days to years, depending on the species. During most of the period of meiotic arrest, the oocytes are about 20 μm in diameter and situation in primordial follicles consisting of a single layer of flattened pregranulosa cells. The mechanism for recruitment of oocytes into the growth phase is not known, but the initiation of oocyte growth and the multiplication of the oocyte's companion somatic cells, the granulosa cells, is usually synchronized. Processes occur during oocyte growth that prepare it for resuming and completing meiosis (oocyte maturation), for fertilization, and for progression through preimplantation development. This chapter will focus on oocyte growth and the acquisition of competence to undergo maturation, fertilization, and embryogenesis. Much pertinent information has been obtained using systems for murine oocyte growth and maturation *in vitro*, and these systems and their contributions will be assessed in detail.

II. GENERAL CHARACTERISTICS OF OOCYTE DEVELOPMENT *IN VIVO*

Oocyte growth is generally correlated with granulosa cell proliferation; as the oocyte grows, the number of layers of granulosa cells around the oocyte increases. In species where a follicular antrum is formed, oocyte growth is nearly completed by the time of antrum formation. Figure 1 shows the size of oocytes isolated from the preantral follicles of mice of increasing ages (8 to 14 days old). In mice, the first wave of oocyte growth and follicular development begins shortly after birth, so oocyte growth and follicular development are directly related to the age of the immature mice.[1-3] Follicles at all stages are developing in mature cycling females, and follicles less developed than those in the first wave are observed in immature mice. Examples of growing oocytes in preantral follicles and a fully grown oocyte in an antral follicle are presented in Figure 2. Atypical follicle development, wherein oocytes grow without concurrent granulosa cell proliferation, has been reported in some strains of mice.[4] The initiation of oocyte growth and folliculogenesis, therefore, are separable processes, but it is still not known whether there is a common signal that is not received or not executed properly by the pregranulosa cells of the incipient granulosa cell-deficient follicle, or whether there are different signals that initiate oocyte growth and follicular development.

Preantral follicles and oocytes develop in the absence of stimulation by gonadotropins. These follicles and oocytes develop in (1) hypophysectomized animals,[5] (2) mice treated with antisera against gonadotropins,[6] and (3) mutant mice having extremely low or undetectable levels of circulating gonadotropins,[7] as shown in Figure 3. Nevertheless, follicle-stimulating hormone (FSH) induces DNA synthesis in granulosa cells of preantral follicles isolated from hamsters,[8,9] and gonadotropins stimulate steroidogenesis in ovaries of infantile mice and rats.[10,11] Accordingly, although preantral follicles can develop in the absence of

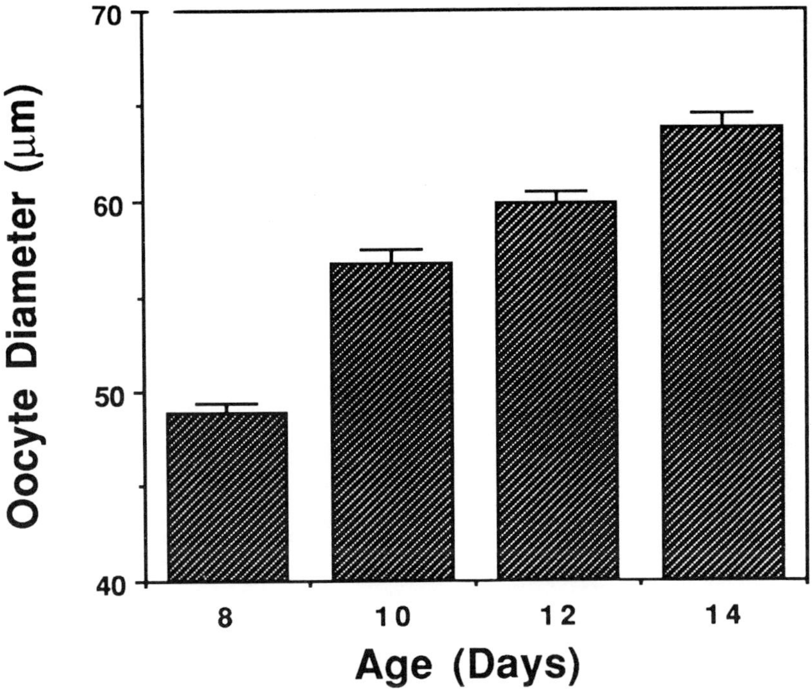

FIGURE 1. Diameter (mean ± SEM) of growing oocytes isolated from mice 8 to 10 days old. Oocytes isolated from ovaries with collagenase and DNase in calcium-free medium were measured, exclusive of the zona pellucida, using an inverted microscope equipped with an ocular micrometer.

gonadotropins, it is possible that these hormones affect the rate of development and follicular function.

The first evidence implicating granulosa cells in oocyte growth and development was the observation of gap junctions at points of apposition between the granulosa cell and oocyte plasma membranes.[12,13] Gap junctions are membrane specializations that allow low molecular weight molecules to diffuse directly from one cell to another. These molecules can participate in the metabolism of the recipient cell (metabolic cooperativity) or in the regulation of the recipient cell's function or development.[14] Heterologous gap junctions are detected between oocytes and granulosa cells in primordial follicles and persist until after the oocyte has become committed to resume meiosis shortly before ovulation.[12,13]

A. GENE EXPRESSION IN OOCYTES

The dictyate oocyte increases in diameter from about 20 µm for resting oocytes in primordial follicles to 70 to 100 µm in well-developed antral follicles, depending on the species. In the mouse, oocytes grow from 20 to 80 µm in 3 to 4 weeks, about a 125-fold increase in volume. Since there is a direct linear relationship between oocyte volume and protein content, there is, therefore, about a 125-fold increase in oocyte protein.[15] As the oocytes grow, there are both qualitative and quantitative changes in the pattern of proteins synthesized, as detected by two-dimensional polyacrylamide gel electrophoresis.[16]

Fully grown mouse oocytes contain about 0.43 ng of RNA, and this amount decreases by about 19% during oocyte maturation.[17] About 19% of the RNA in the immature oocyte is polyadenylated, and this is reduced to about 10% during oocyte maturation *in vitro*.[17] The differential expression of three specific genes throughout oocyte development has been

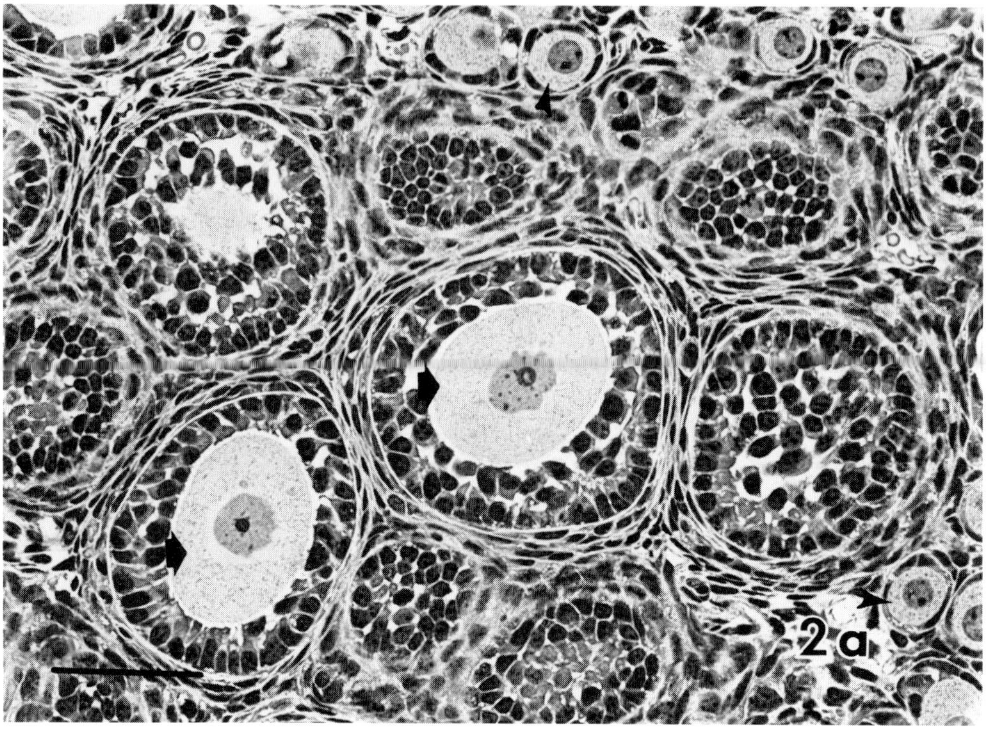

FIGURE 2. (a) Photomicrograph of a section of an 8-day-old mouse ovary. Solid arrows indicate growing oocytes in preantral follicles. Arrow heads point out nongrowing oocytes in primordial follicles. Bar indicates 50 μm. (b) Photomicrograph of a section of a large antral follicle in a mouse ovary. The fully grown oocyte (solid arrow) in the germinal vesicle (arrow head) stage is enclosed by cumulus cells (empty arrow). Bar indicates 100 μm.

established (summarized in Figure 4): the sperm receptor protein of the zona pellucida, *ZP3*; tissue plasminogen activator (*Plat*)*; and the proto-oncogene *Mos*.

1. *ZP3*

The expression and function of *Zp3* is reviewed extensively elsewhere in this volume. *Zp3* mRNA is not found in resting oocytes of primordial follicles, but is detected in growing oocytes and fully grown oocytes through maturation, although the number of transcripts decreases dramatically during oocyte maturation.[18] Synthesis of the *ZP3* glycoprotein begins during oocyte growth and continues in fully grown oocytes, albeit at very low levels.[19]

2. *Plat*

tPA is a serine protease generally considered important in processes of tissue remodeling and cell migration. It is synthesized and secreted by maturing oocytes, but not by germinal vesicle (GV)-stage oocytes.[20] The biological function of tPA during oocyte development is unknown. *Plat* mRNA is detected in growing oocytes, fully grown GV-stage oocytes, and maturing oocytes, but not after fertilization. The transcripts are stable and untranslated before GV breakdown (GVB).[21] Blocking a specific untranslated 3′ region prevents polyadenylation and translation, indicating the importance of this uncoded region in the recruitment of *Plat* mRNA into the translated pool of messages and of polyadenylation in the posttranscriptional control of tPA production during oocyte maturation.[22]

* The enzyme tissue plasminogen activator is abbreviated tPA but the symbol for the gene coding for tPA is *Plat*.

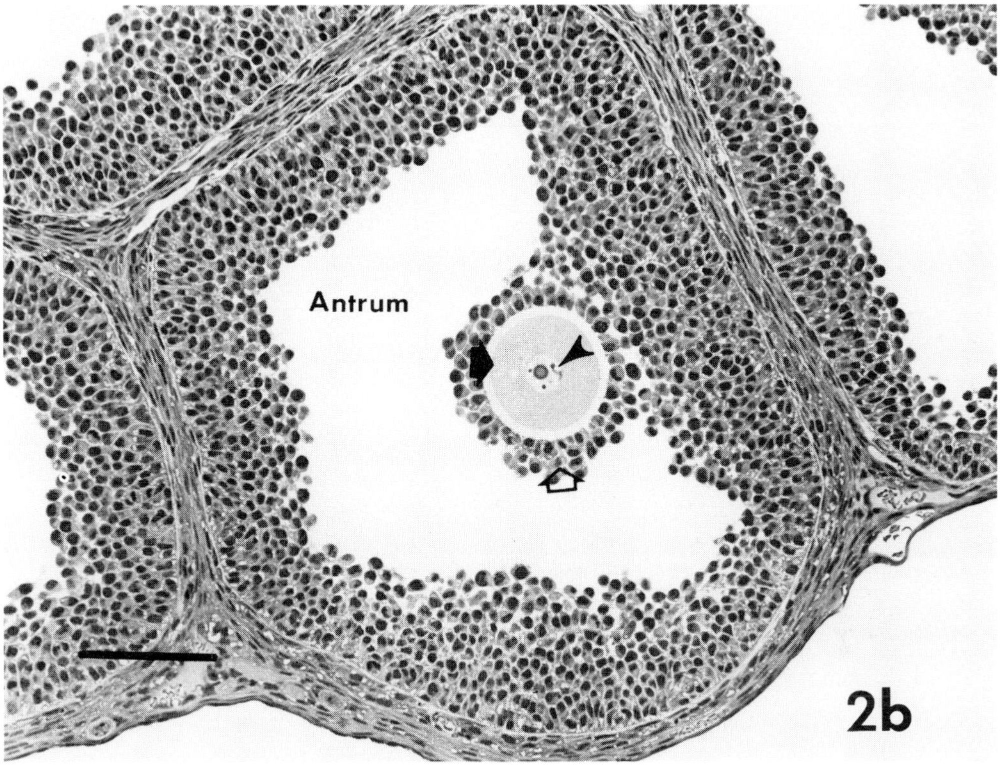

FIGURE 2b.

3. *Mos*

Mos is a proto-oncogene, one of a family of genes which can transform cells when the genes controlling them are altered or when they themselves are modified. *Mos* is the murine cellular homolog of *v-mos*, the transforming gene of Moloney sarcoma virus. *Mos* mRNA, like that of *ZP3* and *Plat,* is not detectable in resting oocytes of primordial follicles, but transcripts are detected in growing oocytes and fully grown oocytes, but not after fertilization.[23,24] The number of transcripts decreases during oocyte maturation. p39^{mos} synthesis is detected in fully grown GV-stage oocytes and maturing oocytes, but, by the same methods, is not found in growing oocytes or after fertilization.[25] As discussed by Schultz elsewhere in this volume, p39^{mos} appears to be required for the progression of oocyte maturation past metaphase I, since microinjection of deoxyoligoribonucleotides antisense to *Mos* mRNA blocks meiosis at or about metaphase I.[25]

4. Summary

As illustrated in Figure 4, mRNA of all three of these genes is detectable at about the same time. Yet, translation seems to be regulated independently; *ZP3* synthesis occurs in growing oocytes, but p39^{mos} or tPA synthesis does not. p39^{mos} synthesis appears in oocytes arrested at the GV stage, but tPA synthesis does not. Synthesis of tPA begins about 4 h after GVB when ZP3 production is undetectable. It will be important to determine what factors regulate translational activity; whether or not the mRNA requires polyadenylation for translation is likely to be one of the keys. There is no evidence of polyadenylation of *Mos* mRNA in GV-stage oocytes where p39^{mos} is detected. In contrast, polyadenylation is an essential step in the translation of *Plat* mRNA.[22]

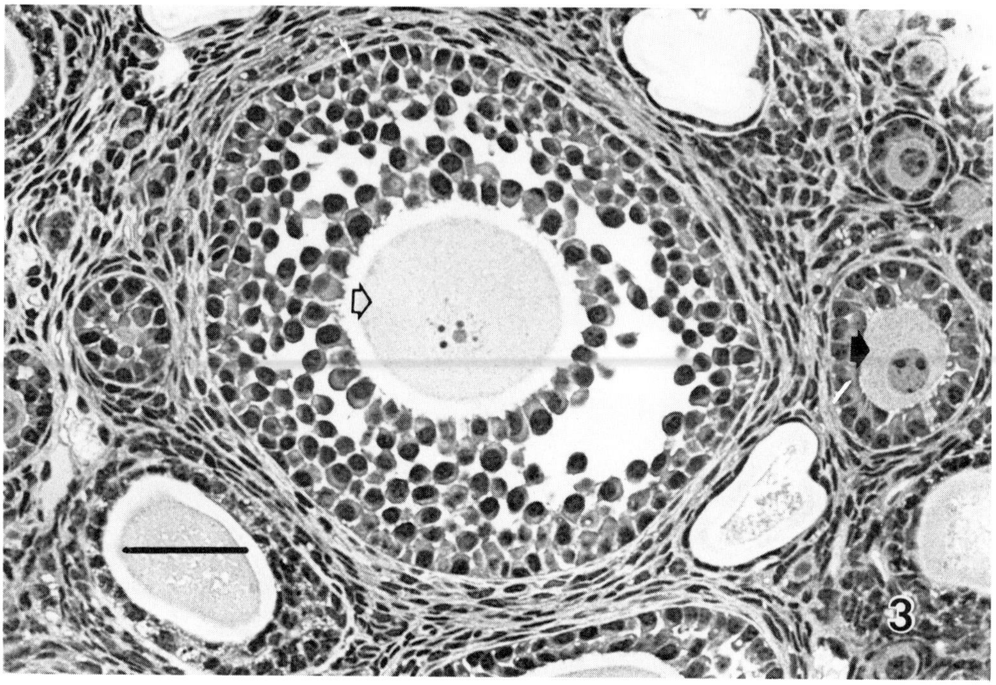

FIGURE 3. Photomicrograph of a small antral follicle in the ovary of a mouse homozygous for the recessive gene hypogonadal (*hpg/hpg*). These mice have undetectable levels of circulating gonadotropins.[7] Follicular development does not progress beyond this stage in hypogonadal mice. The oocyte (empty arrow) is fully grown. A growing oocyte (solid arrow) in a preantral follicle is seen to the right of the antral follicle. Oocyte growth seems to be unaffected by the gonadotropin deficiency. Bar indicates 50 μm.

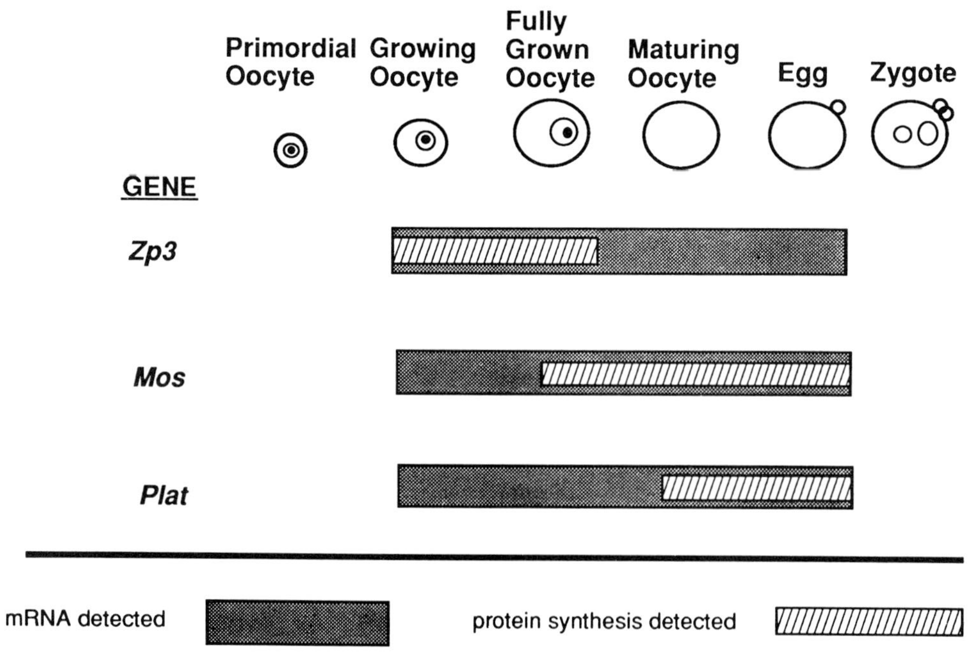

FIGURE 4. Expression of *Zp3*, *Mos*, and *Plat* during oocyte development.

III. OOCYTE DEVELOPMENT *IN VITRO*

The establishment of culture systems for oocytes has had a major impact on our understanding of the mechanisms governing oocyte development. Moreover, these systems have the potential of practical use for human clinical applications, for expanding populations of agriculturally important animals or rare or endangered species, and possibly for the production of transgenic animals.

A. OOCYTE MATURATION *IN VITRO*

Much of our understanding about the mechanisms that control oocyte maturation is derived from studies on oocytes *in vitro*. The initial experiment in this area, conducted by Pincus and Enzmann,[26] demonstrated that rabbit oocytes, isolated from Graafian follicles and cultured, underwent gonadotropin-independent spontaneous maturation, suggesting that components of the follicle maintained the oocyte in meiotic arrest. This study was the foundation for decades of experiments on the mechanism of this follicular action that is discussed in detail by Schultz in Chapter 4. An important problem with the conceptual framework for this work, however, was articulated by Biggers and Powers[27] in 1979. They suggested that spontaneous maturation could be an artifact owing to the shock of removing the oocyte from its natural site and placing it into an artificial environment. If so, this would dismiss the physiological relevance of spontaneous maturation and render the search for the follicular-arresting mechanisms futile.

One argument that seemed to support the notion that spontaneous maturation was actually an artifact of the culture system was that very few of the ova that had undergone spontaneous maturation could be fertilized and undergo embryogenesis.[28-34] Initial studies reported a failure in the formation of the male pronucleus if sperm penetration occurred before the completion of oocyte maturation.[32,33,34] Furthermore, ova that appeared to have completed the nuclear events of oocyte maturation (GVB, meiotic karyokinesis, and polar-body emission) spontaneously *in vitro* also failed to develop the male pronucleus after sperm penetration. It seemed, therefore, that oocytes matured outside the follicle were deficient in some way; although ''nuclear maturation'' was apparently normal, ''cytoplasmic maturation'' was not.[34] The cytoplasmic deficiency seemed to involve a failure to produce a male pronuclear growth factor. Nevertheless, some studies using rodent oocytes showed that a small number of the oocytes that matured *in vitro* could be fertilized and undergo embryogenesis, suggesting that improvements in the culture system might enable the oocyte to undergo a complete maturation.[28] This was verified by studies using oocytes isolated from mice showing equivalent production of live young after oocyte maturation *in vitro* and *in vivo* followed by fertilization *in vitro* and transfer of embryos to pseudopregnant foster mothers.[35,36] These results demonstrated that spontaneous maturation is normal, if carried out using appropriate culture conditions, in the sense that the final derivative of oocyte maturation, a gamete component of fertilization and embryogenesis, is produced. Nevertheless, it is important to realize that although the egg (secondary oocyte) is the same after *in vivo* and *in vitro* maturation, this does not necessarily mean that the mechanisms initiating maturation, i.e., deprivation of the meiosis-arresting factors of the follicle, are the same. This question is discussed in greater detail by Schultz in Chapter 4. The results do, however, support the idea that elements within the follicle maintain the oocyte in meiotic arrest and validate the search for the mechanism of this arrest.

1. Developmental Capacity of Oocytes that Mature *In Vitro*

The conditions of oocyte maturation significantly affect the capacity of the ova to undergo fertilization and embryogenesis. These conditions include the culture medium, the quality of the components of the culture medium, the presence of cumulus cells, serum, and hormones, and the physiological status of the oocyte donor.

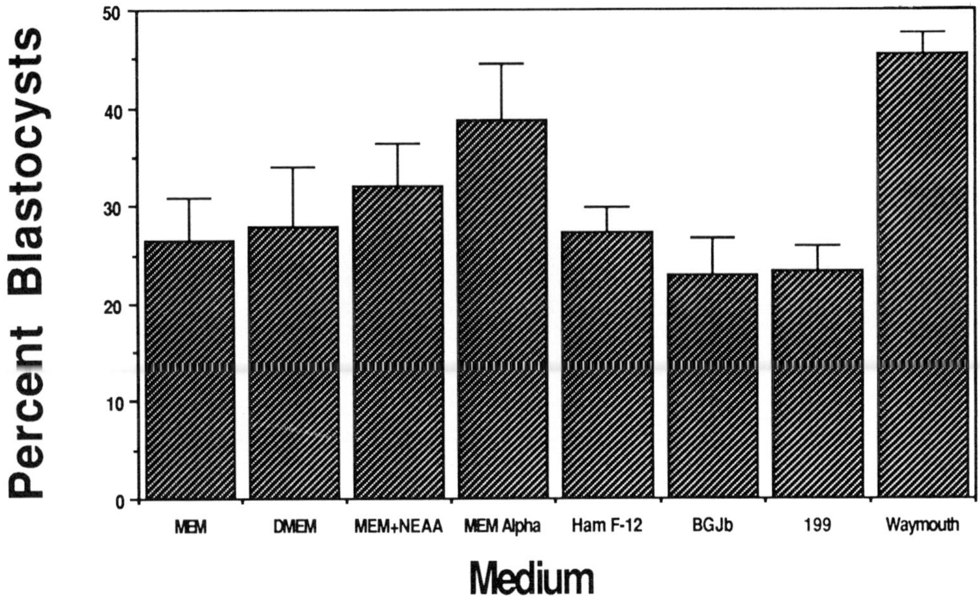

FIGURE 5. Effect of various culture media used for oocyte maturation on the subsequent development of two-cell embryos to the blastocyst stage. Insemination and embryo development occurred in the same medium for all groups. These results show that the culture medium used for oocyte maturation affects the developmental capacity of the embryos. Data adapted fromVan de Sandt et al.[37]

a. Culture Medium

Figure 5 illustrates the developmental capacity of eggs derived from mouse oocytes that matured *in vitro* in one of eight different commercially available culture media. Oocytes that mature in Waymouth medium MB 752/1 have the highest developmental capacity (defined as the percentage of eggs that developed to blastocysts and to live young) of any of the media tested.[37] In addition, the blastocysts derived from oocytes that mature in this medium contain more cells. The reason for the superiority of this medium for oocyte maturation is not known. Nevertheless, these results demonstrate that the processes occurring during spontaneous maturation involving the acquisition of competence for pre- and postimplantation development are affected by the nonhormonal components of the culture medium during the maturation phase. Whether this medium is optimal for the maturation of oocytes of other species remains to be established.

b. Culture Conditions

This need not be discussed in detail here, but it is essential that the entire environment for oocyte maturation and embryo development, including salts, nutrients, sera, water, culture dishes and tubes, gas atmosphere, etc., should be of high quality and free of toxic factors.

c. Serum

The zona pellucida of mouse oocytes undergoes a change when spontaneous maturation occurs in the absence of serum. This change is manifest in two ways that are probably related: increased time for the dissolution of the zona pellucida by α-chymotrypsin (zona hardening) and resistance to fertilization.[38-41] Increasing the concentration of fetal bovine serum (FBS) during oocyte maturation decreases the time required for chymotrypsin-mediated dissolution of the zona pellucida and increases the number of ova that can be fertilized.[39] Dialysis does not remove the factor that prevents zona hardening from FBS, indicating that the factor(s) has a molecular weight greater than 12 kDa, or is bound to a molecule of this size.[39] Mouse serum, calf serum, newborn calf serum, and porcine follicular fluid all contain the zona pellucida-hardening prevention factor, but human fetal cord serum does not.[38,40] The presence of this factor is, therefore, essential during spontaneous maturation to enable the sperm to penetrate the egg.

Zona hardening during oocyte maturation appears to be caused by the precocious release of some cortical granules.[42] The release of cortical granules at the time of sperm penetration modifies *ZP2,* one of the glycoprotein components of the zona pellucida, to $ZP2_f$.[43] This modification apparently is responsible for resistance to further sperm penetration, preventing polyspermy, and to dissolution by acid or α-chymotrypsin. The precocious release of cortical granules occurs during maturation *in vivo* or *in vitro,* but precocious conversion of *ZP2* to $ZP2_f$ does not occur during maturation *in vivo* or *in vitro* in the presence of serum, only during maturation *in vitro* in the absence of serum.[42] It is likely that protease inhibitors present in serum and follicular fluid, therefore, prevent the modification of *ZP2* and preserve fertilizability.

It is not known whether serum provides any other factors that affect the developmental capacity of maturing oocytes. This question is difficult to test because of zona hardening and the failure of sperm penetration after maturation in the absence of serum. Nevertheless, the development of parthenogenetically activated eggs from strain LT/Sv and $B6LTF_1$ mice might provide some informaton. About 10 to 20% of the ovulated eggs in these mice undergo spontaneous parthenogenetic activation, and many of them develop to the blastocyst stage as diploid embryos. LT/Sv and $(C57BL/6J \times LT/Sv)F_1$ (abbreviated $B6LTF_1$) oocytes complete the first meiotic division and become haploid eggs, which become diploid presumably by undergoing karyokinesis without cytokinesis after parthenogenetic activation. Spontaneous parthenogenetic activation of these eggs also occurs after maturation *in vitro.*[44] About 45% of the two-cell parthenotes, derived from $B6LTF_1$ oocytes that undergo spontaneous maturation *in vitro* in the absence of serum, develop to the blastocyst stage.[44] These results suggest that the principal role for serum during oocyte maturation *in vitro*, relative to the developmental capacity of the ova, is associated with the penetrability of the egg by sperm. Drilling a hole in the zona pellucida increased the percentage of rat eggs penetrated by sperm after maturation of cumulus cell-enclosed oocytes *in vitro* in the absence of serum, so that the resulting percentage was similar to that of undrilled eggs after maturation *in vitro* in the presence of serum.[45] However, when a hole was drilled in the zona pellucidae of eggs after maturation of cumulus cell-enclosed oocytes in the presence of serum, the percentage penetrated by sperm was much greater than the undrilled group, indicating changes in the zona pellucida even in the presence of serum. Producing a hole in the zona pellucida of ovulated eggs had no effect on penetration by sperm, and the percentage of eggs penetrated after maturation *in vitro* in the presence of serum followed by zona drilling was equivalent to the percentage of penetrated ovulated eggs.[45] These results emphasize the importance of maintaining the penetrability of the zona pellucida during oocyte maturation *in vitro* and indicate that, in some cases, serum alone cannot prevent some changes in the zona pelludica that affect penetrability.

d. Cumulus Cells

Denuding mouse,[36] rat,[46] cow (bovine),[47,48] or sheep[49] oocytes of their cumulus cells before maturation *in vitro* decreases the percentage of the eggs that form pronuclei or cleave after insemination. Nevertheless, the development of cleaved mouse eggs to the blastocyst stage or to live offspring, after transfer of morula-stage embryos to foster mothers, is equivalent for oocytes that underwent maturation with or without cumulus cells.[36] The zona pellucida of denuded oocytes is harder and more resistant to chymotrypsin digestion than the zona pellucida of cumulus cell-enclosed oocytes after spontaneous maturation in the presence of FBS.[39] Cumulus cells, therefore, participate in the prevention of zona pellucida hardening. The reduction in fertilization of mouse ova after maturation without cumulus cells is most readily explained by the reduced penetrability of the zona pellucida, although other actions of the cumulus cells cannot be excluded. For example, rat oocytes undergo abnormal pronuclear formation when matured in the absence of cumulus cells.[46] Coculture of maturing oocytes with extra cumulus cells, or with mural granulosa cells, has been reported to increase the developmental capacity of the resultant eggs of other species.[49,50] It is not clear, however, whether this beneficial effect is the result of general conditioning of the medium or the production of specific substances by the cumulus or mural granulosa cells.

e. Gonadotropins

Addition of FSH to the medium for oocyte maturation has a beneficial effect on the developmental capacity of maturing mouse oocytes,[51,52] but not of bovine oocytes.[48] Not only does FSH treatment of maturing mouse oocytes increase the percentage of ova that cleave to the two-cell stage after insemination, but it also increases the percentage of two-cell embryos that develop to the blastocyst stage. FSH, therefore, affects the processes occurring during oocyte maturation that relate to the embryonic developmental capacity of the ova. FSH causes a delay in the time of GVB in cumulus cell-enclosed oocytes isolated from pregnant mares' serum gonadotropin (PMSG)-primed mice.[53] In contrast, FSH reduces the total time required for oocyte maturation necessary to produce the highest frequency of cleavage to the two-cell stage after insemination.[52] The possibility that meiosis is reinitiated in oocytes, at least in part, by a positive gonadotropin-dependent stimulus from granulosa/cumulus cells[54] is discussed by Schultz in Chapter 4. Whether this putative maturation-inducing signal, the post-GVB maturation-accelerating signal, or the signal that benefits the developmental capacity of the ova maturing *in vitro* is related remains to be determined. Nevertheless, these observations indicate a positive effect of gonadotropins on oocyte maturational processes.

f. Hormonal Status of the Oocyte Donor

Fully grown GV-stage oocytes, isolated from both PMSG-primed and -unprimed immature mice (22 days old) or from unprimed adult cycling mice at all stages of the estrous cycle, are able to undergo maturation in culture and acquire full developmental capacity.[55] Moreover, GV-stage oocytes isolated from mutant dwarf mice will mature *in vitro* and are competent of fertilization and development to live young.[55] These dwarf females are deficient in growth hormone, prolactin, and thyroid-stimulating hormone, but not in gonadotropins. The antral follicles of these mice are poorly developed and often show atretic degeneration, but the oocytes isolated from them display kinetics of oocyte maturation and developmental capacity identical to that of littermate controls.[55] Growth hormone, prolactin, or thyroid-stimulating hormone, therefore, is not required during mouse oocyte development *in vivo* for oocyte growth, acquisition of competence for spontaneous maturation, or the acquisition of full developmental capacity during maturation *in vitro*.

Oocytes isolated from hypogonadal mutant mice, having undetectable levels of circulating gonadotropins, do not develop a normal capacity for maturation or subsequent embryogenesis.[55] Mice homozygous for the recessive gene hypogonadal (*hpg/hpg*) do not produce gonadotropin-releasing hormone (GnRH) and, therefore, have undetectable levels of circulating gonadotropins.[7] Oocytes isolated from the very few early antral follicles found in the ovaries of these mutant mice exhibit retarded kinetics of GVB, and fewer of the matured oocytes cleave after insemination than in littermate controls. No blastocysts develop from hypogonadal ova after maturation *in vitro*. PMSG priming of hypogonadal females 2 days before isolation of the GV-stage oocytes results in equivalent kinetics of GVB and developmental capacity in the mutant and littermate controls.[55] In conclusion, exposure of oocytes and their companion granulosa cells to gonadotropin *in vivo* before isolation and maturation *in vitro* has a beneficial effect on otherwise developmentally incompetent oocytes.

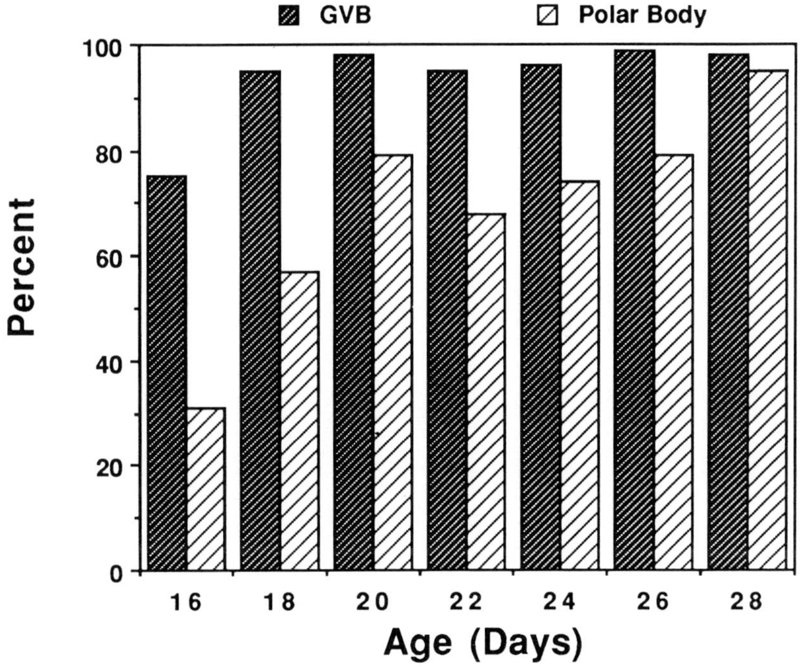

FIGURE 6. Maturation of oocytes isolated from immature mice. Oocytes were isolated from the antral follicles of mice 16 to 28 days old and cultured for 18 h. Less than half of the oocytes that underwent GVB after isolation from 16-day-old mice produced a polar body. In contrast, almost all of the oocytes that underwent GVB after isolation from 28-day-old mice produced a polar body. Data adapted from Eppig and Schroeder.[56]

g. Stage of Follicular Development

Oocytes acquire competence of undergo spontaneous GVB in culture as they near completion of their growth phase, an event that usually occurs about the time of follicular antrum development. Acquisition of competence to undergo GVB, however, does not assume competence to complete the first meiotic division and produce a polar body. Fewer oocytes isolated from the small antral follicles of 16-day-old mice that undergo GVB produce a polar body than oocytes isolated from the larger follicles of older mice (Figure 6).[2] Thus, competence to undergo GVB and then complete the first meiotic division is acquired sequentially. Similarly, the capacity of the oocytes to cleave and complete preimplantation development after maturation *in vitro* is also acquired sequentially. The percentage of the *in vitro*-matured ova that cleave to the two-cell stage after insemination increases with advancing follicular devopment in immature mice (Figure 7).[56] Likewise, the capacity of the two-cell-stage embryos to develop to the blastocyst stage also increases with follicular development (Figure 7).[56] It will be important to resolve the cellular and molecular basis of this sequential acquisition of developmental capacity and whether the changes reside in the cumulus cells, in their ability to regulate oocyte development, an autonomous program in the oocyte, or a combination of these factors.

B. OOCYTE GROWTH *IN VITRO*

Growing oocyte-granulosa cell complexes have been isolated from the ovaries of immature mice and rats, 8 to 12 days old, using proteases, generally crude collagenase.[57-60] This procedure yields oocytes from preantral follicles surrounded by one to three layers of granulosa cells, depending on the age of the donor, but the basal lamina that enclosed the granulosa cells *in situ* is removed. Further treatment with proteases, particularly in the absence of calcium, results in removal of the granulosa cells from the oocyte. Oocytes

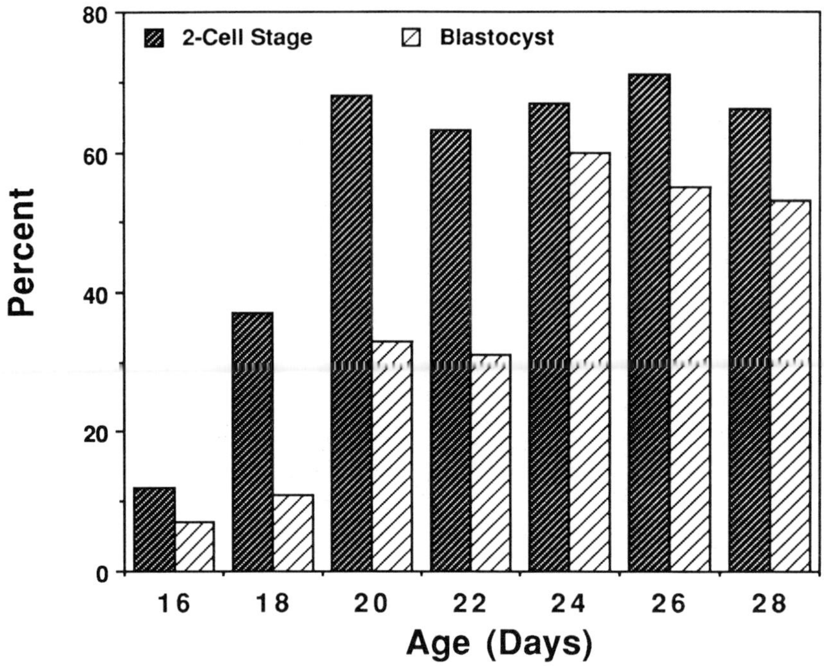

FIGURE 7. Developmental capacity of embryos derived from oocytes that underwent maturation *in vitro* after isolation from the antral follicles of mice 16 to 28 days old. Data indicate the percentage of the mature ova that cleaved to the two-cell stage after maturation *in vitro* and the percentage of the two-cell-stage embryos that developed to the blastocyst stage. Data adapted from Eppig and Schroeder.[56]

isolated from 8- to 12-day-old mice or rats are in mid-growth phase and, without further development, incompetent of undergoing GVB. Figure 8 demonstrates, by scanning electron microscopy, the appearance of an oocyte-granulosa cell complex, isolated from a 12-day-old mouse, after 7 days in culture. During a 12- to 20-day culture period, many granulosa cell-enclosed oocytes grow to nearly full-grown size and 60 to 90% of them acquire competence to undergo spontaneous GVB.[58,59,61]

1. Role of Granulosa Cells in Oocyte Growth and Development

Initial studies on oocyte growth in culture demonstrated that the association of granulosa cells with the oocyte is essential for oocyte growth.[59,60] The culture system must maintain gap junctional association between these two cell types because simple coculture of granulosa cell-denuded oocytes on a monolayer of granulosa cells results in a greatly decreased rate of oocyte growth. Moreover, oocyte growth appears to demand gap junctional communication specifically with granulosa cells, because establishment of gap junctions between oocytes and fibroblasts or Sertoli cells, the male homologue of granulosa cells, fails to support oocyte growth.[62] There is a direct relationship between the number of granulosa cells coupled to the oocyte and the rate of oocyte growth.[63] One function of the gap junctional association of granulosa cells and oocytes is nutrition.[63,64] Granulosa cell-denuded oocytes take up some molecules, such as certain amino acids, sugars, etc., at very low rates, indicating deficiencies in transport systems in the oocyte plasma membrane.[59,65,66] These molecules have a much faster rate of entry into the oocyte when it is coupled to granulosa cells, the gap junctions serving as the conduit for these molecules after they were initially taken up into the granulosa cells. In this way, the granulosa cell membrane compensates for qualitative and quantitative deficiencies in the oocyte plasma membrane.

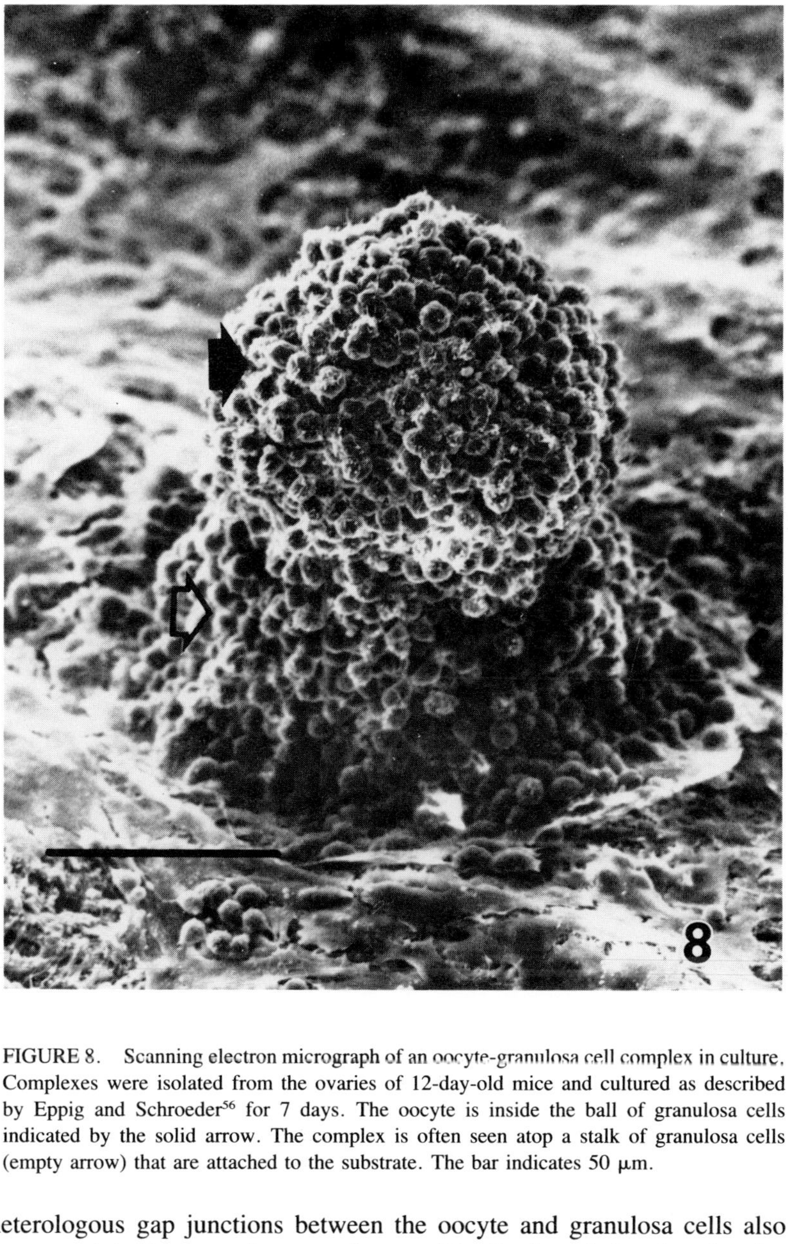

FIGURE 8. Scanning electron micrograph of an oocyte-granulosa cell complex in culture. Complexes were isolated from the ovaries of 12-day-old mice and cultured as described by Eppig and Schroeder[56] for 7 days. The oocyte is inside the ball of granulosa cells indicated by the solid arrow. The complex is often seen atop a stalk of granulosa cells (empty arrow) that are attached to the substrate. The bar indicates 50 μm.

The heterologous gap junctions between the oocyte and granulosa cells also mediate metabolic cooperativity between the two cell types. Distribution of phosphorylated nucleoside derivatives in granulosa cell-enclosed oocytes is different from that in denuded oocytes, but essentially the same as in granulosa cells, suggesting that these somatic cells metabolize the nucleosides after uptake and transfer of the phosphorylated derivatives to the oocyte.[65,66]

Granulosa cells also affect protein phosphorylation in oocytes. The pattern of protein phosphorylation of granulosa cell-denuded oocytes is different from that of granulosa cell-enclosed oocytes.[67] Moreover, the pattern of protein phosphorylation is different in growing oocytes vs. fully grown oocytes. Reestablishment of gap junctional association between the oocytes and granulosa cells, but not with Sertoli cells, restores, to a large extent, the granulosa cell-enclosed pattern of phosphorylation in the oocyte.[67] Granulosa cells, therefore, regulate the pattern of protein phosphorylation in oocytes in a stage- and cell-specific way. Since protein phosphorylation is so critically important in the regulation of cell function and development, these studies underscore the importance of granulosa cell-oocyte interactions for normal oocyte development.

In spite of the significant role of granulosa cells in oocyte development, described above, granulosa cells are not essential for some aspects of oocyte development. Although oocytes cultured in monolayers of fibroblasts do not grow, the viability of the oocytes is sustained by the fibroblasts. Under these conditions, oocytes isolated before they were able to undergo spontaneous GVB *in vitro* acquire competence to resume meiosis and undergo GVB at the same time as age-matched oocytes developing *in vivo*.[68] Growing oocytes, therefore, do not require continued gap junctional association with granulosa cells to acquire competence to undergo spontaneous GVB.

2. Phosphodiesterase Inhibitors

An important component of culture systems for oocyte growth and development in culture is a cyclic adenosine monophosphate-(cAMP)-phosphodiesterase inhibitor such as hypoxanthine or 3-isobutyl-1-methylxanthine (IBMX).[61] These inhibitors appear to function in two ways. First, they maintain the association of granulosa cells with the developing oocytes. In the absence of IBMX or hypoxanthine, many more oocytes lose this association and become denuded of the somatic cells in the culture medium, a condition that greatly retards growth of the oocyte.[61] It is not clear whether the action of the inhibitors is on the granulosa cells, the oocyte, or both; but elevated levels of cAMP are thought to be important for granulosa cell development and function. Second, the phosphodiesterase inhibitors help to maintain the oocyte in meiotic arrest after they acquire competence to undergo GVB. Simply removing the phosphodiesterase inhibitor after oocytes acquire this competence results in the spontaneous maturation of many of them.

3. Comparison of Oocyte Growth *In Vitro* and *In Vivo*

Granulosa cell-enclosed oocytes isolated from 12-day-old mice grow to the size of oocytes isolated from the antral follicles of 16-day-old mice in 10 days of culture.[56] Although this represents considerable growth (about an 80% increase in volume), it appears to fall short of the growth *in vivo* of age-matched oocytes isolated from the antral follicles of 22-day-old mice. Nevertheless, the populations used for comparison may not be exactly comparable, since only oocytes isolated from the antral follicles of mice 16 days old and older were measured for comparison to the oocytes from 12-day-old mice. These antral follicles probably were derived from only the most advanced follicles of 12-day-old mice and represent, therefore, a selected group of the larger oocytes present among the population of oocytes in the 12-day-old mice. Thus, the populations measured are either biased toward the smaller sizes after isolation from 12-day-old-mice or toward the larger sizes after isolation from the older mice.

There is no effect of 17β-estradiol or progesterone on the growth of mouse oocytes *in vitro*. In contrast, testosterone and dihydrotestosterone (DHT) impede oocyte growth. None of these steroid hormones affect the percentage of oocytes that acquire competence to undergo GVB.[69]

4. Developmental Capacity of Oocytes Growth *In Vitro*

When mouse oocyte-granulosa cell complexes are isolated from 12-day-old mice and cultured for 10 days, then removed from medium containing IBMX and stimulated with FSH, the ability of the mature oocytes to cleave to the two-cell stage after insemination is equivalent to that of oocytes that mature *in vitro* after isolation from 22-day-old mice and to that of superovulated ova that matured *in vivo* (Figure 9).[56] Development of the two-cell-stage embryos to blastocysts by the *in vitro* grown oocytes (IVG), however, is equivalent only to that of embryos derived from oocytes isolated from 18-day-old mice (Figure 9). Live young develop from IVG oocytes transferred at the two-cell stage to foster mothers.[56]

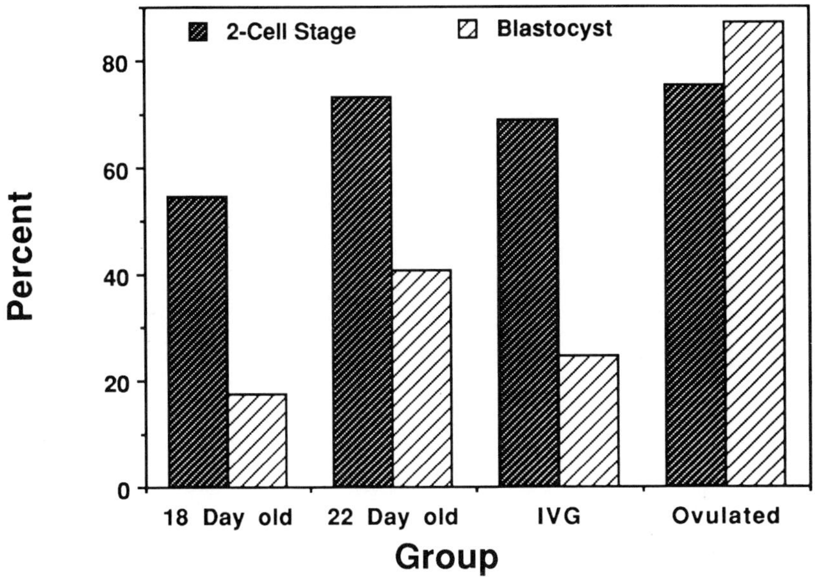

FIGURE 9. Developmental capacity of embryos derived from oocytes that grew and matured *in vitro* (IVG) after isolation from 12-day-old mice and cultured for 10 days. Data indicate the percentage of mature ova that cleaved to the two-cell stage after maturation *in vitro* and the percentage of the two-cell-stage embryos that developed to the blastocyst stage. Data adapted from Eppig and Schroeder.[56]

Although young are produced at a low percentage (about 5%), the results are very promising because they indicate that improved conditions for oocyte development *in vitro* will yield higher frequencies of development to term.

Progress is also being made in the establishment of systems for the development of rat oocytes *in vitro*.[58] Oocyte-granulosa cell complexes isolated from 10- to 11-day-old rats grow from 40 to 45 μm in diameter to an average of about 62 μm diameter in 20-days culture, and 60 to 70% of them are competent of spontaneous GVB. Normal-appearing fertilization occurs in 35 to 45% of the eggs and, of these, about 65% cleave to the two-cell stage. The capacity of the two-cell embryos derived from IVG oocytes to complete embryonic development has not been assessed.

Major improvements in the system for oocyte development *in vitro* are necessary to increase the capacity of the ova to undergo fertilization and embryonic development to term. Because of the intimate association that has been demonstrated between oocytes and granulosa cells, modifications in the culture system will arise from a better understanding of the development and function of granulosa cells and oocyte-granulosa cell interactions as well as oocyte metabolism.

C. USES OF SYSTEMS FOR OOCYTE DEVELOPMENT *IN VITRO*

1. Resolving Basic Mechanisms Governing Oocyte Development

It is well established that culture systems provide a way for studying basic processes in cellular physiology and development utilizing a reductionist philosophy. These systems have already revealed much basic information on the mechanisms of meiotic arrest, the induction of oocyte maturation, and the importance of granulosa cell-oocyte interactions in oocyte and granulosa cell development. Culture systems that produce ova competent of undergoing fertilization and embryogenesis strengthen the physiological significance of experimental findings. Nevertheless, it is critical for investigators to bear in mind that the oocyte is part of a complex structural and functional unit, the follicle, and findings made with isolated oocytes or oocyte-granulosa/cumulus cell complexes must be placed into the context of the follicle and the intact animal.

2. Introduction of Genes

Current technology for introducing genes into the germ line of mice and other species involves either microinjection of pronuclei with appropriate DNA constructs, retroviral infection of preimplantation embryos, or transfection of totipotent embryo-derived cells followed by production of chimeras by introducing the transfected cells into blastocysts. It is possible that another way to introduce genes is by retroviral infection of oocytes in culture. The germ line of SWR/J-RF/J mice becomes infected with ecotropic retroviruses during oocyte development *in vivo*.[70,71] Moreover, inoculation of newborn SWR/J mice with ecotropic murine leukemia virus has produced transgenic mice containing new proviruses.[72] Because of the potential of producing high titers of infectious virus in culture, the culture systems for oocyte development could provide an efficient way to introduce foreign genes into the germ line of mice using retroviral vectors. This method would provide the opportunity for producing transgenic mice without the need for microinjection or micromanipulation protocols. Also, since multiple infections could occur over relatively long incubation times of several days, the chances of integration and subsequent expression may be increased.

3. Expanding Populations of Rare or Endangered Mammals

There is an enormous pool of oocytes (thousands) in the ovary that never becomes available for ovulation because of follicular degeneration and consequent oocyte loss. Assuming that these lost oocytes would have had the capacity for normal development in the absence of follicular atresia, these oocytes could be rescued by appropriate protocols for recovery and culture. Oocytes showing significant post-mortem degenerative changes can, in fact, be removed from deceased mice and the degenerative changes reversed during maturation in culture so that they are competent of fertilization and development to term. Thus, oocytes could be rescued from important individual animals even after unforeseen death. It seems likely that populations of rare or endangered mammals, or of agriculturally important mammals, will be expandable using protocols for oocyte maturation or for both oocyte growth followed by oocyte maturation *in vitro*. In addition, mammals that are reproductively impaired could be assisted by oocyte culture technology.

4. Preservation of Oocytes

a. Cryopreservation

Although protocols for the cryopreservation of embryos and sperm have wide clinical and agricultural application, there are some applications for which cryopreservation of oocytes or eggs would be more advantageous.[73] For example, some of the ethical and legal problems inherent in cryopreservation of human embryos could be resolved. Moreover, because of the possibility of producing many eggs by the techniques of oocyte growth and maturation *in vitro* described in this chapter, it may be necessary to preserve them until suitable sperm become available. For example, if a valuable animal dies unexpectedly and the oocytes are rescued, the eggs could be cryopreserved if appropriate sperm are not on hand. Ovulated eggs from several species can be fertilized after cryopreservation and develop to live young, but success rates are highly variable.[73] GV-stage oocytes survive freeze-thaw protocols and are capable of undergoing spontaneous maturation, but the percentage that undergo fertilization and cleavage to the two-cell stage is very low, and none develop to the blastocyst stage using currently reported protocols.[74] Oocytes that undergo spontaneous maturation *in vitro* can be fertilized, after cryopreservation, and a few of them undergo preimplantation development. Treatment of the maturing cumulus cell-enclosed oocytes with FSH greatly increases the percentage of ova that can undergo fertilization, after freezing and thawing, and embryonic development to term.[74] Mucification of the cumulus oophorus stimulated by FSH during maturation seems to protect the cumulus cells and the ovum during the freeze-thaw protocol.

b. Chemopreservation

Mature ova maintained in culture rapidly lose the capacity for fertilization and development.[52] Oocytes maintained in meiotic arrest *in vitro* with phosphodiesterase inhibitors or cAMP analogues, however, remain competent of undergoing a full maturational program so that ova competent of fertilization and development to term are produced.[51] These meiosis-arresting substances, therefore, can be used to preserve oocytes for limited periods until the oocytes can be cryopreserved or fertilized, after maturation.

IV. CONCLUSION

Culture systems for oocyte growth and maturation in culture have contributed greatly to our understanding of the mechanisms that govern oocyte development. There is certainly much more to be learned and the culture systems will be a primary tool to achieve new information. In addition, the culture systems will be developed for practical uses in the clinic, agriculture, and the expansion of populations of animals essential for research and of animals that are rare or endangered. Evolution of the culture technology is dependent upon the resolution of factors involved in follicular development, the cellular and metabolic processes occurring in the oocyte-granulosa/cumulus cell complex, and the molecular processes that prepare the oocyte for maturation, fertilization, and embryogenesis.

ACKNOWLEDGMENTS

I am very grateful to Drs. Roberto Buccione, Janan Eppig, Allen Schroeder, Richard Schultz, and Barbara Vanderhyden for their helpful suggestions in the preparation of this chapter. I also acknowledge the NIH support for research in my laboratory through grants HD20575, HD21070, and HD23939.

REFERENCES

1. **Mangia, F. and Epstein, C. J.,** Biochemical studies of growing mouse oocytes: preparation of oocytes and analysis of glucose-6-phosphate dehydrogenase and lactate dehydrogenase activities, *Dev. Biol.* 45, 211, 1975.
2. **Sorensen, R. A. and Wassarman, P. M.,** Relationship between growth and meiotic maturation of the mouse oocyte, *Dev. Biol.*, 50, 531, 1976.
3. **Szybek, K.,** In vitro maturation of oocytes from sexually immature mice, *J. Endocrinol.*, 54, 527,1972.
4. **Eppig, J. J.,** Granulosa cell deficient follicles. Occurrence, structure, and relationship to ovarian teratocarcinogenesis in strain LT/Sv mice, *Differentiation,* 12, 111, 1978.
5. **Arendsen de Wolff-Exalto, E.,** Influence of gonadotropins on early follicle cell development and early oocyte growth in the immature rat, *J. Reprod. Fertil.*, 66, 537, 1982.
6. **Purandare, T. V., Munshi, S. R., and Rao, S. S.,** Effect of antisera to gonadotropins on follicular development and fertility of mice, *Biol. Reprod.*, 15, 311, 1976.
7. **Halpin, D. M. G., Jones, A., Fink, G., and Charlton, H. M.,** Postnatal ovarian follicle development in hypogonadal (hpg) and normal mice and associated changes in the hypothalamic-pituitary ovarian axis, *J. Reprod. Fertil.*, 77, 287, 1986.
8. **Roy, S. K. and Greenwald, G. S.,** In vitro steroidogenesis by primary to antral follicles in the hamster during the periovulatory period: effects of follicle-stimulating hormone, luteinizing hormone, and prolactin, *Biol. Reprod.*, 36, 39, 1987.
9. **Roy, S. and Greenwald, G. S.,** In vitro effects of follicle-stimulating hormone, luteinizing hormone, and prolactin on follicular deoxyribonucleic acid synthesis in the hamster, *Endocrinology,* 122, 952, 1988.

10. **Fortune, J. E. and Eppig, J. J.,** Effects of gonadotropins on steroid secretion by infantile and juvenile mouse ovaries in vitro, *Endocrinology,* 105, 760, 1979.
11. **Funkenstein, B., Nimrod, A., and Lindner, H. R.,** The development of steroidogenic capability and responsiveness to gonadotropins in cultured neonatal rat ovaries, *Endocrinology,* 106, 98, 1980.
12. **Anderson, E. and Albertini, D. F.,** Gap junctions between the oocyte and companion follicle cells in the mammalian ovary, *J. Cell Biol.,* 71, 680, 1976.
13. **Gilula, N. B., Epstein, M. L., and Beers, W. H.,** Cell-to-cell communication and ovulation. A study of the cumulus-oocyte complex, *J. Cell Biol.,* 78, 58, 1978.
14. **Lawrence, T. S., Beers, W. H., and Gilula, N. B.,** Transmission of hormonal stimulation by cell-to-cell communication, *Nature,* 272, 501, 1978.
15. **Schultz, R. M. and Wassarman, P. M.,** Biochemical studies of mammalian oogenesis: protein synthesis during oocyte growth and meiotic maturation in the mouse, *J. Cell Sci.,* 24, 167, 1977.
16. **Schultz, R. M., Letourneau, G. E., and Wassarman, P. M.,** Program of early development in the mammal: changes in the patterns and absolute rates of tubulin and total protein synthesis during oocyte growth in the mouse, *Dev. Biol.,* 73, 120, 1979.
17. **Bachvarova, R., De Leon, V., Johnson, A., Kaplan, G., and Payton, B. V.,** Changes in total RNA, polyadenylated RNA, and actin mRNA during meiotic maturation of mouse oocytes, *Dev. Biol.,* 108, 325, 1985.
18. **Philpott, C. C., Ringuette, M. J., and Dean, J.,** Oocyte-specific expression and developmental regulation of ZP3, and sperm receptor of the mouse zona pellucida, *Dev. Biol.,* 121, 568, 1987.
19. **Shimizu, S., Tsuji, M., and Dean, J.,** In vitro biosynthesis of three sulfated glycoproteins of murine zonae pullucidae by oocytes grown in follicle culture, *J. Biol. Chem.,* 258, 5858, 1983.
20. **Huarte, J., Belin, D., and Vassalli, J.-D.,** Plasminogen activator in mouse and rat oocytes: induction during meiotic maturation, *Cell,* 43, 551, 1985.
21. **Huarte, J., Belin, D., Vassalli, A., Strickland, S., and Vassalli, J.,** Meiotic maturation of mouse oocytes triggers the translation and polyadenylation of dormant tissue-type plasminogen activator mRNA, *Genes Dev.,* 1, 1201, 1987.
22. **Strickland, S., Huarte, J., Belin, D., Vassalli, A., Rickles, R. J., and Vassalli, J.-D.,** Antisense RNA directed against the 3′ noncoding region prevents dormant mRNA activation in mouse oocytes, *Science,* 241, 680, 1988.
23. **Keshet, E., Rosenberg, M. P., Mercer, J. A., Propst, F., Vande Woude, G. F., Jenkins, N. A., and Copeland, N. G.,** Developmental regulation of ovarian-specific Mos expression, *Oncogene,* 2, 235, 1988.
24. **Mutter, G. L., Grills, G. S., and Wolgemuth, D. J.,** Evidence for the involvement of the proto-oncogene c-mos in mammalian meiotic maturation and possibly very early embryogenesis, *EMBO J.,* 7, 683, 1988.
25. **Paules, R. S., Buccione, R., Propst, F., Moschel, R.C., Vande Woude, G. F., and Eppig, J. J.,** The mouse *Mos* proto-oncogene is product is present and functions during oogenesis, *Proc. Natl. Acad. Sci. U.S.A.,* 86, 5395, 1989.
26. **Pincus, G. and Enzmann, E. V.,** The comparative behavior of mammalian eggs in vivo and in vitro. I. The activation of ovarian eggs, *J. Exp. Med.,* 62, 655, 1935.
27. **Biggers, J. D. and Powers, R. D.,** Comments on the control of meiotic maturation in mammals, in *Ovarian Follicular Development and Function,* Midgley, A. R. and Sadler, W. A., Eds., 1979, 365.
28. **Cross, P.C. and Brinster, R. L.,** In vitro development of mouse oocytes, *Biol. Reprod.,* 3, 298, 1970.
29. **Fleming, A. D., Evans, G., Walton, E. A., and Armstrong, D. T.,** Developmental capability of rat oocytes matured in vitro in defined medium, *Gamete Res.,* 12, 255, 1985.
30. **Iwamatsu, T. and Chang, M. C.,** Sperm penetration in vitro of mouse oocytes at various times during maturation, *J. Reprod. Fertil.,* 31, 237, 1972.
31. **Leibfried, M. L. and Bavister, B. D.,** Fertilizability of in vitro matured oocytes from golden hamsters, *J. Exp. Zool.,* 226, 481, 1983.
32. **Niwa, K. and Chang, M. C.,** Fertilization of rat eggs in vitro at various times before and after ovulation with special reference to fertilization of ovarian oocytes matured in culture, *J. Reprod. Fertil.,* 43, 435, 1975.
33. **Thibault, C.,** Are follicular maturation and oocyte maturation independent processes?, *J. Reprod. Fertil.,* 51, 1, 1977.
34. **Thibault, C. and Gerard, M.,** Cytoplasmic and nuclear maturation of rabbit oocytes in vitro, *Ann. Biol. Anim. Biochim. Biophys.,* 13, 145, 1973.
35. **Minato, Y. and Toyoda, Y.,** Development to normal young of mouse oocytes matured and fertilized in vitro, *Jpn. J. Zootech. Sci.,* 54, 387, 1983.
36. **Schroeder, A. C. and Eppig, J. J.,** The developmental capacity of mouse oocytes that matured spontaneously in vitro is normal, *Dev. Biol.,* 102, 493, 1984.
37. **Van de Sandt, J. J. M., Schroeder, A. C., and Eppig, J. J.,** Culture media for mouse oocyte maturation affect subsequent preimplantation embryo development, *Mol. Reprod. Dev.,* 25, 164, 1990.

38. **DeFelici, M., Salustri, A., and Siracusa, G.**, "Spontaneous" hardening of the zona pellucida of mouse oocytes during in vitro culture. II. The effect of follicular fluid and glycosaminoglycans, *Gamete Res.*, 12, 227, 1985.
39. **Downs, S. M., Schroeder, A. C., and Eppig, J. J.**, Serum maintains the fertilizability of mouse oocytes matured in vitro by preventing the hardening of the zona pellucida, *Gamete Res.*, 15, 115, 1986.
40. **Eppig, J. J. and Schroeder, A. C.**, Culture systems for mammalian oocyte development: progress and prospects, *Theriogenology*, 25, 97, 1986.
41. **Gianfortoni, J. G. and Gulyas, B. J.**, The effects of short-term incubation (aging) of mouse oocytes on in vitro fertilization, zona solubility, and embryonic development, *Gamete Res.*, 11, 59, 1985.
42. **Ducibella, T., Kurasawa, S., Rangarajan, S., Kopf, G. S., and Schultz, R. M.**, Precocious loss of cortical granules during oocyte meiotic maturation and correlation with an egg-induced modification of the zona pellucida, *Dev. Biol.*, 137, 46, 1990.
43. **Moller, C.C. and Wassarman, P. M.**, Characterization of a proteinase that cleaves zona pellucida glycoprotein ZP2 following activation of mouse eggs, *Dev. Biol.*, 132, 103, 1989.
44. **Eppig, J. J.**, Preimplantation embryonic development of spontaneous mouse parthenotes after oocyte meiotic maturation in vitro, *Gamete Res.*, 4, 3, 1981.
45. **Vanderhyden, B. C., McLaughlin, K. J., Rutledge, J. M., and Armstrong, D. T.**, Zona drilling increases the penetrability of rat oocytes matured in vitro, *Biol. Reprod.*, 40, 953, 1989.
46. **Vanderhyden, B. C. and Armstrong, D. T.**, Role of cumulus cells and serum on the in vitro maturation, fertilization, and subsequent development of rat oocytes, *Biol. Reprod.*, 40, 720, 1989.
47. **Ball, G. D., Leibfried, L. D., Ax, R. L., Bavister, B. D., and First, N. L.**, Factors affecting successful in vitro fertilization of bovine follicular oocytes, *Biol. Reprod.*, 28, 717, 1983.
48. **Sirard, M. A., Parrish, J. J., Ware, C. B., Leibfried-Rutledge, M. L., and First, N. L.**, The culture of bovine oocytes to obtain developmentally competent embryos, *Biol. Reprod.*, 39, 546, 1988.
49. **Staigmiller, R. B. and Moor, R. M.**, Effect of follicle cells on the maturation and developmental competence of ovine oocytes matured outside the follicle, *Gamete Res.*, 9, 221, 1984.
50. **Motlik, J. and Fulka, J.**, Fertilization of rabbit oocytes co-cultured with granulosa cells, *J. Reprod. Fertil.*, 63, 425, 1981.
51. **Downs, S. M., Schroeder, A. C., and Eppig, J. J.**, The developmental capacity of mouse oocytes following maintenance of meiotic arrest in vitro, *Gamete Res.*, 15, 305, 1986.
52. **Schroeder, A. C., Downs, S. M., and Eppig, J. J.**, Factors affecting the developmental capacity of mouse oocytes undergoing maturation in vitro, *Ann. N. Y. Acad. Sci.*, 541, 197, 1988.
53. **Eppig, J. J., Freter, R. R., Ward-Bailey, P. F., and Schultz, R. M.**, Inhibition of oocyte maturation in the mouse: participation of cAMP, steroid hormones, and a putative maturation-inhibitor factor, *Dev. Biol.*, 100, 39, 1983.
54. **Downs, S. M., Daniel, S. A. J., and Eppig, J. J.**, Induction of maturation in cumulus cell-enclosed mouse oocytes by follicle-stimulating hormone and epidermal growth factor: evidence for a positive stimulus of somatic cell origin, *J. Exp. Zool.*, 245, 86, 1988.
55. **Schroeder, A. C. and Eppig, J. J.**, Developmental capacity of mouse oocytes that undergo maturation in vitro: effect of the hormonal state of the oocyte donor, *Gamete Res.*, 24, 1, 1989.
56. **Eppig, J. J. and Schroeder, A. C.**, Capacity of mouse oocytes from preantral follicles to undergo embryogenesis and development to live young after growth, maturation and fertilization in vitro, *Biol. Reprod.*, 41, 268, 1989.
57. **Grob, H. S.**, Growth and endocrine function of isolated ovarian follicles cultivated in vivo, *Biol. Reprod.*, 1, 320, 1969.
58. **Daniel, S. A. J., Armstrong, D. T., and Gore-Langton, R. E.**, Growth and development of rat oocytes in vitro, *Gamete Res.*, 24, 109, 1989.
59. **Eppig, J. J.**, Mouse oocyte development in vitro with various culture systems, *Dev. Biol.*, 60, 371, 1977.
60. **Eppig, J. J.**, A comparison between oocyte growth in coculture with granulosa cells and oocytes with granulosa cell-oocyte junctional contact maintained in vitro, *J. Exp. Zool.*, 209, 345, 1979.
61. **Eppig, J. J. and Downs, S. M.**, The effect of hypoxanthine on mouse oocyte growth and development in vitro: maintenance of meriotic arrest and gonadotropin-induced oocyte maturation, *Dev. Biol.*, 119, 313, 1987.
62. **Buccione, R., Cecconi, S., Tatone, C., Mangia, F., and Colonna, R.**, Follicle cell regulation of mammalian oocyte growth, *J. Exp. Zool.*, 242, 351, 1987.
63. **Herlands, R. L. and Schultz, R. M.**, Regulation of mouse oocyte growth: probable nutritional role for intercellular communication between follicle cells and oocytes in oocyte growth, *J. Exp. Zool.*, 229, 317, 1984.
64. **Brower, P. T. and Schultz, R. M.**, Intercellular communication between granulosa cells and mouse oocytes: existence and possible nutritional role during oocyte growth, *Dev. Biol.*, 90, 144, 1982.
65. **Heller, D. T., Cahill, D. M., and Schultz, R. M.**, Biochemical studies of mammalian oogenesis: metabolic cooperativity between granulosa cells and growing mouse oocytes, *Dev. Biol.*, 84, 455, 1981.

66. **Heller, D. T. and Schultz, R. M.,** Ribonucleoside metabolism by mouse oocytes: metabolic cooperativity between fully grown oocytes and cumulus cells, *J. Exp. Zool.*, 214, 355, 1980.
67. **Colonna, R., Cecconi, S., Tatone, C., Mangia, F., and Buccione, R.,** Somatic cell-oocyte interactions in mouse oogenesis: stage-specific regulation of mouse oocyte protein phosphorylation by granulosa cells, *Dev. Biol.*, 133, 305, 1989.
68. **Canipari, R., Palombi, F., Riminucci, M., and Mangia, F.,** Early programming of maturation competence in mouse oogenesis, *Dev. Biol.*, 102, 519, 1984.
69. **Eppig, J. J., Daniel, S. A. J., and Schroeder, A. C.,** Growth and development of mouse oocytes in vitro, *Ann. N.Y. Acad. Sci.*, 541, 205, 1988.
70. **Jenkins, N. A. and Copeland, N. G.,** High frequency germline acquisition of ecotropic MuLV proviruses in SWR/J-RF/J hybrid mice, *Cell,* 43, 811, 1985.
71. **Lock, L. F., Keshet, E., Gilbert, D. J., Jenkins, N. A., and Copeland, N. G.,** Studies of the mechanism of spontaneous germline ecotropic provirus acquisition in mice, *EMBO J.*, 7, 4169, 1988.
72. **Panthier, J. J., Condamine, H., and Jacob, F.,** Inoculation of newborn SWR/J females with an ecotropic murine leukemia virus can produce transgenic mice, *Proc. Natl. Acad. Sci. U.S.A.*, 85, 1156, 1988.
73. **Frieder, S., Giudice, L. C., and Lamb, E. J.,** Cryopreservation of embryos and ova, *Fertil. Steril.*, 49, 743, 1988.
74. **Schroeder, A.C., Champlin, A. K., Mobraaten, L. E., and Eppig, J. J.,** The developmental capacity of mouse oocytes cryopreserved before and after maturation in vitro and after ovulation, *J. Reprod. Fertil.*, 89, 43, 1990.

Chapter 4

MEIOTIC MATURATION OF MAMMALIAN OOCYTES

Richard M. Schultz

TABLE OF CONTENTS

I. INTRODUCTION

Oogenesis results in the formation of a fully grown oocyte that is present in a preovulatory antral follicle. The oocyte is a highly differentiated cell, and in lower species, e.g., sea urchin, C_0t analysis indicates that the RNA sequence complexity of the oocyte is greater than that at any stage of early development or in differentiated tissues.[21] The oocyte grows while arrested in the first meiotic prophase, and *in vivo,* a hormonal stimulus, which acts on the follicle cells that are coupled by gap junctions to each other and to the oocyte, initiates resumption of meiosis.[3,4] Resumption of meiosis is characterized by germinal vesicle breakdown (GVBD), formation of a metaphase I spindle, movement of the spindle to the cortex, separation of homologous chromosomes with emission of the first polar body, and arrest at metaphase II.[111] In most mammalian species, fertilization triggers completion of meiosis with emission of the second polar body. Resumption of meiosis with arrest at metaphase II is termed meiotic maturation or oocyte maturation, and is a fundamental process in the life cycle of the organism, since only oocytes that mature properly are capable of being fertilized and giving rise to normal development.[111]

During the past 15 years our understanding has increased greatly about the molecular basis for maintenance of meiotic arrest *in vitro,* and most likely *in vivo.* In contrast, the mechanism of gonadotropin-induced resumption of meiosis is still poorly understood, although some new insights have recently come to light. This chapter will focus on our current understanding of maintenance of meiotic arrest and resumption of meiosis and will conclude with a description of specific activities that develop in the cytoplasm of the matured oocyte and that are necessary for early events in egg activation.

II. REGULATION OF OOCYTE MATURATION

A. cAMP AND PROTEIN PHOSPHORYLATION

A seminal observation made over 50 years ago is that oocytes present in explanted antral follicles do not resume meiosis, whereas maturation ensues if the oocyte is removed from its follicle.[94] A reasonable interpretation of these results is that the follicle exerts an inhibitory influence on oocyte maturation. This led, in turn, to a search for inhibitory molecules that are present in the follicle and that can maintain meiotic arrest of oocytes cultured *in vitro.* Much discussion initially focused on whether *in vitro* maturation accurately reflects the events that occur during *in vivo* maturation, i.e., is *in vitro* maturation a valid model system to study oocyte maturation? Resolution of this issue was reached by demonstrating that mouse oocytes matured and fertilized *in vitro* and then transferred to foster mothers, give rise to fertile offspring at the same frequency as oocytes matured *in vivo,* fertilized *in vitro,* and then transferred to foster mothers.[110] It should also be noted that oocytes isolated from either sheep or rabbits that are matured and fertilized *in vitro* and then transferred to foster mothers also give rise to normal preimplantation development to a high extent.

Many types of molecules have been proposed to participate in maintenance of meiotic arrest. Obviously, it is possible for a compound to inhibit oocyte maturation, e.g., GVBD, by a mechanism that has no physiological relevance. An extreme example is that glutaldehyde will inhibit GVBD. No one, however, would propose that this inhibition is physiologically relevant. Thus, several criteria should be met before considering that a molecule participates in maintenance of meiotic arrest. First, the molecule must be present in the follicle and, moreover, in the appropriate compartment within the follicle. Second, the molecule should inhibit maturation at concentrations that are present in the follicle, and this inhibition should be reversible. Last, this inhibition should be reversed by gonadotropin. The nature of the substance may suggest a likely mechanism of action, e.g., activate a protein kinase or bind to a specific class of receptors, and this should be documented as best as is possible. Although this frequently necessitates the use of inhibitors, their judicious use can provide significant insights regarding how a molecule may promote meiotic arrest. Last, it should be noted that most studies of oocyte maturation use GVBD as the indicator for resumption of meiosis, since GVBD is the first easily observed manifestation of resumption of meiosis. GVBD is an early event in oocyte maturation and little is known about controlling mechanisms that govern oocyte maturation subsequent to GVBD and prior to arrest at metaphase II, as well as the molecular basis for arrest at metaphase II.

The first implication of a molecule of known function in maintaining meiotic arrest was the observation that membrane-permeable analogs of cAMP, such as dibutyryl cAMP (dbcAMP) or 8-Br cAMP, and inhibitors of the cAMP phosphodiesterase (PDE), such as theophylline or 3-isobutyl-1-methyl xanthine (IBMX), reversibly inhibit GVBD.[18,22,63,103,119] Thus oocytes initially incubated for several hours in medium containing dbcAMP or a PDE inhibitor are inhibited from undergoing GVBD, but transfer of these oocytes to inhibitor-free medium results in GVBD, emission of the first polar body, and arrest at metaphase II. Moreover, this inhibition is specific, since corresponding concentrations of cGMP analogs do not inhibit GVBD (see below for role of cGMP in oocyte maturation). In these studies and in most of those described below, the surrounding cumulus cells were removed; these oocytes are commonly referred to as either naked or denuded oocytes. The rationale for removing these somatic cells is to simplify the experimental system. It should be noted, however, that this simplification removes an inherent complexity that is likely to play a pivotal role in regulating oocyte maturation (see below). Thus, results of experiments using naked oocytes should be interpreted cautiously.

cAMP, which is present in the oocyte, is a very attractive candidate for participating in maintenance of meiotic arrest, since the mode of cAMP action is known, i.e., cAMP activates a cAMP-dependent protein kinase (PK-A). Based on the observation that membrane-permeable cAMP analogs inhibit oocyte maturation *in vitro,* the following model was put forward to account for maintenance of meiotic arrest and resumption of meiosis (Figure 1).[11] An inhibitory basal level of cAMP is present in the oocyte and this cAMP activates PK-A, which is a heterotetramer composed of two regulatory subunits, which are connected by an intermolecular disulfide bond, and two catalytic subunits; the heterotetramer is inactive. Binding of cAMP to the regulatory subunits promotes the dissociation of the catalytic subunits, which are now active. The activated PK-A phosphorylates a protein X, which is necessary for GVBD, to the form X-P, which is inactive, and this results in maintenance of meiotic arrest. (It should be noted that X-P could actively promote meiotic arrest.) Resumption of meiosis is then triggered by a decrease in the inhibitory level of oocyte cAMP (Figure 2); this decrease may be promoted by the action of a cAMP PDE. This decrease in oocyte cAMP level results in the reassociation of the active catalytic subunits of PK-A with the regulatory subunits, and hence a decrease in PK-A activity. The activity of a protein phosphatase then leads to a net conversion of X-P to the active form X, which is somehow involved in resumption of meiosis.

Several lines of experimental evidence described below are consistent with a role for protein phosphorylation catalyzed by PK-A being necessary for maintenance of meiotic arrest.

1. Mouse oocyte extracts possess a PK-A activity, and photoaffinity labeling of oocyte extracts with [^{32}P] azido cAMP indicates that R_I is the major form of the regulatory subunit.[112]
2. Agents that elevate or maintain oocyte cAMP levels inhibit oocyte maturation *in vitro*. As previously stated, membrane-permeable analogs and PDE inhibitors reversibly inhibit oocyte maturation *in vitro*. Forskolin, which is an activator of adenylate cyclase, can increase oocyte cAMP levels and inhibit GVBD in several species.[24,40,88,98,99,109,115,128] Incubation of rat oocytes with the invasive adenylate cyclase

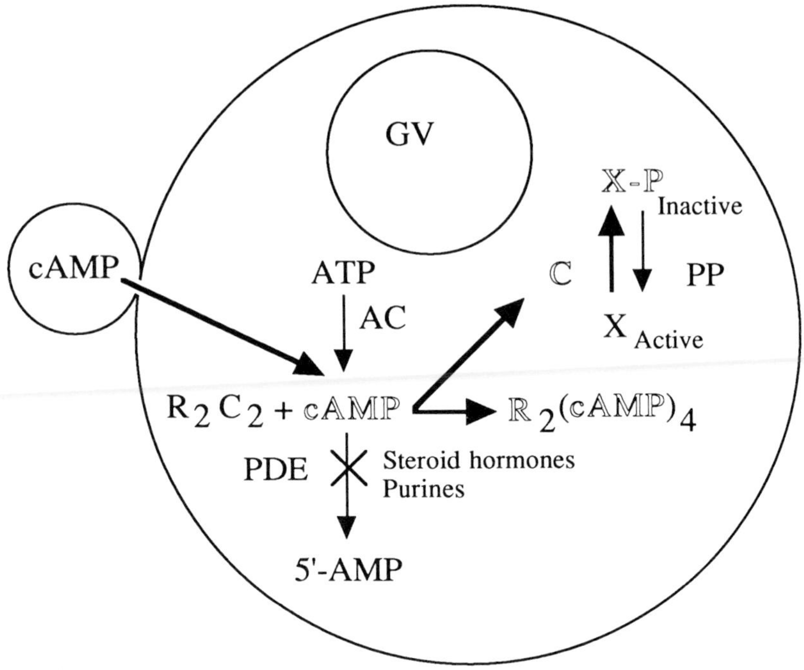

FIGURE 1. Model for maintenance of meiotic arrest by cAMP. Oocyte cAMP is generated by either an oocyte adenylate cyclase or is derived from follicle cells following transfer through the heterologous gap junctions. The cAMP activates a PK-A and the active catalytic subunit C, phosphorylates X, which is active and promotes GVBD, to X-P, which is the inactive form. Steroid hormones and purines may also participate in maintaining inhibitory levels of cAMP by inhibiting the oocyte PDE. R_2C_2, holoenzyme form of PK-A, which is inactive; $R_2(cAMP)_4$, dissociated regulatory subunits of PK-A; C, active catalytic subunit of PK-A; PDE, cAMP phosphodiesterase; PP, protein phosphatase responsible for converting X-P to X. Species depicted in the outline letters represent the dominant forms present.

of *Bordetalla pertussis* results in both an increase in oocyte cAMP levels and inhibition of GVBD.[1]

It should be noted that species differences occur in the ability of these agents to inhibit GVBD. For example, about 0.3 m*M* dbcAMP is required to inhibit mouse oocyte maturation by 80% over a 29-h period, whereas 1 m*M* dbcAMP is required to inhibit porcine oocyte maturation to a similar extent over the same time period.[35,103] At least 10 m*M* dbcAMP is required to inhibit bovine oocytes from undergoing GVBD over a 6-h culture period.[63,119] This high concentration of dbcAMP required to inhibit bovine oocyte maturation provides the likely explanation that earlier studies, which employed significantly lower dbcAMP concentrations, failed to observe dbcAMP inhibition of GVBD.[68] These species differences may reflect differences in (1) membrane permeability of the cAMP analog, (2) ability of the cAMP analog to activate PK-A, or (3) the level of PDE activity present in the oocytes; oocytes with higher levels of PDE activity would require higher external concentrations of the cAMP analog to maintain meiotic arrest.

Species differences are also observed in the concentration dependence of forskolin to inhibit GVBD.[24,40,88,98,99,109,115,128] These differences may reflect either differences in the activitability of adenylate cyclase, or differences in PDE levels. Last, it should be noted that inhibition of GVBD by forskolin is usually transient and may be due to depletion of oocyte ATP levels; the culture medium does not contain direct precursors for ATP synthesis, and moreover, denuded oocytes apparently do not readily synthesize ATP from adenosine.[61]

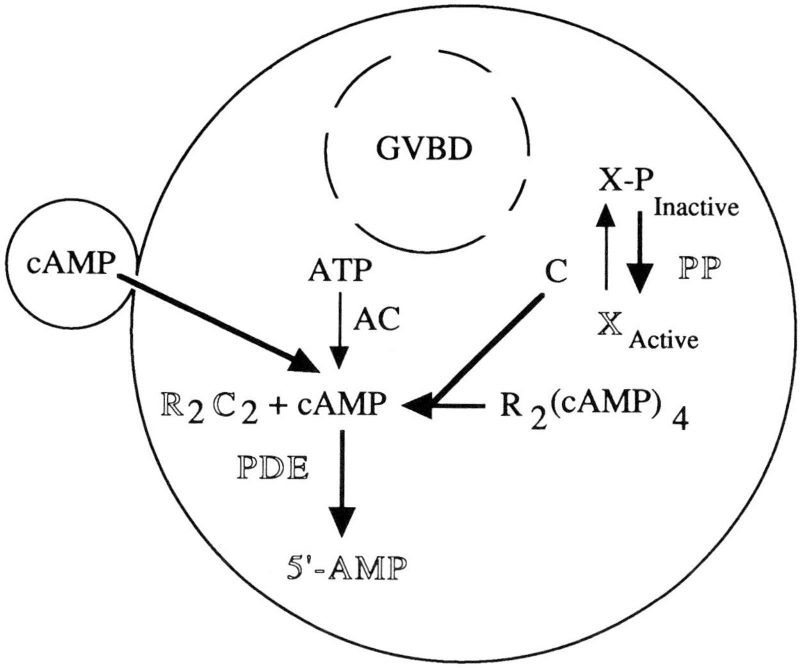

FIGURE 2. Model for resumption of meiosis induced by a decrease in oocyte cAMP. A decrease in oocyte cAMP, which is due in part to the PDE activity, promotes reassociation of the regulatory subunits of PK-A with the catalytic subunits. This results in a decrease in PK-A activity. The protein phosphatase then promotes the conversion of X-P to X, which is somehow involved in GVBD. The symbols are described in the legend to Figure 1.

3. This inhibition effected by cAMP appears mediated via a PK-A.[11] N^6-monobutyryl 2′-deoxy cAMP, which does not activate PK-A, does not inhibit GVBD, whereas N^6-monobutyryl cAMP does inhibit GVBD. Microinjection of mouse oocytes with the catalytic subunit of PK-A inhibits GVBD, whereas microinjection of the protein kinase inhibitor (PKI) of PK-A induces GVBD in oocytes incubated in medium containing concentrations of either dbcAMP or IBMX that inhibit GVBD. The explanation for the inhibitory effect of microinjected catalytic subunit is that microinjected catalytic subunit is present in excess to the amount of endogenous regulatory subunit. Thus, in the absence of activating levels of cAMP, free catalytic subunits are present that will constitutively phosphorylate X to keep it in the inactive state. The explanation for the observation that microinjection of PKI induces maturation under conditions that normally inhibit maturation is that these conditions result in either elevation or maintenance of inhibitory levels of oocyte cAMP and thus activate PK-A by promoting dissociation of the catalytic subunits of PK-A. It is these free, dissociated catalytic subunits that specifically combine with PKI to form an inactive complex. Thus, even though inhibitory levels of cAMP are present, PK-A activity is decreased, and X-P is converted to X.

Three lines of experimental evidence described below are consistent with a decrease in oocyte cAMP providing the trigger for resumption of meiosis.

1. A maturation-associated decrease in mouse oocyte cAMP levels occurs during a period of time in which the oocytes become committed to resume meiosis.[115,128] Commitment is experimentally defined as follows: after a given period of time in culture, oocytes

possessing an intact germinal vesicle are transferred to medium containing either a membrane-permeable cAMP analog or PDE inhibitor and then scored for GVBD at later times. If an oocyte resumes meiosis it is termed "committed". This maturation-associated decrease in oocyte cAMP appears causally related to resumption of meiosis, since PDE inhibitors, which inhibit GVBD, also inhibit the maturation-associated decrease in oocyte cAMP.[115]

It is possible, although unlikely, that fundamental differences exist between different mammalian species in the role for cAMP in regulating oocyte maturation. When differences are observed, they are more likely to be due to differences in experimental design. For example, an increase in oocyte cAMP prior to GVBD is observed in sheep oocytes isolated at various times after gonadotropin administration.[83] An explanation to reconcile this apparent contradictory observation is that the increase in cAMP associated with ovine oocytes is due to cumulus cell remnants that remain attached to the *zona pellucida*. Following gonadotropin administration, cAMP levels dramatically increase in the follicle, and this is reflected in an increase in cumulus cell cAMP. Scanning electron microscopic examination of oocytes obtained by the standard methods for removing cumulus cells reveals that the *zona pellucida* is embedded with remnants of cumulus cell processes.[9] Thus, the apparent increase in oocyte cAMP could be attributed to cAMP present in these cumulus cell remnants. This suggestion is buttressed by the observation that although the amount of radiolabeled actin synthesized and associated with cumulus cell-enclosed mouse oocytes is much greater than that associated with denuded oocytes, this difference is not observed following removal of the *zona pellucida* of the cumulus cell-enclosed oocytes.[17] Moreover, cytochemical labeling of actin with a fluorescently labeled phalloidin reveals *zona pellucida*-associated actin, which is due to the actin-rich cumulus cell processes that penetrate the *zona pellucida*.[17] Measurement of cAMP levels in *zona*-free ovine oocytes could potentially resolve this issue.

2. Microinjection of mouse oocytes, which are incubated in inhibitory concentrations of IBMX, with purified sheep brain PDE undergo GVBD relative to vehicle-injected oocytes or oocytes injected with heat-inactivated PDE.[11] The rationale for this observation is that although the microinjected PDE is inhibited by IBMX, the amount injected is not totally inhibited and the residual activity is sufficient to lower cAMP levels below that necessary to maintain meiotic arrest.
3. Forskolin-induced inhibition of GVBD of denuded mouse oocytes is reversed by addition of 1 m*M* lithium, which in other systems can inhibit adenylate cyclase.[57] In addition, the kinetics of GVBD of oocytes treated with forskolin and lithium are very similar to those of untreated oocytes. Last, lithium cannot reverse inhibition of GVBD by dbcAMP, and this suggests that the site of lithium action is prior to the generation of cAMP.

Although these results are consistent with lithium inhibiting forskolin-induced cAMP synthesis and not acting at a site distal to cAMP synthesis, it should be noted that the effect of lithium on forskolin-induced cAMP accumulation in the oocyte was not measured directly, nor was it demonstrated that lithium treatment permits the maturation-associated decrease in cAMP to occur in these forskolin-treated oocytes.

An oocyte PDE is likely to be responsible, at least in part, for the maturation-associated decrease in oocyte cAMP.[14] Mouse oocyte extracts possess a PDE activity and PDE inhibitors, such as IBMX, inhibit both the cAMP decrease and GVBD. Moreover, consistent with this proposition is the excellent correlation between the dose-dependent inhibition of three different PDE inhibitors (IBMX, theophylline, and Ro 1724/1) to inhibit PDE activity in oocyte extracts and GVBD. Although methyl xanthines such as IBMX and theophylline can perturb calcium fluxes or interact with adenosine receptors, it is unlikely that this is the basis for their ability to inhibit GVBD, since Ro 1724/1, which is structurally dissimilar to IBMX and theophylline and does not exert either the calcium flux perturbations or interact with adenosine receptors, also inhibits GVBD.

The PDE present in mouse oocyte extracts can be modulated by calmodulin, which comprises about 0.3% of the total oocyte protein.[14] The calmodulin antagonists W7 and W13, which are halogenated naphthalene sulfonamides, inhibit GVBD in a concentration-dependent manner, whereas similar concentrations of the less active congeners, W5 and W12, do not.[14] Although this result suggests that a calmodulin-modulated step is involved in GVBD, the target is unlikely to be a calmodulin-modulated PDE activity. The maturation-associated decrease in oocyte cAMP occurs in the presence of concentrations of W7 that inhibit GVBD.[14] Last, no increase in PDE activity is detected in oocyte extracts during the commitment period that precedes GVBD.[14] The sensitivity of the assay, however, would not detect small but significant changes in PDE activity.

The inhibitory effects of calmodulin antagonists suggest that a calcium-dependent step is necessary for GVBD. Consistent with this is the observation that treatment of denuded mouse oocytes, which are incubated in medium containing inhibitory concentrations of dbcAMP, with A23187 results in GVBD.[96] Mouse oocytes incubated in medium containing totally inhibitory concentrations of IBMX and injected with inositol 1,4,5-trisphosphate (IP_3), which is a physiological mediator of release of calcium from intracellular stores, does not result in GVBD.[73] Thus, it remains to be demonstrated if raising intracellular oocyte calcium levels by A23187 reflects an underlying intrinsic process in GVBD or activates secondary processes that can result in resumption of meiosis. Microinjection of *Xenopus* oocytes with IP_3 does not induce maturation in the absence of progesterone, although it does result in activation of *Xenopus* eggs.[92] The temporal and spatial changes in intracellular free calcium during mammalian maturation have not been documented and the role of calcium in maturation is still unresolved.

The model for resumption of meiosis predicts that changes in protein phosphorylation should occur during the commitment period. In fact, a characteristic set of changes in protein phosphorylation occurs during the commitment period in mouse oocytes.[12-14] A basic phosphoprotein (M_r = 60,000) undergoes an apparent dephosphorylation and a set of phosphoproteins (M_r = 24,000, 29,000, and 36,000) exhibits an apparent increase in phosphorylation. In addition, oocytes incubated in medium containing dbcAMP and microinjected with PKI manifest during the commitment period a similar set of changes in protein phosphorylation.[11] Thus, cAMP-independent protein kinases presumably catalyze the phosphorylation of the proteins whose apparent phosphorylation increases, since these increases are observed in oocytes microinjected with PKI. Changes in the pattern of phosphoprotein synthesis also occur during ovine oocyte maturation.[19,20]

Changes in the relative distribution of phosphoamino acids also occur during the commitment period in mouse oocytes.[12] The relative amount of phosphoserine significantly decreases, whereas that of phosphotyrosine doubles. The relative increase in the amount of phosphotyrosine is interesting, since tyrosine phosphorylation is correlated with cell proliferation in other systems, and resumption of meiosis is a specialized form of cell division. Treatment of polyacrylamide gels with alkalai, which preferentially hydrolyzes phosphoserine and phosphothreonine, indicates that tyrosine phosphorylation does not underlie the increase in the apparent phosphorylation of the three proteins whose apparent phosphorylation increases during the commitment period.[12]

The observation that the increase in apparent phosphorylation of the three phosphoproteins occurs in oocytes incubated in medium containing dbcAMP but microinjected with PKI suggests that cAMP-independent protein kinases catalyze the phosphorylation of these proteins. This in turn suggests that other types of protein kinases may be involved in regulation of oocyte maturation. Accordingly, the effects of protein kinase C (PK-C), which is the calcium, phospholipid-dependent protein kinase and is activated by diacylglycerol or phorbol diesters, on oocyte maturation have been examined. Biologically active phorbol diesters or a diacylglycerol, which is a more physiological PK-C activator, inhibit GVBD in mouse oocytes in a concentration-dependent manner; biologically inactive phorbol diesters do not inhibit GVBD.[13,127] Retinoic acid, which is known to reverse the effects of phorbol diesters in other systems, also overcomes the inhibitory effect of phorbol diesters on mouse oocyte maturation.[127] It should be noted, however, that although these results are consistent with the mode of action of these drugs to inhibit PK-C, to date, PK-C activity has not been measured directly in mammalian oocytes.

The mechanism by which these agents inhibit oocyte maturation differs from that of cAMP.[13] First, elevation of oocyte cAMP levels after GVBD does not inhibit polar body emission, whereas addition of PK-C activators after GVBD does inhibit polar body emission. Second, PK-C activators do not inhibit the maturation-associated decrease in oocyte cAMP. Consistent with PK-C activators acting at a step distal to that of PK-A is that GVBD does not occur in oocytes microinjected with PKI and incubated in medium that contains a PK-C activator. Although the PK-C activators do not inhibit the apparent dephosphorylation of the 60,000-Da phosphoprotein, their ability to inhibit GVBD correlates with inhibition of the apparent increases in phosphorylation that occur during the commitment period. It is also interesting to note that the calmodulin antagonist W7, which inhibits GBVD but not the maturation-associated decrease in oocyte cAMP nor the apparent dephosphorylation of the 60,000-Da phosphoprotein, inhibits the apparent increase in phosphorylation of the phosphoprotein of M_r = 36,000. The corresponding concentration of the less active congener W5, which does not inhibit GVBD, does not inhibit any of the changes in phosphoprotein synthesis that occur during the commitment period.

Results of these experiments suggest that the maturation-associated changes in protein phosphorylation that occur during the commitment period may be causally related to GVBD. This proposal gains further credence from studies with meiotically incompetent mouse oocytes.[12] During its growth phase, the oocytes acquire the ability to resume meiosis and the fraction of oocytes capable of undergoing GVBD increases with increasing oocyte diameter.[120] For example, only about 10% of mouse oocytes 60 μm in diameter will undergo GVBD, whereas >90% of oocytes larger than 75 μm in diameter can undergo GVBD. Using oocytes about 60 μm in diameter, the oocytes that do not undergo GVBD do not manifest the maturation-associated changes in protein phosphorylation, whereas the 10% that do undergo GVBD do display these changes.[12] Moreover, microinjection of meiotically incompetent oocytes with PKI does not induce GVBD, although microinjected PKI does induce the apparent dephosphorylation of the basic phosphoprotein (M_r = 60,000) but not the increase in apparent phosphorylation of the other set of maturation-associated phosphoproteins. Thus, the dephosphorylation of the M_r = 60,000 protein is not sufficient to induce GVBD and meiotic incompetence may be due, in part, to an inability of the oocyte to catalyze the phosphorylation of these other proteins. It still remains to be demonstrated, however, if these changes in protein phosphorylation are causally related to GVBD or are simply biochemical markers for this process.

Little attention has been paid to the role of protein phosphatases in mammalian oocyte maturation. Protein dephosphorylation is implicit in the model for resumption of meiosis, and apparent protein dephosphorylation occurs during the commitment period to resume meiosis (see above). Results from lower species implicate protein phosphatases in oocyte maturation. For example, injection of *Xenopus* oocytes with the protein inhibitor of protein phosphatase-1 inhibits progesterone-induced maturation.[48,67]

It should be emphasized that the role of cAMP and protein phosphorylation in regulating oocyte maturation in mammalian oocytes may not be generalized to all species. Although a decrease cAMP is thought to be involved in resumption of meiosis in oocytes of lower species, e.g., *Xenopus*, an increase in cAMP initiates maturation in brittle-star oocytes and oocytes of several hydrozoans.[49,80,81,134] In addition, in contrast to the inhibitory effect of

PK-C activators on mammalian oocyte maturation, PK-C activators induce maturation in *Xenopus* and *Spisula* oocytes.[38,123]

B. PURINES AND STEROID HORMONES

Although cAMP plays a role in maintenance of meiotic arrest, other molecules that are present in follicle are also likely to participate in meiotic arrest. To date, two other classes of molecules, purines and steroid hormones, have been shown to inhibit oocyte maturation at concentrations present in follicular fluid. In addition, both purines and steroid hormones are likely to inhibit maturation by modulating oocyte cAMP levels.

Hypoxanthine and adenosine are present in m*M* concentrations in follicular fluid of several species; serum contains much lower levels of these purines.[33,44] At concentrations present in follicular fluid—precautions to minimize the artifactual generation of purines during isolation of follicular fluid were employed and thus these concentrations are likely to approximate *in situ* levels present in follicular fluid—these purines transiently inhibit maturation of denuded mouse oocytes, although the inhibition by hypoxanthine is much more pronounced than that of adenosine.[29,30,32,33,42] Moreover, this purine-induced inhibition of GVBD is markedly enhanced when either dbcAMP or IBMX is present at concentrations that by themselves do not result in significant inhibition of GVBD.[30,35,124] These purines may inhibit oocyte PDE and thus potentiate dbcAMP- or IBMX-induced inhibition of GVBD, since a correlation exists between the ability of hypoxanthine or adenosine to inhibit GVBD and oocyte extract PDE activity; hypoxanthine is the more potent inhibitor.[34] (Guanosine, which is not detected in murine follicular fluid, is the most effective inhibitor of GVBD and PDE activity in oocyte extracts.) Thus, purines would potentiate dbcAMP-mediated inhibition of GVBD by raising the effective intracellular concentration of the cAMP analog or would augment IBMX-induced inhibition of GVBD by further inhibiting endogenous PDE activity. Moreover, the ability of microinjected PKI to induce GVBD in oocytes incubated in inhibitory concentrations of hypoxanthine indicates that purines mediate maintenance of meiotic arrest via a cAMP-dependent mechanism.[34]

Compared to mouse denuded oocytes, the inhibitory effect of purines on GVBD is increased in cumulus cell-enclosed oocytes.[32] A likely basis for this observation is that functional heterologous gap junctions are present between cumulus cells and the oocyte, and this communication pathway, which provides the major route of entry of most metabolites into the oocyte, facilitates the uptake and metabolism of both hypoxanthine and adenosine.[15,32,61,62] The inhibitory effect of hypoxanthine on GVBD in cumulus cell-enclosed oocytes, however, is likely to be more complex than that produced in denuded oocytes. Both bredinin and mycophenolic acid, which are inhibitors of IMP dehydrogenase and thereby inhibit metabolism of hypoxanthine to either xanthyl or guanyl compounds, inhibit the ability of hypoxanthine to maintain meiotic arrest of cumulus cell-enclosed oocytes.[32] These IMP dehydrogenase inhibitors, however, fail to overcome inhibition of meiotic arrest elicited by hypoxanthine or adenosine on denuded oocytes. Moreover, injection of these inhibitors into mice induces resumption of meiosis in antral follicles by a mechanism that does not entail release of endogenous gonadotropins or atresia.[31] The cumulus cells in follicles in which oocytes resumed meiosis do not manifest any signs of mucification and no pycnotic granulosa cells are observed. Further specificity for induction of maturation is that injection of mice with inhibitors of adenylosuccinate synthetase, which is involved in the conversion of hypoxanthine to adenyl derivatives, does not induce resumption of meiosis. Thus, the inhibitory effect of hypoxanthine on GVBD appears mediated by cumulus cell metabolism of hypoxanthine to guanyl derivatives. Although the guanyl derivatives may act directly on the oocyte by entering the oocyte via the communication pathway and inhibiting oocyte PDE—guanosine is the most potent inhibitor of PDE activity in oocyte extracts—it is also possible that guanyl derivatives play an additional role in maintenance of meiotic arrest. Guanyl derivatives such as GTP could activate guanine nucleotide binding proteins, which are central modulators of adenylate cyclase as well as other enzymes. This modulatory role of GTP could function at either the level of the cumulus cells or the oocyte, or both.

The situation with adenosine is also complex. Although adenosine is metabolized by cumulus cells, the inhibitory effect of adenosine on maturation of cumulus cell-enclosed oocytes does not appear to necessitate adenosine metabolism.[32,107] 2-Chloroadenosine is not metabolized and yet its ability to interact synergistically with hypoxanthine to maintain meiotic arrest of cumulus cell-enclosed oocytes is similar to that of adenosine. In addition, adenosine synergizes with forskolin to inhibit GVBD in denuded oocytes. Inhibition of adenosine uptake by dipyridamole does not prevent this synergistic interaction with forskolin to inhibit maturation in denuded oocytes, and moreover, this potentiation of forskolin-induced inhibition of GVBD is also observed with several nonmetabolizable adenosine analogs, e.g., N^6-phenylisopropyl adenosine, 2-chloro adenosine, or 5′-*N*-ethylcarboxamide. Thus, some of the inhibitory effects of adenosine on GVBD may be attributed to adenosine's interaction with adenosine receptors on cumulus cells and/or the oocyte.

Steroid hormones, such as progesterone, testosterone, and 17β-estradiol, are present in follicular fluid at concentrations in the μg/ml range. Such concentrations of these steroid hormones can inhibit GVBD in denuded mouse oocytes.[69] This inhibition is transient, however, and provides the likely explanation that many early studies, which examined GVBD after an overnight culture, failed to report an inhibitory effect. Similar to purines, these steroid hormones can also potentiate the inhibition of maturation of denuded oocytes by low concentrations of dbcAMP.[43,103,116]

The likely basis for this transient inhibition and synergistic interaction with dbcAMP to inhibit maturation is that steroid hormones inhibit oocyte PDE activity.[69] Concentrations of steroid hormones present in follicular fluid inhibit PDE activity in mouse oocyte extracts. This inhibition is not complete and hence accounts for the transient nature of inhibition of GVBD; a similar transient inhibition of GVBD is observed at concentrations of known PDE inhibitors, as well as purines that only partially inhibit oocyte extract PDE. The ability of steroid hormones to inhibit oocyte PDE thus accounts for their ability to potentiate inhibition of GVBD by low concentrations of dbcAMP. Consistent with the proposal that the steroid hormones, in fact, inhibit oocyte PDE *in situ* is that concentrations of steroid hormones that inhibit mouse oocyte extract PDE activity to the same extent as a known PDE inhibitor, Ro 20 1724/1, potentiate inhibition of GVBD by low concentrations of dbcAMP to the same extent as the PDE inhibitor.[69]

Taken *in toto,* results of these studies suggest that several different types of molecules that are present in the follicle and follicular fluid may play a role in maintenance of meiotic arrest and may do so in a hierarchial manner, since molecules that regulate either directly or indirectly cAMP levels in the oocyte will participate in maintenance of meiotic arrest. These molecules are likely to play a modulatory role in maintenance of meiotic arrest rather than a primary role, since cumulus cell-oocyte complexes grafted to follicle wall hemisections do not resume meiosis in the absence of follicular fluid and the concentrations of these molecules in follicular fluid do not decrease following administration of gonadotropin.[47,77] It is also interesting to note that when oocytes become competent to resume meiosis, the antrum starts to appear. Possibly steroid hormones and purines present in this nascently forming follicular fluid play a modulatory role to maintain meiotic arrest of the now meiotically competent oocyte.

C. OOCYTE MATURATION INHIBITOR AND MULLERIAN-INHIBITING SUBSTANCE

Much discussion has centered on a substance called oocyte maturation inhibitor or OMI.[111] Earlier studies indicated that porcine follicular fluid, but not the same concentration of serum, and granulosa cell-conditioned medium contain a substance(s) that can inhibit GVBD in cumulus cell-enclosed oocytes; little inhibition of GVBD is observed when denuded oocytes are used in the assay. Follicular fluid or granulosa cell-conditioned medium obtained from small antral follicles possesses higher levels of this inhibitory activity than follicular fluid or granulosa cell-conditioned medium derived from large antral follicles. Trypsin treatment of follicular fluid results in a loss of activity, but charcoal treatment of follicular fluid does not. The inhibition promoted by follicular fluid is reversible, and addition of luteinizing hormone overcomes follicular fluid-mediated inhibition of GVBD. Thus, the aforementioned criteria for an inhibitor of ooyte maturation appear satisfied.

OMI, which is proposed to be a low molecular polypeptide, has been partially purified using a bioassay that quantifies the effect of follicular fluid and purified fractions on their ability to inhibit GVBD.[95] The inhibitory activity separates into several components during purification, and thus it is not clear if multiple inhibitors that are chemically distinct are present or whether these differences reflect multiple forms of the same molecule. Conflicting results are obtained following protease treatment of these partially purified fractions—sometimes the inhibitory activity is lost and sometimes it is not.[95] There is also no explanation for these contradictory observations and the initial observation that trypsin treatment of follicular fluid results in a loss of the inhibitory activity.

The recent isolation of purines as the principal component from follicular fluid that is responsible for inhibiting GVBD complicates the issue of OMI as a low molecular polypeptide.[33] In addition, many aspects of inhibition of GVBD by concentrations of purines that are present in follicular fluid—cumulus cell-enclosed oocytes are more sensitive than denuded oocytes, inhibitory levels of purines are present in follicular fluid but not serum, inhibition of GVBD is transient and not species specific, purines augment the inhibitory effects of low dbcAMP or IBMX concentrations on GVBD—are similar to those reported for OMI. Thus, although it is possible that OMI is a low molecular polypeptide that participates in maintenance of meiotic arrest, OMI will have to be purified and characterized, and its mechanism of action characterized in order to establish its biological significance. An explanation for the failure to observe an inhibitory effect of follicular fluid on oocyte maturation in several studies is that end point assays were used and the kinetics of maturation were not monitored, i.e., GVBD was determined after a fixed interval of time.[46,77,100] Thus, any transient inhibition on the kinetics of GVBD would not have been observed.

Mullerian-inhibiting substance (MIS), which has recently been isolated and is related to TGF-β, is responsible for regression of the Mullerian duct. MIS is present in the ovary throughout fetal and adult life and is present in granulosa cells and follicular fluid.[95,125] MIS, which is obtained from conditioned medium prepared from transfected CHO cells with a human MIS gene, can inhibit GVBD of either denuded or cumulus cell-enclosed rat oocytes, and antibodies to MIS overcome MIS inhibition of GVBD. Low concentrations of the nonionic detergent NP-40 must be present to observe the inhibitory effect of highly purified MIS preparations (>90% pure), but the extent of inhibition of GVBD is low when compared to that elicited by dbcAMP, IBMX, or purines, which can totally inhibit GVBD. In contrast, results of another study that used immunopurified MIS could not demonstrate any inhibition of GVBD in cumulus cell-enclosed rat oocytes.[126] Extremely high concentrations of MIS (9 μg/ml) led to an increased level of oocyte degeneration. Thus, although MIS may play a regulatory role in maintenance of meiotic arrest, further experimentation is obviously required to strengthen this proposal.

D. PROTEIN SYNTHESIS

In rodents, GVBD occurs when protein synthesis is inhibited, but the oocytes arrest at the circular bivalent stage.[28,39,132,133] Addition of protein synthesis inhibitors shortly after GVBD occurs permits the oocytes to emit a polar body and arrest at metaphase II.[133] Interestingly, a number of changes in the pattern of protein synthesis occur subsequent to GVBD, and these may be necessary for continued nuclear progression.[113,133]

Short-lived proteins may also be involved in resumption of meiosis.[39] Rat or mouse oocytes incubated in medium containing a protein synthesis inhibitor (puromycin or cycloheximide) to inhibit protein synthesis, and either dbcAMP or IBMX to inhibit GVBD, do not undergo GVBD when transferred to medium containing the protein synthesis inhibitor but not dbcAMP or IBMX. In contrast, oocytes incubated for the same period of time in medium containing dbcAMP or IBMX do undergo GVBD following transfer to dbcAMP/IBMX-free medium. Thus, short-lived proteins that are degraded during the initial incubation in medium containing the protein synthesis inhibitor may be required for resumption of meiosis. Their half-life is long enough, however, so that addition of protein synthesis inhibitors at the outset of the culture period does not inhibit GVBD. There are no reports to date that characterize protein turnover in GV-intact oocytes or oocytes at different stages of maturation. Short-lived proteins may also be involved in maintenance of arrest at metaphase II, since incubation of metaphase II-arrested eggs in medium containing protein synthesis inhibitors results in egg activation.[118]

Protein synthesis is required, however, for GVBD in porcine and ovine oocytes.[52,79] A possible explanation for these species differences is that in rodents, GVBD starts to occur within 45 to 60 min following oocyte isolation, whereas in these larger animals, GVBD takes hours to occur following oocyte isolation. Thus, if a similar set of short-lived proteins is involved in GVBD in all of these species, addition of protein synthesis inhibitors at the start of the culture period will not inhibit GVBD in rodent oocytes but may in porcine or ovine oocytes.

E. MATURATION PROMOTING FACTOR AND CELLULAR ONCOGENES

Maturation promoting factor (MPF), which was first described in amphibian oocytes, is an activity present in both meiotic and mitotic cells and is capable of promoting a G2- to M-phase transition.[71,81] A more appropriate name for the MPF acronym is M-phase promoting factor.[58,60] The activity is centrally involved in cell cycle regulation—it apparently regulates chromosome condensation and nuclear envelope breakdown—and is not species specific.

Mammalian oocytes possess an MPF activity. Fusion of meiotically competent oocytes that have undergone GVBD with meiotically incompetent oocytes results in GVBD of the nucleus derived from the meiotically incompetent oocyte.[5,6,54] In addition, microinjection of cytoplasm obtained from mouse oocytes that have undergone GVBD into *Xenopus* or starfish oocytes induces GVBD in these recipient oocytes, which normally require a steroid hormone or 1-methyl adenine, respectively, to initiate resumption of meiosis.[72,121] Cytoplasm obtained from mouse oocytes that have not undergone GVBD is not capable of inducing GVBD in the recipient *Xenopus* or starfish oocytes.

In contrast with the fusion of a mouse oocyte at metaphase I with a single GV-intact oocyte, fusion of an oocyte at metaphase I with multiple GV-intact oocytes results in a delay in both chromosome condensation and GVBD.[51,54] Moreover, in the giant cells that are derived from these multiple fusions that result in a "pearl chain", nuclei closest to the region of fusion with the metaphase I oocyte undergo GVBD, whereas nuclei farthest from this region of fusion do not undergo GVBD. These differences may be due to differences in dilution and distribution of MPF following fusion of the donor oocyte with the recipient oocytes.

The generation of mouse oocyte MPF activity does not require the contents of the germinal vesicle.[6] This was shown by bissecting a meiotically competent GV-intact oocyte into a nucleate and anucleate fragment. After the nucleate fragment has undergone GVBD, the anucleate fragment is fused with a two-cell blastomere, and chromosome condensation is almost immediately observed in the interphase nucleus in the two-cell blastomere. It should be noted that GV material is required for other activities to develop during maturation (see Section IV).

The amount of mouse oocyte MPF can be quantified by removing oocyte cytoplasm

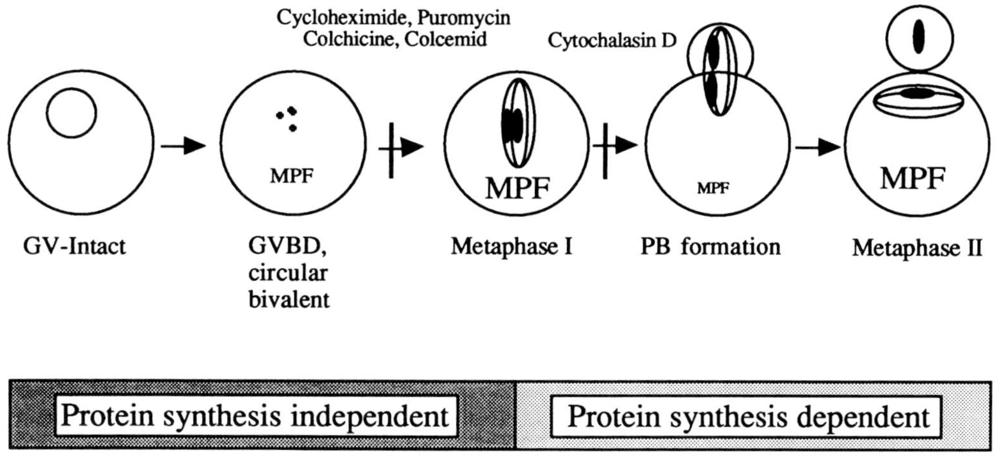

FIGURE 3. Schematic representation of changes in MPF activity during oocyte matuation. The different stages of oocyte maturation are shown and the levels of MPF activity are proportional to the size of the MPF typescript. Also indicated are the points of meiotic arrest in the presence of certain inhibitors, as well as the period of time when protein synthesis is required for MPF production.

and microinjecting it into recipient *Asterina pectinifera* oocytes, which are then scored for GVBD.[60] Using this bioassay, it was demonstrated that mouse oocyte MPF activity increases following GVBD and reaches a peak level at metaphase I. The amount of MPF activity then decreases during polar body emission, but increases by the time the oocyte arrests at metaphase II. Thus, MPF activity peaks with each metaphase, which is similar to that in lower species. These changes in MPF activity, as well as other aspects of regulation of MPF activity that are described below, are schematically depicted in Figure 3.

As in starfish oocytes, the initial appearance of mouse oocyte MPF does not require protein synthesis.[27,55,60,93] GVBD occurs when protein synthesis is inhibited and MPF levels initially increase, but then decline in the continued presence of a protein synthesis inhibitor; the first meiotic cell cycle arrests somewhere between GVBD and formation of the first metaphase spindle.[60] This effect is reversible, since restoration of protein synthesis subsequent to GVBD results in both the formation of the metaphase I spindle and the reappearance of MPF activity. These oocytes ultimately emit a polar body and arrest at metaphase II. Thus, protein synthesis is not required for the initial production of MPF, but is required for its continued production and changes in activity during maturation. Interestingly, addition of protein synthesis inhibitors at either metaphase I or II results in chromosome decondensation and nuclear envelope formation.[18a] This is consistent with protein synthesis being required for the continued production of MPF, which is involved in maintaining chromosomes in a condensed state.

Amplification of MPF occurs in *Xenopus* and starfish oocytes.[71,81] For example, injection of a small volume of cytoplasm from a matured *Xenopus* oocyte into a recipient GV-intact oocyte induces GVBD, and the level of MPF activity in the injected oocyte's cytoplasm increases to a value similar to that present in a progesterone-induced matured oocyte. Amplification of mammalian oocyte MPF apparently requires protein synthesis.[55] Following fusion of either mouse or pig oocytes that have undergone GVBD with several homologous GV-intact oocytes, the incidence of GVBD of all of the nuclei is about 90%, whereas only about 7% of all of the nuclei undergo GVBD when cycloheximide is present. Studies concerning the role of c-*mos* in oocyte maturation may provide additional insights regarding the protein synthesis requirement for MPF generation and activity (see below).

Nuclear progression beyond GVBD is not necessary for the appearance of MPF activity.[60] Treatment of oocytes with cytochalasin arrests nuclear progression at metaphase I, whereas treatment with colcemid inhibits formation of the metaphase I spindle. Neither of these agents inhibits the appearance of MPF activity in the mouse oocyte, and in each of these cases, chromosomes condense and MPF activity rises to levels found at metaphase I. The level of MPF, however, does not decrease as is the case when nuclear progression occurs beyond metaphase I. Thus, generation of high levels of MPF activity is independent of spindle formation but is highly correlated with the presence of condensed chromosomes. This may reflect an association of MPF with condensed chromosomes or components associated with the chromosomes. Consistent with this proposal are results of experiments using physically bissected GV-intact oocytes.[7] Nucleate fragments cultured for either 2 to 3 or 15 to 17 h, or anucleate fragments cultured for 2 to 3 h prior to fusion with interphase blastomeres from two-cell embryo induce premature chromosome condensation. In contrast, anucleate fragments cultured for 15 to 17 h prior to blastomere fusion results in a marked reduction in the incidence of premature chromosome condensation. In addition, inhibiting protein synthesis in cytochalasin-treated metaphase I-arrested oocytes results in a loss of MPF activity that correlates with chromosome movement to the spindle poles and chromosome decondensation, with the appearance of two nucleus-like structures.[60] Thus, as is the case in other cell types, the loss of MPF activity may signal a metaphase to anaphase transition.

It is also interesting to note that meiotically incompetent oocytes possess an anti-MPF activity.[51,53] Whereas fusion of a meiotically competent oocyte that has undergone GVBD with a meiotically incompetent oocyte results in the GVBD of the GV present in the meiotically incompetent oocyte, fusion of meiotically competent GV-intact oocyte with meiotically incompetent oocytes results in inhibition of GVBD of the GV derived from the meiotically competent oocyte. Thus, decreases in the level of this anti-MPF activity may be involved in acquisition of meiotic competence, and it will be of interest to ascertain if this anti-MPF activity inhibits the maturation-associated changes in protein phosphorylation (see above). In addition, it will be of interest to determine if this anti-MPF activity present in meiotically incompetent oocytes is similar to a factor present in somatic cells and *Xenopus* and starfish oocytes that inhibits MPF activity.[2,58]

Recently MPF has been purified from *Xenopus* eggs, and one of the two subunits (M_r = 34,000) was shown to be the cell homologue of the fission yeast cell cycle control protein encoded by the $cdc2^+$ gene.[37,56,78] Genetic studies indicate that the $cdc2^+$ gene product is involved with entry into mitosis, and biochemical studies indicate that its activity is modulated by protein phosphorylation. Moreover, the $cdc2^+$ protein is a major substrate for tyrosine phosphorylation. Although the relative amount of phosphotyrosine increases during mouse oocyte maturation, alkalai treatment of gels reveals no increase in a phosphotyrosine-containing protein whose molecular weight would correspond to that of the 34,000-Da subunit.[12]

Proteins encoded by specific cellular oncogenes, which are likely to play central roles in cell cycle regulation, are also likely to be involved in oocyte maturation. A role for c-*mos* proto-oncogene was recently described in *Xenopus* maturation.[97,105] c-*Mos* is a phosphoprotein of M_r = 39,000 and sequence analysis suggests that it may be a serine protein kinase. c-*Mos* RNA is present in the oocyte, but is not translated until maturation is induced by progesterone; the mechanism for the specific mobilization of *mos* mRNA is not known. Microinjection of *mos* antisense oligonucleotides, but not sense oligonucleotides, inhibits both the synthesis of the *mos* protein and GVBD that is promoted by either progesterone or insulin. Moreover, the appearance of the c-*mos* translation product appears prior to the appearance of MPF, and during the period of time in which protein synthesis is required for GVBD. All of these results are consistent with c-*mos* being translated from a maternal mRNA and modulating MPF activity, possibly by a phosphorylation mechanism.

In situ hybridization experiments demonstrate that mouse oocytes possess c-*mos* transcripts, and metabolic labeling studies indicate that low levels of c-*mos* synthesis can be detected in the GV-intact oocyte.[70,86,87,89] The level of c-*mos* synthesis increases following GVBD, remains elevated in the ovulated egg, but cannot be detected in one-cell embryo.[89] Consistent with these changes in c-*mos* synthesis is that *in situ* hybridization studies indicate that the level of c-*mos* transcripts does not change significantly during oocyte maturation, but that c-*mos* transcripts cannot be detected in the one-cell embryo.[70,86,87]

Microinjection of mouse oocytes with c-*mos* antisense oligonucleotides, but not sense oligonucleotides, does not inhibit GVBD, but does inhibit polar body emission.[89] Cytological analysis of the oocytes reveals that oocytes are arrested at the circular bivalent stage and that mitochondria are clustered around the metaphase I chromosomes. The reason that GVBD may not be inhibited in mouse oocytes by microinjected *mos* antisense oligonucleotides, as is the case with *Xenopus* oocytes, may be that low levels of c-*mos* synthesis are detected in the GV-intact mouse oocyte and that the *mos* protein has a half-life that is longer than the time it takes to undergo GVBD. Initially incubating the oocytes in an inhibitor of protein synthesis, which would allow degradation of the existing c-*mos*, followed by microinjection of *mos* antisense oligonucleotides, may afford conditions that ablate c-*mos* function and result in inhibiting GVBD.

Results of these experiments have identified additional proteins that are involved in regulating meiotic maturation. These proteins are also highly conserved between species and this should pave the way toward elucidating molecular aspects of mammalian oocyte maturation and constructing a sequence of events that leads to GVBD.

III. MECHANISM OF GONADOTROPIN-INDUCED RESUMPTION OF MEIOSIS

A. TERMINATION OF INTERCELLULAR COMMUNICATION

It has been known for many years that oocytes present within an explanted preovulatory antral follicle do not resume meiosis, but do so in response to added gonadotropin or removal of the oocyte from the follicle. A network, of gap junctions, which mediate the direct passage of small molecules between cells, exists between the different cell types present in the follicle. Thus, each of the three cell types present in the follicle—granulosa cells, cumulus cells, and the oocyte—are in principle capable of communicating with each other via this syncytium.[111] This syncytium of gap junctions appears necessary to maintain meiotic arrest in the follicle, since grafting cumulus cell-enclosed oocytes to follicle wall hemisections or granulosa cell cultures results in reestablishment of intercellular communication and maintenance of meiotic arrest.[47,77] Denuded oocytes, which are not capable of establishing intercellular communication with these somatic cells, resume meiosis when added to follicle wall hemisections. Moreover, addition of luteinizing hormone (LH) to follicle wall hemisections to which cumulus cell-enclosed oocytes are grafted results in resumption of meiosis.

An attractive model that accounts for these observations emerged over 10 years ago and has dominated the focus of research about the mechanism of gonadotropin-induced resumption.[23,59] Briefly, the model proposes that cAMP is involved in maintenance of meiotic arrest. The oocyte, however, lacks an adenylate cyclase activity but does possess a PDE activity. The cAMP that is necessary to maintain meiotic arrest is derived from the surrounding follicle cells, which do possess an active adenylate cyclase, and enters the oocyte via the syncytium of gap junctions. This transfer of cAMP from follicle cells to the oocyte inhibits resumption of meiosis of follicle-enclosed oocytes. Gonadotropins induce resumption of meiosis by terminating gap junction-mediated intercellular communication between the cumulus cells and the oocyte. Deprived of its exogenous source of cAMP, the oocyte is no longer able to maintain an inhibitory level of cAMP, since the oocyte cannot generate its own cAMP and the oocyte's PDE activity induces a cAMP decrease that results in resumption of meiosis. This model provides a simple explanation for why oocytes removed from their follicle resume meiosis, and also provides an attractive explanation for the paradoxical finding that while cAMP inhibits oocyte maturation in culture, agents that elevate cAMP levels in the follicle, such as gonadotropins, induce resumption of meiosis of follicle-enclosed oocytes.

A number of experimental observations are consistent with this model.There is a substantial body of evidence that implicates cAMP in promoting maintenance of meiotic arrest and that a decrease in oocyte cAMP is involved in resumption of meiosis (see above). The observation that treatment of denuded rat or mouse oocytes with cholera toxin (CT), which activates adenylate cyclase by catalyzing an ADP-ribosylation of the α_s subunit of the guanine nucleotide binding protein G_s, does not increase oocyte cAMP levels nor inhibit GVBD suggests that oocytes may not possess an adenylate cyclase.[23] Oocytes, however, do possess an adenylate cyclase activity, since forskolin treatment results in an increase in cAMP levels in several species, although the increase is quite low when compared to either cumulus or granulosa cells.[24,40,88,98,99,109,115,128] In addition, inclusion of a PDE inhibitor is frequently necessary to observe this effect. The increase in cAMP levels in response to forskolin is not due to the presence of cumulus cell remnants that remain embedded in the *zona pellucida* following denudation and that do contain an active adenylate cyclase, since forskolin treatment of *zona*-free oocytes, which are obtained by either enzymatic or chemical treatment and which removes the cumulus cell remnants, results in an elevation of oocyte cAMP to nearly the same level as in *zona*-intact denuded oocytes.[9]

The lack of an effect of cholera toxin on increasing mouse and rat oocyte cAMP levels and inhibiting GVBD may result from an inability of cholera toxin to be translocated into the oocyte and thus be capable of ADP-ribosylating the α_s subunit, or that the activity of the catalytic subunit of the cyclase is only loosely coupled with the G_s subunit, which in other systems stimulates adenylate cyclase activity. In contrast, cholera toxin treatment results in both an increase in sheep oocyte cAMP levels and a transient inhibition of GVBD.[19] This effect, however, may attribute to cumulus cell remnants associated with the *zona pellucida*; examining the effect of CT treatment on cAMP levels of *zona*-free sheep oocytes could resolve this apparent species difference.

Although oocytes do possess an adenylate cyclase and hence are not obligatorily dependent on an exogenous cAMP source, the level of this activity may be too low to generate inhibitory levels of cAMP to maintain meiotic arrest. Thus, the presence of an oocyte adenylate cyclase does pose a serious problem for the termination of intercellular communication model.

There is indirect evidence that cAMP made in follicle cells can be transmitted to the oocyte. Oocytes obtained from cumulus cell-oocyte complexes treated with either follicle-stimulating hormone (FSH) or cholera toxin contain more cAMP than do their denuded counterparts that are similarly treated—cumulus cell adenylate cyclase is activated by either FSH or CT—and correlated with this increase in oocyte cAMP is a transient inhibition in GVBD in several species.[10,98,99,108] Thus, cAMP made in cumulus cells may enter the oocyte via the communication pathway and result in inhibiting GVBD. It should be noted, however, that the transfer of cAMP from the follicle cells to the oocyte has only been observed under these nonphysiological conditions and has not been observed under basal or resting conditions.

Resumption of meiosis *in vitro* may be due in part to termination of follicle cell cAMP to the oocyte.[10] Although the oocyte possesses an adenylate cyclase and can therefore generate endogenous cAMP, the high level of oocyte PDE activity may not allow accumulation of cAMP to levels that inhibit GVBD. The follicle cell contribution of cAMP to the oocyte may generate sufficient levels of oocyte cAMP necessary to maintain meiotic arrest. Thus, removal of the oocyte from its follicle disrupts basal movement of cAMP from the follicle cells to the oocyte, and the oocyte's PDE would now result in a decrease in oocyte cAMP and resumption of meiosis.

Central to the communication uncoupling model for gonadotropin-induced resumption of meiosis is that a reduction in intercellular communication between the cumulus cells and the oocyte occurs prior to GVBD. Intercellular communication between cumulus cells and the oocyte is frequently measured by what is termed a coupling assay that is performed as follows: oocyte-cumulus cell complexes, which are obtained from ovaries after discrete periods of time following administration of gonadotropin, are incubated in medium containing a radiolabeled compound such as [^{3}H]uridine. After the incubation, radioactivity present in the denuded oocytes obtained from these complexes and radioactivity present in the intact cumulus cell-oocyte complex are determined and expressed as a ratio of the number of cpm present in the denuded oocyte to that in the intact complex; expressing the data as this ratio corrects for changes in uptake of the radiolabeled precursor by the cumulus cells, which could result in changes in the amount of radioactivity present in the oocyte without reflecting changing in the extent of metabolic cooperativity. It should be noted that under these conditions, denuded oocytes poorly take up the radiolabeled precursor and that the major route of entry of the precursor into a cumulus cell-enclosed oocyte is via the communication pathway, i.e., the compound is first taken up by the cumulus cells and then enters the oocyte via the heterologous gap junctions that are present between the cumulus cells and the oocyte. Thus, a decrease in the coupling ratio is indicative of a decrease in the extent of intercellular communication between the cumulus cells and the oocyte.

Results of experiments from several studies using oocyte-cumulus cell complexes obtained from mice, hamsters, or sheep reveal contradictory results. Some investigators observe that a reduction in the extent of metabolic coupling occurs prior to GVBD and this is consistent with the termination of communication model.[101,102] In contrast, however, other investigators observe that GVBD occurs prior to any significant reduction in the extent of intercellular communication.[41,50,84,85,98,106] In addition, mouse oocytes become committed to resume meiosis prior to any observable reduction in intercellular communication between cumulus cells and the oocyte, and moreover, at a time when the cAMP content of the cumulus cell-oocyte complex is increasing.[42a,115] These observations cast serious reservations regarding the validity of the termination of intercellular communication model as previously presented. Although species differences may exist and thus account for these differences, it seems unlikely that such a major difference would exist between these species for such an important and fundamental process.

Recent observations concerning gap junction dynamics during ovulation attempt to provide a possible explanation as to how GVBD could occur prior to uncoupling of cumulus cells and the oocyte and still retain the general spirit of the termination of communication model.[74,75] Gonadotropin-induced resumption of meiosis results in a rapid down-regulation of the number of homologous gap junctions between cumulus cells that correlates with GVBD in rat oocytes, and this decrease precedes the decrease in the number of heterologous gap junctions present between cumulus cells and the oocyte (Figure 4). This reduction in the extent of communication between the cumulus cells and the granulosa cells would effectively isolate the oocyte from the follicle cell mass, which grafting experiments show is essential for maintenance of meiotic arrest (see above). The result would be equivalent to liberating the oocyte from its follicle and a reduced flux of cAMP to the oocyte from the follicle cell mass would then result in resumption of meiosis (see above).

This modified version of the termination of intercellular communication model is very attractive. It incorporates many of the key features of the termination of communication model and can resolve apparent discrepancies, e.g., how resumption of meiosis could occur when no reduction in the extent of intercellular communication is observed between cumulus cells and the oocyte. Neither it nor its predecessor, however, addresses the crucial issue of demonstrating that the reduction in the number of gap junctions between the cumulus cells occurs prior to or concomitant with the time of oocyte commitment. Although results of a recent study indicate that a decrease in the number of gap junctions correlates with the percentage of hamster oocytes committed to resume meiosis, establishing a causal correlation is confounded by the fact that it is impossible to measure both the gap junctional status of

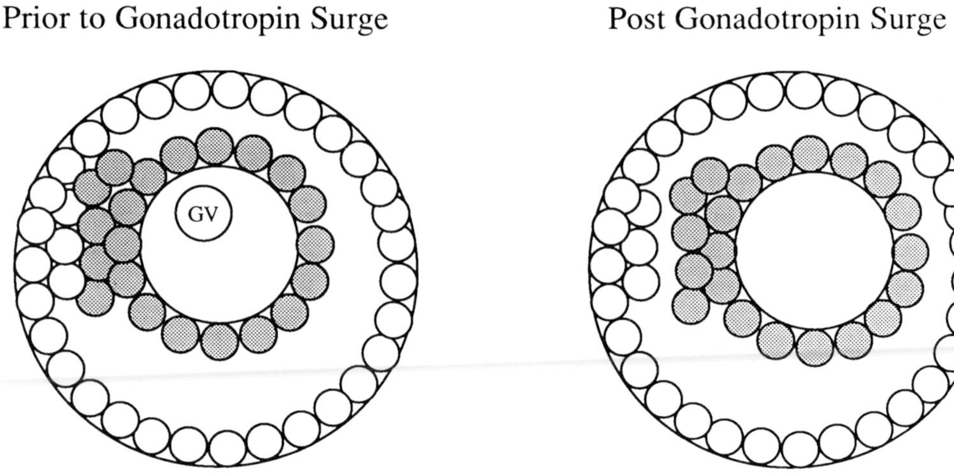

FIGURE 4. Schematic representation of gonadotropin-induced reduction in gap junctions between cumulus and granulosa cells. Prior to the gonadotropin surge, a synctium of gap junctions is present between the three different cell types present in the follicle. Following the gonadotropin surge, a reduction in the number of gap junctions between the granulosa cells and cumulus cells occurs. Although this is depicted as a physical separation of the two cell types in the figure, they are still physically associated. The significance of this reduction in intercellular communication, however, is that the cumulus cell-oocyte complex is functionally uncoupled from the granulosa cell mass, which appears necessary to maintain meiotic arrest. A reduction in intercellular communication between the oocyte and cumulus cells occurs subsequently. Open circles, granulosa cells; shaded circles, cumulus cells; large circle; oocyte/egg.

cumulus cells surrounding a given oocyte and if that oocyte is committed to resume meiosis, since measurement of one parameter excludes measurement of the other parameter.[99a] For example, the measurement of gap junctions must be performed on a GV-intact oocyte and this measurement entails fixation of the sample. Measurement of commitment, however, requires observing if the oocyte in that complex will undergo GVBD at a later time, and this is prevented by fixing the sample for analysis of the number of gap junctions. Although it is possible to measure these parameters on populations of complexes, the data obtained are by necessity correlative and do not provide pertinent insight regarding the commitment status of an oocyte in a given complex with the status of the gap junctions in that complex. Thus, the role of a gonadotropin-induced reduction in the extent of intercellular communication in resumption of meiosis is still unresolved. If this reduction in communication does not play a primary role in initiating resumption of meiosis, it may play a secondary role in insuring that the oocyte that is committed to resuming meiosis will successfully complete meiotic maturation.

B. GENERATION OF POSITIVE STIMULUS OF FOLLICLE CELL ORIGIN

The assumption in the termination of intercellular communication model for gonadotropin-induced resumption of meiosis is that the trigger for initiating maturation is the reduction in concentration of the same molecule(s) that is responsible for maintenance of meiotic arrest, e.g., cAMP. It is also possible, however, that a positive stimulus of follicle cell origin is generated in response to gonadotropin and that this signal either promotes a reduction in oocyte cAMP or bypasses cAMP-mediated maintenance of meiotic arrest. For example, this positive signal could activate an oocyte PDE and thereby effect a decrease in oocyte cAMP concentration or the signal could either directly or indirectly bypass the inhibitory effect promoted by PK-A. The ability of microinjected PKI to overcome inhibition of maturation induced by dbcAMP or IBMX provides a conceptual framework for this possibility.

Interestingly, a positive stimulus generated by somatic cells that acts on the oocyte to initiate meiotic maturation is frequently observed in lower species. Neither starfish nor *Xenopus* oocytes resume meiosis when placed in culture, but do so when given either 1-methyl adenine or progesterone, respectively. In *Xenopus*, the gonadotropin stimulates the synthesis and release of progesterone from the follicle cells. Progesterone then interacts with the oocyte and inhibits adenylate cyclase activity. The consequent reduction in oocyte cAMP levels is then thought to trigger resumption of meiosis.[80,81]

Although results from several earlier studies were interpreted as being consistent with the reduction in communication model, they may also be interpreted as being consistent with a gonadotropin-induced generation of a positive stimulus that initiates resumption of meiosis. Two early cases will be discussed. In the first instance, it was observed that IBMX markedly inhibits GVBD in oocytes present in cumulus cell-oocyte complexes, whereas incubation of these complexes in medium containing both IBMX and LH results in a marked increase in the incidence of GVBD.[22] LH is also able to stimulate the incidence of GVBD in oocytes present in cumulus cell-oocyte complexes incubated in medium containing inhibitory concentrations of dbcAMP. This increase in GVBD occurs at a time prior to cumulus cell mucification, which results in the physical disruption of intercellular communication, and therefore cannot be attributed to a reduced extent of communication. Thus, the data are compatible with an LH-induced generation of a positive stimulus of follicle cell origin that is involved in resumption of meiosis.

In the second instance, a cAMP-dependent acceleration of GVBD of cumulus cell-enclosed oocytes was observed relative to denuded oocytes.[50] In a complex protocol, cumulus cell-oocyte complexes are incubated in IBMX to prevent GVBD but treated with agents such as dbcAMP, FSH, or LH, each of which increases cumulus cell cAMP levels. Following removal of the attached cumulus cells and transfer to IBMX-free medium, it is observed that the rate of GVBD of these treated oocytes is significantly faster than a corresponding group of denuded oocytes treated in a similar fashion. Within the detection limits of the metabolic coupling assay for intercellular communication, these treatments do not result in a reduction in the extent of intercellular communication. Again, these data are consistent with a positive stimulus that is generated in the cumulus cells in response to cAMP and that is involved in resumption of meiosis.

More recent studies are also consistent with the generation of a positive stimulus in the follicle cells.[42] Incubation of complexes with FSH and a concentration of dbcAMP that is not totally inhibitory results first in a decrease in the extent of GVBD, but then an increase in the extent of GVBD when compared to the extent of GVBD in complexes incubated in just the dbcAMP. Moreover, this increase in the incidence of GVBD occurs during a time in which there is no detectable decrease in the extent of intercellular communication as measured by the coupling assay. The initial inhibitory effect on GVBD is attributed to the movement of cumulus cell cAMP to the oocyte. This additional cAMP would augment the inhibitory effect of the dbcAMP present in the medium. The increased incidence of GVBD at later times, however, is attributed to a positive stimulus that somehow overcomes the inhibitory effect of dbcAMP.

Subsequent studies provide additional evidence for such a positive stimulus. Cumulus cell-enclosed oocytes or denuded oocytes are incubated in medium containing either a concentration of dbcAMP, IBMX, or hypoxanthine that permits a modest extent of GVBD.[35] Treatment of these oocytes with FSH results in an increased incidence of GVBD in the cumulus cell-enclosed oocytes compared to the denuded oocytes. In fact, this difference can be observed by simply pulsing the complexes or denuded oocytes that are incubated in medium containing IBMX with dbcAMP and then reducing the concentration of dbcAMP in the subsequent incubation period. Since the only difference between these two goups is the presence of cumulus cells, the enhanced extent of GVBD in oocytes present in complexes may be attributed to a signal generated in response to an elevation in cumulus cell cAMP;

this signal somehow overcomes the inhibitory effect of either IBMX, dbcAMP, or hypoxanthine on maturation.

Taken together, the results described above suggest that gonadotropin-induced resumption of meiosis may entail the production of a positive stimulus of follicle cell origin that is transmitted to the oocyte via the communication pathway. Thus, this model is diametrically opposed in conceptual framework to the uncoupling model, since the new model requires the presence of a functional communication pathway between the oocyte and the follicle cells to promote resumption of meiosis. It will be of great interest to ascertain the nature and mode of action of the putative positive stimulus, although this will likely be a most difficult task. In addition, the formulation of the positive stimulus model makes it all the more imperative to resolve the temporal aspects of when the reduction in intercellular communication occurs following gonadotropin administration.

C. cGMP

LH induces an increase in follicle cGMP levels, but no clear role of cGMP in oocyte maturation has emerged. Several studies report interesting effects of perturbing cGMP levels on oocyte maturation.[64-66] For example, high concentrations of Br-cGMP (6 m*M*) inhibit maturation of cumulus cell-enclosed hamster oocytes, but not denuded oocytes, and LH can overcome this inhibition. Similarly, dbcGMP transiently inhibits maturation of cumulus cell-enclosed mouse oocytes. In contrast, FSH-induced inhibition of GVBD is partially overcome by elevation of cumulus cell cGMP levels, and this correlated with lower levels of cAMP present in the FSH-treated complexes. dbcGMP can also partially overcome hypoxanthine-mediated inhibition of GVBD in mouse cumulus cell-oocyte complexes. Addition of dbcGMP to mouse cumulus cell-oocyte complexes cultured in the presence of inhibitory concentrations of guanosine or dbcAMP, however, has either no effect on GVBD or augments the extent of inhibition of GVBD, respectively. There is currently no satisfactory explanation to account for these differences.

Perturbing follicle cGMP levels also produces interesting results with regard to LH-induced maturation.[66] LH induction of maturation of hamster follicle-enclosed oocytes is enhanced by the presence of dbcGMP, or nitroprusside or atrial natriuretic factor, each of which activates guanylate cyclase and increases follicle cGMP levels. Nitroprusside treatment also reduces the LH-induced increase in follicle cAMP levels. In addition, these treatments result in an increase in the level of cumulus cell, but not oocyte, PDE activity. Thus, the enhanced effect of cGMP on oocyte maturation may be to elicit a shorter transient rise in follicle cAMP, which may somehow provide the physiological trigger for gonadotropin-induced maturation.

The issue that is unresolved in these studies is the temporal relationship between these changes in cGMP in the different ovarian compartments and the time of oocyte commitment to maturation. Thus, for these changes in cGMP to play a role in the resumption of meiosis, the changes must occur prior to or concomitant with the commitment period. If they occur after the commitment period, a role for cGMP in resumption of meiosis is very ambiguous, although it may play a role in later events in oocyte maturation. If the changes in cGMP correlate with commitment to resume meiosis, the development of reagents that specifically perturb cGMP metabolism and/or the cGMP-dependent protein kinase may provide the necessary tools to approach the physiological role of cGMP in oocyte maturation.

D. GROWTH FACTORS AND GnRH

Resumption of meiosis of follicle-enclosed oocytes by several different types of molecules has been reported, although their mechanism of action and physiological relevance are still unclear. Physiological concentrations of epidermal growth factor (EGF) induce resumption of meiosis in oocytes present in both large and small follicles.[26] Consistent with this observation is that EGF treatment of cumulus cell-oocyte complexes, which are cultured in concentrations of either dbcAMP, IBMX, or hypoxanthine that do not totally inhibit GVBD, results in a significant increase in the extent of GVBD compared to similarly treated denuded oocytes.[29,35] EGF-induced resumption of meiosis may occur by a mechanism similar to that elicited by gonadotropin. It should be noted, however, results of another study did not observe an EGF-induced elevation of cumulus cell cAMP.[45] Although the statistical analysis of the data did not indicate any significant differences in cAMP levels, examination of the data suggests that such an increase may occur and that results of further experiments would indicate that the differences are significant. An explanation for the differences in cAMP levels in response to EGF in these two studies may be that the inclusion of IBMX, which was done in the study that detected significant cAMP differences, may enhance cAMP accumulation by inhibiting PDE activity.

In the context of the positive stimulus model, the increase in cumulus cell cAMP would generate a signal that overrides the inhibitory effect of the cAMP analog or PDE inhibitor. It should be noted, however, that treatment of follicles with EGF also results in a reduction of intercellular communication between cumulus cells and the oocyte. Regrettably, a precise correlation of this reduction in communication with commitment to resume meiosis was not determined.[26] Such information is pivotal to interpret data in the context of the termination of intercellular communication model.

Transforming growth factor-β (TGF-β) induces maturation of follicle-enclosed oocytes.[45] Physiological concentrations of TGF-β (p*M*) increase the rate of GVBD of follicle-enclosed oocytes by about a factor of two. (It should be noted, however, that the rate of GVBD in untreated follicles was very high during the culture period of 8 h when compared to that reported by others. Thus, it is possible that the culture conditions employed in this study resulted in a high incidence of an atresia-like induction of maturation and that TGF-β may somehow stimulate this process.) Similarly, treatment of cumulus cell-oocyte complexes also results in an increased rate of GVBD when compared to untreated complexes or denuded oocytes.[45] Results of another study did not reveal that TGF-β increased the extent of GVBD in cumulus cell-enclosed oocytes cultured in the presence of subinhibiting concentrations of hypoxanthine.[29] These apparent differences may be due to the fact that GVBD was assessed after a 21- to 22-h culture period in this latter study, and thus any transient acceleration of maturation would not have been observed.

Members of the insulin family of growth factors also can stimulate GVBD in cumulus cell-oocyte complexes. Insulin-like growth factor I or II (IGF I/II), but not insulin, stimulates maturation, and addition of TGF-β has no additive effect.[45] In addition, neither TGF-β, IGF-I, nor II increases cAMP levels in cumulus cell-oocyte complexes. Although TGF-β can modulate the effect of EGF on granulosa cell differentiation and EGF by itself can stimulate maturation, the stimulatory effect of EGF is inhibited by low concentrations (10 p*M*) of TGF-β. Results of another study indicated that IGF-I and II, platelet-derived growth factor, bombesin, and nerve growth factor do not stimulate maturation of cumulus cell-enclosed oocytes cultured in the presence of subinhibiting concentrations of hypoxanthine.[29] Again, the differences reported in these two studies may be attributed to differences in the time at which GVBD was assessed (see above).[29]

GnRH in the n*M* range induces maturation of follicle-enclosed oocytes, but the kinetics of maturation are slower than those induced by LH.[25] GnRH binds to a receptor distinct from that of LH, since GnRH antagonists inhibit GnRH- but not LH-induced resumption of meiosis of follicle-enclosed oocytes. In addition, GnRH is likely to induce maturation by a mechanism different than that of LH, since treatment of follicles with GnRH does not result in an increase in cAMP levels.

A possible mode of action for GnRH is to activate PK-C.[25] Consistent with this is that treatment of follicles with activators of PK-C or phospholipase C, which will generate diacylglycerol that is also an activator of PK-C, induces resumption of meiosis. In addition, GnRH-induced maturation is more sensitive to inhibition by 1-(5-isoquinolinylsulfonyl)-2-

methylpiperazine (H7) than by N-[2-(methylamino)ethyl]-5-isoquinolinesulfonamide (H8); H7 is a more potent inhibitor of PK-C than is H8. These results must also be interpreted with caution, since the mode of action of these inhibitors was not documented and it is possible that the inhibition effected by these compounds is at a step distal to the initial events of GnRH action and common to LH-induced maturation. Nevertheless, these results raise the possibility that these different stimulatory signals may initiate distinct pathways that ultimately converge and lead to resumption of meiosis. Demonstrating that LH-induced maturation, which is known to be mediated by elevating cAMP, is more sensitive to inhibition by H8 than by H7 would lend further credence to a GnRH-induced PK-C cascade that leads to maturation. It should be noted that in *Xenopus* oocytes, progesterone- or insulin-induced maturation apparently activate different second messenger cascades that converge to a common pathway that leads to resumption of meiosis.[80,104,122]

In summary, results of these studies indicate that several other types of molecules can initiate resumption of meiosis of follicle-enclosed oocytes. Although the physiological significance of maturation induced by these agents is still not clear, elucidating the mechanism of action should promote insights regarding the mechanism of gonadotropin-induced resumption of meiosis.

IV. GENERATION OF CYTOPLASMIC ACTIVITIES DURING OOCYTE MATURATION

Oocyte maturation is accompanied by dramatic changes in the pattern of protein synthesis.[113] Most of these changes occur subsequent to GVBD, and although some may require mixing of nucleoplasm and cytoplasm, most of these changes in the pattern of protein synthesis occur in anucleate oocyte fragments.[114] Some of these changes are necessary for nuclear progression subsequent to GVBD (see above), but others are apparently required for cortical granule exocytosis and development of the ability to decondense the head of the fertilizing sperm.

Oocyte maturation is accompanied by a maturation-associated decrease (about 30%) in the number of cortical granules; fertilization or egg activation results in essentially a total loss of the remaining cortical granules.[36] The release of cortical granule material is thought to be responsible for modifications of the mouse egg's *zona pellucida* that constitute the *zona pellucida* block to polyspermy.[82,130,131] The reduction in the number of cortical granules requires GVBD, since it is prevented by dbcAMP, and moreover, the ability of oocytes to respond to the calcium ionophore A23187 with a reduction in the number of cortical granules occurs subsequent to GVBD and develops between 7 and 12 h after initiation of maturation. Last, GV-intact oocytes microinjected with IP_3, which elicit egg-induced modifications of the *zona pellucida* when microinjected into metaphase II-arrested eggs, does not result in egg-induced modifications of the *zona pellucida*.[73a]

The replacement of histones by protamines, which are basic proteins, during spermatogenesis results in a highly condensed nucleus.[135] The protamines contain high amounts of cysteine that are oxidized during sperm maturation. Reduction of these protamines in the egg's cytoplasm is necessary for sperm nuclear decondensation and transformation into the male pronucleus after fertilization. This reduction is apparently accomplished by glutathione, which is the major physiologically relevant thiol-reducing agent.[16,90,91,136] Treatment of eggs with diamide, which is an antioxidant of glutathione, reversibly inhibits the decondensation of microinjected hamster sperm nuclei. Treatment of mice *in vivo* with an inhibitor of glutathione synthesis, L-buthionine-S,R-sulfoxamine (BSO), inhibits sperm decondensation of mouse sperm nuclei following *in vitro* fertilization. The inability of the sperm nucleus to decondense is apparently due to inadequate amounts of reducing power present in the egg, since disulfide-poor sperm nuclei decondense when microinjected into either GV-intact oocytes or eggs.

The ability of sperm nuclei to decondense in the egg cytoplasm is acquired during oocyte maturation; the decondensing activity is barely detectable in GV-intact eggs and becomes maximal in metaphase II-arrested eggs.[90] Correlated with the ability to decondense the sperm nucleus is a maturation-associated increase in the amount of reduced glutathione, and moreover, inhibiting this maturation-associated increase in glutathione levels with BSO results in a decreased ability of the matured oocyte to decondense microinjected sperm nuclei.

Sperm nuclei in a GV-intact oocyte will not decondense, whereas sperm nuclei in an oocyte that has undergone GVBD will decondense. Moreover, since dithiothreitol-treated hamster sperm nuclei will decondense in the cytoplasm of GV-intact oocytes, the contents of the GV may be necessary for the development of this decondensing activity, i.e., sufficient reducing levels of glutathione. It was not demonstrated, however, if anucleate fragments obtained by bisection of GV-intact oocytes and allowed to mature for a similar time as the nucleate fragments develop the ability to decondense microinjected sperm nuclei.[91] The contents of the GV may be necessary, since GVBD is apparently required for development of cytoplasmic activities that control male and female pronuclei formation.[8,135] Fertilization of nucleate fragments, which are obtained from bissected oocytes and allowed to undergo GVBD, results in transformation of the sperm head into a pronucleus. Fertilization of anucleate fragments, which are also obtained from bissected oocytes and cultured for the same length of time as the nucleate fragments, does not result in this transformation.

REFERENCES

1. **Aberdam, E., Hanski, E., and Dekel, N.,** Maintenance of meiotic arrest in isolated rat oocytes by the invasive adenylate cyclase of *Bordetalla Pertussis, Biol. Reprod.*, 36, 530, 1987.
2. **Adlakha, R. C., Sahasrabuddhe, C. G., Wright, D. A., and Rao, P. N.,** Evidence for the presence of inhibitors of mitotic factors during G1 period in mammalian cells, *J. Cell Biol.*, 97, 1701, 1983.
3. **Albertini, D. F. and Anderson, E.,** The appearance and structure of intercellular connections during the ontogeny of the rabbit ovarian follicle with particular reference to gap junctions, *J. Cell Biol.*, 63, 234, 1974.
4. **Anderson, E. and Albertini, D. F.,** Gap junctions between the oocyte and companion follicle cells in the mammalian ovary, *J. Cell Biol.*, 71, 680, 1976.
5. **Balakier, H.,** Induction of maturation in small oocytes from sexually immature mice by fusion with meiotic or mitotic cells, *Exp. Cell Res.*, 112, 137, 1978.
6. **Balakier, H. and Czolowska, R.,** Cytoplasmic control of nuclear maturation in mouse oocytes, *Exp. Cell Res.*, 110, 466, 1977.
7. **Balakier, H. and Masui, Y.,** Chromosome condensation activity in the cytoplasm of anucleate and nucleat fragments of mouse oocytes, *Dev. Biol.*, 113, 155, 1986.
8. **Balakier, H. and Tarkowski, A. K.,** The role of germinal vesicle karyoplasm in the development of the male pronucleus in the mouse, *Exp. Cell Res.*, 128, 79, 1980.
9. **Bornslaeger, E. A. and Schultz, R. M.,** Adenylate cyclase activity in *zona*-free mouse oocytes, *Exp. Cell Res.*, 156, 277, 1985.
10. **Bornslaeger, E. A. and Schultz, R. M.,** Regulation of mouse oocyte maturation: effect of elevating cumulus cell cAMP on oocyte cAMP levels, *Biol. Reprod.*, 33, 698, 1985.
11. **Bornslaeger, E. A., Mattei, P. M., and Schultz, R. M.,** Involvement of cAMP-dependent protein kinase and protein phosphorylation in regulation of mouse oocyte maturation, *Dev. Biol.*, 114, 453, 1986.
12. **Bornslaeger, E. A., Mattei, P. M., and Schultz, R. M.,** Protein phosphorylation in meiotically competent and incompetent mouse oocytes, *Mol. Reprod. Dev.*, 1, 19, 1988.
13. **Bornslaeger, E. A., Poueymirou, W. T., Mattei, P., and Schultz, R. M.,** Effects of protein kinase C activators on germinal vesicle breakdown and polar body emission of mouse oocytes, *Exp. Cell Res.*, 165, 507, 1986.
14. **Bornslaeger, E. A.,Wilde, M. W., and Schultz, R. M.,** Regulation of mouse oocyte maturation: involvement of cyclic AMP phosphodiesterase and calmodulin, *Dev. Biol.*, 105, 488, 1984.
15. **Brower, P. T. and Schultz, R. M.,** Intercellular communication between granulosa cells and mouse oocytes: existence and possible nutritional role during oocyte growth, *Dev. Biol.*, 90, 144, 1982.

16. **Calvin, H. I., Grosshans, K., and Blake, E.,** Estimation and manipulation of glutathione levels in prepubertal mouse ovaries and ova: relevance of sperm nucleus transformations in the fertilized egg, *Gamete Res.,* 14, 265, 1986.
17. **Canipari, R., Bevilacqua, A., Colonna, R., De Felici, M., and Mangia, F.,** Actin synthesis is not regulated by granulosa cells in mouse growing and preovulatory oocytes, *Gamete Res.,* 20, 115, 1988.
18. **Cho, W. K., Stern, S., and Biggers, J. D.,** Inhibitory effect of dibutyryl cAMP on mouse maturation *in vitro, J. Exp. Zool.,* 187, 383, 1974.
18a. **Clarke, H. J. and Masui, Y.,** The induction of reversible and irreversible chromosome decondensation by protein synthesis inhibition during meiotic maturation of mouse oocytes, *Dev. Biol.,* 97, 291, 1983.
19. **Crosby, I. M., Moor, R. M., Heslop, J. P., and Osborn, J. C.,** cAMP in ovine oocytes: localization of synthesis and its actions on protein synthesis, phosphorylation, and meiosis, *J. Exp. Zool.,* 234, 307, 1985.
20. **Crosby, I. M., Osborn, J. C., and Moor, R. M.,** Changes in protein phosphorylation during the maturation of mammalian oocytes *in vitro, J. Exp. Zool.,* 229, 459, 1984.
21. **Davidson, E. H.,** *Gene Activity in Early Development,* 3rd ed., Academic Press, New York, 1986, chap. 2.
22. **Dekel, N. and Beers, W. H.,** Rat oocyte maturation *in vitro*: relief of cyclic AMP inhibition by gonadotropins, *Proc. Natl. Acad. Sci. U.S.A.,* 75, 4369, 1978.
23. **Dekel, N. and Beers, W. H.,** Development of the rat oocyte *in vitro:* inhibition and induction of maturation in the presence or absence of the cumulus oophorus, *Dev. Biol.,* 75, 247, 1980.
24. **Dekel, N., Aberdam, E., and Sherizly, I.,** Spontaneous maturation *in vitro* of cumulus-enclosed rat oocytes is inhibited by forskolin, *Biol. Reprod.,* 31, 244, 1984.
25. **Dekel, N., Aberdam, E., Goren, S., Feldman, B., and Shalgi, R.,** Mechanism of action of GnRH-induced oocyte maturation, *J. Reprod. Fertil. Suppl.,* 37, 319, 1989.
26. **Dekel, N. and Sherizly, I.,** Epidermal growth factor induces maturation of rat follicle-enclosed oocytes, *Endocrinology,* 116, 406, 1985.
27. **Doree, M.,** Protein synthesis is not involved in initiation or amplification of the maturation factor (MPF) in starfish oocytes, *Exp. Cell Res.,* 139, 127, 1982.
28. **Downs, S. M.,** Protein synthesis inhibitors prevent both spontaneous and hormone-induced maturation of isolated mouse oocytes *in vitro, Mol. Reprod. Dev.,* in press.
29. **Downs, S. M.,** Specificity of epidermal growth factor action on maturation of the murine oocyte and cumulus oophorus *in vitro, Biol. Reprod.,* 47, 371, 1989.
30. **Downs, S. M. and Eppig, J. J.,** Cyclic adenosine monophosphate and ovarian follicular fluid act synergistically to inhibit mouse oocyte maturation, *Endocrinology,* 114, 418, 1984.
31. **Downs, S. M. and Eppig, J. J.,** Induction of mouse oocyte maturation *in vivo* by perturbants of purine metabolism, *Biol. Reprod.,* 36, 431, 1987.
32. **Downs, S. M., Coleman, D. L., and Eppig, J. J.,** Maintenance of murine meiotic arrest: uptake and metabolism of hypoxanthine and adenosine by cumulus cell-enclosed and denuded oocytes, *Dev. Biol.,* 117, 174, 1986.
33. **Downs, S. M., Coleman, D. L., Ward-Bailey, P. F., and Eppig, J. J.,** Hypoxanthine is the principal inhibitor of oocyte maturation in a low molecular weight fraction of porcine follicular fluid, *Proc. Natl. Acad. Sci. U.S.A.,* 82, 454, 1985.
34. **Downs, S. M., Daniel, S. A. J., Bornslaeger, E. A., Hoppe, P. C., and Eppig, J. J.,** Maintenance of meiotic arrest in mouse oocytes by purines: modulation of cAMP levels and cAMP phosphodiesterase activity, *Gamete Res.,* 23, 323, 1989.
35. **Downs, S. M., Daniel, S. A. J., and Eppig, J. J.,** Induction of maturation in cumulus cell-enclosed mouse oocytes by follicle-stimulating hormone and epidermal growth factor: evidence for a positive stimulus of somatic cell origin, *J. Exp. Zool.,* 245, 86, 1988.
36. **Ducibella, T., Kurasawa, S., Rangarajan, S., Kopf, G. S., and Schultz, R. M.,** Precocious loss of egg cortical granules during meiotic maturation and correlation with an egg-induced modification of the *zona pellucida, Dev. Biol.,* 137, 46, 1990.
37. **Dunphy, W. G., Brizuela, L., Beach, D., and Newport, J. W.,** The *Xenopus* cdc2 protein is a component of MPF, a cytoplasmic regulator of mitosis, *Cell,* 54, 423, 1988.
38. **Eckberg, W. R., Szuts, E. Z., and Carroll, A. G.,** Protein kinase C activity, protein phosphorylation and germinal vesicle breakdown in *Spisula* oocytes, *Dev. Biol.,* 124, 57, 1987.
39. **Eckholm, C. and Magnusson, C.,** Rat oocyte maturation: effects of protein synthesis inhibitors, *Biol. Reprod.,* 21, 1287, 1979.
40. **Eckholm, C., Hillensjo, T., Magnusson, C., and Rosberg, S.,** Stimulation and inhibition of rat oocyte meiosis by forskolin, *Biol. Reprod.,* 30, 537, 1984.
41. **Eppig, J. J.,** The relationship between cumulus cell-oocyte coupling, oocyte meiotic maturation, and cumulus expansion, *Dev. Biol.,* 89, 268, 1982.

42. **Eppig, J. J. and Downs, S. M.,** The effect of hypoxanthine on mouse oocyte growth and development *in vitro*: maintenance of meiotic arrest and gonadotropin-induced oocyte maturation, *Dev. Biol.*, 119, 313, 1987.
42a. **Eppig, J. J. and Downs, S. M.,** Gonadotropin-induced murine oocyte maturation *in vivo* is not associated with decreased cyclic adenosine monophosphate in the oocyte-cumulus cell complex, *Gamete Res.*, 20, 125, 1988.
43. **Eppig, J. J., Freter, R. R., Ward-Bailey, P. F., and Schultz, R. M.,** Inhibition of oocyte maturation in the mouse: participation of cAMP, steroid hormones, and a putative maturation-inhibitory factor, *Dev. Biol.*, 100, 39, 1983.
44. **Eppig, J. J., Ward-Bailey, P. F., and Coleman, D. L.,** Hypoxanthine and adenosine in murine ovarian follicular fluid: concentrations and activity in maintaining meiotic arrest, *Biol. Reprod.*, 33, 1041, 1985.
45. **Feng, P., Catt, K. J., and Knecht, M.,** Transforming growth factor-β stimulates meiotic maturation of the rat oocyte, *Endocrinology*, 122, 181, 1988.
46. **Fleming, A. D., Khalil, W., and Armstrong, D. T.,** Porcine follicular fluid does not inhibit rat oocytes *in vitro, J. Reprod. Fertil.*, 69, 665, 1983.
47. **Foote, W. D. and Thibault, C.,** Recherches experimentales sur la maturation *in vitro* des oocytes de trui et de veau, *Ann. Biol. Anim. Biochem. Biophys*, 9, 329, 1969.
48. **Foulkes, J. G. and Maller, J. L.,** *In vivo* actions of protein phosphatase inhibitor-2 in *Xenopus* oocytes, *FEBS Lett.*, 150, 155, 1982.
49. **Freeman, G. and Ridgway, E. B.,** The role of cAMP in oocyte maturation and the role of germinal vesicle contents in mediating maturation and subsequent developmental events in hydrozoans, *Roux's Arch. Dev. Biol.*, 197, 197, 1988.
50. **Freter, R. R. and Schultz, R. M.,** Regulation of murine oocyte meiosis: evidence for a gonadotropin-induced, cAMP-dependent reduction in a maturation inhibitor, *J. Cell Biol.*, 98, 1119, 1984.
51. **Fulka, J., Jr., Motlik, J., Fulka, J., and Crozet, N.,** Inhibition of nuclear maturation in fully grown porcine and mouse oocytes after their fusion with growing porcine oocytes, *J. Exp. Zool.*, 235, 255, 1985.
52. **Fulka, J., Jr., Motlik, J., Fulka, J., and Jilek, F.,** Effect of cycloheximide on nuclear maturation of pig and mouse oocytes, *J. Reprod. Fertil.*, 77, 281, 1986.
53. **Fulka, J., Jr.,** Maturation-inhibiting activity in growing mouse oocytes, *Cell Differ.*, 17, 45, 1985.
54. **Fulka, J., Jr., Motlik, J., Fulka, J., and Crozet, N.,** Activity of maturation promoter factor in mammalian oocytes after its dilution by single and multiple fusions, *Dev. Biol.*, 118, 176, 1986.
55. **Fulka, J., Jr., Flechon, J.-E., Motlik, J., and Fulka, J.,** Does autocatalytic amplification of maturation-promoting factor (MPF) exist in mammalian oocytes?, *Gamete Res.*, 21, 185, 1988.
56. **Gautier, J., Norbury, C., Lohka, M., Nurse, P., and Maller, J.,** Purified maturation-promoting factor contains the product of a *Xenopus* homolog of the fission yeast cell cycle control gene $cdc2^+$, *Cell*, 54, 433, 1988.
57. **Gavin, A.-C. and Schorderet-Slatkine, S.,** The interaction of lithium with forskolin-inhibited meiotic maturation of denuded mouse oocytes, *Exp. Cell Res.*, 179, 298, 1988.
58. **Gerhart, J., Wu, M., and Kirschner, M.,** Cell cycle dynamics of an M-phase-specific cytoplasmic factor in *Xenopus laevis* oocytes and eggs, *J. Cell Biol.*, 98, 1247, 1984.
59. **Gilula, N. B., Epstein, M. L., and Beers, W. H.,** Cell-to-cell communication and ovulation. A study of the cumulus cell-oocyte complex, *J. Cell Biol.*, 78, 58, 1978.
60. **Hashimoto, N. and Kishimoto, T.,** Regulation of meiotic metaphase by a cytoplasmic maturation-promoting factor during mouse oocyte maturation, *Dev. Biol.*, 126, 242, 1988.
61. **Heller, D. H. and Schultz, R. M.,** Ribonucleoside metabolism by mouse oocytes: metabolic cooperativity between the fully-grown oocyte and cumulus cells, *J. Exp. Zool.*, 214, 355, 1980.
62. **Heller, D. H., Cahill, D. M., and Schultz, R. M.,** Biochemical studies of mammalian oogenesis: metabolic cooperativity between granulosa cells and growing mouse ooytes, *Dev. Biol.*, 84, 455, 1981.
63. **Homa, S. T.,** Effects of cyclic AMP on the spontaneous meiotic maturation of cumulus-free bovine oocyte cultured in chemically defined medium, *J. Exp. Zool.*, 248, 222, 1988.
64. **Hubbard, C. J.,** Cyclic AMP changes in the component cells of Graafian follicles: possible influences on maturation in the follicle-enclosed oocytes of hamsters, *Dev. Biol.*, 118, 343, 1986.
65. **Hubbard, C. J. and Price, J.,** The effects of follicle-stimulating hormone and cyclic guanosine 3′,5′-monophosphate on cyclic adenosine 3′,5′-monophosphate diesterase and resumption of meiosis in hamster cumulus-oocyte complexes, *Biol. Reprod.*, 39, 829, 1988.
66. **Hubbard, C. J. and Terranova, P. F.** Inhibitory action of cyclic guanosine 5′-phosphoric acid (GMP) on oocyte maturation: dependence on an intact cumulus, *Biol. Reprod.*, 26, 628, 1982.
67. **Huchon, D., Ozon, R., and Demaille, J. G.,** Protein phosphatase-1 is involved in *Xenopus* oocyte maturation, *Nature (London)*, 294, 358, 1981.
68. **Jagiello, G., Ducayen, M. B., and Goonan, W. D.,** A note on the inhibition of *in vitro* meiotic maturation of mammalian oocytes by dibutyryl cyclic AMP, *J. Exp. Zool.*, 218, 309, 1981.

69. **Kaji, E., Bornslaeger, E. A., and Schultz, R. M.,** Inhibition of mouse oocyte cyclic nucleotide phosphodiesterase by steroid hormones: a possible mechanism for steroid hormone inhibition of oocyte maturation, *J. Exp. Zool.*, 243, 489, 1987.
70. **Keshet, E., Rosenberg, M. P., Mercer, J. A., Propst, F., Vande Woude, G. G., Jenkins, N. A., and Copland, N. G.,** Developmental regulation of ovarian-specific *Mos* expression, *Oncogene*, 2, 235, 1988.
71. **Kishimoto, T.,** Regulation of metaphase by a maturation-promoting factor, *Dev. Growth Differ.*, 30, 105, 1988.
72. **Kishimoto, T., Yamazaki, K., Kato, Y., Koide, S. S., and Kanatani, H.,** Induction of starfish oocyte maturation by maturation-promoting factor of mouse and surf clam oocytes, *J. Exp. Zool.*, 231, 293, 1984.
73. **Kurasawa, S., Kopf, G. S., and Schultz, R. M.,** unpublished data, 1989.
73a. **Kurasawa, S., Schultz, R. M., and Kopf, G. S.,** Egg-induced modifications of the *zona pellucida* of the mouse egg: effects of microinjected inositol 1,4,5-trisphosphate, *Dev. Biol.*, 133, 295, 1989.
74. **Larsen, W. J., Wert, S. E., and Brunner, G. D.,** A dramatic loss of cumulus cell gap junctions is correlated with germinal vesicle breakdown in rat oocytes, *Dev. Biol.*, 113, 517, 1986.
75. **Larsen, W. J., Wert, S. E., and Brunner, G. D.,** Differential modulation of rat follicle cell gap junction populations at ovulation, *Dev. Biol.*, 122, 61, 1987.
76. **Leibfried, L. and First, N. L.,** Effect of bovine and porcine follicular fluid and granulosa cells on maturation of oocyte *in vitro, Biol. Reprod.*, 23, 699, 1980.
77. **Leibfried, L. and First, N. L.,** Follicular control of meiosis in the porcine oocyte, *Biol. Reprod.*, 23, 705, 1980.
78. **Lohka, M. J., Hayes, M. K., and Maller, J. L.,** Purification of maturation-promoting factor, an intracellular regulator of early mitotic events, *Proc. Natl. Acad. Sci. U.S.A.*, 85, 3009, 1988.
79. **Moor, R. M. and Crosby, I. M.,** Protein synthesis requirements for germinal vesicle breakdown in ovine oocytes, *J. Embryol. Exp. Morphol.*, 94, 207, 1986.
80. **Maller, J. L.,** Interaction of steroids with cyclic nucleotide systems in amphibian oocytes, *Adv. Cyclic Nucleotide Res.*, 15, 295, 1983.
81. **Maller, J. L.,** Regulation of amphibian oocyte maturation, *Cell Differ.*, 16, 211, 1985.
82. **Moller, C. C. and Wassarman, P. M.,** Characterization of a proteinase that cleaves *zona pellucida* glycoprotein ZP2 following activation of mouse eggs, *Dev. Biol.*, 132, 103, 1989.
83. **Moor, R. M. and Heslop, J. P.,** Cyclic AMP in mammalian follicle cells and oocytes during maturation, *J. Exp. Zool.*, 216, 205, 1981.
84. **Moor, R. M., Osborn, J. C., Cran, D. G., and Walters, D. E.,** Selective effect of gonadotropins on cell coupling, nuclear maturation and protein synthesis in mammalian oocytes, *J. Embryol. Exp. Morphol.*, 61, 347, 1981.
85. **Moor, R. M., Smith, M. W., and Dawson, R. M. C.,** Measurement of intercellular coupling between oocytes and cumulus cells using intracellular markers, *Exp. Cell Res.*, 126, 15, 1980.
86. **Mutter, G. L. and Wolgemuth, D. J.,** Distinct developmental patterns of c-*mos* protooncogene expression in female and male germ cells, *Proc. Natl. Acad. Sci. U.S.A.*, 84, 5301, 1987.
87. **Mutter, G. L., Grills, G. S., and Wolgemuth, D. J.,** Evidence for the involvement of the proto-oncogene c-*mos* in mammalaian meiotic maturation and possibly very early embryogenesis, *EMBO J.*, 7, 683, 1988.
88. **Olsiewski, P. A. and Beers, W. H.,** cAMP synthesis in the rat oocyte, *Dev. Biol.*, 100, 287, 1983.
89. **Paules, R. S., Buccione, R., Propst, F., Moschel, R. C., Vande Woude, G. F., and Eppig, J. J.,** The mouse c-*mos* proto-oncogene product is present and functions during oogenesis, *Proc. Natl. Acad. Sci. U.S.A.*, 86, 5395, 1989.
90. **Perreault, S. D., Barbee, R. R., and Slott, V. L.,** Importance of glutathione in the acquisition and maintenance of sperm nuclear decondensing activity in maturing hamster oocytes, *Dev. Biol.*, 125, 181, 1988.
91. **Perreault, S.D., Wolff, R. A., and Zirkin, B. R.,** The role of disulfide bond reduction during mammalian sperm nuclear decondensation *in vivo, Dev. Biol.*, 101, 160, 1984.
92. **Picard, A., Giraud, F., Le Bouffant, F., Sladeczek, F., Le Peuch, C., and Dorre, M.,** Inositol 1,4,5-trisphosphate microinjection triggers activation, but not meiotic maturation in amphibian and starfish oocytes, *FEBS Lett.*, 182, 446, 1985.
93. **Picard, A., Peaucellier, G., Le Bouffant, F., Le Peuch, C., and Dorre, M.,** Role of protein synthesis and proteases in production and inactivation of maturation-promoting activity during meiotic maturation of starfish oocytes, *Dev. Biol.*, 109, 311, 1985.
94. **Pincus, G. and Enzmann, E. V.,** The comparative behaviour of mammalian eggs *in vivo* and *in vitro, J. Exp. Med.*, 62, 665, 1935.
95. **Pomerantz, S. H.,** Oocyte maturation inhibitor: several candidates for the title, in *Nonsteroidal Gonadal Factors: Physiological Roles and Possibilities in Contraceptive Development,* Hodgen, G. D., Rosenwaks, Z., and Spieler, J. M., Eds., The Jones Institute Press, Norfolk, 1988, 265.
96. **Powers, R. D. and Paleos, G. A.,** Combined effects of calcium and dibutyryl cAMP on germinal vesicle breakdown in the mouse oocyte, *J. Reprod. Fertil.*, 66, 1, 1982.

97. **Probst, F., Rosenberg, M. P., and Vande Woude, G. F.,** Proto-oncogene expression in germ cell development, *Trends Genet.,* 4, 183, 1988.
98. **Racowsky, C.,** Effect of forskolin on the spontaneous maturation and cyclic AMP content of rat oocyte-cumulus complexes, *J. Reprod. Fertil.,* 72, 107, 1984.
99. **Racowsky, C.,** Effect of forskolin on the spontaneous maturation and cyclic AMP content of hamster oocyte-cumulus complexes, *J. Exp. Zool.,* 234, 87, 1985.
99a. **Racowsky, C., Baldwin, K. V., Larabell, C. A., DeMarais, A. A., and Kazilek, C. J.,** Down-regulation of membrana granulosa cell gap junctions is correlated with irreversible commitment to resume meiosis in gold Syrian hamster oocytes, *Eur. J. Cell Biol.,* 49, 244, 1989.
100. **Racowsky, C. and McGaughey, R. W.,** Further studies of the effects of follicular fluid and membrana granulosa cells on the spontaneous maturation of pig oocytes, *J. Reprod. Fertil.,* 66, 505, 1982.
101. **Racowsky, C. and Satterlie, R. A.,** Metabolic, fluorescent dye and electrical coupling between hamster oocytes and cumulus cells during meiotic maturation *in vivo* and *in vitro, Dev. Biol.,* 108, 191, 1985.
102. **Racowsky, C. and Satterlie, R. A.,** Decreases in heterologous metabolic and dye coupling, but not in electrical coupling accompany meiotic resumption in hamster oocyte-cumulus complexes, *Eur. J. Cell Biol.,* 43, 283, 1987.
103. **Rice, C. and McGaughey, R. W.,** Effect of testosterone and dibutyryl cAMP on the spontaneous maturation of pig oocytes, *J. Reprod. Fertil.,* 62, 245, 1981.
104. **Sadler, S. E. and Maller, J. L.,** *In vivo* regulation of cyclic AMP phosphodiesterase in *Xenopus* oocytes. Stimulation by insulin and insulin-like growth factor 1, *J. Biol. Chem.,* 262, 10644, 1987.
105. **Sagata, N., Oskarsson, M., Copeland, T., Brumbaugh, J., and Vande Woude, G. F.,** Function of c-*mos* proto-oncogene product in meiotic maturation in *Xenopus* oocytes, *Nature (London),* 335, 519, 1988.
106. **Salustri, A. and Siracusa, G.,** Metabolic coupling, cumulus expansion, and meiotic resumption in mouse cumuli oophori cultured *in vitro* in the presence of FSH or dbcAMP, or stimulated *in vivo* by hCG, *J. Reprod. Fertil.,* 68, 335, 1983.
107. **Salustri, A., Petrungaro, S., Conti, M., and Siracusa, G.,** Adenosine potentiates forskolin-induced delay of meiotic maturation by mouse denuded oocytes: evidence for an oocyte surface site of adenosine action, *Gamete Res.,* 21, 157, 1988.
108. **Salustri, A., Petrungaro, S., De Felici, M., Conti, M., and Siracusa, G.,** Effect of follicle-stimulating hormone on cyclic adenosine monophosphate level and on meiotic maturation in mouse cumulus cell-enclosed oocytes cultured *in vitro, Biol. Reprod.,* 33, 797, 1985.
109. **Sato, E. and Koide, S. S.,** Forskolin and mouse oocyte maturation *in vitro, J. Exp. Zool.,* 230, 125, 1984.
110. **Schroeder, A. C. and Eppig, J. J.,** The developmental capacity of mouse oocytes that matured spontaneously *in vitro* is normal, *Dev. Biol.,* 102, 493, 1984.
111. **Schultz, R. M.,** Molecular aspects of oogenesis, in *Experimental Approaches to Mammalian Embryonic Development,* Rossant, J. and Pedersen, R. A., Eds., Cambridge University Press, New York, 1988, 195.
112. **Schultz, R. M. and Richards, J. S.,** unpublished data, 1989.
113. **Schultz, R. M. and Wassarman, P. M.,** Specific changes in the pattern of protein synthesis during meiotic maturation of mammalian oocytes *in vitro, Proc. Natl. Acad. Sci. U.S.A.,* 74, 538, 1977.
114. **Schultz, R. M., Letourneau, G. E., and Wassarman, P. M.,** Meiotic maturation of mouse oocytes *in vitro*: protein synthesis in nucleate and anucleate oocyte fragments, *J. Cell Sci.,* 30, 251, 1978.
115. **Schultz, R. M., Montgomery, R. R., and Belanoff, J. R.,** Regulation of mouse oocyte maturation: implication of a decrease in oocyte cAMP and protein dephosphorylation in commitment to resume meiosis, *Dev. Biol.,* 97, 264, 1983.
116. **Schultz, R. M., Montgomery, R. R., Ward-Bailey, P. F., and Eppig, J. J.,** Regulation of oocyte maturation in the mouse: possible roles of intercellular communication, cAMP, and testosterone, *Dev. Biol.,* 95, 294, 1983.
117. **Simanis, V. and Nurse, P.,** The cell cycle control gene cdc2+ of fission yeast encodes a protein kinase potentially regulated by phosphorylation, *Cell,* 45, 261, 1986.
118. **Siracusa, G., Whittingham, D. G., Molinaro, M., and Vivarelli, E.,** Parthenogenetic activation of mouse oocytes induced by inhibitors of protein synthesis, *J. Embryol. Exp. Morphol.,* 43, 157, 1978.
119. **Sirard, M. A. and First, N. L.,** *In vitro* inhibition of oocyte nuclear maturation in the bovine, *Biol. Reprod.,* 39, 229, 1988.
120. **Sorensen, R. A. and Wassarman, P. M.,** Relationship between growth and meiotic maturation of the mouse oocyte, *Dev. Biol.,* 50, 531, 1976.
121. **Sorensen, R. A., Cyert, M. S., and Pedersen, R. A.,** Active maturation-promoting factor is present in mature mouse oocytes, *J. Cell Biol.,* 100, 1637, 1985.
122. **Stith, B. J. and Maller, J. L.,** The effect of insulin on intracellular pH and ribosomal protein S6 phosphorylation in oocytes of *Xenopus laevis, Dev. Biol.,* 102, 79, 1984.
123. **Stith, B. J. and Maller, J. L.,** Induction of meiotic maturation in *Xenopus* oocytes by 12-O-tetradecanoylphorobol 13-acetate, *Exp. Cell Res.,* 169, 514, 1987.

124. **Suzuki, K., Kurasawa, S., Kitai, H., Oba, M., Komatsu, S., Yoda, K., and Iizuka, R.,** Cooperative inhibitory effect of follicular fluid and cAMP on hamster oocyte maturation, *Experientia,* 42, 795, 1986.
125. **Takahashi, M., Koide, S., and Donahoe, P. K.,** Mullerian inhibiting substance as oocyte meiosis inhibitor, *Mol. Cell Endocrinol.,* 47, 225, 1986.
126. **Tsafriri, A., Picard, J.-Y., and Josso, N.,** Immunopurified anti-Mullerian hormone does not inhibit spontaneous resumption of meiosis in vitro of rat oocytes, *Biol. Reprod.,* 38, 481, 1988.
127. **Urner, F. and Schorderet-Slatkine, S.,** Inhibition of denuded mouse oocyte meiotic maturation by tumor-promoting phorbol esters and its reversal by retinoids, *Exp. Cell Res.,* 154, 600, 1984.
128. **Urner, F., Herrmann, W. L., Baulieu, E. E., and Schorderet-Slatkine, S.,** Inhibition of denuded mouse oocyte meiotic maturation by forskolin, an activator of adenylate cyclase, *Endocrinology,* 113, 1170, 1983.
129. **Vivarelli, E., Conti, M., De Felici, M., and Siracusa, G.,** Meiotic resumption and intracellular cAMP levels in mouse oocytes treated with compounds which act on cAMP metabolism, *Cell Differ.,* 12, 271, 1983.
130. **Wassarman, P. M.,** Early events in mammalian fertilization, *Annu. Rev. Cell Biol.,* 3, 100, 1987.
131. **Wassarman, P. M.,** The biology and chemistry of fertilization, *Science,* 235, 553, 1987.
132. **Wassarman, P. M., Josefowicz, W. J., and Letourneau, G. E.,** Meiotic maturation of mouse oocytes *in vitro*: inhibition of maturation at specific stages of nuclear progression, *J. Cell Sci.,* 22, 431, 1976.
133. **Wassarman, P. M., Schultz, R. M., Letourneau, G. E., LaMarca, M. J., Josefowicz, W. J., and Bleil, J. D.,** Meiotic maturation of mouse oocytes *in vitro,* in *Ovrian, Follicular, and Corpus Luteum Function,* Channing, C. P., Marsh, J. M., and Sadler, W. A., Eds., Plenum Press, New York, 1979, 251.
134. **Yamashita, M.,** Involvement of cAMP in initiating maturation of the brittle-star *Amphipholis kochi* oocytes: induction of oocyte maturation by inhibitors of cyclic nucleotide phosphodiesterase and activators of adenylate cyclase, *Dev. Biol.,* 125, 109, 1988.
135. **Yanagimachi, R.,** Mammalian fertilizaton, in *The Physiology of Reproduction,* Vol. 1, Knobil, E., Neill, J., et al., Eds., Raven Press, New York, 1988, 135.
136. **Zirkin, B. R., Soucek, D. A., Chang, T. S. K., and Perreault, S. D.,** *In vitro* and *in vivo* studies of mammalian sperm nuclear decondensation, *Gamete Res.,* 11, 349, 1985.

Chapter 5

PROGRESS TOWARD UNDERSTANDING THE MOLECULAR BASIS OF CAPACITATION

Harvey M. Florman and Donner F. Babcock

TABLE OF CONTENTS

I. INTRODUCTION

A. OVERVIEW

Mammalian sperm, examined immediately following natural mating, fail to fertilize eggs. While these cells are motile and appear structurally mature, fertility can only be demonstrated following additional maturation either within the female reproductive tract or under selected conditions *in vitro*.[1-4] Historically, the term "capacitation" has been defined as the time interval (and, by extension, the alterations produced therein) encompassing this additional maturation.[5-7] Modifications of sperm motility, metabolism, and cell surface features all apparently are associated with such acquisition of fertilizing ability. Assigning causal relationships between such alterations and the ability to fertilize, however, remains a difficult problem, primarily because of the limited range of analytical tools available for examination of those few sperm in a population that actually achieve fertilization. We will consider briefly the methods available for determination of fertile competence in sperm and the assumptions that underlie them (Section II).

The major portion of this review is organized around the concept that for many mammalian species the predominant and physiologically relevant acquisition of sperm fertility may progress through a limited number of intermediary stages. These develop in response to identifiable instructive signals and proceed by definable and general mechanisms. Recent advances in understanding the molecular basis for interaction between mammalian sperm and egg thereby provide an additional opportunity to examine the requirements for functional maturation of sperm. Section III discusses these advances and their significance. Finally, we consider how different functional states of sperm are maintained and how the transitions between states may be accomplished during the acquisition of fertilizing ability (Section IV).

Readers interested in the more general topic of capacitation may refer to previous comprehensive reviews.[8-11] Methodology for the capacitation of sperm *in vitro* and *in vivo* also were considered previously.[11-14] In addition, specific reviews have examined the role of exogenous lipids and of the modifications to sperm membranes,[15-18] as well as that of metabolic alterations,[19,20] associated with capacitation.

B. FUNCTIONAL STATES OF SPERM: DEFINITIONS

Either surgical intervention or ejaculation can release sperm from storage in the cauda epididymis and initiate a progression of events that leads invariably either to fertilization or to cell death. Because capacitation is defined with fertilization as an end point, it contains no inherent criteria for assessment of physiological relevance. Thus, the term capacitation may be applied legitimately to an ionophore treatment regimen that, for instance, produces mouse or bovine sperm capable of penetrating zona-free hamster eggs. Nevertheless, such sperm almost certainly would be incapable of the physiologically relevant signal transduction processes that occur at the zona pellucida. As a further example, similar imprecision results from application of the term "uncapacitated" both to cauda epididymal sperm suspended in defined media and to ejaculated sperm in seminal fluids. Such usage implies a functional equivalence that has yet to be established. Much confusion is avoided by focusing instead upon that set of activity states that may represent the progression between mating and fertilization.

1. The *storage state* for mammalian sperm is here identified as the native condition (*circa* 10^9 cells/ml and presumably both immotile and incapable of fertilization) of sperm in cauda epididymal fluids *in situ*. Perforce, such epididymal sperm usually are studied following dilution *in vitro*, with accompanying initiation of motility (and perhaps other alterations, but still incapable of fertilization). Such diluted epididymal sperm will be defined as in a *resting state*. During the resting state, sperm are still incapable of

fertilizing eggs. Yet, the relationship between cells in the storage and resting states must be established.

2. The bulk of available evidence indicates that the admixture of epididymal sperm with seminal fluids, as occurs during natural mating, results in increased requirements for the manifestation of fertility by such ejaculated or by treated epididymal sperm.[12,21-23] Freshly ejaculated sperm, or epididymal sperm treated with seminal fluids *in vitro*, will be considered to be in a *repressed state*.
3. During appropriate incubation, *in vitro* or *in vivo,* sperm progress to a *fertile state,* in which they are functionally mature. By these terms, we are specifying those sperm that are able to undergo a zona-induced acrosome reaction and that additionally can complete the subsequent fertilization of zona-intact eggs.
4. Several published results indicate that the postmating transition(s) from the repressed state to the fully mature, fertile state may be at least partially reversible. Specifically, treatment of populations of fertile sperm with seminal or epididymal fluids results in a transient loss of fertilizing ability.[6,24,25] We will refer to such sperm as being in an *inactivated state*. The relationship of the inactivated, repressed, and resting states is not well characterized, since complete reversibility of the fertile state has not been demonstrated.

It must be emphasized that these functional states are referenced specifically to the fertility of sperm. They represent only a conceptual framework (of the authors' construction) assembled solely to support an examination of selected regions of current knowledge. These terms are not intended to reflect additional aspects of cellular function, such as motility and metabolism.

II. DETERMINATION OF THE FERTILE STATE: ASSAY METHODS

A number of systems are available for the determination of the fertile state. Initial assays were based on the fertilization of eggs (zona pellucida-encased or -denuded), which can result from simple coincubation with sperm populations *in vitro*. However, preliminary incubation of sperm alone decreases the time required to fertilize eggs that are added subsequently. Comparison of the rates of fertilization by sperm subjected to control and to treatment incubation conditions thereby provides a measure of the acquisition of fertility (and satisfies an accepted criteria for the completion of capacitation). However, such determinations of fertilization *in vitro* are cumbersome, frequently report on the fertility of a fraction of the cell population (due to the typical use of sperm/egg ratios >100:1), and necessitate additional controls before any failure of fertilization can be assigned to the sperm.

A second assay strategy is based on sperm-acrosomal exocytosis (the sperm acrosome reaction), a generally acknowledged requirement for both fertilization of eggs and of the completion of capacitation. Assessments of acrosomal status are much simpler than direct examination of fertility. For some rodent species (particularly the guinea pig and hamster), occurrence of the acrosome reaction can be determined by simple microscopic examination. Various staining procedures aid assessment of exocytosis in other species where the acrosomal organelle is less prominant[26-28] and sometimes discriminate between intermediate states of unknown significance.[29] The exposure of previously inaccessible acrosomal components also allows assays based upon associated enzymatic[30] or antigenic activity.[31,32]

Mammalian sperm display both agonist-independent ("spontaneous") and agonist-dependent acrosome reactions. Spontaneous exocytosis increases as populations of sperm attain the fertile state, and, in some cases, particularly with rodents, the rates are accelerated sufficiently as to pemit assessment of the fertile state.[12,14,33-35] Yet, the general utility of spontaneous exocytosis as an assay is questionable, since such acceleration varies across species and may not always provide a sufficient signal (as in the case of fertile-state bull sperm).[36,37] In much the same fashion, induction of exocytosis by exogenous, nonphysiological agents (Ca^{2+}/H^+ ionophores; lysophospholipids) may increase as sperm are incubated under conditions that support the acquisition of the fertile state.[37-39] Determination of responses to such agents could provide another assay scheme, although the association between fertility and exocytosis in these cases is not well established.

The incidence of zona-dependent exocytosis also increases as sperm enter the fertile state. A physiological agonist of the acrosome reaction of mouse sperm has been isolated from the zona pellucida, and strong evidence indicates that similar zona agonists are widely distributed among mammalian species (see Section III.A). Agonist-induced exocytosis is first evident (in mouse and bovine) when sperm enter the fertile state, as demonstrated by examination of time courses and by stimulators and inhibitors of the fertile state.[29,36] While both zona-promoted and (in some species) spontaneous acrosome reactions are potentiated by transition to the fertile state, the relationship between these two processes is not understood. For the purposes of this review, we will consider the ability to undergo acrosomal exocytosis in response to zona agonists as a hallmark of the fertile state.

Finally, the development of the fertile state is associated with alterations in a variety of sperm surface features. Alterations in lectin[40,41] and antibody[32,42-45] binding sites occur during appropriate incubations. Again, the precise relationship between these modifications and fertility is not certain.

In conclusion, a number of assay systems are available for detection of the fertile state, Perforce, the direct proof of functional maturation requires documented fertilization of eggs, despite the difficulties entailed. Yet, efforts to define the molecular events regulating transition to the fertile state have led to the development of a number of assays that are indirect and, in some cases, correlate to fertility in uncertain fashion. In addition, most of these assays are invasive, single-observation designs, and there is a need for the development of a vital labeling technique permitting determinations in living sperm. This is underscored by the recent applications of biophysical and biochemical methods to single spermatozoa[46-48] and by the refinement of gamete coincubation systems that allow fertilization with limited numbers of sperm.[14] Surface-directed reagents (e.g., antibodies)[32] or membrane probes (e.g., chlortetracycline) may provide the opportunity to correlate activity state with biochemical alterations in single cells.

III. CONTROL OF FERTILIZATION BY FUNCTIONAL MATURATION OF SPERM

A. INTERACTION OF FUNCTIONALLY MATURE SPERM WITH EGGS

1. Penetration to the Zona Pellucida Surface

Animal eggs are encapsulated in one or more investments which must be negotiated by fertilizing sperm.[11,49,50] For mammals, the zona pellucida, an acellular glycoprotein coat, immediately surrounds the egg. The zona pellucida of ovulated eggs is itself surrounded by cumulus oophorus cells, embedded in an extracellular matrix. In most mammalian species, the cumulus is retained through the time of fertilization.[10,11,50-52] Sperm penetrate these investments, and fusion of sperm with eggs occurs in multiple steps. In this section, we consider this reaction sequence and the associated interactions between functionally mature sperm and eggs. Subsequent sections address the mechanisms by which sperm may actively acquire these abilities. Certain aspects of these topics are considered in greater detail in other chapters in this volume.

Sperm penetration of the extracellular matrix of the cumulus oophorus can be considered the initial step in the cellular interactions of fertilization. Early studies focused on an essential role of acrosomal hyaluronidase in penetration of this matrix, and suggested that acrosomal exocytosis begins at its outer margin.[53-57] More recent reexamination of this question has led to the conclusion that acrosome-reacted sperm associate with cumulus cells, thereby becoming immobilized, whereas sperm with intact acrosomes readily penetrate to the surface of the zona pellucida.[51,56,58-60] The role of acrosomal hyaluronidase therefore remains obscure.

For many species,[31,61-66] though possibly not in all,[67-69] functionally mature sperm adhere by means of their apical plasma membrane to zonae prior to the initiation of acrosomal exocytosis. This initial gamete adhesion of the most studied sperm, those of the mouse, results from the interaction of serine/threonine-linked (O-linked) oligosaccharides on the zona pellucida glycoprotein, ZP3, with a galactose-binding molecule on the sperm surface.[31,49,50,70,71] This carbohydrate recognition component of sperm has not been unequivocally identified, and possible candidates include glycosyltransferases,[72-75] proteases or protease-inhibitor binding sites,[76-80] and other oligosaccharide recognition proteins.[10,11,81] None of these are yet known to interact specifically with ZP3.

2. Initiation of Acrosomal Exocytosis

It appears that zona-derived glycoproteins function as primary agonists for the acrosome reaction of many mammalian species during natural mating. It remains uncertain whether secondary agonists (perhaps present in the reproductive milieu) or nonagonist-induced pathways (perhaps related to the spontaneous acrosome reaction *in vitro*) may serve as alternative, auxillary, or fail-safe mechanisms. Such questions of physiological relevance likely will be debated for some time to come. It is apparent that development of the fertile state is associated with enhanced rates of spontaneous[10,11,35] and zona-promoted exocytosis.[23,29]

In recent years the signal transducing pathways associated with agonist-induced responses (including exocytosis) of somatic cells have been investigated intensively. Certain common themes have emerged from such studies. These include the presence of high affinity, transmembrane receptors that in some cases themselves have associated, agonist-responsive enzymatic activity (i.e., the protein-tyrosine kinase activity of the receptors for insulin and for several growth factors[82] and the guanylate cyclase activity of the receptors for ANF and speract[83,84]), or are themselves ion-conducting channels (i.e., nicotinic cholinergic and GABA receptors[85]). In some instances part or all of the actions of the agonist are transmitted by the dissociation of G-proteins affected by receptor occupancy and resulting in alterations in the activity of downstream targets.[86]

a. Transduction of the Zona Agonist Signal

Following initial gamete adhesion, sperm undergo the acrosome reaction in response to specific agonists such as the ZP3 component of the mouse zona pellucida. A variety of evidence indicates that functionally homologous elements in the zonae of other species initiate exocytosis.[23,36,59,62,64,87,88] A GTP-binding regulatory protein (G-protein) apparently mediates an early step in the agonist-activated signal transduction pathway. The acrosome reaction induced by structurally intact zonae, by soluble extracts of zonae, and by purified agonist are all inhibited by pertussis toxin (islet-activating protein),[48,89,90] a relatively specific inhibitor of some G-proteins.[91] These toxin targets include those of the G_i class known to be components of mammalian sperm, as demonstrated by toxin-catalyzed ADP-ribosylation *in vitro*, by protease mapping, and by immunoreactivity methods.[89,90,92,93]

Available evidence indicates that G-proteins function in transduction of the instructive signal provided by occupancy of mammalian sperm receptors for the zona-agonist of acrosomal exocytosis.[48,89,90] The immediate targets of activated (dissociated) G-proteins, as well as those of other putative signal transduction pathways, are not yet functionally characterized and knowledge of other intermediate steps is, at best, fragmentary. Rather than catalog the incomplete (and often contradictory) evidence, we will consider instead recent work from this and other laboratories focused upon those early events responsible for elevations of Ca_i and pH_i. Comparisons between the mammalian and sea urchin systems are

TABLE 1
Effects of Organic and Inorganic Calcium Channel Antagonists on the Agonist-Induced Acrosome Reaction of Bull Sperm

Compound	IC_{50}
Nitrendipine	500 nM
Nisoldapine	500 nM
Verapamil	700 nM
Ni^{2+}	50 μM
Co^{2+}	100 μM

justified, as demonstrated by several indicators of similarity: such comparison will be drawn where possible.

b. Central Role of Ca^{2+} and Ca^{2+} Channels

A nearly complete chain of evidence indicates that the acrosome reaction promoted by the primary zona-agonist is a consequence of elevations of Ca_i resulting from induced entry of external Ca^{2+}. Links in this chain include (1) a dependence upon external Ca^{2+} for acrosome reaction induced by Ca^{2+}-selective ionophores.[30,94-96] It is generally assumed that agonist-induced exocytosis also requires influx of external Ca^{2+}; (2) a blockade of exocytosis by specific inhibitors of Ca^{2+} entry (see below); (3) an elevation of Ca_i in individual sperm destined to subsequently undergo the acrosome reaction in response to treatment with solubilized zona components; and (4) a suppression of both exocytosis and Ca_i elevations induced by pertussis toxin.[48] Missing yet is documentation of similar response to purified agonist *in vitro* and perhaps ultimately of elevations of Ca_i in the fertilizing sperm observed *in situ*, following natural mating.

Similar evidence indicates a central role of Ca_i in the agonist-induced acrosome reaction of marine invertebrate sperm.[47,97-99] Briefly, an obligatory entry of external Ca^{2+} in response to a (partially characterized) fucose-sulfate glycopolymer of the egg surface also results in elevation of Ca_i required for the acrosome reaction.[99,100]

Both mammalian and sea urchin sperm apparently possess voltage-gated Ca^{2+} channels with properties resembling those of the L-type Ca^{2+} channel of vertebrate cardiac and skeletal muscle. In sperm of both the ram[101] and bull[102] elevation of external K^+ promotes an elevation of Ca_i that (1) is prolonged in duration, but reversible upon return to media of more physiological ionic composition; (2) depends upon both external pH and $[Ca^{2+}]$; (3) displays characteristic pharmacological sensitivities to dihydropyridine, phenylalkylamine, and benzothiazapine inhibitors; and (4) may require large depolarizations (estimated from external $[K^+]$ dependence). For sea urchin sperm, binding sites of relatively high affinity for Ca^{2+} channel antagonists were found in plasma membrane preparations[103-105] and antagonist-sensitive Ca^{2+} conductances have been demonstrated by reconstitution into lipid bilayers.[47] A dihydropyridine-sensitive Ba^{2+} conductance recently was reconstituted from boar sperm membranes.[106] Further application of the tools of electrophysiology may be expected to yield more complete characterization of these and other ion channels present in the surface membranes of sperm.

Pharmacological evidence implicates voltage-gated Ca^{2+} channels of sperm as regulatory elements involved in elevation of Ca_i during agonist-induced acrosomal exocytosis. The acrosome reaction of bull sperm responding to solubilized zonae (Table 1) and of sea urchin sperm responding to egg jelly[99,103,105,105a] are both inhibited by preliminary treatment either with potent Ca^{2+} channel antagonists or with transition metal ion inhibitors.

In other circumstances, however, dihydropyridine[107] or phenylalkylamine[108] antagonists are ineffective inhibitors of spontaneous acrosomal exocytosis. Higher concentrations of phenylalkylamines (D600, verapamil) paradoxically induce acrosomal Ca^{2+} deposition[109] and exocytosis in rodent sperm.[107,110] Because verapamil inhibited Na^+-dependent Ca^{2+} release from ram sperm membrane vesicles,[111] a putative Na^+/Ca^{2+} exchange mechanism was suggested to mediate entry of Ca^{2+} during the verapamil-induced acrosome reaction.[109] It should be noted, however, that a variety of tertiary amines (including the related Ca^{2+} antagonist, D600; the local anesthetic dibucaine; and the calmodulin antagonist, trifluoperazine), when present at concentrations approaching 1 m*M*, all induce similar alterations to bull sperm motility and metabolism.[110] These effects are probably mediated by Ca^{2+}, but the underlying mechanism, although remaining unclear, is unlikely to be highly selective in nature.

Other putative cellular components involved in maintenance of sperm Ca^{2+} homeostasis also could conceivably participate in regulatory elevations of Ca_i. Mechanisms operating at the plasma membrane include a Ca^{2+}-ATPase activity[112-115] that possibly represents the major cellular transporter responsible for extrusion of Ca^{2+}.[114,116] Additional Ca^{2+} transport mechanisms may exist, including a Na^+-coupled antiporter.[111,117] Recently, an interdependence of proton and Ca^{2+} transport in bull sperm preparations treated with urea and EGTA was interpreted as evidence of direct $Ca^{2+}/2H^+$ exchange activity,[118] although other interpretations can be applied as well.

Regulation of Ca_i also might involve sequestration or mobilization from internal stores, which are restricted in the sperm cell to mitochondria and the acrosome. The single mitochondrion at the base of the sea urchin sperm accumulates Ca^{2+} following the egg jelly-induced acrosome reaction.[119] Because such accumulation continues even after sperm become immotile and infertile, it is presumably a consequence of the loss of surface membrane permeability barriers. This mitochondrial accumulation has received little attention and no physiological role is apparent. In other tissues such accumulation is often, but not always, irreversible and is associated with senescent or pathological changes.[120]

In mammalian sperm, as in most somatic cells, mitochondria will accumulate Ca^{2+} when plasma membrane permeability barriers are compromised by hypotonic lysis[121] or by treatment with cholesterol-directed agents, such as digitonin.[116,122] Mitochondrial accumulation of Ca^{2+} has not been examined following the agonist-induced acrosome reaction of mammalian sperm. Yet, such Ca^{2+} uptake might not be expected since motility is maintained in these cells during the postexocytotic phase of penetration through the zona. This indicates that flagella membranes remain intact during this period. Ca^{2+} is accumulated during incubations that produce the spontaneous acrosome reaction of guinea pig sperm,[123,124] but the relationship between mitochondrial sequestration and exocytosis is unknown.

The most unusual feature of the sequestration of Ca^{2+} by sperm mitochondria is its occurrence in apparently normal populations of resting-state epididymal sperm.[116,122,125] However, it is difficult to exclude the possibility of selective accumulation by a damaged subpopulation of cells. Although electron microanalysis of individual sperm indicated that mitochondrial accumulation was uniformily distributed,[126] only a limited number of cells were examined.

In contrast to epididymal sperm, ejaculated sperm of the bull accumulate Ca^{2+} much less avidly.[118,127] Although the physiological significance of this developmental alteration in the ability to sequester Ca^{2+} remains to be established, the responsible seminal peptide, caltrin, has been characterized.[117,128] Immunologically cross-reactive peptide components of seminal plasma also inhibit mitochondrial sequestration by sperm of at least one rodent species.[129] However, ejaculated sperm of the ram accumulate as much Ca^{2+} as reported for epididymal bull sperm.[130] Thus, further studies of sperm from a number of species are required in order to determine the universality of this developmental alteration.

Mammalian sperm are not known to possess reticular structures that might represent mobilizable depots for Ca^{2+}. However, it has been speculated that the acrosome might itself accumulate Ca^{2+} by active processes.[10,11,35] No direct evidence exists of either uptake or release of Ca^{2+} from this organelle.

In summary, agonist-induced entry of Ca^{2+} is a central obligatory event in the acrosome reaction of both mammalian and echinoderm sperm. Current available evidence also indicates that activation of voltage-gated Ca^{2+} channels is an early component of the signal transduction pathways of both sea urchin and mammalian sperm responding to egg agonists of exocytosis. Very recent evidence indicates that alternatively, ligand-gated, mechanisms also may mediate Ca^{2+} entry into both mammalian[129a,129b,191] and sea urchin[105a] sperm. It is probable that G-protein activation, possibly in association with activation of other putative signal-transducing pathways,[131] leads directly, or indirectly, to alterations in the cellular machinery controlling the balance between influx and efflux of Ca^{2+}.

c. *Regulation of pH_i in the Absence and in the Presence of Agonists*

Elevation of the pH_i of echinoderm sperm also apparently is associated with, and obligatory for, the acrosome reaction promoted by its primary agonist.[99] Early work employing ion substitution and ionophore treatments[132] indicated that such cellular alkalinization is necessary for the actin polymerization that produces acrosomal filament extension (formation of the ''acrosomal process'') and, therefore, for the completion of the acrosome reaction. Moreover, treatment of sperm with a monoclonal antibody to a major 210-kDa membrane glycoprotein increases Ca_i while having no effect on pH_i. Acrosome reactions fail to occur in response to this Ca_i elevation unless pH_i is also elevated by treatment with NH_4^+.[133,134] In contrast, NH_4^+ alone induces exocytosis when applied at sufficiently high concentrations.[135] Such acrosome reactions, as well as those promoted by alkaline (pH 9) seawater,[136] are each inhibited by Ca^{2+} channel antagonists and therefore apparently depend upon secondary activation of Ca^{2+} channels. Thus, obligatory but separable polymerization and membrane fusion events, initiated by elevations of pH_i and Ca_i, apparently are both required for the acrosome reaction of sea urchin sperm.

Elevations of mammalian sperm pH_i also apparently are associated with the agonist-induced acrosome reaction. Solubilized zona-agonists induce elevations of bull sperm pH_i[48] and loss of pH gradients across the mouse acrosomal membrane.[90,137] Preliminary video imaging[277] indicates alkalinization of all cellular compartments is associated with agonist-induced exocytosis in bull sperm. In addition, it is widely assumed (despite only limited supporting data)[138-140] that the sperm acrosome, in analogy to other lysosomal organelles, maintains an acidic interior. Although mammalian sperm do not form an acrosomal filament, and are not known to possess pools of soluble actin, various alternative regulatory consequences of acrosomal alkalinization have been proposed. These include autocatalytic activation of protease precursors.[10]

For most somatic cells it is thought that Na^+/H^+ and Cl^-/HCO_3^- exchange activities, respectively, maintain resting pH_i in the face of acidic and alkaline challenge.[141] In addition, these pathways apparently are regulatory targets of both receptor-tyrosine kinase and G-protein-linked signal transduction pathways that alter pH_i as required for various cellular responses.[142] Although the deduced sequence of the human growth factor-activatable Na^+/H^+ exchanger and of the Cl^-/HCO_3^- exchanger each contain several possible phosphorylation sites in their putative cytoplasmic C-terminal domains, the molecular basis for regulation remains to be established.[143,144]

Several studies employing ion replacement and tracer flux measurements indicate that Na^+/H^+ exchange probably also functions both in pH_i homeostasis and in the *egg peptide*-induced pH_i elevation for sea urchin sperm.[145-147] However, the *egg jelly glycoconjugate* agonist of the acrosome reaction elevates pH_i in the absence of external Na^+,[140] indicating that other pH-regulatory mechanisms also function in echinoderm sperm. Among other possibilities, it was suggested[148] that the Na^+/K^+ ATPase may contribute to pH_i regulation.

The spectral properties of intracellular carboxyfluorescein, or of the related BCECF probe, indicate a relatively acidic pH_i (6.3 to 6.7) in epididymal or ejaculated sperm of the bull,[48,149] ram,[101] and pig.[150] These estimates were supported by noninvasive ^{31}P-NMR determinations of a similar pH_i in dense suspensions of bull sperm.[151] However, it was noted recently[277] that a new, seminaphthofluorescein indicator (CSNAFL-2) reports bull sperm pH_i nearly 0.5 pH units higher than previously reported values. The basis for these discrepant findings is still under investigations but apparently results, at least in part, from different selective intracellular compartmentation of the newer and older indicators.

The previous acidic determinations of the pH_i ($\approx$6.5) of sperm examined in physiological media (pH_o 7.4 to 7.5) are approximately consistent with electrochemical equilibrium of H^+_i and H^+_o, as predicted from estimates of sperm membrane potential (≈ -60 mV). However, mammalian sperm actively regulate pH_i since (1) in mildly acidic media ($pH_o \approx$ 6.5) both bull and ram sperm maintain pH_i at values distant from electrochemical equilibrium;[101,149] (2) a large H^+ permeability is observed in membrane vesicles isolated from bull sperm;[149] and (3) the more recent alkaline estimates of pH_i argue against passive control.

The mechanisms that control pH_i in mammalian sperm have not been characterized. In particular, the involvement of Na^+/H^+ exchange in this regulation is not clear. For example, epididymal rat sperm are immotile in Na^+-free media but reinitiate motility upon addition of Na^+, with an associated, massive proton release. Thus, it was inferred that Na^+/H^+ exchange promoted a required elevation of pH_i.[152] However, Li^+ failed to substitute for Na^+ in reactivation, in contrast to the properties of the Na^+/H^+ exchanger of somatic cells[142] or with the putative Na^+/H^+ exchanger of echinoderm sperm.[153,154]

In other circumstances, pH_i of sperm is altered by mechanisms that also display features inconsistent with operation of Na^+/H^+ exchange. Epididymal sperm of the ram and bull do not recover rapidly from an acidic load imposed by treatment with permeant weak acids[149] or by transient exposure to ammonia.[278] Resting pH_i is sensitive to alterations of the membrane potential of both mammalian[101,149] and sea urchin[153,154] sperm. The resultant alterations in pH_i are insensitive to amiloride and to related inhibitors of the major Na^+/H^+ exchanger of somatic cells.[101,154] Moreover, for bull and ram sperm as determined by spectral probes,[101,155] and for ram sperm, as determined by weak base distribution[279] techniques, resting pH_i is affected little, if at all, by adjustments of transmembrane Na^+ and pH gradients. Thus, it is unlikely that Na^+/H^+ exchange is the principle regulator of pH_i, at least in the absence of agonist.

The role of alternative mechanisms for sperm pH_i regulation is not understood. Bicarbonate is required for the agonist-induced acrosome reaction of mouse[156] and bull[36] sperm. Bicarbonate also plays some role in the spontaneous acrosome reaction of guinea pig sperm.[108,157] It is uncertain whether these observations relate to involvement of HCO_3^-/Cl^- exchange or to the apparent direct stimulatory action of HCO_3^- on sperm adenylate cyclase.[150,158,159] Examination of anion flux activity in porcine sperm[160] indicates that external Cl^- partially inhibits HCO_3^- uptake, as do several inhibitors of the anion exchanger of somatic cells. Reported estimates of V_{max} ($\approx$133 pmol $SO_4^{-2} \times 10^{-8}$ cells $\times$ min^{-1}) of the putative anion exchanger of sperm can be calculated to correspond to $\approx$15 pmol $\times$ $cm^{-2} \times s^{-1}$, comparable to the values reported for both erythrocytes and ascites tumor cells.[161] Most published examinations of sperm pH_i were conducted in nominally HCO_3^--free media. The effects upon sperm pH_i of the systematic replacement of external Cl^- have not been reported. The lack of this and other more direct evidence indicates that control of sperm pH_i by HCO_3^-/Cl^- exchange is still speculative.

In addition to pH_i regulation at the plasma membrane, it is generally believed that acid equivalents are transported across the acrosomal membrane. The weak base, aminoacridine, is selectively accumulated in acidic compartments, but also binds to nucleic acids and other cellular components. Correction of measured distributions for such binding allowed estimations of pH_i of sea urchin sperm and examination by fluorescence microscopy detected a small, apparently acidic, acrosomal compartment. A similar localization of fluorescence was noted in the head of guinea pig and mouse sperm.[140] The acridine fluorescence of hamster and mouse sperm heads decreased following treatment with either ammonia or with the ionophore nigericin, presumably as a consequence of discharging transmembrane proton gradients.[137-139] From determined distributions and estimated volumes (but without corrections for binding), an acidic intraacrosomal pH (pH <5) was calculated for hamster sperm. Similar pH_i estimates were provided from distributions of the weak base, methylamine, which also presumably was accumulated preferentially in an acidic cellular compartment.

The ability of a variety of relatively nonselective inhibitors to discharge aminoacridine fluorescence in hamster sperm[162] was interpreted as evidence that the widely distributed vacuolar H^+-ATPase, responsible for the acidification of lysosomes and other organelles in somatic cells,[163] might also function to maintain an acidic acrosomal compartment in sperm. Although this remains an appealing possibility, more definitive evidence is not yet available.

Thus, pH_i is regulated both in the absence of agonist and in agonist-treated sperm. A number of lines of evidence indicate that a Na^+/H^+ exchange pathway is absent or, if present, then functions either as a minor contributor to pH_i control or displays a number of unique features not observed in its somatic cell counterpart. The nature of other regulatory transport systems has not been determined.

d. Cross-Talk between pH_i and Ca_i

In sperm, pH_i and Ca_i are controlled by distinct but interrelated mechanisms. For sea urchins, egg jelly promotes sperm pH_i elevation in Ca^{2+}-free seawater, even though exocytosis fails to occur. This alkalinization is transient in complete seawater but the effect persists in Ca^{2+}-free conditions.[140] In addition, the monoclonal antibody J10-14 elevates Ca_i without producing either an internal alkalinization or acrosomal exocytosis.[133,134] It is clear then that in sea urchin sperm Ca_i does not control pH_i. In contrast, an uptake of Ca^{2+} (and presumably an elevation of Ca_i) is promoted in these cells by $NH_4{}^+$-induced alkalinization.[135] It is still unclear whether this effect is exerted directly or indirectly upon putative Ca^{2+} channels of sperm or upon other components of Ca_i homeostasis.

A similar, regulatory action of pH_i on mammalian sperm Ca_i was noted recently. The rate of the Ca_i increase that followed depolarization of bull sperm was rapidly and reversibly increased or decreased when pH_i was intentionally elevated or depressed with permeant weak bases or acids.[102] The relationship of agonist-induced elevations of pH_i and Ca_i, such as are necessary for exocytosis, is not yet known, but may require that a permissive pH_i be achieved prior to elevation of Ca_i. There are two critical and related consequences of these observations. First, the modulating action of pH_i on Ca_i homeostasis may also be important in the spontaneous acrosome reaction, which is often examined in media whose bicarbonate, permeant-weak acid or ionophore content may affect pH_i. Such alterations of pH_i may produce unanticipated ramifications. Second, the effects of J10-14 on sea urchin sperm suggest that, at least in those species, an elevation of Ca_i, in the absence of other associated alterations, is not sufficient to induce exocytosis. Whether a similar circumstance applies to mammalian sperm is uncertain.

e. Sperm Membrane Potential

An inside negative, K^+-dominated membrane potential was determined for sea urchin sperm by both isotope distribution[140,164] and fluorescence[100,165,167] techniques. Early application of these methods to mammalian sperm was less successful. Cyanine dye fluorescence determinations in suspensions of guinea pig sperm[168] indicated an inside positive potential ($\Delta\psi \approx +13$mV) that perhaps reflected dye toxicity. More recent cyanine dye fluorescence determinations[278] in bull epididymal sperm suspensions indicate somewhat higher resting potentials (≈ -70 mV). This resting potential was affected by transmembrane $[K^+]$ gradients, although a mitochondrial contribution most likely is present. Measurements of permeant cation distributions in epididymal sperm of the ram indicate resting potentials of ≈ -40 mV[279] that are consistent with this interpretation.

Both isotopic probes[169] and the cyanine[165] and bisoxonol[100] dyes indicate that sea urchin sperm depolarize in response to egg jelly, in agreement with the earlier observations of ionic movements during the agonist-induced acrosome reaction.[99,145,170,171] In addition, the more rapidly responding cyanine indicators also detected an earlier, transient hyperpolarization of sea urchin sperm treated with agonist.[135,136] Because the agonist-induced hyperpolarization requires the presence of external Ca^{2+}, it may involve opening of Ca^{2+}-activated K^+ channels. The egg jelly peptide, speract, also promotes a hyperpolarization of isolated flagellar membrane vesicles. This hyperpolarization is suppressed by elevations of external K^+. However, the K^+ movements promoted by speract are resistant to the K^+ channel blocker, TEA^+; are enhanced by GTP; and are partially mimicked by GTP γ-S[47] This suggests that G-proteins may mediate K^+ channel opening in the speract signal transduction pathway.

In summary, most available evidence is consistent with the interpretation that for echinoderm sperm the earliest responses to egg jelly occur within seconds. This results in alterations in membrane potential that secondarily elevate pH_i and Ca_i, probably by actions on voltage-dependent exchangers and channels. The involvement of voltage-gated Ca^{2+} channels in the acrosome reaction of mammalian sperm permits speculation concerning a preceding, zona-induced depolarization. As yet, no actions of zona agonists on mammalian sperm membrane potential have been described.

f. Downstream Events in Sperm Signal Transduction

The presence of calmodulin in sea urchin and mammalian sperm is well documented[172-177] and is likely to mediate many of the consequences of any elevation of Ca_i. A number of potential targets for the Ca^{2+}-calmodulin complex probably are present in sperm. These include adenylate cyclase,[178,179] cyclic nucleotide phosphodiesterase,[180-182] protein phosphatase,[183,184] Ca^{2+}-ATPase,[185] phospholipase A_2,[186] unidentified flagellar elements controlling flagellar waveform,[187] and perhaps sperm Ca^{2+} channels themselves.[100]

However, pharmacological treatments directed toward calmodulin action have produced conflicting evidence for the involvement of calmodulin-mediated processes in acrosomal exocytosis. In sea urchin sperm calmodulin antagonists (W7 and trifluoperazine) block the elevation of Ca_i and the acrosome reaction induced by egg jelly, apparently by preventing opening of Ca^{2+} channels.[100] The ability of calmodulin antagonists to block the agonist-induced acrosome reaction of mammalian sperm has not been examined. However, the most selective of these agents, calmidazolium, blocks fusion of acrosome-reacted human sperm to denuded eggs,[188] suggesting that calmodulin may have alternative or additional roles in fertilization. Although multiple calmodulin-binding peptides are present in mammalian sperm,[188,189] their functions are unknown.

Other potential effector systems are present in sperm and conceivably also may participate in signal transduction leading to sperm exocytosis. Induction of acrosome reactions in mammalian sperm, either by application of the divalent cation/proton ionophore, A23187, or by addition of nonzona agonists, causing hydrolysis of phosphatidylinositides and the associated production of polyphosphoinositols.[96,190,191] Production of these potential second messengers[192] in sperm is dependent upon prior entry of Ca^{2+},[96,191] yet their role in the signal transduction process is undetermined.

In summary, the fragmentary nature of current knowledge does not allow construction of an integrative picture of the downstream pathways that couple Ca_i elevation to exocytosis. Recent characterization of a permeable sperm model system that undergoes Ca^{2+}-induced swelling of the acrosomal matrix and release of contained peptides[193] may facilitate investigation of the possible roles of Ca^{2+} and calmodulin in such downstream events.

B. REGULATION OF GAMETE INTERACTION BY FUNCTIONAL MATURATION OF SPERM

1. Early Events: Cumulus Penetration and Zona Pellucida Adhesion

Attainment of fertility by functional maturation of sperm is not accomplished in a single step, but rather by events that separately affect different components of the gamete interaction sequence. Regulation is first apparent during sperm penetration of the cumulus oophorus matrix. Sperm that either are not functionally mature, or have already completed the acrosome reaction, are apparently unable to penetrate the cumulus matrix.[51,56,59,60,194] Thus, the cumulus may act as a physiological, although not a physical, filter and admit only a subpopulation of sperm that are capable of interacting with eggs. The selectivity mechanism has not been described, due in part to a limited understanding of the physical and biochemical milieu of the cumulus matrix. It has been speculated that the expression of ability to penetrate the cumulus oophorus reflects cell surface modifications associated with functional maturation.[51,195]

The species-selective, initial adhesion of sperm to the zona pellucida also apparently develops during the transitions that follow mating. This acquisition is best documented for cauda epididymal sperm of rodents. Such sperm, initially examined in the resting state immediately upon release *in vitro*, establish only weak, reversible contacts with the zona pellucida that are nonspecific in nature.[64,196-199] The capacity for specific adhesion is acquired during incubation, also presumably as a consequence of functional maturation.

Yet, adhesion competence and fertilizing ability can be differentiated, at least in the mouse, based on (1) kinetics, as zona adhesion develops more rapidly than fertility ($t_{1/2}$ of 5 to 10 min and 1 h, respectively); and (2) incubation conditions, as adhesion depends primarily on Ca_o (although without absolute specificity), whereas fertility displays more complex media requirements.[200-203] Thus, expression of adhesion activity is regulated in sperm and can be considered a component, though not a rate-limiting one, of the functional maturation process.

The mechanisms that account for the suppression of adhesion competence in the resting state and for its eventual expression during maturation are not completely understood, but may involve inhibitors present in epididymal fluids. Epididymal fluid, or material dissociated from resting-state mouse sperm, can block the adhesion of epididymal sperm to zonae.[204-206] At least one active factor may be an epididymal fluid oligosaccharide, characterized in preliminary fashion as poly-*N*-acetyllactosamine, that binds reversibly to carbohydrate binding proteins on the sperm.[205,206] Such exogenous ligands could compete at sperm surface adhesion sites, blocking access of zona pellucida oligosaccharides. Adhesion activity is also regulated either directly or indirectly by Ca_o, which remains relatively low until sperm reach the oviduct.[207] Ca^{2+}-dependent, carbohydrate-recognition proteins exist elsewhere[208] and such a component might mediate sperm interaction with the oligosaccharide adhesion sites on ZP3.[70]

2. Regulation of Exocytosis During Functional Maturation

The rates of both spontaneous and agonist-induced sperm exocytosis are influenced by functional state. For example, low rates of exocytosis are observed in resting or repressed-state cells. Subsequent maturation enhances the rate of spontaneous acrosome reaction by as much as 10-fold in some (though not all) species,[14,35] and further amplifies the rate of zona-evoked exocytosis[23,29] by as much as 50-fold.[33] Agonist-induced exocytosis and fertility *in vitro* increase with similar time courses during incubation and are similarly effected by stimulatory and inhibitory agents.[23,29] Enhanced spontaneous exocytosis and agonist sensitivity increased coincidently with fertility, suggesting that these are late, possibly rate-limiting, events in the maturation process.[14,23,36]

The mechanisms that account for increased response to agonist by fertile sperm have not been determined, although two general (and nonexclusive) models can be considered.

First, fertility may be suppressed through the action of endogenous competitive antagonists that bind to, and hence occlude, agonist receptors on the sperm surface. In the simplest case, such antagonists would act exclusively at the sperm surface to prevent the initiation of signal transduction. A possible role of competitive inhibitors in the expression of zona adhesion competency in mouse sperm has been discussed. It is a plausible speculation that a similar mechanism may control agonist-induced exocytosis. Such adsorption and release of seminal and epididymal fluid components to sperm will be considered in detail (Section IV).

A second potential means of regulation is through alterations of key intracellular components necessary for effective response to zona agonists. Phosphorylation by (cAMP-dependent and by other) protein kinases alters signal transduction in a variety of somatic systems[213-215] and in the response of sea urchin sperm to chemotactic peptides.[83,84] There are several indications that increased adenylate cyclase activity is associated with the acquisition of fertility *in vivo* and *in vitro,*[209-212] although these must be considered in light of the inharmonious reports of the effects of permeant cAMP analogs on the spontaneous acrosome reaction.[10,11,20] Thus, mediation of the zona-induced acrosome reaction by phosphorylation mechanisms has not yet been demonstrated.

Sperm Ca_i may mediate alternate sites of regulation during the acquisition of fertility. Accumulation of Ca^{2+} is observed during incubations *in vitro* that lead to spontaneous exocytosis,[123,124] although the relative contribution of mitochondrial accumulation to this process is uncertain.[126] The development of hyperactivated motility in resting-state hamster sperm is associated with an elevated Ca_i, as determined by quin 2 fluorescence.[217] Fertility was not determined directly in this study, although the observed alterations in motility are associated with acquisition of fertility.[10,11] In contrast, no elevation of Ca_i, reported by quin 2 fluorescence, accompanies the repressed/mature transition of rabbit sperm.[218] The basis for these differences is unclear and at present conclusions about regulation of Ca_i are premature.

Zona agonists produce elevations of Ca_i in fertile-state bull sperm, whereas resting-state cells failed to display such increases.[48] Although the apparent resting Ca_i of these cells did not differ significantly, agonist regulation of Ca_i is altered as a consequence of the acquisition of fertility. Further work is necessary to identify the site at which signal transduction fails in infertile, immature cells.

Other intracellular ions may also promote exocytosis, either by causing secondary rises in Ca_i or by interacting synergistically with Ca^{2+}. Such a possible role of internal alkalinization in the agonist-promoted acrosome reaction of bull and echinoderm sperm, as well as the potential relationship between Ca_i and pH_i, was considered above. A number of studies have examined pH_i regulation during transitions to the fertile state, and its possible mediatory role in spontaneous exocytosis. Fluorescence indicator dyes report alkalinization during maturation-supporting incubations of resting-state hamster sperm.[138,139,217] The role of pH_i elevations in maturation, or in supporting exocytosis, remains controversial. However, it is worth noting that for somatic cells, stimulus-related alkalinizations or acidifications may reflect *in vitro* incubation conditions.[141,219]

Nevertheless, it may be speculated that maturation-associated elevations of sperm pH_i play an important supporting role in exocytosis. In agonist-induced exocytosis Ca_i is regulated by voltage-gated Ca^{2+} channels.[102] Potential cross-talk between Ca_i and pH_i was discussed previously. In this regard, it is interesting that changes have also been reported in sperm membrane potential during maturation.[220] It may be that the relatively acidic pH_i in resting and repressed-state sperm[48,101,149] prevents premature exocytosis by minimizing channel openings.

Finally, mediatory roles of HCO_3^-, Na^+, K^+, Mg^{2+}, and Zn^{2+} have been proposed in spontaneous or zona-evoked exocytosis[156,221-226] and reviewed elsewhere.[10,11,14] In general, these arguments are based on studies of ion depletion, ionophore-promoted ionic movement, and on ionic accumulation into unspecified compartments. At present, a direct demonstration of a mediatory role for these ions in exocytosis has not been provided. Thus, there is evidence that a number of ionic species can function as regulators of sperm exocytosis, although direct association with the acrosome reaction is available only in the case of the agonist-induced process and only then for Ca_i and pH_i.

In summary, the mechanisms that account for enhanced rates of spontaneous and agonist-induced exocytosis in fertile sperm are not well understood. Considerable attention has focused on modulations of intracellular mediators that may occur during maturation and may produce spontaneous acrosome reactions.[10,11,16,20] Yet, such hypotheses remain speculative. In addition, there is evidence that early events in the agonist-activated signal-transducing pathway, such that occur prior to the elevation of sperm Ca_i and pH_i, may be regulated during acquisition of fertile competency in (bull) sperm.

IV. SPERM ACTIVITY STATES: MECHANISMS OF STABILIZATION AND OF TRANSITION

In this section we will briefly consider how activity states are established and maintained in sperm and how transitions between states are driven. Focus will be on the limited number of activity states defined in this review: readers concerned with other areas of sperm capacitation, such as the development of "hyperactivated" motility, may be guided by more general reviews.[5,10,11]

A. ESTABLISHMENT OF THE RESTING STATE

Sperm are extensively modified during posttesticular progress and storage in the cauda epididymis. Structural alterations include changes in cellular (presumably membrane) lipid[16,227-231] and protein composition, the latter apparently due to the combined effects of dissociation to and adsorption from epididymal fluids as well as to processing of surface proteins.[208,232-238] This may account for the divese and general functional changes observed as sperm progress through the epididymis, including altered motility,[239] metabolism,[240,241] membrane protein[242] and lipid diffusion rates,[243] and the acquisition of the potential to interact with eggs.[244-248]

Surface restructuring may also form the basis for the establishment and maintenance of the resting state, as specific components may be inserted or functionally altered. We have previously considered one example of this type of modification: the reversible association of poly-*N*-acetyllactosaminyl oligosaccharides from epididymal plasma with sperm surface sites. Such molecules may occlude sites and competitively inhibit sperm recognition of the zona pellucida.[73,206] Consider also the case of acrosome stabilizing factor, or ASF. ASF is a 260-kDa holoprotein, containing two copies each of a large (92 kDa) and small (38 kDa) subunit. Asparagine residues on each subunit are decorated with both high mannose and complex N-linked oligosaccharides.[249,250] The glycoprotein is synthesized by principle cells in the corpus region of the epididymis and is subsequently secreted into the epididymal lumen. ASF reversibly suppresses spontaneous exocytosis when added to sperm that were previously incubated under conditions supporting functional maturation; presumably similar inhibitory consequences ensue from its addition to epididymal sperm *in vivo*. As such, it could help establish the resting state. The assumed mechanism of action is by adsorption of ASF to sperm during transit through the corpus epididymis, thereby inhibiting constitutive exocytosis during later sperm storage in the cauda epididymis. Distal flow of epididymal plasma results in an apparent three- to four-fold concentration of ASF in cauda epididymal fluids, relative to levels in the corpus region.[250] This could drive reassociation of ASF with sperm surface binding sites exposed during prolonged storage. The identity of these sperm surface receptors and the mechanism whereby ASF suppresses exocytosis remain unknown. Other negative modulators may also be present in epididymal fluids. For example, guinea pig epididymal plasma contains a low molecular weight factor that suppresses spontaneous exocytosis.[251]

In addition to a functional role of protein components, the ionic composition of epididymal fluid may contribute to the establishment and stabilization of the resting state. As sperm migrate along the epididymis the concentrations of Na^+ and K^+ vary greatly: Na^+ declines from >100 m*M* in caput fluid to <50 m*M* in the caudal storage region, whereas K^+ shows an opposite pattern, rising from 15 to 20 m*M* in the caput to approximately 40 m*M* in the cauda.[252] These shifts in the relative concentrations of Na^+ and K^+ might be expected to depolarize membrane potentials determined predominantly by K^+ permeability, and hence activate voltage-gated channels such as may participate in sperm acrosome reactions (see above). Such premature exocytosis may be prevented by the relatively low pH of 6.85 prevailing in cauda epididymal fluid.[253] As was discussed previously, such acidic conditions may limit the opening of sperm Ca^{2+} channels[102] and, consequently, of acrosome reactions. Additionally, the low concentrations of Ca^{2+} in epididymal fluids[240,252] may minimize spontaneous exocytosis.

Sperm acquire the capability of fertilizing eggs as they move through the epididymis. Yet, they are faced with the prospects of prolonged storage prior to mating. It is essential that they be maintained in a functionally inhibited state during this time, thereby limiting the spontaneous completion of events (e.g., acrosomal exocytosis) that would preclude fertilization. These cells, incapable of new gene expression, apparently achieve this end by the creation of an inhibitory microenvironment, containing specific proteins as well as a profile of inorganic ions that, acting in concert, impose and maintain a storage (and resting) state condition. Maturation is only possible following release from this microenvironment, as will be considered later.

B. THE NATURE OF THE REPRESSED STATE AND OF RESTING/REPRESSED TRANSITIONS

Addition of seminal fluids to cauda epididymal sperm propels these cells from the resting state into a repressed state. This transition is associated with additional alterations of sperm surface proteins, as a consequence both of adsorptive and dissociative processes[42,237,254-256] as well as in lipid composition.[16] In addition, changes in metabolic and motility patterns occur after mixture of sperm with seminal fluids.[11] It is not known whether this transition can be reversed.

Sperm in the repressed state are more resistant to functional maturation then are resting-state cells.[11] Surface modifications account for at least a portion of the inhibitory activity. One well-characterized example involves adsorption of a negative regulator from bovine seminal fluids. Resting-state sperm of this species accumulate extracellular Ca^{2+} during incubation *in vitro*. Flux through this uptake pathway, which may be Na^+-dependent,[117] is decreased by seminal fluids,[127] or by a small peptide (5411 M_r), caltrin, purified from those fluids.[117,128,256,257]

Caltrin may serve a number of functions during postepididymal sperm development. The inhibitory effects on Ca^{2+}-transport may contribute to the suppression of spontaneous acrosome reactions in repressed-state sperm. Ca^{2+} concentrations in seminal fluids may be 10 to 20-fold higher than in epididymal plasma: typical values are 7 to 10 and 0.25 to 0.85 m*M*, respectively.[240,241,252] An elevated drivng force for Ca^{2+} influx might be sufficient to trigger acrosomal exocytosis (see above): inhibition of that uptake could minimize such elevated spontaneous secretion. At least a portion of this inwardly directed driving force will be relieved by citrate, a Ca^{2+}-chelator present in high concentrations in seminal fluids (*circa* 20 m*M*).[240,241] Caltrin also could participate, by blocking uptake. In this regard, caltrin has been localized by indirect immunofluorescence to the distal flagellum and to the acrosomal region of the bovine sperm head,[256] consistent with a role in the suppressing spontaneous exocytosis. Inhibition of this uptake route may account for the slightly lower Ca_i

of cells in the repressed state, as compared to resting-state cells.[48] (A second role of caltrin will be considered later.) The sperm surface may additionally be modified by adsorption of other seminal fluid polypeptides,[237] as well as by enzymatic resculpturing, but the role of these processes in the establishment of the repressed state is uncertain.

Inorganic ions may also regulate the resting/repressed transition. Na^+ and K^+ are more concentrated in seminal fluids, at 80 to 90 m*M* each,[207,240] than in the epididymis.[252] The significance of the high concentrations of K^+ are not well understood. Divalent cations are also enriched in seminal plasma. Ca^{2+} levels are elevated, relative to the epididymal environment and to serum.[207,240,252] In addition, Zn^{2+} is present at >1 m*M* concentrations and may either be free in seminal plasma, accumulated by spermatozoa, or bound to a class of Zn^{2+}-binding proteins.[14,240,241] This cation may be significant in establishing the repressed state, as it will inhibit fertilization of mouse eggs *in vitro* when added to gamete cocultures.[14,258] In the hamster, inclusion of (nonspecific) Zn^{2+}-chelators in sperm incubation media supports resting/fertile-state transitions,[226] although it was not demonstrated that these agents acted as chelators.

Finally, it should be noted that functionally mature sperm can be subsequently inactivated by treatment with seminal or epididymal fluids.[11,24,204] It is generally presumed, though not demonstrated, that these sperm are driven back to the repressed state.

C. TRANSITIONS TO THE FERTILE STATE

Sperm can enter a fertile state from either the resting or the repressed state. This typically occurs in the female reproductive tract or in an appropriate environment *in vitro*. A number of convergent mechanisms drive these transitions.

First, negative regulators adsorbed to the sperm surface dissociate, or are otherwise neutralized. The loss of cell surface proteins during maturation *in vitro* and *in vivo* has been extensively documented. Moreover, shedding of specific macromolecules is reversible and may be associated with maturation status: that is, dissociation can be correlated with onset of fertility and, in some cases, the reassociation of released materials signals a loss of the fertile state.[43-45,204,255,259-262] Dissociation may expose relevant surface features, such as agonist receptors, similar to the mechanisms suggested to regulate zona adhesion activity in sperm.[205,206]

Other changes also occur at the sperm surface associated with maturation. Alterations occur in lipid composition.[16,264-268] lectin binding site distribution,[40,41,269,270] diffusion rates of membrane proteins[17,242] and lipids,[46] and in intramembranous particle distribution.[271,272] This reorganization is not generalized, but rather involves specific components. In addition to exposing surface sites, these modifications may alter the ability of sperm membranes to engage in membrane fusion reactions.

Alterations in membrane lipid composition, and particularly in the cholesterol/phospholipid molar ratio, may enhance the ability of sperm plasma membranes to fuse with the underlying domains of the acrosomal membrane. Postulated mechanisms for these effects include an alteration of Ca^{2+} permeability, the generation of bilayer-destabilizing lysophospholipids, and an alteration in membrane "fluidity". Changes of these types might be expected to influence the efficiency of membrane fusion. In addition, some experimental evidence supports a role of membrane lipids in functional maturation, although the mechanism by which they may act or even the obligatory nature of the modifications has not been defined.[11,16,265,266,273]

Second, positive regulatory elements are present on the sperm surface and their activity must be expressed. Caltrin, the seminal fluid peptide that regulates Ca^{2+} sequestration by bovine sperm, is an example of an extrinsic positive regulator. Cauda epididymal bovine sperm are relatively insensitive to agonists in the zona pellucida, requiring >25 mg zona protein per milliliter for a half-maximal response, whereas ≈20 μg/ml suffices with fertile-state, ejaculated sperm.[36] Signal transduction in epididymal sperm fails at an early step, prior to the elevation of pH_i and Ca_i by agonist.[48] High affinity, agonist response can be induced by incubating cauda epididymal sperm with seminal fluids[36] and is associated with a recoupling of agonist to pathways that elevate Ca_i (unpublished observations). Furthermore, the active constituent has been purified from seminal fluids: this activity comigrates with caltrin during gel electrophoresis and displays a similar ability to bind to dansyl-calmodulin (unpublished observations). It is therefore likely that caltrin plays a second role in sperm maturation, functioning as a positive regulator of the agonist-promoted acrosome reaction. The action of caltrin is stoichiometric rather than catalytic, consistent with adsorption of the peptide to the sperm surface.[36] Further details concerning the mechanism of action are not known.

While caltrin is apparently essential for agonist-promoted signaling in the specific case of bull sperm, the generality of this mechanism remains to be demonstrated. Cauda epididymal sperm from the mouse[62,64] and hamster[59] are unlike bull sperm in that they undergo acrosome reactions following treatment with the homologous agonist. It has been hypothesized[23] that molecules similar to caltrin are present in the rodent, but are synthesized and applied to sperm within the epididymis rather than upon addition of seminal fluids. Recently, an immunologically related (though, in general, structurally distinct) peptide was detected in guinea pig seminal vesicle extracts using antisera produced against bovine caltrin. It is uncertain whether similar peptides are also present in epididymal tissues.[129] Whether functionally similar peptides are present in rodent epididymal fluids, and can account for the agonist sensitivity of sperm from those regions, must be resolved. In addition, unrelated positive regulators may be present and added to sperm either from epididymal fluid, from seminal plasma, or from the female reproductive tract. For example, oviductal or follicular glycosaminoglycans may bind to sperm of some species and drive transitions into the functionally mature state.[37,274]

Posttranslational modification of sperm membrane components can also contribute to the development of fertility. The monoclonal antibody M42, that recognizes a 185/200-kDa doublet in resting-state mouse sperm, inhibits agonist-induced exocytosis.[238,275,276] During transit through the epididymis this antigen is altered from a high M_r precursor (195/210 kDa). When sperm from different regions of the epididymis were incubated (under conditions that support acquisition of fertility by resting state—i.e., cauda epididymal—sperm) it was found that the development of agonist sensitivity occurs at the roughly the same stage of epididymal transit as does the modification of M42 antigen. This study suggests that early events, such as those occurring within the epididymis (or, by extension, in the testis) may be essential for eventual development of fertility, but the positive effects of these alterations are transiently masked by negative regulators. Other forms of posttranslational modification include changes in protein phosphorylation. Recently,[131] it has been suggested that a limited set of proteins serve as substrates for sperm tyrosine kinases and that phosphorylation of these proteins may increase during the resting/fertile transition of mouse sperm. Further studies are required to demonstrate a causal relationship between these posttranslational modifications and fertility.

The inorganic ionic composition of the female reproductive tract may additionally regulate transition to the fertile state. There is a progressive rise in Na^+ to levels approximating that of serum (140 to 150 m*M* in the mouse) and a decline in K^+ from seminal fluid levels (15- to 20-fold that of serum) to concentrations in the oviduct only four to five times the serum representation.[14,207,240]

D. SUMMARY

A variety of strategies are used to regulate activity-state transitions in mammalian sperm. First, extrinsic negative control elements act to suppress or inhibit certain cellular functions. These agents may act as simple competitive inhibitors, as has been suggested for the mouse epididymal lactosaminoglycan that may modulate zona adhesion ability in sperm.[73,206] Additional or alternative mechanisms are plausible for other negative regulators. While active agents are being identified (i.e., ASF), the mechanism of action remains uncertain. Second, extrinsic positive regulatory elements are present and play an essential, supportive role in specific aspects of sperm-egg interaction. For example, caltrin functions as an essential modulator of the zona-induced acrosome reaction of bovine sperm. Again, further studies are required to illuminate mechanisms. Finally, the ionic composition of the extracellular fluids is highly specialized and may provide instructional signals to sperm.

V. CONCLUSIONS

We have attempted to review briefly our current understanding of the regulation of sperm fertility, focusing (although not exclusively) on the control of signal transducing pathways during sperm maturation. We have side-stepped many of the complexities of this area by addressing a set of specific activity states and have attempted to highlight the role of membrane and intracellular mediators of exocytosis.

The ability of sperm to fertilize eggs is regulated during final maturation at a number of points in the gamete interaction sequence, rather than at a single sensitive site. In addition, competence to accomplish individual components of the gamete interaction sequence (adhesion to zona; zona-induced exocytosis) is acquired separately, indicating that these processes are individually regulated. The mechanisms that mediate transitions to the fertile state are not completely understood. Nevertheless, progress toward such an understanding is apparent in at least three regards.

First, our knowledge of the events of gamete interaction has advanced greatly by the realization that a signal transduction system is present and highly conserved in sperm from mammalian and invertebrate species. This pathway, as it functions in fertile-state sperm, includes a central and obligatory elevation of Ca_i, most likely through G-protein-promoted activation of voltage-gated Ca^{2+} channels. Associated elevations of pH_i may also be essential for exocytosis. With this level of understanding, it becomes feasible to discuss potential mechanisms for the regulation of signal transduction during maturation. Second, grasp of the fundamental biochemical and physiological processes as applied to the sperm cell has also enlarged. Thus, demonstration and characterization of homeostatic pathways in sperm (e.g., ion transport proteins; second messenger-responsive effectors) provide a framework for examining the perturbations encompassed by functional maturation. Finally, in some cases putative instructive molecules have been isolated (ASF, caltrin) and studies have been initiated into their mechanisms of action. The identification of such regulatory molecules may provide the key to understanding certain component processes of sperm maturation.

REFERENCES

1. **Austin, C. R.,** Observations on the penetration of the sperm into the mammalian egg, *Aust. J. Sci. Res. Ser. B*, 4, 581, 1951.
2. **Austin, C. R.,** The "capacitation" of the mammalian sperm, *Nature (London)*, 170, 326, 1952.
3. **Chang, M. C.,** Fertilizing capacity of spermotozoa deposited in fallopian tubes, *Nature (London)*, 168, 997, 1951.
4. **Yanagimachi, R. and Chang, M. C.,** Fertilization of hamster eggs in vitro, *Nature (London)*, 200, 281, 1963.
5. **Austin, C. R.,** Sperm maturation in the male and female genital tracts, in *Biology of Fertilization*, Vol. 2, Metz, C. B. and Monroy, A., Eds., Academic Press, New York, 1985, 121.
6. **Chang, M. C., Austin, C. R., Bedford, J. M., Brackett, B. G., Hunter, R. H. F., and Yanagimachi, R.,** Capacitation of spermatozoa and fertilization in mammals, in *Frontiers in Reproduction and Fertility Control. A Review of Reproductive Sciences and Contraceptive Development*, Greep, R. O. and Koblinsky, M. A., Eds., MIT Press, Cambridge, 1977, 434.

7. **Chang, M. C.,** The meaning of sperm capacitation, *J. Androl.,* 5, 45, 1984.
8. **Fraser, L. R.,** Mechanisms controlling mammalian fertilization, in *Oxford Reviews of Reproductive Biology,* Vol. 6, Finn, C. A., Ed., Clarendon Press, Oxford, 1984, 174.
9. **Yanagimachi, R.,** Specificity of sperm-egg interaction, in *Immunobiology of Gametes,* Edinin, M. and Johnson, M. H., Eds., Cambridge University Press, London, 1977, 255.
10. **Yanagimachi, R.,** Mechanisms of fertilization in mammals, in *Fertilization and Embryonic Development In Vitro,* Mastroianni, L. and Biggers, J. D., Eds., Plenum Press, New York, 1981, 81.
11. **Yanagimachi, R.,** Mammalian fertilization, in *The Physiology of Reproduction,* Knobil, E. and Neill, J., Eds., Raven Press, New York, 1988, 135.
12. **Rogers, B. J.,** Mammalian sperm capacitation and fertilization in vitro, *Gamete Res.,* 1, 165, 1978.
13. **Moore, H. D. M. and Bedford, J. M.,** The interaction of mammalian gametes in the female, in *Mechanism and Control of Animal Fertilization,* Hartmann, J. F., Ed., Academic Press, New York, 1983, 453.
14. **Bavister, B. D.,** Animal in vitro fertilization and embryo development, in *Developmental Biology,* Vol. 4, Browder, L. W., Ed., Plenum Press, New York, 1986, 81.
15. **Bedford, J. M. and Cooper, G. W.,** Membrane fusion events in fertilization of vertebrate eggs, in *Cell Surface Reviews (Membrane Fusion),* Vol. 5, Poste, G. and Nicolson, G. L., Eds., North-Holland, Amsterdam, 1978, 65.
16. **Langlais, J. and Roberts, K. D.,** A molecular membrane model of sperm capacitation and the acrosome reaction of mammalian spermatozoa, *Gamete Res.,* 12, 183, 1985.
17. **O'Rand, M. G.,** Changes in sperm surface properties corollated with capacitation, in *The Spermatozoa,* Fawcett, D. W. and Bedford, J. M., Eds., Urban and Schwartzenberg, Baltimore, 1979, 195.
18. **O'Rand, M. G.,** Modification of the sperm membrane during capacitation, *Ann. N.Y. Acad. Sci.,* 383, 392, 1982.
19. **Garbers, D. L. and Kopf, G. S.,** The regulation of spermatozoa by calcium and cyclic nucleotides, *Adv. Cyclic Nucleotide Res.,* 13, 251, 1980.
20. **Fraser, L. R. and Ahuja, K. K.,** Metabolic and surface events in fertilization, *Gamete Res.,* 20, 491, 1988.
21. **Shalgi, R., Kaplan, R., Nebel, L., and Kraicer, P. F.,** The male factor in fertilization of rat eggs in vitro, *J. Exp. Zool.,* 217, 399, 1981.
22. **Nagai, T., Niwa, K., and Iritani, Y.,** Effect of sperm concentration during preincubation in a defined medium on fertilization in vitro of pig follicular oocytes, *J. Reprod. Fertil.,* 70, 271, 1984.
23. **Florman, H. M. and First, N. L.,** The regulation of acrosomal exocytosis. II. The zona pellucida-induced acrosome reaction of bovine spermatozoa is controlled by extrinsic positive regulatory elements, *Dev. Biol.,* 128, 464, 1988.
24. **Chang, M. C.,** A detrimental effect of seminal plasma on the fertilizing capacity of sperm, *Nature (London),* 179, 258, 1957.
25. **Thomas, T. S., Reynolds, A. B., and Oliphant, G.,** Evaluation of the site of synthesis of rabbit sperm acrosome stabilizing factor using immunocytochemical and metabolic labelling techniques, *Biol. Reprod.,* 30, 693, 1984.
26. **Bryan, J. H. B. and Akruk, S. R.,** A naphthol yellow S and erythrosin B staining procedure for use in studies of the acrosome reaction of rabbit sperm, *Stain Technol.,* 52, 47, 1977.
27. **Cross, N. L., Morales, P., Overstreet, J. W., and Hanson, F. W.,** Two simple methods for detecting acrosome reacted human sperm, *Gamete Res.,* 15, 213, 1986.
28. **Didoin, B. A., Dobrinsky, J. R., Giles, J. R., and Graves, C. N.,** Staining procedure to detect viability and the true acrosome reaction in spermatozoa of various species, *Gamete Res.,* 22, 51, 1989.
29. **Ward, C. R. and Storey, B. T.,** Determination of the time course of capacitation in mouse spermatozoa using a chlortetracycline fluorescence assay, *Dev. Biol.,* 104, 287, 1984.
30. **Triana, L. R., Babcock, D. F., Lorton, S. P., and Lardy, H. A.,** Release of acrosomal hyaluronidase follows increased membrane permeability to calcium in the presumptive capacitation sequence for spermatozoa of the bovine and other mammalian species, *Biol. Reprod.,* 23, 47, 1980.
31. **Florman, H. M., Bechtol, K. B., and Wassarman, P. M.,** Enzymatic dissection of the functions of the mouse egg's receptor for sperm, *Dev. Biol.,* 106, 243, 1984.
32. **Blach, E. L., Amman, R. P., Bowen, R. A., Sawyer, H. R., and Hermenet, M. J.,** Use of monoclonal antibody to evaluate integrity of the plasma membrane of stallion sperm, *Gamete Res.,* 21, 233, 1988.
33. **Austin, C. R.,** Capacitation of spermatozoa, *Int. J. Fertil.,* 12, 25, 1967.
34. **Bedford, J. M.,** Sperm capacitation and fertilization in mammals, *Biol. Reprod. (Suppl.),* 2, 128, 1970.
35. **Meizel, S.,** The mammalian sperm acrosome reaction. A biochemical approach, in *Development in Mammals,* Johnson, M. H., Ed., North-Holland, New York, 1978, 1.
36. **Florman, H. M. and First, N. L.,** The regulation of acrosomal exocytosis. I. Sperm capacitation is required for the induction of acrosome reactions by the bovine zona pellucida in vitro, *Dev. Biol.,* 128, 453, 1988.
37. **Parrish, J. J., Susko-Parrish, J. L., Winer, M. A., and First, N. L.,** Capacitation of bovine sperm by heparin, *Biol. Reprod.,* 38, 1171, 1988.

38. **Fleming, A. D. and Yanagimachi, R.,** Effects of various lipids on the acrosome reaction and fertilizing capacity of guinea pig spermatozoa with special reference to the possible involvement of sperm membrane lysophospolipids in the acrosome reaction of guinea pig spermatozoa, *J. Exp. Zool.*, 229, 485, 1981.
39. **Byrd, W. and Wolf, D. P.,** Acrosomal status in fresh and capacitated human ejaculated sperm, *Biol. Reprod.*, 34, 859, 1987.
40. **Koehler, J. K.,** Lectins as probes of the spermatozoon surface, *Arch. Androl.*, 6, 197, 1981.
41. **Ahuja, K. K.,** Carbohydrate determinants involved in mammalian fertilization, *Am. J. Anat.*, 174, 207, 1985.
42. **Oliphant, G. and Brackett, B. G.,** Immunological assessment of surface changes of rabbit sperm undergoing capacitation, *Biol. Reprod.*, 9, 404, 1973.
43. **Vernon, R. B., Hamilton, M. S., and Eddy, E. M.,** Effects of in vivo and in vitro fertilization environments on the expression of a surface antigen of the mouse sperm tail, *Biol. Reprod.*, 32, 669, 1985.
44. **Okabe, M., Takada, K., Adachi, T., Kohama, Y., Miura, T., and Aonuma, S.,** Studies on sperm capacitation using monoclonal antibody—disappearance of an antigen from the anterior part of the mouse sperm head, *J. Pharmacol. Dyn.*, 9, 55, 1986.
45. **Okabe, S., Takada, K., Adachi, T., Kohama, Y., and Miura, T.,** Inconsistent reactivity of an anti-sperm monoclonal antibody and its relationship to sperm capacitation, *J. Reprod. Immunol.*, 9, 67, 1986.
46. **Wolf, D. E., Hagopian, S. S., and Isojima, S.,** Changes in sperm plasma membrane lipid diffusibility after hyperactivation during in vitro capacitation in the mouse, *J. Cell Biol.*, 102, 1372, 1986.
47. **Liévano, A., Vega-SaenzdeMiera, E. C., and Darszon, A.,** Ca^{2+} channels from the sea urchin sperm plasma membrane, *J. Gen. Physiol.*, 95, 273, 1990.
48. **Florman, H. M., Tombes, R. M., First, N. L., and Babcock, D. F.,** An adhesion-associated agonist from the zona pellucida activates G protein-promoted elevations of internal Ca^{2+} and pH that mediate mammalian sperm acrosomal exocytosis, *Dev. Biol.*, 135, 133, 1989.
49. **Wassarman, P. M.,** Early events in mammalian fertilization, *Annu. Rev. Cell Biol.*, 2, 109, 1987.
50. **Wassarman, P. M.,** The biology and chemistry of fertilization, *Science*, 235, 553, 1987.
51. **Talbot, P.,** Sperm penetration through oocyte investments in mammals, *Am. J. Anat.*, 174, 331, 1985.
52. **Talbot, P. and DiCartantonio, G.,** The oocyte-cumulus complex: ultrastructure of the extracellular components in hamster and mice, *Gamete Res.*, 10, 127, 1984.
53. **McClean, D. and Rowlands, I. W.,** Role of hyaluronidase in fertilization, *Nature (London)*, 150, 627, 1942.
54. **Leonard, S. L. and Kurzrok, R.,** A study of hyaluronidase: effects on the follicle cells of ovulated rats, *Endocrinology*, 37, 171, 1945.
55. **Austin, C. R. and Bishop, M. W. H.,** Role of the rodent perforatorium and acrosome in fertilization, *Proc. R. Soc. London Ser. B.*, 149, 241, 1958.
56. **Austin, C. R.,** Capacitation and the release of hyaluronidase from spermatozoa, *J. Reprod. Fertil.*, 3, 310, 1960.
57. **Hartree, E. F. and Srivastava, P. N.,** Chemical composition of the acrosomes of ram sperm, *J. Reprod. Fertil.*, 9, 47, 1965.
58. **Storey, B. T., Lee, M. A., Muller, C. R., and Wirtschafter, D. C.,** Binding of mouse spermatozoa to the zonae pellucidae of mouse eggs in cumulus: evidence that the acrosomes remain substantially intact, *Biol. Reprod.*, 31, 1119, 1984.
59. **Cherr, G. N., Lambert, H., Meizel, S., and Katz, D. F.,** In vitro studies of the golden hamster sperm acrosome reaction: completion on the zona pellucida and induction by solubilized zonae pellucidae, *Dev. Biol.*, 114, 119, 1986.
60. **Cummins, J. M. and Yanagimachi, R.,** Development of ability to penetrate the cumulus oophorus by hamster spermatozoa capacitated in vitro, in relation to the timing of the acrosome reaction, *Gamete Res.*, 15, 187, 1986.
61. **Saling, P. M., Sowinski, J., and Storey, B. T.,** An ultrastructural study of epididymal mouse spermatozoa binding to zonae pellucidae in vitro: sequential relationship to the acrosome reaction, *J. Exp. Zool.*, 209, 229, 1979.
62. **Florman, H. M. and Storey, B. T.,** Mouse gamete interactions: the zona pellucida is the site of the acrosome reaction leading to fertilization in vitro, *Dev. Biol.*, 91, 121, 1982.
63. **Florman, H. M., Saling, P. M., and Storey, B. T.,** Fertilization of mouse eggs in vitro: time resolution of the reactions preceeding penetration of zona pellucida, *J. Androl.*, 3, 373, 1982.
64. **Bleil, J. D. and Wassarman, P. M.,** Sperm-egg interactions in the mouse: sequence of events and induction of the acrosome reaction by a zona pellucida glycoprotein, *Dev. Biol.*, 95, 317, 1983.
65. **Crozet, N.,** Ultrastructural aspects of in vivo fertilization in the cow, *Gamete Res.*, 10, 241, 1984.
66. **Crozet, N. and Dumont, M.,** The site of the acrosome reaction during in vivo penetration of the sheep oocyte, *Gamete Res.*, 10, 97, 1984.
67. **Kuzan, F., Fleming, A. D., and Siedel, G.,** Successful fertilization in vitro of fresh intact oocytes by perivitelline (acrosome-reacted) spermatozoa in the rabbit, *Fertil Steril.*, 41, 766, 1984.

68. **Myles, D. G., Hyatt, H., and Primakoff, P.,** Binding of both acrosome-intact and acrosome-reacted guinea pig sperm to the zona pellucida, *Dev. Biol.,* 121, 559, 1987.
69. **Morales, P., Cross, N. L., Overstreet, J. W., and Hanson, F. W.,** Acrosome intact and acrosome-reacted human sperm can initiate binding to the zona pellucida, *Dev. Biol.,* 133, 385, 1989.
70. **Florman, H. M. and Wassarman, P. M.,** O-linked oligosaccharides of mouse egg ZP3 account for its sperm receptor activity, *Cell,* 41, 313, 1985.
71. **Bleil, J. D. and Wassarman, P. M.,** Galactose at the nonreducing terminus of O-linked oligosaccharides of mouse egg zona pellucida glycoprotein ZP3 is essential for the glycoprotein's sperm receptor activity, *Proc. Natl. Acad. Sci. U.S.A.,* 85, 6778, 1988.
72. **Shur, B. D. and Hall, N. G.,** A role for mouse sperm surface galactosyltransferase in sperm binding to the egg zona pellucida, *J. Cell Biol.,* 95, 574, 1982.
73. **Macek, M. B. and Shur, B. D.,** Protein-carbohydrate complementarity in mammalian gamete recognition, *Gamete Res.,* 20, 93, 1988.
74. **Shur, B. D. and Neely, C. A.,** Plasma membrane association, purification, and partial characterization of mouse sperm β1,4-galactosyltransferase, *J. Biol. Chem.,* 263, 17706, 1988.
75. **Cardullo, R. A., Armant, D. R., and Millette, C. F.,** Characterization of fucosyltransferase activity during mouse spermatogenesis: evidence for a cell surface fucosyltransferase, *Biochemistry,* 28, 1611, 1989.
76. **Hartmann, J. F. and Hutchison, C. F.,** Nature of the prepenetration contact interactions between hamster gametes in vitro, *J. Reprod. Fertil.,* 36, 49, 1974.
77. **Saling, P. M.,** Involvement of trypsin-like activity in binding of mouse spermatozoa to zona pellucida, *Proc. Natl. Acad. Sci. U.S.A.,* 76, 6231, 1981.
78. **Poirier, G. R., Robinson, R., Richardson, R., Hinds, K., and Clayton, D.,** Evidence for a binding site on the sperm plasma membrane which recognizes the murine zona pellucida, *Gamete Res.,* 14, 235, 1986.
79. **Robinson, R., Richardson, R., Hinds, K., Clayton, D., and Poirier, G. R.,** Features of a seminal proteinase inhibitor-zona pellucida binding component on murine sperm, *Gamete Res.,* 16, 217, 1987.
80. **Benau, D. A. and Storey, B. T.,** Zona-binding site sensitive to trypsin inhibitors, *Biol. Reprod.,* 32, 282, 1987.
81. **Tulsiani, D. R. P., Skudlarek, M. D., and Orgebin-Crist, M.-C.,** Novel α-D-mannosidase of rat sperm plasma membranes: characterization and potential role in sperm-egg interactions, *J. Cell Biol.,* 109, 1257, 1989.
82. **Silbey, D. R., Benovic, J. L., Caron, M. C., and Lefkowitz, R. J.,** Regulation of transmembrane signalling by receptor phosphorylation, *Cell,* 48, 913, 1987.
83. **Garbers, D. L.,** Guanylate cyclase, a cell surface receptor, *J. Biol. Chem.,* 264, 9103, 1989.
84. **Schulz, S., Chinkers, M., and Garbers, D. L.,** The guanylate cyclase/receptor family of proteins, *FASEB J.,* 3, 2026, 1989.
85. **Miller, C.,** Genetic manipulation of ion channels: a new approach to structure and mechanism, *Neuron,* 2, 1195, 1989.
86. **Ross, E. M.,** Signal sorting and amplification through G protein-coupled receptors, *Neuron,* 3, 141, 1989.
87. **O'Rand, M. G. and Fisher, S. J.,** Localization of zona pellucida binding sites on rabbit spermatozoa and induction of the acrosome reaction by solubilized zonae, *Dev. Biol.,* 119, 551, 1987.
88. **Cross, N. L., Morales, P., Overstreet, J. W., and Hanson, F. W.,** Induction of acrosome reactions by the human zona pellucida, *Biol. Reprod.,* 38, 235, 1988.
89. **Endo, Y., Lee, M. A., and Kopf, G. S.,** Evidence for the role of a guanine nucleotide-binding regulatory protein in the zona pellucida-induced mouse sperm acrosome reaction, *Dev. Biol.,* 119, 210, 1987.
90. **Endo, Y., Lee, M. A., and Kopf, G. S.,** Characterization of an islet-activating protein-sensitive site in mouse sperm that is involved in the zona pellucida-induced acrosome reaction, *Dev. Biol.,* 129, 12, 1988.
91. **Moss, J. and Vaughan, M.,** ADP-ribosylation of guanyl nucleotide-binding regulatory proteins by bacterial toxins, *Adv. Enzymol.,* 61, 303, 1988.
92. **Bentley, J. K., Garbers, D. L., Domino, S. E., Noland, T. D., and Van Dop, C.,** Spermatozoa contain a guanine nucleotide-binding protein ADP-ribosylated by pertussis toxin, *Biochem. Biophys. Res. Commun.,* 138, 728, 1986.
93. **Kopf, G. S., Woolkalis, M. J., and Gerton, G. L.,** Evidence for a guanine nucleotide-binding regulatory protein in invertebrate and mammalian sperm: identification by islet-activating protein-catalyzed ADP-ribosylation and immunochemical methods, *J. Biol. Chem.,* 261, 7327, 1986.
94. **Yanagimachi, R.,** Acceleration of the acrosome reaction and activation of guinea pig spermatozoa by detergents and other reagents, *Biol. Reprod.,* 13, 519, 1975.
95. **Green, D. L. P.,** The induction of the acrosome reaction in guinea pig sperm by the divalent metal cation ionophore A23187, *J. Cell Sci.,* 32, 137, 1978.
96. **Roldan, E. R. S. and Harrison, R. A. P.,** Polyphosphoinositide breakdown and subsequent exocytosis in the Ca^{2+}/ionophore-induced acrosome reaction of mammalian spermatozoa, *Biochem. J.,* 259, 397, 1989.

97. **Shapiro, B. M., Schackmann, R. W., Tombes, R. M., and Kazazoglou, T.,** Coupled ionic and enzymatic regulation of sperm behavior, *Curr. Top. Cell. Regul.,* 26, 97, 1986.
98. **Trimmer, J. S. and Vacquier, V. D.,** Activation of sea urchin gametes, *Annu. Rev. Cell Biol.,* 2, 1, 1986.
99. **Schackmann, R. W.,** Ionic regulation of the sea urchin sperm acrosome reaction by egg-derived peptides, in *The Cell Biology of Fertilization,* Schatten, H. and Schatten, G., Eds., Academic Press, New York, 1989, 3.
100. **Guerrero, A. and Darszon, A.,** Egg jelly triggers a calcium influx which inactivates and is inhibited by calmodulin antagonists in sea urchin spermatozoa, *Biochim. Biophys. Acta,* 980, 109, 1989.
101. **Babcock, D. F. and Pfeiffer, D. R.,** Independent elevation of cytosolic [Ca^{2+}] and pH of mammalian sperm by voltage-dependent and pH-sensitive mechanisms, *J. Biol. Chem.,* 262, 15041, 1987.
102. **Babcock, D. F.,** Cross talk between voltage-dependent components of the signal transduction machinery of bovine spermatozoa, *Biol. Reprod.,* 38 (Suppl. 1), 60, 1988.
103. **Kazazoglou, T., Schackmann, R. W., Fosset, M., and Shapiro, B. M.,** Calcium channel antagonists inhibit the acrosome reaction and bind to plasma membranes of sea urchin sperm, *Proc. Natl. Acad. Sci. U.S.A.,* 82, 1460, 1985.
104. **Toowicharanont, P. and Shapiro, B. M.,** Regional differentiation of the sea urchin sperm plasma membrane, *J. Biol. Chem.,* 263, 6877, 1988.
105. **Garcia-Soto, J., Moiurelle, M., Varga, I., De la Torre, L., Ramirez, E., Lopez-Colome, A. M., and Darszon, A.,** Sea urchin sperm head plasma membranes: characteristics and egg jelly induced Ca^{2+} and Na^{+} uptake, *Biochim. Biophys. Acta,* 944, 1, 1988.
105a. **Guerrero, A. and Darszon, A.,** Evidence for the activation of two different Ca^{2+} channels during the egg jelly-induced acrosome reaction of sea urchin sperm, *J. Biol. Chem.,* 264, 19593, 1989.
106. **Cox, T. and Peterson, R. N.,** Identification of calcium conducting channels in isolated boar sperm plasma membranes, *Biochem. Biophys. Res. Commun.,* 161, 162, 1989.
107. **Roldan, E. R. S., Shibata, S., and Yanagimachi, R.,** Effect of Ca^{2+} regulation during capacitation and the acrosome reaction of guinea pig spermatozoa, *Gamete Res.,* 13, 281, 1986.
108. **Hyne, R. V.,** Bicarbonate and calcium dependent induction of rapid guinea pig acrosome reaction by monovalent ionophore, *Biol. Reprod.,* 31, 312, 1984.
109. **Ruknudin, A.,** Cytochemical study of intracellular calcium in hamster spermatozoa during the acrosome reaction, *Gamete Res.,* 22, 375, 1989.
110. **Singh, J. P., Babcock, D. F., and Lardy, H. A.,** Induction of accelerated acrosome reaction in guinea pig sperm, *Biol. Reprod.,* 22, 566, 1980.
111. **Bradley, M. P. and Forrester, I. T.,** A sodium-calcium exchange mechanism in plasma membrane vesicles isolated from ram sperm flagella, *FEBS Lett.,* 121, 15, 1980.
112. **Roldan, E. R. S. and Fleming, A. D.,** Is a Ca^{2+}-ATPase involved in Ca^{2+} regulation during capacitation and the acrosome reaction of guinea pig spermatozoa?, *J. Reprod. Fertil.,* 85, 297, 1989.
113. **Breitbart, H., Darshan, R., and Rubinstein, S.,** Evidence for the presence of ATPase-dependent calcium pump and ATPase activities in bull sperm head membranes, *Biochem. Biophys. Res. Commun.,* 122, 479, 1984.
114. **Breitbart, H., Rubinstein, S., and Nass-Arden, L.,** The role of calcium and Ca^{2+}-ATPase in maintaining motility in ram spermatozoa, *J. Biol. Chem.,* 260, 11548, 1985.
115. **Breitbart, H., Stern, B., and Rubinstein, S.,** Calcium transport and Ca^{2+}-ATPase activity in ram sperm plasma membrane vesicles, *Biochim. Biophys. Acta,* 728, 349, 1983.
116. **Babcock, D. F., First, N. L., and Lardy, H. A.,** Transport mechanism for succinate and phosphate localized in the plasma membrane of bovine spermatozoa, *J. Biol. Chem.,* 250, 6488, 1975.
117. **Rufo, G. A., Schoff, P. K., and Lardy, H. A.,** Regulation of calcium content in bovine spermatozoa, *J. Biol. Chem.,* 259, 2547, 1984.
118. **Rigoni, F., Dell Antone, P., and Deana, R.,** Evidence for a ΔpH-driven Ca^{2+}-uptake in EGTA-treated bovine spermatozoa, *Eur. J., Biochem.,* 169, 417, 1988.
119. **Cantino, M. E., Schackmann, R. W., and Johnson, D. E.,** Changes in subcellular elemental distribution accompanying the acrosome reaction in sea urchin sperm, *J. Exp. Zool.,* 226, 255, 1983.
120. **Somlyo, A. P. and Himpens, B.,** Cell calcium and its regulation in smooth muscle, *FASEB J.,* 3, 2266, 1989.
121. **Storey, B. T.,** Energy metabolism of spermatozoa. IV. Effect of calcium on respiration of mature epididymal sperm of the rabbit, *Biol. Reprod.,* 13, 1, 1975.
122. **Babcock, D. F., First, N. L., and Lardy, H. A.,** Action of ionophore A23187 at the cellular level. Separation of effects at the plasma and mitochondrial membranes, *J. Biol. Chem.,* 251, 3881, 1976.
123. **Singh, J. P., Babcock, D. F., and Lardy, H. A.,** Increased calcium ion flux is a component of capacitation of spermatozoa, *Biochem. J.,* 172, 549, 1978.
124. **Coronel, C. E. and Lardy, H. A.,** Characterization of Ca^{2+} uptake by guinea pig epididymal spermatozoa, *Biol. Reprod.,* 37, 1097, 1987.

125. **Vijayaraghavan, S., Bhattachjaryya, A., and Hoskins, D. D.**, Calcium uptake by bovine epididymal spermatozoa is regulated by the redox state of the mitochondrial pyridine nucleotides, *Biol. Reprod.*, 40, 744, 1989.
126. **Babcock, D. F., Stammerjohn, D., and Hutchins, T.**, Calcium redistribution in individual cells correlated with ionophore action on motility, *J. Exp. Zool.*, 204, 391, 1978.
127. **Babcock, D. F., Singh, J. P., and Lardy, H. A.**, Alteration of membrane permeability to calcium ions during maturation of bovine spermatozoa, *Dev. Biol.*, 69, 85, 1979.
128. **Rufo, G. A., Singh, J. P., Babcock, D. F., and Lardy, H. A.**, Purification and characterization of a calcium transport inhibitor protein from bovine seminal plasma, *J. Biol. Chem.*, 257, 4627, 1982.
129. **Coronel, C. E., San Agustin, J., and Lardy, H. A.**, Identification and partial characterization of caltrin-like proteins in the reproductive tract of the guinea pig, *Biol. Reprod.*, 38, 713, 1988.
129a. **Thomas, P. and Meizel, S.**, An influx of extracellular calcium is required for initiation of the human sperm acrosome reaction induced by human follicular fluid, *Gamete Res.*, 20, 397, 1988.
129b. **Blackmore, P. F., Beebe, S. J., Danforth, D. R., and Alexander, N.**, Progesterone and 17-alpha-hydroxyprogesterone. Novel stimulators of calcium influx in human sperm, *J. Biol. Chem.*, 265, 1376, 1990.
130. **Zarca, A., Rubinstein, S., and Breitbart, H.**, Transport mechanism for calcium and phosphate in ram spermatozoa, *Biochim. Biophys. Acta*, 944, 351, 1988.
131. **Leyton, L. and Saling, P.**, 95 kD sperm proteins bind ZP3 and serve as tyrosine kinase substrates in response to zona binding, *Cell*, 57, 1130, 1989.
132. **Tilney, L., Kiehart, D., Sardet, C., and Tilney, M.**, The polymerization of actin. IV. The role of Ca^{++} and H^+ in the assembly of actin and in membrane fusion in the acrosomal reaction of echinoderm sperm, *J. Cell Biol.*, 77, 536, 1978.
133. **Trimmer, J. S., Schackmann, R. W., and Vacquier, V. D.**, Monoclonal antibodies increase intracellular Ca^{2+} in sea urchin spermatozoa, *Proc. Natl. Acad. Sci. U.S.A.*, 83, 9055, 1986.
134. **Trimmer, J. S., Ebina, Y., Schackmann, R. W., Meinhof, C.-G., and Vacquier, V. D.**, Characterization of a monoclonal antibody which induces the acrosome reaction of sea urchin sperm, *J. Cell Biol.*, 105, 1120, 1986.
135. **Garcia-Soto, J., Gonzalez-Martinez, M., De la Torre, L., and Darszon, A.**, Internal pH can regulate Ca^{2+} uptake and the acrosome reaction in sea urchin sperm, *Dev. Biol.*, 120, 112, 1987.
136. **Garcia-Soto, J. and Darszon, A.**, High pH-induced acrosome reaction and Ca^{2+} uptake in sea urchin sperm suspended in Na^+-free sea water, *Dev. Biol.*, 110, 338, 1985.
137. **Lee, M. A. and Storey, B. T.**, Evidence for plasma membrane impermeability to small ions in acrosome-intact mouse spermatozoa bound to mouse zonae pellucidae, using an aminoacridine fluorescent probe: time course of the zona-induced acrosome reaction monitored by both chlortetracycline and pH probe fluorescence, *Biol. Reprod.*, 33, 235, 1985.
138. **Meizel, S. and Deamer, D. W.**, The pH of the hamster sperm acrosome, *J. Histochem. Cytochem.*, 26, 98, 1978.
139. **Working, P. K. and Meizel, S.**, Correlation of increased intraacrosomal pH with the hamster sperm acrosome reaction, *J. Exp. Zool.*, 227, 97, 1983.
140. **Christen, R., Schackmann, R. W., and Shapiro, B. M.**, Interaction between sperm and sea urchin egg jelly, *Dev. Biol.*, 98, 1, 1983.
141. **Thomas, R. C.**, Bicarbonate and pH_i response, *Nature (London)*, 337, 601, 1989.
142. **Grinstein, S., Rotin, D., and Mason, M. J.**, Na^+/H^+ exchange and growth factor-induced cytosolic pH changes. Roles in cellular proliferation, *Biochim. Biophys. Acta*, 998, 73, 1989.
143. **Sardet, C., Franchi, A., and Pouyssegur, J.**, Molecular cloning, primary structure and expression of the human growth factor-activatable Na^+/H^+ antiporter, *Cell*, 56, 271, 1989.
144. **Kopito, R. R. and Lodish, H. F.**, Primary structure and transmembrane orientation of the murine anion exchange protein, *Nature (London)*, 316, 234, 1985.
145. **Hansbrough, R. K. J. and Garbers, D. L.**, Sodium-dependent activation of rapid guinea pig sperm acrosome reactions by monovalent ionophores, *J. Biol. Chem.*, 256, 1447, 1981.
146. **Schackmann, R. W. and Chock, P. B.**, Alteration of intracellular $[Ca^{2+}]$ in sea urchin sperm by the egg peptide speract. Evidence that increased intracellular $[Ca^{2+}]$ is coupled to Na^+ entry and increased intracellular pH, *J. Biol. Chem.*, 261, 8719, 1986.
147. **Lee, H. C.**, Internal GTP stimulates the speract receptor mediated voltage changes in sea urchin spermatozoa membrane vesicles, *Dev. Biol.*, 126, 91, 1988.
148. **Gatti, J. L. and Christen, R.**, Regulation of internal pH of sea urchin sperm. A role for the Na/K pump, *J. Biol. Chem.*, 260, 7599, 1985.
149. **Babcock, D. F.**, Examination of the internal ionic environment and ionophore action by null point measurements employing the fluorescein chromatophore, *J. Biol. Chem.*, 258, 6380, 1983.
150. **Tajima, Y., Okamura, N., and Sugita, Y.**, The activating effects of bicarbonate on sperm motility and respiration at ejaculation, *Biochim. Biophys. Acta*, 924, 519, 1987.

151. **Smith, M. B., Babcock, D. F., and Lardy, H. A.,** A ^{31}P-NNR study of the epididymis and epididymal sperm of the bovine and hamster, *Biol. Reprod.*, 33, 1029, 1985.
152. **Wong, P. Y. D., Lee, W. M., and Tsang, A. Y. S.,** The effects of extracellular sodium on acid release and motility initiation in rat caudal epididymal. spermatozoa in vitro, *Exp. Cell Res.*, 131, 97, 1981.
153. **Lee, H. C.,** A membrane potential-sensitive Na^+-H^+ exchange system in flagella isolated from sea urchin spermatozoa, *J. Biol. Chem.*, 259, 15315, 1984.
154. **Lee, H. C.,** The voltage-sensitive Na^+/H^+ exchange in flagella isolated from sea urchin sperm, *J. Biol. Chem.*, 260, 10794, 1985.
155. **Babcock, D. F., Rufo, G. A., and Lardy, H. A.,** Potassium-dependent increases in cytosolic pH stimulate metabolism and motility of mammalian sperm, *Proc. Natl. Acad. Sci. U.S.A.*, 80, 1327, 1983.
156. **Lee, M. A. and Storey, B. T.,** Bicarbonate is essential for fertilization of mouse eggs: mouse sperm require it to undergo the acrosome reaction, *Biol. Reprod.*, 34, 349, 1986.
157. **Bhattacharyya, A. and Yanagimachi, R.,** Synthetic organic pH buffers can support fertilization of guinea pig eggs, but not as efficiently as bicarbonate buffer, *Gamete Res.*, 19, 123, 1988.
158. **Okamura, N., Tajima, Y., Soejima, A., Masudo, H., and Sugita, Y.,** Sodium bicarbonate stimulates the motility of mammalian spermatozoa through direct activation of adenylate cyclase, *J. Biol. Chem.*, 260, 9699, 1985.
159. **Garty, N. B. and Salomon, Y.,** Stimulation of partially purified adenylate cyclase from bull sperm by bicarbonate, *FEBS Lett.*, 218, 148, 1987.
160. **Okamura, N., Tajima, Y., and Sugita, Y.,** Decrease in bicarbonate transport activities during epididymal maturation of porcine sperm, *Biochem. Biophys. Res. Commun.*, 157, 1280, 1988.
161. **Hoffman, E. K.,** Anion transport systems in the plasma membrane of vertebrate cells, *Biochim. Biophys. Acta*, 864, 1, 1986.
162. **Working, P. K. and Meizel, S.,** Evidence that an ATPase functions in the maintenance of the acidic pH of the hamster sperm acrosome, *J. Biol. Chem.*, 256, 4708, 1983.
163. **Forgac, M.,** Structure and function of vacuolar class of ATP-driven proton pumps, *Physiol. Rev.*, 69, 769, 1989.
164. **Schackmann, R. W., Christen, R., and Shapiro, B. M.,** Membrane potential depolarization and increased pH_i accompany the acrosome reaction of sea urchin sperm, *Proc. Natl. Acad. Sci. U.S.A.*, 78, 6066, 1981.
165. **Gonzalez-Martinez, M. and Darszon, A.,** A fast transient hyperpolarization occurs during the sea urchin sperm acrosome reaction induced by egg jelly, *FEBS Lett.*, 218, 247, 1987.
167. **Lee, H. C. and Garbers, D. L.,** Modulation of the Na^+-H^+ exchange system in sea urchin spermatozoa through membrane potential changes induced by the egg peptide speract, *J. Biol. Chem.*, 261, 16026, 1986.
168. **Rink, T. J.,** Membrane potential of guinea-pig spermatozoa, *J. Reprod. Fertil.*, 51, 155, 1977.
169. **Schackmann, R. W., Christen, R., and Shapiro, B. M.,** Measurement of plasma membrane and mitochondrial potentials in sea urchin sperm. Changes upon activation and induction of the acrosome reaction, *J. Biol. Chem.*, 259, 13914, 1984.
170. **Schackmann, R. W., Eddy, E. M., and Shapiro, B. M.,** The acrosome reaction of *Stronglyocentrotus purpuratus* sperm. Ion requirements and movements, *Dev. Biol.*, 65, 483, 1978.
171. **Schackmann, R. W. and Shapiro, B. M.,** A partial sequence of ionic changes associated with the acrosome reaction of *Stronglyocentrotus purpuratus*, *Dev. Biol.*, 81, 145, 1981.
172. **Garbers, D. L., Hansbrough, R. J., Radany, E. W., Hyne, R. V., and Kopf, G. S.,** Purification and characterization of calmodulin from sea urchin sperm, *J. Reprod. Fertil.*, 59, 377, 1980.
173. **Jones, H. P., Lenz, R. W., Palevitz, B. A., and Cormier, M. J.,** Calmodulin localization in mammalian spermatozoa, *Proc. Natl. Acad. Sci. U.S.A.*, 77, 2772, 1980.
174. **Feinberg, J. M. F., Weinman, J. S., Weinman, S. J., Walsh, M. P., Harricane, M. C., Gabrion, J., and Demaille, J. G.,** Immunocytochemical and biochemical evidence for the presence of calmodulin in bull sperm flagellum. Isolation and characterization of sperm calmodulin, *Biochim. Biophys. Acta*, 673, 303, 1981.
175. **Moore, P. B. and Dedman, J. R.,** Calmodulin, a calmodulin acceptor protein, and calcimedins: unique antibody localization in hamster sperm, *J. Cell Biochem.*, 25, 99, 1984.
176. **Camatini, M., Anelli, G., and Casale, A.,** Immunocytochemical localization of calmodulin in intact and acrosome-reacted boar sperm, *Eur. J. Cell Biol.*, 41, 89, 1986.
177. **Weinman, S., Ores-Carton, C., Escaia, F., Feinberg, J., and Puskin, S.,** Calmodulin immunoelectron microscopy: redistribution during ram spermatogenesis and epididymal maturation, *J. Histochem. Cytochem.*, 34, 1181, 1986.
178. **Kopf, G. S. and Vacquier, V. D.,** Characterization of a calmodulin-stimulated adenylate cyclase from abalone spermatozoa, *J. Biol. Chem.*, 259, 7590, 1984.
179. **Gross, M. K., Toscano, D. G., and Toscano, W. A.,** Calmodulin-mediated adenylate cyclase from mammalian sperm, *J. Biol. Chem.*, 262, 8672, 1987.
180. **Chaudry, P. S. and Casillas, E. R.,** Calmodulin-stimulated cyclic nucleotide phosphodiesterases in plasma membranes of bovine epididymal spermatozoa, *Arch. Biochem. Biophys.*, 262, 439, 1988.

181. **Geremia, R., Rossi, R., Mocini, D., Pezzotti, R., and Conti, M.,** Characterization of a calmodulin-dependent high-affinity cAMP and cGMP phosphodiesterase from male mouse germ cells, *Biochem. J.*, 217, 693, 1984.
182. **Wasco, W. M. and Orr, G. A.,** Function of calmodulin in mammalian sperm: presence of a calmodulin-dependent cyclic nucleotide phospodiesterase associated with demembranated rat caudal epididymal sperm, *Biochem. Biophys. Res. Commun.*, 118, 636, 1984.
183. **Tang, F. Y. and Hoskins, D. D.,** Phosphoprotein phosphatase of bovine epididymal sperm, *Biochem. Biophys. Res. Commun.*, 62, 328, 1975.
184. **Tash, J. S., Krinks, M., Patel, J., Means, R. L., Klee, C. B., and Means, A. R.,** Identification, characterization, and functional correlation of calmodulin-dependent protein phosphatase in sperm, *J. Cell Biol.*, 106, 1625, 1988.
185. **Ashraf, M., Peterson, R. N., and Russell, L. D.,** Characterization of (Ca^{2+} + Mg^{2+}) adenosine triphosphatase activity and calcium transport in boar sperm plasma membrane vesicles and their relation to phosphorylation of plasma membrane proteins, *Biol. Reprod.*, 31, 1061, 1984.
186. **Thakkar, J. K., East, J., and Franson, R. C.,** Modulation of phospholipase A_2 activity associated with human sperm membranes by divalent cations and calcium antagonists, *Biol. Reprod.*, 30, 679, 1984.
187. **Brokaw, C. J. and Nagayama, S. M.,** Modulation of the asymmetry of sea urchin sperm flagellar bending by calmodulin, *J. Cell Biol.*, 100, 1875, 1983.
188. **Aitken, R. J., Clarkson, J. S., Hulme, M. J., and Henderson, C. J.,** Analysis of calmodulin acceptor proteins and the influence of calmodulin antagonists on human spermatozoa, *Gamete Res.*, 21, 93, 1988.
189. **Noland, T. D., Van Eldik, L. J., Garbers, D. L., and Burgess, W. H.,** Distribution of calmodulin and calmodulin-binding proteins in membranes from bovine epididymal spermatozoa, *Gamete Res.*, 11, 297, 1985.
190. **Nikolopoulou, M., Soucek, D. A., and Vary, J. C.,** Modulation of lipid composition of boar sperm plasma membranes during an acrosome reaction in vitro, *Arch. Biochem. Biophys.*, 250, 30, 1987.
191. **Thomas, P. and Meizel, S.,** Phosphatidylinositol 4,5-bisphosphate hydrolysis in human sperm stimulated with follicular fluid or progesterone is dependent on Ca^{2+} influx, *Biochem. J.*, 264, 539, 1989.
192. **Berridge, M. J. and Irvine, R. F.,** Inositol phosphates and cell signaling, *Nature (London)*, 341, 197, 1989.
193. **Noland, T. D. and Olson, G. E.,** Calcium-induced modification of the acrosomal matrix in digitonin-permeabilized guinea pig spermatozoa, *Biol. Reprod.*, 40, 1057, 1989.
194. **Talbot, P., DiCartantonio, G., Zao, P., Penkala, J., and Haimo, L. T.,** Motile cells lacking hyaluronidase can penetrate the hamster oocyte cumulus complex, *Dev. Biol.*, 108, 387, 1985.
195. **Zao, P. Z. R., Meizel, S., and Talbot, P.,** Release of hyaluronidase and β-acetylhexosaminidase during in vitro incubation of hamster sperm, *J. Exp. Zool.*, 234, 63, 1985.
196. **Hartmann, J. F., Gwatkin, R. B. L., and Hutchison, C. F.,** Early contact interactions between mammalian gametes in vitro: evidence that the vitellus influences adherence between sperm and zona pellucida, *Proc. Natl. Acad. Sci. U.S.A.*, 69, 2767, 1972.
197. **Hartmann, J. F. and Hutchison, C. F.,** Involvement of two carbohydrate containing components in the binding of uncapacitated spermatozoa to eggs of the golden hamster in vitro, *J. Exp. Zool.*, 201, 383, 1977.
198. **Inoue, M. and Wolf, D. P.,** Sperm binding characteristics of the murine zona pellucida, *Biol. Reprod.*, 13, 340, 1975.
199. **Gwatkin, R. B. L.,** *Fertilization Mechanisms in Man and Mammals*, Plenum Press, New York, 1977.
200. **Saling, P. M., Wolf, D. P., and Storey, B. T.,** Calcium-dependent binding of mouse epididymal spermatozoa to the zona pellucida, *Dev. Biol.*, 65, 515, 1978.
201. **Heffner, L. J., Saling, P. M., and Storey, B. T.,** Separation of calcium effects on motility and zona binding ability in mouse spermatozoa, *J. Exp. Zool.*, 212, 53, 1980.
202. **Heffner, L. J. and Storey, B. T.,** Cold lability of mouse sperm binding to zona pellucida, *J. Exp. Zool.*, 219, 155, 1982.
203. **Yanagimachi, R.,** Requirements of extracellular calcium ions for various stages of fertilization and fertilization-related phenomena in the hamster, *Gamete Res.*, 5, 323, 1982.
204. **Oliphant, G. and Brackett, B. G.,** Capacitation of mouse spermatozoa in media with elevated ionic strength and reversible decapacitation with epididymal extracts, *Fertil. Steril.*, 24, 948, 1973.
205. **Shur, B. D. and Hall, N. G.,** Sperm surface galactosyltransferase activities during in vitro capacitation, *J. Cell Biol.*, 95, 567, 1982.
206. **McLaughlin, J. D. and Shur, B. D.,** Binding of caput epididymal mouse sperm to the zona pellucida, *Dev. Biol.*, 124, 557, 1987.
207. **Borland, R. M., Hazra, S., Biggers, J. D., and Lechene, C. P.,** The elemental composition of the environments of the gametes and preimplantation embryo during initiation of pregnancy, *Biol. Reprod.*, 16, 147, 1977.
208. **Drickamer, K.,** Two distinct classes of carbohydrate-recognition domains in animal lectins, *J. Biol. Chem.*, 263, 9557, 1988.

209. **Berger, T. and Clegg, E. D.,** Adenylate cyclase activity in porcine sperm in response to female reproductive tract secretions, *Gamete Res.*, 7, 169, 1983.
210. **Morton, B. and Albagli, L.,** Modification of hamster spem adenyl cyclase by capacitation in vitro, *Biochem. Biophys. Res. Commun.*, 50, 695, 1973.
211. **Stein, D. M. and Fraser, L. R.,** Cyclic nucleotide metabolism in mouse epididymal spermatozoa during capacitation in vitro, *Gamete Res.*, 10, 283, 1984.
212. **Stein, D. M., Fraser, L. R., and Monks, N. J.,** Adenosine and Gpp(NH)p modulate mouse sperm adenylate cyclase, *Gamete Res.*, 13, 151, 1986.
213. **Lefkowitz, R. J. and Caron, M. G.,** Adrenergic receptors. Models for the study of receptors coupled to guanine nucleotide regulatory proteins, *J. Biol. Chem.*, 263, 4993, 1988.
214. **Imagawa, T., Leung, A. T., and Campbell, K. P.,** Phosphorylation of the 1,4-dihydropyridine receptor of the voltage-dependent Ca^{2+} channel by an intrinsic protein kinase in isolated triads from rabbit skeletal muscle, *J. Biol. Chem.*, 262, 8333, 1987.
215. **Catterall, W. A., Seager, M. J., and Takahashi, M.,** Molecular properties of dihydropyridine-sensitive calcium channels in skeletal muscle, *J. Biol. Chem.*, 263, 3535, 1988.
217. **White, D. R. and Aitken, R. J.,** Relationship between calcium, cyclic AMP, ATP and intracellular pH and the capacity of hamster spermatozoa to express hyperactivated motility, *Gamete Res.*, 22, 163, 1989.
218. **Mahanes, M. S., Ochs, D. L., and Eng, L. A.,** Cell calcium of ejaculated rabbit spermatozoa before and following in vitro capacitation, *Biochem. Biophys. Res. Commun.*, 134, 664, 1986.
219. **Ganz, M. B., Boyarsky, G., Bernd Sterzel, R., and Boron, W. F.,** Arginine vasopressin enhances pH_i regulation in the presence of HCO_3^- by stimulating three acid-base transport systems, *Nature (London)*, 337, 648, 1989.
220. **Calzada, L., Bernal, A., and Loustaunau, E.,** Effect of steroid hormones and capacitation on membrane potential of human spermatozoa, *Arch. Androl.*, 21, 121, 1988.
221. **Mrsny, R. J. and Meizel, S.,** Potassium ion influx and Na^+, K^+-ATPase activity are required for the hamster sperm acrosome reaction, *J. Cell Biol.*, 91, 77, 1981.
222. **Mrsny, R. J., Siiteri, J. E., and Meizel, S.,** Hamster sperm Na^+, K^+-adenosine triphosphatase: increased activity during capacitation in vitro and its relationship to cyclic nucleotides, *Biol. Reprod.*, 30, 573, 1984.
223. **Hyne, R. V.,** Bicarbonate and calcium dependent induction of rapid guinea pig acrosome reaction by monovalent ionophore, *Biol. Reprod.*, 31, 312, 1984.
224. **Hyne, R. V., Edwards, K. P., and Smith, J. D.,** Changes in guinea pig sperm intracellular sodium and potassium content during capacitation and treatment with monovalent ionophore, *Gamete Res.*, 12, 65, 1985.
225. **Hyne, R. V., Higginson, R., Kohlman, D., and Lopata, A.,** Sodium requirement for capacitation membrane fusion during guinea pig acrosome reaction, *J. Reprod. Fertil.*, 70, 83, 1984.
226. **Andrews, J. C. and Bavister, B. D.,** Capacitation of hamster spermatozoa with the divalent cation chelators D-penicillamine, L-histidine, and L-cysteine in a protein-free culture medium, *Gamete Res.*, 23, 159, 1989.
227. **Dawson, R. M. C. and Scott, T.W.,** Phospholipid composition of epididymal spermatozoa prepared by density gradient centrifugation, *Nature (London)*, 202, 292, 1964.
228. **Poulos, A., Voglmayr, J. K., and White, I. G.,** Phospholipid changes in spermatozoa during passage through the genital tract of the bull, *Biochim. Biophys. Acta*, 306, 194, 1973.
229. **Scott, T. W., Voglmayr, J. K., and Setchell, B. P.,** Lipid composition and metabolism in testicular and ejaculated ram spermatozoa, *Biochem. J.*, 102, 456, 1967.
230. **Parks, J. E. and Hammerstedt, R. H.,** Developmental changes occurring in the lipids of ram epididymal spermatozoa plasma membrane, *Biol. Reprod.*, 32, 653, 1985.
231. **Hammerstedt, R. H., Keith, A. D., Hay, S., Deluca, N., and Amman, R. P.,** Changes in ram sperm membranes during epididymal transit, *Arch. Biochem. Biophys.*, 196, 7, 1979.
232. **Olson, G. E. and Danzo, B. J.,** Surface changes in rat spermatozoa during epididymal transit, *Biol. Reprod.*, 24, 431, 1981.
233. **Brown, C. R., von Glos, K. I., and Jones, R.,** Changes in plasma membrane glycoproteins of rat spermatozoa during maturation in the epididymis, *J. Cell Biol.*, 96, 256, 1983.
234. **Zaheb, R. and Orr, G. A.,** Characterization of a maturation-associated glycoprotein on the plasma membrane of rat caudal epididymal sperm, *J. Biol. Chem.*, 259, 839, 1984.
235. **Volgmayr, J. K., Fairbanks, G., Jakowitz, M. A., and Colella, J. R.,** Post-testicular developmental changes in the ram sperm cell surface and their relationship to luminal fluid proteins of the reproductive tract, *Biol. Reprod.*, 22, 655, 1980.
236. **Voglmayr, J. K., Fairbanks, G., and Lewis, R. G.,** Surface glycoprotein changes in ram spermatozoa during epididymal maturation, *Biol. Reprod.*, 29, 767, 1983.
237. **Russell, L. D., Peterson, R. N., Hunt, W., and Strack, L. E.,** Post-testicular surface modifications and contributions of reproductive tract fluids to the surface polypeptide composition of boar spermatozoa, *Biol. Reprod.*, 30, 959, 1984.

238. **Lakoski, K. A., Carron, C. P., Cabot, C. L., and Saling, P. M.,** Epididymal maturation and the acrosome reaction in mouse sperm: response to zona pellucida develops coincident with modifications of M42 antigen, *Biol. Reprod.*, 38, 221, 1988.
239. **Acott, T. S., Katz, D. F., and Hoskins, D. D.,** Movement characteristics of bovine epididymal spermatozoa: effects of forward motility protein and epididymal maturation, *Biol. Reprod.*, 29, 389, 1983.
240. **Mann, T.,** *The Biochemistry of Semen and of the Male Reproductive Tract,* Methuen, New York, 1960.
241. **Mann, T. and Lutwak-Mann, C.,** *Male Reproductive Function and Semen,* Springer-Verlag, New York, 1981.
242. **Myles, D. G., Primakoff, P., and Koppel, D. E.,** A localized surface protein of guinea pig sperm exhibits free diffusion in its domain, *J. Cell Biol.*, 98, 1905, 1984.
243. **Wolf, D. E. and Voglmayr, J. K.,** Diffusion and regionalization in membranes of maturing ram spermatozoa, *J. Cell Biol.*, 98, 1678, 1984.
244. **Blandau, R. J. and Rumery, R. E.,** The relationship of swimming movements of epididymal spermatozoa to their fertilizing capacity, *Fertil. Steril,* 15, 571, 1964.
245. **Bedford, J. M.,** Development of the fertilizing ability of spermatozoa in the epididymis of the rabbit, *J. Exp. Zool.*, 163, 319, 1966.
246. **Orgebin-Crist, M. C.,** Sperm maturation in the rabbit epididymis, *Nature (London),* 216, 816, 1967.
247. **Saling, P. M.,** Development of the ability to bind to zonae pellucidae during epididymal maturation: reversible immobilization of mouse spermatozoa by lanthanum, *Biol. Reprod.*, 26, 429, 1982.
248. **Fournier-Delpech, S., Courtens, J. L., Pisselet, C. L., Delaleu, B., and Courot, M.,** Acquisition of zona binding by ram spermatozoa during epididymal passage, as revealed by interaction with rat oocytes, *Gamete Res.*, 5, 403, 1982.
249. **Thomas, T. S., Wilson, W. L., Reynolds, A. B., and Oliphant, G.,** Chemical and physical characterization of rabbit sperm acrosome stabilizing factor, *Biol. Reprod.*, 35, 691, 1986.
250. **Reynolds, A. B., Thomas, T. S., Wilson, W. L., and Oliphant, G.,** Concentration of acrosome stabilizing factor (ASF) in rabbit epididymal fluid and species specificity of anti-ASF antibodies, *Biol. Reprod.*, 40, 673, 1989.
251. **Hyne, R. V. and Garbers, D. L.,** Inhibition of the guinea pig sperm acrosome reaction by a low molecular weight factor(s) in epididymal fluid and serum, *J. Reprod. Fertil.*, 64, 151, 1982.
252. **Jenkins, A. D., Lechene, C. P., and Howard, S. S.,** Concentrations of seven elements in the intraluminal fluids of the rat seminiferous tubules, rete testis, and epididymis, *Biol. Reprod.*, 23, 981, 1980.
253. **Levine, N. and Kelly, H.,** Measurement of pH in the rat epididymis in vivo, *J. Reprod. Fertil,* 52, 333, 1978.
254. **Johnson, W. L. and Hunter, A. G.,** Seminal antigens: the alteration in the genital tract of female rabbits and during partial in vitro capacitation with β-amylase and β-glucuronidase, *Biol. Reprod,* 7, 332, 1972.
255. **Koehler, J. K., Nudelman, E. D., and Hakomori, S.,** A collagen-binding protein on the surface of ejaculated rabbit spermatozoa, *J. Cell Biol.*, 86, 529, 1980.
256. **San Agustin, J. T., Hughes, P., and Lardy, H. A.,** Properties and function of caltrin, the calcium-transport inhibitor of bull seminal plasma, *FASEB J.*, 1, 60, 1987.
257. **Lewis, R. V., San Agustin, J. T., Kruggel, W., and Lardy, H. A.,** The structure of caltrin, the calcium transport inhibitor of bovine seminal plasma, *Proc. Natl. Acad. Sci. U.S.A.*, 82, 6490, 1985
258. **Aonuma, S., Okabe, M., Kawaguchi, M., and Kishi, Y.,** Zinc effects on mouse spermatozoa and in vitro fertilization, *J. Reprod. Fertil.*, 63, 463, 1981.
259. **Johnson, M. H.,** The macromolecular organization of membranes and its bearing on events leading up to fertilization, *J. Reprod. Fertil.*, 44, 167, 1975.
260. **Schill, W. B., Heimburger, N., Schliessler, H., Stolla, R., and Fritz, H.,** Reversible attachment and localization of the acid-stable seminal plasma acrosin-trypsin inhibitors on boar spermatozoa as revealed by the indirect immunofluorescent staining technique, *Hoppe Seylers Z. Physiol. Chem.*, 356, 1473, 1975.
261. **Eng, L. A. and Oliphant, G.,** Rabbit sperm reversible decapacitation by membrane stabilization with a highly purified glycoprotein from seminal plasma, *Biol. Reprod.*, 38, 235, 1978.
262. **Oliphant, G. and Singhas, C. A.,** Iodination of rabbit sperm plasma membrane: relationship of specific surface proteins to epididymal function and sperm capacitation, *Biol. Reprod.*, 21, 937, 1979.
263. **Voglmayr, J. K. and Sawyer, F. R.,** Surface transformation of ram spematozoa in uterus, oviduct and cauda epididymal fluids in vitro, *J. Reprod. Fertil.*, 78, 315, 1986.
264. **Snider, D. R. and Clegg, E. D.,** Alteration of phospholipids in porcine spermatozoa during in vivo uterus and oviduct incubation, *J. Anim. Sci.*, 40, 269, 1975.
265. **Davis, B. K.,** Timing of fertilization in mammals: sperm cholesterol/phospholipid ratio as a determinant of the capacitation interval, *Proc. Natl. Acad. Sci. U.S.A.*, 78, 7560, 1981.
266. **Go, K. J. and Wolf, D. P.,** The role of sterols in sperm capacitation, *Adv. Lipid Res.*, 20, 317, 1982.
267. **Go, K. J. and Wolf, D. P.,** Albumin-mediated changes in the sperm sterol contents during capacitation, *Biol. Reprod.*, 32, 145, 1985.
268. **Llanos, M. N. and Meizel, S.,** Phospholipid methylation increases during capacitation of golden hamster sperm in vitro, *Biol. Reprod.*, 28, 1043, 1983.

269. **Gordon, M., Dandekar, P. V., and Bartoszewicz, W.,** The surface coat of epididymal, ejaculated, amd capacitated sperm, *J. Ultrastruct. Res.*, 50, 199, 1975.
270. **Talbot, P. and Franklin, L. E.,** Surface modification of guinea pig sperm during in vitro capacitation: an assessment using lectin induced agglutination of living sperm, *J. Exp. Zool.*, 203, 1, 1978.
271. **Koehler, J. K. and Gaddum-Rosse, P.,** Media induced alterations of the membrane associated particles of the guinea pig sperm tail, *J. Ultrastruct. Res.*, 51, 106, 1975.
272. **Friend, D. S., Orci, L., Perrelet, A., and Yanagimachi, R.,** Membrane particle changes attending the acrosome reaction in guinea pig spermatozoa, *J. Cell Biol.*, 74, 561, 1977.
273. **Fayrer-Hosken, R. A., Brackett, B. G., and Brown, J.,** Reversible inhibition of rabbit sperm-fertilizing ability by cholesterol sulfate, *Biol. Reprod.*, 36, 878, 1987.
274. **Handrow, R. R., Boehm, S. K., Lenz, R. W., Robinson, J. A., and Ax, R. L.,** Specific binding of the glycosaminoglycan ^{3}H-heparin to bull, monkey, and rabbit spermatozoa in vitro, *J. Androl.*, 5, 51, 1984.
275. **Saling, P. M.,** Mouse sperm antigens that participate in fertilization. IV. A monoclonal antibody prevents zona penetration by inhibition of the acrosome reaction, *Dev. Biol.*, 117, 511, 1986.
276. **Leyton, L., Robinson, A., and Saling, P.,** Relationship between the M42 antigen of mouse sperm and the acrosome reaction induced by ZP3, *Dev. Biol.*, 132, 174, 1989.
277. **Babcock, D. F. and Florman, H. M.,** unpublished results.
278. **Babcock, D. F.,** unpublished observations.
279. **Christen, R.,** personal communication.

Chapter 6

SPERM RECEPTORS OF MAMMALIAN EGGS

Jeffrey D. Bleil

TABLE OF CONTENTS

I. INTRODUCTION

The basic approach to understanding the molecular basis of mammalian sperm-egg interaction has involved identifying and characterizing sperm and egg proteins having presumptive functions in fertilization and finally determining those functions using the tools of biochemistry and cell biology. Considerable effort has been directed toward identifying the egg's zona pellucida proteins which serve as sperm receptors, and these studies will be emphasized in this review. This review focuses on studies concerning fertilization in mouse, the mammal in which fertilization is best characterized, although pertinent information concerning fertilization in other mammals will be discussed.

II. THE GAMETES

A. THE EGG

The mouse egg is approximately 80 μm in diameter. It has grown as an oocyte from a diameter of about 10 μm within the ovary over a period of 16 days. During its growth, the oocyte's nucleus is arrested in diplotene of meiosis. At the time of ovulation the oocyte "matures" into an egg: it loses tight contact with surrounding follicle cells, the oocyte nucleus dissociates, and the chromosomes progress through meiosis. At the time of fertilization, chromosomes of the egg are arrested in Metaphase I of meiosis, the homologues of which have been extruded from the cell as a small "polar body".[1]

Surrounding the unfertilized mammalian egg is the zona pellucida, a porous, extracellular glycoprotein coat of approximately 5 μm in thickness, which mediates sperm-egg recognition, entrance of the sperm into the egg, and the slow block to polyspermy. Because of the important role the zona pellucida plays in mediating fertilization and because of the relative ease with which this extracellular coat can be purified, the zona pellucida has been intensively studied and represents the best characterized of all components involved in mammalian fertilization. In the mouse, the zona pellucida is composed of three glycoproteins, designated ZP1, ZP2, and ZP3, having apparent molecular weights on SDS-polyacrylamide gels of 200, 120, and 83 kDa, respectively.[2] All three proteins contain both N- and O-linked oligosaccharide chains and are synthesized and secreted by the oocyte during its growth.[3-8] All three of the proteins have been purified, characterized, and their functions in fertilization and zona pellucida structure largely determined.[2,3,9-16] The structure of the zona pellucida is made up of filaments, the basic units of which are ZP2-ZP3 dimers, and these filaments appear to be cross-linked by ZP1.[17,18] ZP1 exists in the zona pellucida as a disulfide-linked homodimer composed of monomers which have an apparent molecular weight of 130 kDa.[2] As described below, ZP3 functions at both the "primary receptor" for sperm and "inducer of the acrosome reaction". Also described below are experiments which suggest that ZP2 functions as the "secondary receptor" for sperm.

The ovulated mouse egg is released into the ampulla of the oviduct within a mass of follicle (cumulus) cells. Shortly after ovulation, these cells migrate away from one another, tethered by long cross-links composed largely of hyaluronic acid. This cumulus mass, approximately 0.3 mm in diameter, fills the oviduct lumen. It is the cumulus mass with which sperm are first presented at the time of fertilization *in vivo*.

B. THE SPERM

Sperm development and differentiation, or spermatogenesis, takes place in the seminiferous tubules of the testes. The stem cells for spermatogenesis, spermatogonia, give rise to spermatocytes. Spermatocytes are the undifferentiated spermatogenic cells that perform the meiotic divisions and growth phase of spermatogenesis, each generating four haploid spermatids. During the haploid spermatid stage, the spermatogenic cell differentiates (spermiogenesis) into the spermatozoon or sperm, forming the organelles which will be necessary for motility, recognition, and penetration of the zona pellucida, and fusion with the egg plasma membrane. Because spermiogenesis occurs during the spermatid stage, many of the proteins required for sperm interaction with the egg are synthesized during this stage. Thus, each sperm's ability to fertilize an egg may, at least in part, depend upon gene expression from a haploid set of chromosomes (see below).

The fully differentiated sperm in mammals is a highly elongated cell composed of a head, midpiece, and tail.[19,20] The vast majority of spermatid cytoplasm has been pinched off as a ''cytoplasmic body'' during the cell's differentiation. The sperm tail or flagellum is similar to flagellae of other cells, having the classic arrangement of 9 + 1 pairs of microtubules. Providing energy for flagellar movement are the spirally arranged mitochondria of the middle piece. Within the mitochondria are the distal centriole, from which the axial filament of the flagellum has arisen, and the proximal part of the axial filament (tail). The sperm head contains, at its most posterior section, the proximal centriole, which will serve as the male pronucleus centrosome following fusion of sperm and egg. The remainder of the sperm head is taken up by the sperm nucleus, containing the highly compacted chromatin, and the acrosome.

The acrosome is a large vesicle containing proteolytic enzymes that are apparently involved in penetration of sperm through the zona pellucida.[21,22] The acrosome arises in the spermatid from vesicles associated with the Golgi.[19] Fusion of these vesicles during spermatid differentiation results in a large lysosome which spreads out over the front half of the spermatid nucleus. The most posterior portion of the stretched acrosome is called the ''equatorial segment''. Although its size and structure vary considerably in different mammals, the acrosome of mouse sperm takes up about half the volume of the head and is tightly wrapped around the sperm nucleus. The acrosome is confined by one continuous membrane, functionally differentiated into the inner acrosomal membrane, apposed to the nucleus, and the outer acrosomal membrane, apposed to the sperm head plasma membrane. During fertilization, the outer acrosomal membrane and plasma membrane of the head will undergo calcium-dependent fusion at numerous points of contact, and these two membranes will be released as hybrid vesicles.[23-25] The inner acrosomal membrane will then be contiguous with the plasma membrane at the equatorial segment.[26-28] This ''acrosome reaction'' will thus expose any acrosomal enzymes associated with the inner acrosomal membrane, providing a means by which the sperm can penetrate the zona pellucida.

The entire surface of the sperm is covered with a plasma membrane, which appears to be regionally differentiated on the basis of localization of certain surface antigens.[29-31] Such regional differentiation of the sperm surface has been interpreted to reflect the specialized functions of different parts of the plasma membrane. As will be seen below, at least for the plasma membrane overlying the acrosomal region of the sperm head, there exists clear evidence that this region of the membrane is specialized for recognition of the zona pellucida of the egg and for those events associated with fusion of plasma and outer acrosomal membranes.

1. Maturation

Differentiated sperm released from the testes traverse the epididymus. In mouse, as sperm pass through the epididymus, they undergo what is termed ''maturation''. Maturation is characterized by a group of cellular modifications that include changes in sperm morphology and gaining of forward motility.[32-34] Modification of the sperm surface morphology is of significance with respect to the important role of the sperm surface in egg recognition. Some of these changes associated with maturation have been characterized, namely, changes in net charge, changes in lectin binding sites on the plasma membrane, and, perhaps the cause of the above changes, accumulation of epididymal-secreted proteins on the sperm surface.[26,35] Prior to maturation (before the sperm traverse the entire epididymus), sperm are immotile and will not, therefore, fertilize eggs. Motility is progressively developed as the sperm traverse the epididymus, and, upon entrance into the cauda epididymes, sperm are fully motile.[32] The mechanism whereby motility is gained is not known.

2. Capacitation

Whether isolated *in vitro* or following ejaculation, mature cauda epididymal or vas deferens sperm are not immediately capable of binding to the zona pellucida of eggs, and, consequently, cannot fertilize eggs.[36] Such sperm must reside for some time in the female genital tract or in culture medium before they are capable of binding to the zona pellucida. The time-dependent requirement for fertilizing capability has been termed "capacitation".[26]

A number of changes in the sperm have been correlated with the temporal requirements for capacitation. During capacitation, sperm change in their motility, originally moving in a linear fashion, and finally developing a sinusoidal movement common to flagellae.[37] Uncapacitated sperm have very different histochemical and ultrastructural characteristics than capacitated sperm. For example, at the ultrastructural level, plasma membrane-associated particles of the sperm head undergo rearrangement during capacitation.[34,38] Finally, uncapacitated sperm are covered with an extracellular coat of glycoproteins (probably acquired during transit through the epididymus), and this coat is lost during capacitation.[39-41]

The simplest explanation for these observations is that capacitation, at the molecular level, is the removal of the extracellular glycoprotein coat covering the sperm plasma membrane. Newly ejaculated sperm (or *in vitro*-isolated caudal and vas deferens sperm) covered with a thick glycoprotein coat would be restricted in normal flagellar movement and prevented from recognizing the zona pellucida via plasma membrane-associated binding sites on the sperm head. With time in culture or in the female genital tract, these glycoproteins are released from the sperm surface, allowing for normal flagellar movement and exposing zona pellucida recognition sites on the sperm head plasma membrane. Such sperm would be "capacitated."

One prediction of the above model would be that the surface-associated glycoproteins of uncapacitated sperm, released from sperm during capacitation, might, at high concentrations, prevent capacitated sperm from binding to eggs. Indeed, such surface-associated glycoproteins have been isolated and do prevent capacitated sperm from binding to eggs. These glycoproteins have been termed "decapacitation factors".[39,41]

During capacitation *in vitro* sperm exhibit an increased susceptibility to undergo the acrosome reaction.[42] It was at one time concluded that the acrosome reaction was an indicator of the completion of capacitation.[26,43] However, acrosome-reacted mouse sperm will not bind to the zona pellucida.[10,11,44] Mouse sperm bind to the zona pellucida via the plasma membrane overlying the head, and the fertilizing sperm undergoes the acrosome reaction at the zona pellucida surface following binding.[11,45-48] Thus, capacitation is completed prior to degenerative acrosome reactions, and acrosome reaction of a mouse sperm prior to zona pellucida binding is a dead-end event. Such acrosome reactions, in culture, are possibly the consequence of increased fluidity (and therefore fusibility) of the plasma membrane. It is reasonable to argue that the surface coat over uncapacitated sperm serves to limit plasma membrane fluidity and possibly restrict movement of plasma membrane proteins whose interaction might be required for initiation of membrane fusion. The sperm is thus delayed from undergoing degenerative acrosome reactions by the presence of the glycoprotein coat. It has been suggested that the time required for completion of capacitation *in vivo* corresponds to the time required for the ejaculated sperm to reach the ampulla of the mouse oviduct, where ovulated, fertilizable eggs reside.[49] Hence, the ejaculated sperm would only be susceptible to undergoing the acrosome reaction at or about the time of approaching the egg.

In conclusion, a proper definition of capacitation is the acquired ability of the sperm head plasma membrane to bind the zona pellucida. This appears to be accomplished, at least in part, by loss of the sperm's surface-associated glycoprotein coat.

A more complete review of capacitation is presented by Florman in Chapter 5.

III. SEQUENCE OF EVENTS IN SPERM-EGG INTERACTION

A. PENETRATION OF THE CUMULUS MASS

The ovulated egg of mammals is released within a mass of follicle cells through which the fertilizing sperm must pass in order to reach the zona pellucida. Since sperm represent an excellent source of hyaluronidase and because acrosome reactions release hyaluronidase (presumably stored within the acrosome), it was originally thought that the acrosome reaction was a prerequisite to sperm penetration through the cumulus mass.[50] However, experiments in mouse and hamster have shown that sperm traverse this apparent barrier prior to the acrosome reaction. In cumulus-intact culture, virtually all sperm were found bound to egg zonae pellucidae with intact acrosomes.[51,52] Since these experiments were performed at low sperm concentration and prior to any discernible cumulus dispersal, it can be concluded that the sperm of mouse and hamster penetrate the cumulus mass intact. As stated above, this conclusion is required by the finding that only acrosome-intact sperm bind to the mouse egg zona pellucida. Whether the acrosome-intact fertilizing sperm progresses through the cumulus layer by releasing hyaluronidase is not known. Although the cumulus layer does not, apparently, mediate specific interactions between sperm and egg, it may function to restrict the sperm's movements during interaction with the zona pellucida, thus ensuring fertilization (see below).

Having traversed the cumulus mass, the acrosome-intact sperm then interacts with the egg zona pellucida. In summary, the sperm first "attaches" and then "binds" to the zona pellucida via the plasma membrane overlying the head. Second, the sperm undergoes the acrosome reaction, binds again, and penetrates the zona pellucida, gaining access to the egg plasma membrane. The sequence of sperm interactions with the zona pellucida is illustrated in Figure 1 and described below.

B. ATTACHMENT AND BINDING OF SPERM TO THE EGG'S ZONA PELLUCIDA

Immediately following mixing of capacitated sperm and cumulus-free eggs *in vitro,* sperm first associate loosely and nonspecifically with the zona pellucida via the head plasma membrane ("attachment"). Attachment has been historically defined as a loose association between sperm and unfertilized egg zona pellucida, which is no tighter than the association between sperm and fertilized egg zona pellucida.[53,54] In a mixture of capacitated sperm, unfertilized eggs, and two-cell embryos (or fertilized one-cell embryos or artificially activated eggs), attached sperm are defined as those sperm that are removed from unfertilized eggs when all sperm have been removed from the two-cell embryos after the mixture has been gently pipetted.[12] Because the zona pellucida of fertilized eggs have undergone the "zona reaction" (the slow block to polyspermy which destroys specific sperm-egg recognition), attachment is thought of as a loose, nonspecific interaction between sperm and egg, analogous to loose attachment of virus to host surface prior to specific binding.[55,56]

Attached sperm subsequently bind tightly and specifically to the unfertilized egg zona pellucida (but not to the fertilized or artificially activated egg zona pellucida).[11] Bound sperm are very tightly associated with the zona pellucida and are defined as those sperm remaining associated with the unfertilized egg zona pellucida after gentle pipetting conditions which remove all sperm from fertilized eggs.[12] It should be noted that both attached and bound sperm are found associated with unfertilized eggs.[10-12] That these two types of association represent two steps in sperm interaction with the zona pellucida was shown by the fact that sperm attached to the zona pellucida of mouse eggs could be quantitatively "chased" into bound sperm.[11]

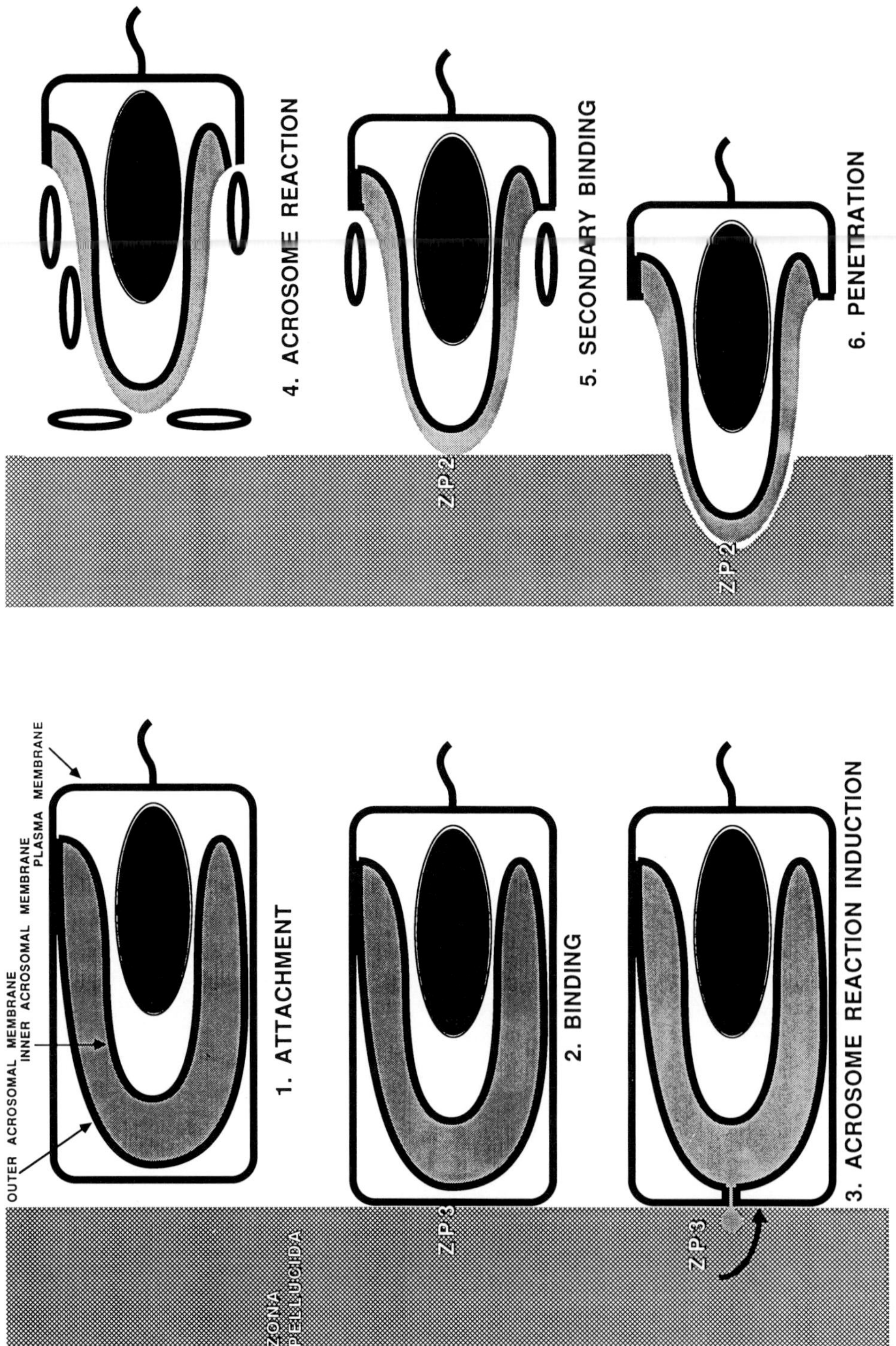

FIGURE 1. Sequence of interactions between the sperm head and the egg's zona pellucida.

Sperm first bind to the zona pellucida of unfertilized mouse eggs via the plasma membrane overlying the head. This has been shown by both electron microscopy and light microscopy. Within short time periods after mixing mouse sperm and eggs *in vitro* (e.g., 5 min), virtually all sperm bound to the zona pellucida are found to be acrosome-intact.[10,44] These same acrosome-intact, bound sperm have been shown to arise from previously attached sperm.[12] Binding of acrosome-intact sperm to the zona pellucida of mouse eggs follows classic saturation kinetics and reaches its maximum within 10 to 15 min under the *in vitro* conditions normally employed (between 10^5 and 10^6 sperm per milliliter).[10,45] A maximal level of about 2000 sperm can bind per egg, about the number of expected for sperm that associate head-on with the zona pellucida and individually cover a surface of 10 μm^2.[12]

Binding of sperm to the egg zona pellucida represents specific interaction between the plasma membrane of the sperm head and a sperm receptor in the zona pellucida. A number of observations lead to this conclusion: (1) sperm will bind to the zona pellucida of unfertilized eggs but not to the zona pellucida of fertilized or artificially activated eggs (i.e., the "zona reaction" results in modification or destruction of the egg's sperm receptor in the zona pellucida);[12,57] (2) sperm will not bind to egg zonae pellucidae treated with proteases, some lectins, or some antibodies directed against the zona pellucida;[35,58-60] (3) binding of mammalian sperm to eggs is partially species-specific (e.g, guinea pig sperm do not bind to the zona pellucida of mouse eggs);[61,62] (4) solubilized, unfertilized mouse egg zonae pellucidae evidence "receptor activity" for sperm in a "competition assay".[12] Soluble egg zonae pellucidae block sperm binding to eggs in a concentration-dependent manner (50% inhibition occurring at about 0.5 zona equivalents per microliter or about 10 n*M* ZP2 and ZP3 and about 1 n*M* ZP1), and this inhibition is not observed if solubilized zonae pellucidae from fertilized eggs are tested. In a similar competition assay, the receptor activity of solubilized egg, but not embryo zonae pellucidae, was first demonstrated in hamster.[57]

C. INDUCTION OF THE ACROSOME REACTION

Following binding of mouse sperm to the egg zona pellucida, the sperm undergoes the acrosome reaction at the zona surface. This statement derives from a number of lines of evidence: (1) in time-course experiments, those same acrosome-intact sperm found bound to the zona pellucida (shortly after mixing sperm and eggs) have been successfully "chased" into acrosome-reacted sperm;[11,44,45,47,48] (2) in the presence of QNB, a specific inhibitor of the zona-induced acrosome reaction, sperm bound to the zona pellucida and underwent the acrosome reaction only following dilution of QNB;[46] (3) acrosome-intact hamster sperm have been directly observed associating with the zona pellucida, subsequently undergoing the acrosome reaction, and finally penetrating the zona pellucida;[52] (4) finally, the zona pellucida has been shown to be a catalyst for the acrosome reaction: solubilized zonae pellucidae, in solution, increased the rate of acrosome reactions in capacitated sperm.[11,45,47,63] As described below, ZP3 has been shown to be the "inducer of the acrosome reaction" in mouse.

The time-course for sperm acrosome reactions at the zona pelludica surface follows a sigmoid curve and has a half-maximal time of between 3 and 8 min following primary binding.[10,11]

A more complete review of the acrosome and acrosome reaction is presented by Kopf in Chapter 7.

D. SECONDARY BINDING AND PENETRATION OF THE ZONA PELLUCIDA

In order to remain associated with zona pellucida following the acrosome reaction, contact between the sperm plasma membrane and the zona pellucida surface is relayed to a new interaction between the inner acrosomal membrane and the zona pellucida.[15,16,44] In pulse-chase experiments,[10] eggs were incubated with capacitated sperm for a period of 5 min. At the end of the 5-min "pulse", all unbound sperm were removed from eggs. Eggs with bound sperm were incubated for increasing periods of time in the absence of free-swimming sperm ("chase") and washed once again to remove any sperm that had been bound at 0-min chase but subsequently became loosely associated ("secondary attachment"). Following the 5-min pulse, virtually all bound sperm were found to be acrosome-intact. During the 30-min chase period, most (about 70%) of the sperm became loosely associated with the zona pellucida: all of these sperm were acrosome reacted. In addition, approximately 20% of sperm that were bound via their plasma membrane at 0-min chase had again successfully bound to the zona pellucida via their inner acrosomal membrane. The remaining 10% of sperm remained bound to the zona pellucida via their plasma membrane. These results are interpreted to mean that the transition from plasma membrane to inner acrosomal membrane interaction with the zona pellucida following the acrosome reaction is accompanied by a loose secondary attachment, followed by a tight "secondary binding".[10] Secondary attachment appears to be coincident with the acrosome reaction and probably reflects the loss of plasma membrane association with ZP3 followed by association between inner acrosomal membrane and zona pellucida. It is likely that secondary binding and the early stages of sperm penetration of the zona pellucida are one and the same event.

It has been estimated that the half-time of secondary binding is approximately 5 to 9 min following primary binding.[10] It should be noted that two, temporally separated binding events, designated B1 and B2, have been observed for hamster sperm-egg interaction, and these results were consistent with B1 and B2 corresponding to binding via plasma and inner acrosomal membranes, respectively.[64]

The above results immediately suggest a possible role for the cumulus mass in successful fertilization. The cumulus mass may serve as a support for the plasma membrane-bound sperm, preventing it from swimming away following the acrosome reaction and during the period of secondary attachment, assuring rebinding (secondary binding) via the inner acrosomal membrane. In fact, analogous pulse-chase experiments with hamster gametes revealed that a higher number of acrosome-reacted sperm (following acrosome reaction at the zona pellucida surface) were found on the surface of the zona pellucida of cumulus-intact than on cumulus-free eggs.[52]

The mechanism of penetration of sperm through the zona pellucida is not well understood. As stated above, the acrosome in mammalian sperm contains proteases, which are thought to be involved in penetration.[22,65] The finding that mouse sperm undergo the acrosome reaction at the zona pellucida surface and that acrosome-reacted sperm release acrosin, a trypsin-like protease found in the acrosome, suggests that the acrosome reaction provides the sperm with a means by which to bore through the zona pellucida and gain access to the egg plasma membrane.[23,27] Since sperm bore an oblique hole through the zona pellucida as they penetrate, and since the forward force provided by the whip-like movement of the sperm tail is alone insufficient to allow the sperm to penetrate the zona pellucida, an inner acrosomal membrane-associated "penetration protease" would provide the means for successful penetration of the sperm through the zona pellucida.[26,27] The recent finding that secondary binding (via the inner acrosomal membrane) appears to be mediated by ZP2 and is inhibited by trypsin inhibitors (see below) makes it tempting to speculate that secondary binding merely represents the early stages of penetration and that the main substrate for the penetration protease of sperm is ZP2.

In hamster, sperm are found in the perivitelline space within 30 min after mixing sperm and eggs.[52] Assuming similar kinetics for primary and secondary binding in hamster and mouse, the period of time required for penetration through the zona pellucida is under 25 min following secondary binding.

E. FUSION AND EGG ACTIVATION

Having successfully penetrated the zona pellucida, the sperm gains entrance to the perivitelline space and access to the egg plasma membrane. Fusion of sperm with the egg plasma membrane "activates" the egg. Egg activation is characterized by progression of the egg chromosomes from Metaphase I to Metaphase II of meiosis (formation of the female pronucleus), and development of both "fast" and "slow" blocks to polyspermy. The fast block to polyspermy in mouse occurs at the level of the plasma membrane and is thought to parallel that in lower organisms (i.e., to involve depolarization of the plasma membrane).[66,67] The second or slow block to polyspermy is due to alterations of the zona pellucida, called the "zona reaction".[9,68] The zona reaction is apparently caused by molecules released from cortical granules that have fused with egg plasma membrane.[9,49,69,70] Functional consequences of the zona reaction are, first, free-swimming sperm will no longer bind to the fertilized egg zona pellucida.[49,71] Second, sperm which are penetrating the zona pellucida will penetrate no further following the zona reaction.[69,71] All three proteins of the mouse egg zona pellucida persist following the zona reaction, but both ZP2 and ZP3 undergo structural modifications (see below).

IV. FUNCTION OF ZONA PELLUCIDA PROTEINS AS SPERM RECEPTORS

In order to determine the functions of zona pellucida proteins in mediating sperm recognition of and penetration through the mouse egg zona pellucida, the sequence of sperm interactions with the zona pellucida has been studied in cumulus-free culture. In this way, purified zona pellucida proteins, fragments thereof, or antibodies against the proteins could be added to sperm-egg mixtures and tested for their ability to perturb specific events along the fertilization pathway.

A. ZP3 IS THE PRIMARY RECEPTOR FOR SPERM

In order to identify the primary receptor for sperm in the mouse egg zona pellucida, each of the zona pellucida glycoproteins, ZP1, ZP2, and ZP3, was purified and tested for receptor activity in the "competition assay".[12] Of the three proteins, only ZP3 blocked sperm binding to eggs, and this inhibition quantitatively accounted for inhibition of sperm binding by total soluble zonae pellucidae. (Identical results were also obtained for hamster ZP3.[72]) Although ZP3 persists in the fertilized egg zona pellucida (to which sperm will not bind), this protein, designated ZP3f, did not evidence receptor activity at the concentrations tested. Thus, in the competition assay, ZP3 has the properties expected of a sperm receptor, and ZP3f, as expected, has lost these properties. Further evidence that ZP3 is the zona pellucida glycoprotein recognized by the sperm head plasma membrane is provided by the fact that ^{125}I-labeled ZP3 competitively binds to the heads of acrosome-intact, but not acrosome-reacted, sperm,[73] and that sperm specifically associate with silica beads to which purified ZP3 is covalently bound.[74] The sperm head plasma membrane contains an estimated 10 to 50,000 binding sites for ZP3, but no binding sites are apparent on the sperm midpiece or tail.[73]

An O-linked oligosaccharide of ZP3 is responsible for its receptor activity. The first evidence that a stable, relatively immobile part of the glycoprotein was involved in sperm recognition came from the following observations. First, a variety of denaturing conditions (100°C, 2% SDS, 8 *M* urea, or treatment at pH 2.5) did not diminish the receptor activity of ZP3.[12] Indeed, sperm will bind to zonae pellucidae of live, ovulated eggs to the same extent as to zonae pellucidae of eggs that have been fixed in glutaraldehyde or in formaldehyde.[61,62,75] Second, ZP3f, purified from fertilized eggs or artificially activated eggs, although indistinguishable from ZP3 on SDS polyacrylamide gels, size-exclusion chromatography, or isoelectric focusing gels, has lost its receptor activity.[12,76] Thus, the transition from ZP3 to ZP3f appeared to be a subtle change, probably not involving alteration of the polypeptide chain. Finally, Florman et al. demonstrated that the polypeptide chain of ZP3 was unnecessary for receptor activity: ZP3 glycopeptides, generated by extensive pronase digestion, retained receptor activity associated with the intact glycoprotein.[13] In fact, a purified O-linked oligosaccharide of ZP3, or about 3900 Da, was found to account for the receptor activity of the protein alone and demonstrated to be the only ZP3 oligosaccharide with affinity for mouse sperm.[14,75] The ZP3 "receptor oligosaccharide" contains at least one terminal α-linked galactose residue (at the nonreducing end of the oligosaccharide chain) which is required for receptor activity.[75] Either removal of this terminal galactose residue by α-galactosidase or modification by galactose oxidase destroys the receptor activity of the purified oligosaccharide. Significantly, the oligosaccharide's receptor activity, destroyed by treatment with galactose oxidase (which oxidizes the C-6 hydroxyl of galactose to an aldehyde), can be restored by subsequent reduction with sodium borohydride, which reforms the C-6 hydroxyl.

Recently, experiments in our laboratory (in preparation) indicate that loss of receptor activity at the time of egg activation (transition from ZP3 to ZP3f) is correlated with modification of the receptor oligosaccharide of ZP3. Radiolabeled receptor oligosaccharide (3900 Da), detected on ion suppression HPLC, is largely lost following fertilization or artificial activation of unfertilized eggs, and this is accompanied by appearance of a new, lower molecular weight oligosaccharide. One possible explanation for this observation is that, at the time of egg activation, a cortical granule glycosidase removes one or more sugars from the receptor oligosaccharide chain, thereby removing all or part of the site recognized by sperm.

There exists a number of cases in which oligosaccharide chains are involved in sperm-egg recognition. In sea urchins, sperm first recognize a fucose-sulfate polysaccharide in the jelly coat surrounding the egg.[77,78] Interestingly, this polysaccharide of sea urchins, like ZP3 in mice, is involved in induction of the sperm acrosome reaction (see below). Also, an oligosaccharide containing terminal *N*-acetyl glucosamine has been suggested to mediate sperm recognition of Ascidian eggs.[79,80]

B. ZP3 MEDIATES THE SPERM ACROSOME REACTION FOLLOWING PRIMARY BINDING

As stated above, solubilized zonae pellucidae block sperm binding to eggs in a competition assay, and this receptor activity is 50% maximal at a concentration of about 0.5 zona pellucida equivalent per microliter. ZP3 (or the receptor oligosaccharide of the protein) has been shown to be responsible for this activity. At higher concentrations, solubilized zonae pellucidae induce the acrosome reaction of capacitated sperm (50% maximum acrosome reaction at a concentration of approximately 1.5 to 2 zona pellucida equivalents per microliter).[11] In turn, ZP3 has been shown to be responsible for this induction of sperm acrosome reactions.[11] (Identical results have been obtained with hamster egg ZP3.[72]) It is of interest to note that, whereas the concentration dependence of ZP3's receptor activity increases in a hyperbolic manner, the concentration dependence of ZP3's ability to induce the acrosome reaction increases in a sigmoidal manner.[11] Also, neither the O-linked oligosaccharide of ZP3 nor glycopeptides of the protein evidence capacity to induce sperm acrosome reactions, indicating that the ability of the protein to induce the acrosome reaction requires an intact polypeptide chain.[13,14] Therefore, ZP3 has both receptor activity and ability to induce the sperm acrosome reaction, but these two functions evidence different forms of concentration dependency, and are physically separable.

One possible explanation for these observations is that two or more identical receptor oligosaccharides must be physically linked via polypeptide chain in order to initiate the acrosome reaction. Soluble ZP3 would then interact in a di- or multivalent manner with the sperm plasma membrane, leading to cross-linking of ZP3 binding sites on the sperm head ("patching"), and this, in turn, leading to induction of the acrosome reaction in a manner similar to other extracellular signaling systems. Since the intact zona pellucida essentially presents to the sperm a gel composed of ZP3 receptor oligosaccharides, it is easy to see how multiple receptor sites on the zona pellucida might lead to patching of ZP3 recognition sites on the sperm head at the point of contact between sperm and zona pellucida. Indeed, patching of soluble ZP3 on the sperm head has been implied by immunological studies.[81,82] Proteolytic fragments of ZP3 which have receptor activity but are incapable of inducing the acrosome reaction were incubated with capacitated mouse sperm. If these sperm (presumably bearing bound ZP3 fragments) were then exposed to IgG directed against the ZP3 fragments, acrosome reactions were induced. However, Fab fragments derived from this same IgG were incapable of inducing acrosome reactions. Finally, when sperm were incubated with ZP3 proteolytic fragments, followed by anti-ZP3 Fab, acrosome reactions were successfully induced only after addition of secondary antibody IgG directed against the Fabs. Thus, glycopeptides of ZP3 can induce the sperm's acrosome reaction, but only after they have been cross-linked. In fact, apparent time-dependent patching of ^{125}I-ZP3 on the heads of acrosome-intact sperm has been observed in our laboratory (unpublished observations).

The hypothesis[83] that ZP3 interaction with the sperm head leads to induction of the acrosome reaction by multivalent association, followed by patching of the ZP3 recognition protein on the sperm head, followed by signal transduction, and, ultimately, calcium-dependent membrane fusion, is inferred by other observations. If it is true that the ZP3 polypeptide chain bears two or more identical "receptor oligosaccharides", it might be expected that higher concentrations of the purified receptor oligosaccharide would be required in order to obtain the same level of receptor activity as intact ZP3 (i.e., ZP3 would have higher avidity for the sperm head than the univalent oligosaccharide). Indeed, unpublished observations in our laboratory suggest that this is the case. Consistent with this observation, other unpublished work in our laboratory indicates that ZP3 does associate more tightly with the sperm head than do ZP3 glycopeptides.

The patching hypothesis described above predicts that selective removal of all but one receptor oligosaccharide of ZP3 would result in a "defective" ZP3 molecule which would evidence receptor activity in the competition assay, but would be incapable of inducing the sperm's acrosome reaction. Such a defective ZP3 molecule appears to have been produced by partial cortical granule release. An incomplete transition of ZP3 to ZP3f can be obtained if ovulated mouse eggs are treated with phorbol esters (12-*O*-tetradecanoyl phorbol-13-acetate [TPA]). Such treatment results in incomplete cortical granule release and alteration of ZP3 such that ZP3 *retains* receptor activity but has apparently *lost* ability to induce sperm acrosome reactions at the concentrations tested.[84-86] Importantly, if this defective ZP3 in soluble form is incubated with sperm at concentrations normally capable of inducing acrosome reactions, subsequent addition of anti-ZP3 IgG to the sperm results in successful induction of the acrosome reaction.[82] Thus, it might be argued that ZP3, bearing at least two identical receptor oligosaccharides, loses, on average, one of those receptor "sites" following TPA-induced partial cortical granule release. Consequently, ZP3 retains receptor activity, but has lost ability to "patch" on the sperm surface because it bears, on average, a monovalent receptor oligosaccharide. Subsequent IgG-induced patching of monovalent ZP3 molecules on the sperm surface would result in acrosome reaction. If this interpretation of these results is correct, it would be expected that, in the competition assay, defective ZP3 generated by TPA-induced partial cortical granule release retains receptor activity, but at a considerably higher soluble concentration than ovulated egg ZP3. In turn, soluble defective ZP3 might evidence ability to induce the sperm acrosome, but only at relatively high concentrations.

It follows from the patching hypothesis that the sperm surface protein which recognized ZP3 would likely exist as a multimer, thus providing the means to form a cross-linked complex on the sperm head. It remains to be seen whether sperm proteins which have been suggested to play a role in primary binding of sperm to zona pellucida (see below) have multimeric properties.

C. EVIDENCE THAT ZP2 IS THE SECONDARY RECEPTOR FOR SPERM

The nature of inner acrosomal membrane association with the zona pellucida following the acrosome reaction was first indicated by results of whole-mount autoradiographic experiments similar to those performed using ^{125}I-ZP3. ^{125}I-ZP2 was found to have affinity for the heads of acrosome-reacted sperm, but not the heads of acrosome-intact sperm.[73] In pulse-chase binding experiments, a monoclonal antibody (or its Fab fragments) directed against ZP2, when bound to ZP2 in the intact zona pellucida, blocked secondary binding (via inner acrosomal membrane) but had no effect upon primary binding (via plasma membrane) of sperm to the zona pellucida.[73] Moreover, these same antibodies, directed against ZP2, blocked association of soluble ^{125}I-ZP2 with the inner acrosomal membrane of acrosome-reacted sperm. Hence, ZP2 appears to be a likely candidate as "secondary receptor" for sperm in the zona pellucida, acting as the protein to which newly acrosome-reacted sperm bind via their inner acrosomal membrane.

The possibility that secondary binding, as observed in the light microscope, represents the first stages of sperm penetration through the zona pellucida is supported by the finding that trypsin inhibitors (soybean trypsin inhibitor [SBTI] and α 1 antitrypsin) block secondary, but not primary, binding of mouse sperm to eggs.[10] These results are consistent with previous work, in which effects on primary and secondary binding were not assessed.[87] SBTI inhibition of secondary binding appears to block association of the sperm inner acrosomal membrane with ZP2, since ^{125}I-SBTI, like ^{125}I-ZP2, associates specifically with the heads of acrosome-reacted sperm.[73] These results are consistent with the finding that *p*-aminobenzamidine, a synthetic trypsin inhibitor, was found to inhibit B2 but not B1 during hamster sperm-egg interaction *in vitro*.[64] One interpretation of these results is that secondary binding represents the tight association between sperm and zona pellucida during protease-dependent penetration of the zona pellucida and that ZP2 serves as the substrate for the sperm inner acrosomal protease. A likely candidate for such a protease is acrosin, a trypsin-like protease associated with the acrosome of most mammalian sperm (see Urch, Chapter 9). However, no direct evidence that acrosin (or any other acrosomal protease) is directly involved in secondary binding or penetration exists to date. Indeed, whether acrosin resides on the inner acrosomal membrane is in dispute.[65]

Inhibitors of trypsin have been shown by other investigators to block sperm binding to eggs, and this was originally interpreted as a consequence of inhibition of a sperm plasma membrane-associated trypsin-like protease involved in primary binding. A serine protease-active site titrant, 4-methylumbelliferyl-*p*-guanidinobenzoate (MUGB), which is hydrolyzed by mouse sperm, inhibited association of mouse sperm with the zona pellucida.[88] In order to minimize the effect of this inhibition upon secondary binding, only uncapacitated sperm, which have been shown to be incapable of undergoing the zona-induced acrosome reaction, were used in this study. However, since uncapacitated sperm cannot bind to the zona pellucida of mouse eggs,[36] it is difficult to interpret these results.

As stated above, only capacitated, acrosome-intact sperm will bind to the mouse egg zona pellucida. Nevertheless, once having bound to the zona pellucida via its head plasma membrane and having undergone the acrosome reaction at the zona pellucida surface, a sperm will bind again and remain bound during penetration of the zona pelludica. Since the acrosome-reacted sperm would appear to have ZP2 recognition sites associated with the inner acrosomal membrane, why do not free-swimming, acrosome-reacted sperm bind to the zona pellucida? This is, at present, unclear, but it is possible that for secondary binding to occur, inner acrosomal membrane-associated sites that recognize ZP2 must be presented to the zona pellucida during the acrosome reaction. One way this requirement might be fulfilled would be if such secondary binding sites were present at high number but of low affinity. These sites, presented to the zona pellucida in a concerted manner during loss of hybrid vesicles (acrosome reaction), and along with flagellar force directing the head into the zona pellucida at that time, would be allowed to bind. However, free-swimming, acrosome-reacted sperm hitting the zona pellucida, usually in a glancing blow, would not have the time or proper orientation for tight binding (i.e., multiple site interactions) to develop. In fact, the number of ZP2 binding sites on the inner acrosomal membrane appears to be much larger than the number of ZP3 binding sites on the plasma membrane,[73] and a measured low affinity of the inner acrosomal membrane for ZP2 might be expected, since dissociation of a protease and proteolytic products is quite rapid.

At the time of egg activation and zona reaction, the polypeptide chain of ZP2 is cleaved by a protease, presumably released from cortical granules.[9,68] The proteolytic fragments remain held together by disulfide cross-links, and so ZP2 persists following fertilization and the zona pellucida remains intact. (ZP2 from hamster zonae pellucidae is cleaved in an apparently identical manner.[72]) It is possible that the specific proteolytic cleavage of ZP2 results in folding of the protein in such a way that the substrate for the presumptive sperm penetration protease becomes inaccessible. This would account for penetrating sperm being blocked from further penetration following fertilization. However, no direct evidence for the relationship between ZP2 cleavage and penetration blockade exists to date.

V. SPERM PROTEINS THAT MAY FUNCTION IN FERTILIZATION

As described above, the proteins of the mouse egg pellucida have been identified and characterized and their roles in sperm-egg interaction have been largely elucidated. The next obvious step in understanding the molecular basis for sperm interaction with the zona pellucida would be the identification of the sperm proteins which recognize ZP3 and ZP2. Some sperm proteins have been proposed as having roles in sperm recognition and penetration of the zona pellucida, and these are described below.

A. SPERM PROTEINS THAT MAY FUNCTION IN PRIMARY BINDING

Because considerable information exists concerning the primary receptor for sperm in the mouse zona pellucida (ZP3), certain predictions can be made about the nature of the sperm's ZP3 recognition protein. Some of the most obvious predictions are (1) a sperm ZP3 recognition protein should reside on the plasma membrane overlying the sperm head; (2) this protein should be capable of associating specifically with ZP3 and specifically with the receptor oligosaccharide of ZP3 (that is, it should be a lectin-like surface protein); (3) since a terminal α-linked galactose residue is required for sperm to recognize the receptor oligosaccharide, the ZP3 recognition protein should have measurable affinity for galactose. Of sperm proteins proposed to have possible roles in primary binding, those that fulfill at least some of these criteria are the following.

1. Mouse Sperm Galactosyl Transferase

Sperm galactosyl transferase may play a role in binding of the mouse sperm plasma membrane to the zona pellucida. Some T/t mouse mutants evidence segregation distortion (transmission of alleles to progeny in non-Mendelian ratios), and this has been ascribed to overexpression of galactosyl transferase on sperm bearing the t haplotype. Heterozygous +/t males transmit the t haplotype to as many as 99% of their offspring.[89] These more successful sperm have been shown to express a higher level of galactosyl transferase, leading to the hypothesis that galactosyl transferase is responsible for sperm binding to the zona pellucida and that stronger binding results in higher fertility.[89-91] One hypothesis which explains segregation distortion of t haplotype is that galactosyl transferase is expressed during the haploid stage of spermatogenesis and that genetically haploid spermatids give rise to phenotypically haploid sperm: thus, t sperm bearing high levels of galactosyl transferase would bind more efficiently to the zona pellucida.[89] Although spermatids develop within a common syncytium, it would appear that neither galactosyl transferase nor its mRNA are shared among cells of the syncytium: immunological evidence suggests (1) that galactosyl transferase is located on the plasma membrane overlying the sperm head, placing the protein in the expected location for interaction with ZP3; and (2) that the endoplasmic reticulum is not shared among spermatids.[92,93] This is clearly not the case for all membrane-associated proteins expressed in the haploid spermatid and may be unique to galactosyl transferase.[94]

It has been proposed that mouse sperm surface galactosyl transferase binds to a terminal glcNAc moiety of ZP3 and, in the absence of substrate UDPgal, the enzyme functions essentially as a lectin for the sperm receptor.[90,91] In support of the hypothesis, it has been demonstrated that sperm are prevented from binding to the zona pellucida of eggs that had been treated with milk galactosyl transferase and UDPgal.[90,91] Presumably, galactose is transferred to the receptor oligosaccharide of ZP3 in such experiments, preventing subsequent recognition by sperm. It should be noted that the requirement for a terminal α-linked galactose residue on the receptor oligosaccharide of ZP3 does not exclude a role for sperm galactosyl transferase in primary binding.

Direct evidence for galactosyl transferase functioning as the sperm's ZP3 recognition protein does not exist. Specific interaction between ZP3 or its receptor oligosaccharide and mouse sperm galactosyl transferase has not, as yet, been demonstrated. Importantly, there may be no relationship between the fact that t-bearing sperm from +/t heterozygotes evidence positive segregation distortion and the observation that these t-bearing sperm bear greater levels of galactosyl transferase. The "sibling" +-bearing sperm from +/t heterozygotes are largely acrosome reacted and cannot, therefore, bind to the zona pellucida or fertilize eggs.[95]

2. A 95-kDa Mouse Sperm Protein

Purified ^{125}I-ZP3 has been shown to bind a 95-kDa band on nitrocellulose blots containing sperm proteins separated by SDS-polyacrylamide gel electrophoresis.[96] Antibodies directed against phosphotyrosine also bind to a 95-kDa protein on such blots and, interestingly, localize to the acrosomal region of mouse sperm. This result is consistent with evidence that calcium-dependent membrane fusion, leading to the acrosome reaction, is mediated by G-proteins.[97-101] These authors have postulated that a 95-kDa sperm surface protein binds ZP3, and subsequent patching of the sperm protein leads to the acrosome reaction via kinase-activated, G-protein-mediated membrane fusion. If the 95-kDa sperm protein is the ZP3 recognition protein, specific association with ZP3 and with its receptor oligosaccharide should be demonstrable.

3. A-56 kDa Mouse Sperm Protein

A mouse sperm protein, which runs at 56 kDa on SDS-polyacrylamide gels, has many of the properties expected of the sperm's ZP3 recognition protein. It has specific affinity for ZP3, since it has been found to bind to affinity columns bearing purified ZP3, but not ZP2.[102] As expected of the sperm protein that recognizes the receptor oligosaccharide of ZP3, the 56-kDa protein binds to affinity columns bearing galactose (but not *N*-acetyl glucosamine). The 56-kDa protein has been shown to be a surface protein on the basis of its susceptibility, in intact sperm, to proteolytic digestion, and its modification by surface-specific radioiodination techniques. Most importantly, cross-linking experiments demonstrated that ZP3 specifically associates with this protein on the surface of the sperm head. Although this protein fulfills many of the criteria expected of the ZP3 recognition protein, it remains to be demonstrated that it has attenuated affinity for ZP3f, and whether this protein, as hypothesized, plays a direct role in the ZP3-induced acrosome reaction.

It should be noted that a 54-kDa rat testes protein having affinity for galactose and related to asialoglycoprotein receptor has been identified.[103] Whether this protein is identical to the 56-kDa protein of mouse is not known.

B. ACROSIN: A MOUSE SPERM PROTEIN THAT MAY FUNCTION IN PENETRATION

As stated above, there exists the possibility that secondary binding of sperm to the zona pellucida, mediated by ZP2, may simply represent the first stages of sperm penetration. In turn, ZP2 exists as a likely candidate for the substrate of the sperm penetration protease (see above). Acrosin, the most abundant acrosomal protease, has some of the properties expected of the penetration protease. First, acrosin has trypsin-like substrate specificity,[23,104,105] meaning that it would be inhibited by protease inhibitors that block secondary binding (see above). Second, nitrocellulose blots containing total sperm proteins (separated by SDS-polyacrylamide gel electrophoresis) bind ^{125}I-labeled, soluble zona pellucida proteins.[106,107] The band to which one or more of the zona pellucida proteins binds appears to be acrosin.[59,103,104] Whether ZP2, as expected, is the only zona pellucida protein that binds acrosin on such blots is not known.

Although acrosin interaction with ZP2 can be inferred from the types of results described above, no direct evidence for participation of the protease in sperm interaction with the zona pellucida exists to date.

A more complete review of acrosin is presented by Urch in Chapter 9.

VI. CONCLUSION

Fertilization in mammals involves a discreet series of interactions between sperm and egg, beginning with recognition (''binding'') of the egg by the sperm and ultimately resulting in fusion of egg and sperm and formation of the diploid zygote. These interactions involve specific molecules of the sperm and of the egg. Glycoproteins of the zona pellucida, which mediates sperm-egg recognition, sperm penetration, and the slow block to polyspermy, have been identified and characterized, and many of their functions in fertilization determined. Further biochemical analysis of these glycoproteins will undoubtedly serve to answer many of the questions that remain concerning fertilization in mammals. Molecular characterization of sperm molecules which interact with the zona pellucida represents another area of study which will certainly provide greater insight into the biochemistry of fertilization.

REFERENCES

1. **Wassarman, P. M. and Josefowicz, W. J.,** Oocyte development in the mouse: an ultrastructural composition of oocytes isolated at various stages of growth and meiotic competence, *J. Morphol.,* 156, 209, 1978.
2. **Bleil, J. D. and Wassarman, P. M.,** Structure and function of the zona pellucida: identification and characterization of the proteins of the mouse oocyte's zona pellucida, *Dev. Biol.,* 76, 185, 1980b.
3. **Bleil, J. D. and Wassarman, P. M.,** Synthesis of zona pellucida proteins by denuded and follicle-enclosed mouse oocytes during culture in vitro, *Proc. Natl. Acad. Sci. U.S.A.,* 77, 1029, 1980c.
4. **Greve, J. M., Salzmann, G. S., Roller, R. J., and Wassarman, P. M.,** Biosynthesis of the major zona pellucida glycoprotein secreted by oocytes during mammalian oogenesis, *Cell,* 31, 749, 1982.
5. **Roller, R. J., Kinloch, R. A., Hiraoka, B. Y., Li, S. S.-L., and Wassarman, P. M.,** Gene expression during mammalian oogensis and early embryogenesis: quantification of three messenger RNAs abundant in fully grown mouse oocytes, *Development,* 106, 251, 1989.
6. **Roller, R. J. and Wassarman, P. M.,** Role of asparagine-linked oligosaccharides in secretion of glycoproteins of the mouse egg's extracellular coat, *J. Biol. Chem.,* 258, 13243, 1983.
7. **Philpott, C. C., Ginguette, M. J., and Dean, J.,** Oocyte-specific expression and development regulation of ZP3, the sperm receptor of the mouse zona pellucida, *Dev. Biol.,* 121, 568, 1987.
8. **Salzmann, G. S., Greve, J. M., Roller, R. J., and Wassarman, P. M.,** Biosynthesis of the sperm receptor during oogenesis in the mouse, *EMBO J.,* 2, 1451, 1983.

9. **Bleil, J. D., Beall, C. F., and Wassarman, P. M.,** Mammalian sperm-egg interaction: fertilization of mouse eggs triggers modification of the major zona glycoprotein, *Dev. Biol.,* 86, 189, 1981.
10. **Bleil, J. D., Greve, J. M., and Wassarman, P. M.,** Identification of a secondary sperm receptor in the mouse egg zona pellucida: role in maintenance of binding acrosome-reacted sperm, *Dev. Biol.,* 128, 376, 1988.
11. **Bleil, J. D. and Wassarman, P. M.,** Sperm-egg interactions in the mouse: sequence of events and induction of acrosome reaction by a zona pellucida glycoprotein, *Dev. Biol.,* 95, 317, 1983.
12. **Bleil, J. D. and Wassarman, P. M.,** Mammalian sperm-egg interaction: identification of a glycoprotein in mouse egg zonae pellucidae possessing receptor activity for sperm, *Cell,* 20, 873, 1980a.
13. **Florman, H. M., Bechtol, K. B., and Wassarman, P. M.,** Enzymatic dissection of the functions of the mouse egg's receptor for sperm, *Dev. Biol.,* 106, 243, 1984.
14. **Florman, H. M. and Wassarman, P. M.,** O-Linked oligosaccharides of mouse egg ZP3 account for its sperm receptor activity, *Cell,* 41, 313, 1985.
15. **Wassarman, P. M.,** The biology and chemistry of fertilization, *Science,* 235, 553, 1987.
16. **Wassarman, P. M.,** Zona pellucida glycoproteins, *Annu. Rev. Biochem.,* 57, 415, 1988.
17. **Greve, J. M. and Wassarman, P. M.,** Mouse egg extracellular coat is a matrix of interconnected filaments possessing a structural repeat, *J. Mol. Biol.,* 181, 253, 1985.
18. **Greve, J. M. and Wassarman, P. M.,** unpublished observations.
19. **Fawcett, D. W.,** The mammalian spermatozoa, *Dev. Biol.,* 44, 394, 1975.
20. **Eddy, E. M.,** The spermatozoon, in *The Physiology of Reproduction,* Vol. 1, Knobil, E. and Neill, J. D., Eds., Raven Press, New York, 1988, 27.
21. **Allison, A. C. and Hartree, E. F.,** Lysosomal enzymes in the acrosome and their possible role in fertilization, *J. Reprod. Fertil.,* 21, 501, 1970.
22. **McRorie, R. A. and Williams, W. L.,** Biochemistry of mammalian fertilization, *Annu. Rev. Biochem.,* 43, 777, 1974.
23. **Green, D. P. L. and Hockaday, A. R.,** The histochemical localization of acrosin in guinea pig sperm after the acrosome reaction, *J. Cell Sci.,* 32, 177, 1978.
24. **Flaherty, S. P. and Olson, G. E.,** Membrane domains in guinea pig sperm and their role in membrane fusion events of the acrosome reaction, *Anat. Rec.,* 220, 267, 1988.
25. **Flechon, J.-E., Harrison, R. A. P., Flechon, B., and Escaig, J.,** Membrane fusion events in the Ca^{2+} ionophore-induced acrosome reaction of ram spermatozoa, *J. Cell Sci.,* 81, 43, 1986.
26. **Yanagimachi, R.,** Mammalian fertilization, in *Physiology of Reproduction,* Knobil, E. and Neill, J., Eds., Raven Press, New York, 1988, 135.
27. **Yanagimachi, R. and Noda, Y. D.,** Scanning electron microscopy of golden hamster spermatozoa before and during fertilization, *Experientia,* 28, 69, 1972.
28. **Flechon, J.-E.,** Sperm surface changes during the acrosome reaction as observed by freeze-fracture, *Am. J. Anat.,* 174, 239, 1985.
29. **Aguas, A. P. and Pinto Da Silva, P.,** Regionalization of transmembrane glycoproteins in the plasma membrane of boar sperm is revealed by fracture-label, *J. Cell Biol.,* 97, 1356, 1983.
30. **Holt, W. V.,** Membrane heterogeneity in the mammalian spermatozoa, *Int. Rev. Cytol.,* 87, 159, 1984.
31. **Peterson, R. N., Gillot, M., Hunt, W., and Russell, L. D.,** Organization of the boar spermatozoan plasma membrane: evidence for separate domain (subdomains) of integral membrane proteins in the plasma membrane overlying the principal segment of the acrosome, *J. Cell Sci.,* 88, 343, 1987.
32. **Saling, P. M.,** Development of the ability to bind to the zona pellucida during epididymal maturation: reversible immobilization of mouse sperm by lanthanum, *Biol. Reprod.,* 26, 429, 1982.
33. **Cooper, G. W. and Bedford, J. M.,** Acquisition of surface charge by the plasma membrane of mammalian spermatozoa during epididymal maturation, *Anat. Rec.,* 169, 300, 1971.
34. **Saxena, N. K., Russell, L. D., Saxena, N., and Peterson, R. N.,** Immunofluorescence antigen localization on boar sperm plasma membranes. Monoclonal antibodies reveal apparent new domains and apparent redistribution of surface antigens during maturation and ejaculation, *Anat. Rec.,* 214, 238, 1986.
35. **Nicholson, G. L., Usui, N., Yanagimachi, R., Yanagimachi, H., and Smith, J. R.,** Lectin binding sites on the membranes of rabbit spermatozoa. Changes in surface receptors during epididymal maturation and after ejaculation, *J. Cell Biol.,* 74, 950, 1977.
36. **Saling, P. M., Storey, B. T., and Wolf, D. P.,** Calcium-dependent binding of mouse spermatozoa to the zona pellucida, *Dev. Biol.,* 65, 515, 1978.
37. **Yanagimachi, R.,** The movement of golden hamster spermatozoa before and after capacitation, *J. Reprod. Fertil.,* 23, 193, 1970.
38. **Hartmann, J. F. and Gwatkin, R. B. L.,** Alteration of sites on the mammalian sperm surface following capacitation, *Nature,* 234, 479, 1971.
39. **Aunomas, S., Mayumi, T., Suzuki, K., Noguchi, T., Iwai, M., and Okabe, M.,** Studies on sperm capacitation. I. The relationship between a given sperm-coating antigen and a sperm capacitation phenomenon, *J. Reprod. Fertil.,* 35, 425, 1973.

40. **Oliphant, G.,** Removal of sperm seminal plasma components as a prerequisite to induction of the acrosomal reaction, *Fertil. Steril.*, 27, 28, 1976.
41. **Reddy, J. M., Stark, R. A., and Zaneveld, J. D.,** A high molecular weight antifertility factor from human seminal plasma, *J. Reprod. Fertil.*, 60, 126, 1980.
42. **Friend, D. S., Orci, L., Perrelet, A., and Yanagimachi, R.,** Membrane particle changes attending the acrosome reaction in guinea pig spermatozoa, *J. Cell Biol.*, 74, 561, 1977.
43. **Yanagimachi, R., Nicholson, G. L., Noda, Y. D., and Fujimoto, M.,** Electron microscopic observations of the distribution of acidic anionic residues on hamster spermatozoa and eggs before and during fertilization, *J. Ultrastruct. Res.*, 43, 344, 1973.
44. **Saling, P. M., Sowinksy, J., and Storey, B. T.,** An ultrastructural study of epididymal mouse spermatozoa binding to zonae pellucidae in vitro: sequential relationship to the acrosome, reaction, *J. Exp. Zool.*, 209, 229, 1979.
45. **Florman, H. M., Saling, P. M., and Storey, B. T.,** Fertilization of mouse eggs in vitro. Time resolution of the reactions preceding penetration of the zona pellucida, *J. Androl.*, 3, 373, 1982.
46. **Florman, H. M. and Storey, B. T.,** Inhibition of in vitro fertilization of mouse eggs: 3-quinuclidynyl benzilate specifically blocks penetration of zonae pellucidae by mouse spermatozoa, *J. Ep. Zool.*, 216, 159, 1981.
47. **Florman, H. M. and Storey, B. T.,** Mouse gamete interactions: the zona pellucida is the site of the acrosome reaction leading to fertilization in vitro, *Dev. Biol.*, 91, 121, 1982.
48. **Saling, P. M. and Storey, B. T.,** Mouse gamete interactions during fertilization in vitro. Chlortetracycline as a fluorescent probe for the mouse acrosome reaction, *J. Cell Biol.*, 83, 544, 1979.
49. **Wolf, D. P.,** The mammalian egg's block to polyspermy, in *Fertilization and Embryonic Development In Vitro,* Mastroianni, L. and Biggers, J. D., Eds., Plenum Press, New York, 1981, 185.
50. **Ready, J. M., Joyce, C., and Zaneveld, J. D.,** Role of hyaluronidase in fertilization: the antifertility activity of myocrosin, a nontoxic hyaluronidase inhibitor, *J. Androl.*, 1, 28, 1980.
51. **Storey, B. T., Lee, M. A., Muller, C., Ward, C. R., and Wirtschafter, D. G.,** Binding of mouse spermatozoa to the zona pellucida mouse eggs in cumulus: evidence that the acrosomes remain substantially intact, *Biol. Reprod.*, 3, 1119, 1984.
52. **Cherr, G. N., Lambert, H., Meizel, S., and Katz, D. F.,** In vitro studies of the golden hamster sperm acrosome reaction: completion of the zona pellucida and induction by homologous soluble zonae pellucidae, *Dev. Biol.*, 114, 119, 1986.
53. **Hartmann, J. F., Gwatkin, R. B. L., and Hutchison, C. F.,** Early contact interactions between mammalian gametes in vitro: evidence that the vitellus influences adherence between sperm and zona pellucida, *Proc. Natl., Acad. Sci. U.S.A.*, 69, 2767, 1972.
54. **Inoue, M. and Wolf, D. P.,** Fertilization associated changes in the murine zona pellucida: a time sequence study, *Biol. Reprod.*, 13, 546, 1975.
55. **Crowell, R. L. and Lonberg-Holm, K.,** *Virus Attachment and Entry into Cells,* American Society of Microbiology, Washington, D.C., 1986.
56. **Lindberg, A. A.,** Bacteriophase receptors, *Annu. Rev. Microbiol.*, 27, 205, 1973.
57. **Gwatkin, R. B. L. and Williams, D. T.,** Receptor activity of the hamster and mouse zona pellucida before and after the zona reaction, *J. Reprod. Fertil.*, 49, 55, 1976.
58. **Jilek, F. and Pavlok, A.,** Antibodies against mouse ovaries and their effect on fertilization in vitro and in vivo in the mouse, *J. Reprod. Fertil.*, 42, 377, 1975.
59. **Oikawa, T., Nicolson, G. L., and Yanagimachi, R.,** Wheat germ agglutinin blocks mammalian fertilization, *Nature (London),* 241, 256, 1973.
60. **Shivers, C. A., Dudkiewiocz, A. B., Franklin, L. E., and Fussell, E. N.,** Inhibition of sperm-egg interaction by specific antibody, *Science,* 178, 1211, 1972.
61. **Gulyas, B. J. and Schmell, E. D.,** Sperm-egg recognition and binding in mammals, in *Bioregulators of Reproduction,* Jagiello, G. and Vogel, H. J., Eds., Academic Press, New York, 1981, 499.
62. **Schmell, E. G. and Gulyas, B. J.,** Mammalian sperm-egg recognition and binding in vitro. I. Specificity of sperm interactions with live and fixed eggs in homologous inseminations of hamster, mouse, and guinea pig oocytes, *Biol. Reprod.*, 23, 1075, 1985.
63. **Uto, N., Yoshimatrsu, N., Lopata, A., and Yanagimachi, R.,** Zona-induced acrosome reaction of hamster spermatozoa, *J. Exp. Zool.*, 248, 113, 1988.
64. **Hartmann, J. F. and Hutchison, C. F.,** Nature of the prepenetration contact interactions between hamster gametes in vitro, *J. Reprod. Fertil.*, 36, 49, 1974.
65. **Castellani-Ceresa, L., Berruti, G., and Colombo, R.,** Immunocytochemical localization of acrosin in boar spermatozoa, *J. Exp. Zool.*, 227, 297, 1983.
66. **Epel, D. and Vacquier, V. D.,** Membrane fusion events during inverterbrate fertilization, in *Cell Surface Reviews,* Vol. 5, Poste G. and Nicolson, G. L., Eds., Elsevier/North-Holland, Amsterdam, 1978, 1.
67. **Jaffe, L. A. and Gould, M.,** Polyspermy-preventing mechanisms, in *Biology of Fertilization,* Vol. 3, Metz, C. B. and Monroy, A., Eds., Academic Press, New York, 1985, 223.

68. **Moller, C. C. and Wassarman, P. M.,** Characterization of a proteinase that cleaves zona pellucida glycoprotein ZP2 following activation of mouse eggs, *Dev. Biol.*, 132, 103, 1989.
69. **Gulyas, B. J.,** Cortical granules of mammalian eggs, *Int. Rev. Cytol.*, 63, 357, 1980.
70. **Wolf, D. P. and Hamada, M.,** Induction of zona and oolemmal blocks to sperm penetration with cortical granule exudate, *Biol. Reprod.*, 17, 350, 1977.
71. **Sato, K.,** Polyspermy preventing mechanisms in mouse eggs fertilized in vitro, *J. Exp. Zool.*, 210, 350, 1979.
72. **Moller, C. C., Bleil, J. D., and Wassarman, P. M.,** Structural and functional relationships between mouse and hamster zona pellucida glycoproteins, *Dev. Biol.*, 137, 276, 1990.
73. **Bleil, J. D. and Wassarman, P. M.,** Autoradiographic visualization of the mouse egg's sperm receptor bound to sperm, *J. Cell Biol.*, 102, 1363, 1986.
74. **Vazquez, M. H., Phillips, D. M., and Wassarman, P. M.,** Interaction of mouse sperm with purified sperm receptors covalently linked to silica beads, *J. Cell Sci.*, 92, 713, 1989.
75. **Bleil, J. D. and Wassarman, P. M.,** Galacotse at the nonreducing terminus of O-linked oligosaccharides of mouse egg zona pellucida glycoprotein ZP3 is essential for the glycoprotein's sperm receptor activity, *Proc. Natl. Acad. Sci. U.S.A.*, 85, 6778, 1988.
76. **Bleil, J. D. and Wassarman, P. M.,** unpublished observation.
77. **Lopo, A. C., Glabe, C. G., Lennarz, W. J., and Vacquier, V. D.,** Sperm-egg binding events during sea urchin fertilization, *Ann. N. Y. Acad. Sci.*, 383, 405, 1982.
78. **Segall, G. K. and Lennarz, W. J.,** Chemical characterization of the component of the jelly coat from sea urchin egg responsible for induction of the acrosome reaction, *Dev. Biol.*, 71, 33, 1979.
79. **Lambert, C. C.,** Fertilization-induced modification of chorion N-acetyl glucosamine groups blocks polyspermy in Ascidians, in *Biology of Fertilization*, Vol. 2, Metz, C. B. and Monroy, A., Eds., Academic Press, New York, 1985, 361.
80. **Rosati, F.,** Sperm-egg interactions in Ascidians, in *Biology of Fertilization*, Vol. 2, Metz, C. B. and Monroy, A., Eds., Academic Press, New York, 1985, 361.
81. **Leyton, L., Robinson, A., and Saling, P.,** Relationship between the M42 antigen of mouse sperm and the acrosome reaction induced by ZP3, *Dev. Biol.*, 132, 174, 1989.
82. **Leyton, L. and Saling, P.,** Evidence that aggregation of mouse sperm receptors by ZP3 triggers the acrosome reaction, *J. Cell Biol.*, 108, 2163, 1989.
83. **Wassarman, P. M., Bleil, J. D., Florman, H. M., Greve, J. M., Roller, R. J., Salzmann, G. S., and Samuels, F. G.,** The mouse egg's receptor for sperm: what is it and how does it work?, *Cold Spring Harbor Symp. Quant. Biol.*, 50, 11, 1985.
84. **Endo, Y., Mattei, P., Kopf, G. S., and Schultz, R. M.,** Effects of a phorbol ester on mouse eggs: dissociation of sperm receptor activity from acrosome reaction-inducing activity of the mouse zona pellucida protein, ZP3, *Dev. Biol.*, 123, 574, 1987.
85. **Endo, Y., Schultz, R. M., and Kopf, G. S.,** Effects of phorbol esters and a diacylglycerol on mouse eggs: inhibition of fertilization and modification of the zona pellucida, *Dev. Biol.*, 119, 199, 1987.
86. **Kurasawa, S., Schultz, R. M., and Kopf, G. S.,** Egg-induced modifications of the zona pellucida of mouse eggs: effects of microinjection inositol 1,4,5-trisphosphate, *Dev. Biol.*, 133, 295, 1989.
87. **Saling, P. M.,** Involvement of trypsin-like activity in binding of mouse spermatozoa to zona pellucida, *Proc. Natl. Acad. Sci. U.S.A.*, 78, 6231, 1981.
88. **Benau, G. and Storey, B. S.,** Characterization of the mouse sperm plasma membrane zona-binding site sensitive to trypsin inhibitors, *Biol. Reprod.*, 36, 282, 1987.
89. **Shur, B. D. and Bennett, D.,** A specific defect in galactosyl transferase regulation on sperm bearing mutant alleles of the T/t locus, *Dev. Biol.*, 71, 243, 1979.
90. **Shur, B. D. and Hall, N. G.,** A role for mouse sperm surface galactosyltransferase in sperm binding to the egg zona pellucida, *J. Cell Biol.*, 95, 574, 1982.
91. **Shur, B. D. and Hall, N. G.,** Sperm surface galactosyl transferase activities during in vitro capacitation, *J. Cell Biol.*, 95, 567, 1982.
92. **Lopez, L. C., Bayna, E. M., Litoff, D., Shaper, N. L., Shaper, J. H., and Shur, B. D.,** Receptor function of mouse sperm galactosyl transferase during fertilization, *J. Cell Biol.*, 101, 1501, 1985.
93. **Shur, B. D. and Neely, C. A.,** Plasma membrane association, purification, and partial characterization of mouse sperm beta-1,4-galactosyltransferase, *J. Biol. Chem.*, 263, 17706, 1988.
94. **Braun, R. E., Behringer, R. R., Peschon, J. J., Brinster, R. L., and Palmiter, R. D.,** Genetically haploid spermatids are phenotypically diploid, *Nature*, 337, 373, 1989.
95. **Brown, J., Cebra-Thomas, J. A., Bleil, J. D., Wassarman, P. M., and Silver, L. M.,** A premature acrosome reaction is programmed by mouse t haplotypes during sperm differentiation and could play a role in transmission ratio distortion, *Development*, 106, 769, 1989.
96. **Leyton, L. and Saling, P.,** 95 Kd sperm proteins bind ZP3 and serve as tyrosine kinase substrates in response to zona binding, *Cell*, 57, 1123, 1989.
97. **Casey, P. J. and Gilman, A. G.,** G protein involvement in receptor-effector coupling, *J. Biol. Chem.*, 263, 2577, 1988.

98. **Endo, Y., Lee, M. A., and Kopf, G. S.,** Evidence for the role of a guanine nucleotide-binding regulatory protein in the zona pellucida-induced mouse sperm acrosome reaction, *Dev. Biol.,* 119, 210, 1987.
99. **Endo, Y., Lee, M. A., and Kopf, G. S.,** Characterization of an islet-activating protein-sensitive site in mouse sperm that is involved in the zona pellucida-induced acrosome reaction, *Dev. Biol.,* 129, 12, 1988.
100. **Florman, H. M., Tombes, R. M., First, N. L., and Babcock, D. F.,** An adhesion-associated agonist from the zona pellucida activates G protein-promoted elevations in Ca^{2+} and pH that mediate mammalian sperm acrosomal exocytosis, *Dev. Biol.,* 135, 133, 1989.
101. **Kopf, G. S., Woolkalis, M. J., and Gerton, G. L.,** Evidence for a guanine nucleotide-binding regulatory protein in invertebrate and mammalian sperm, *J. Biol. Chem.,* 261, 7327, 1986.
102. **Bleil, J. D. and Wassarman, P. M.,** Identification of a mouse sperm protein that recognizes ZP3, *J. Cell Biol.,* 109, 125a, 1989.
103. **Abdullah, M. and Kierszenbaum, A. L.,** Identification of rat testis galactosyl receptor using antibodies to liver asialoglycoprotein receptor: purification and localization on surfaces of spermatogenic cells and sperms, *J. Cell Biol.,* 108, 367, 1989.
104. **Topfer-Peterson, E. and Henschen, A.,** Zona pellucida binding and fucose binding on boar sperm acrosin is not correlated with proteolytic activity, *Biol. Chem. Hoppe Seyler,* 369, 69, 1988.
105. **Tesarik, J., Drahorad, J., and Peknikova, J.,** Subcellular immunochemical localization of acrosin in human spermatozoa during the acrosome reaction and zona pellucida penetration, *Fertil. Steril.,* 50, 133, 1988.
106. **Brown, C. R. and Jones, R.,** Binding of zona pellucida proteins to a boar sperm polypeptide of Mr 53,000 and identification of zona moieties involved, *Development,* 99, 333, 1987.
107. **Jones, R. and Brown, C. R.,** Identification of a zona-binding protein from boar spermatozoa as proacrosin, *Exp. Cell Res.,* 171, 503, 1987.
108. **Topfer-Peterson, E. and Henschen, A.,** Acrosin shows zona and fucose binding, novel properties for a serine proteinase, *FEBS Lett.,* 226, 38, 1987.

Chapter 7

THE MAMMALIAN SPERM ACROSOME AND THE ACROSOME REACTION

Gregory S. Kopf and George L. Gerton

TABLE OF CONTENTS

I. INTRODUCTION

The mammalian sperm acrosome is a membrane-limited organelle which overlies the nucleus of the mature spermatozoon. This organelle is a product of the Golgi complex and is synthesized and assembled during spermiogenesis.[1-4] Although acrosome morphology varies tremendously from species to species, its basic structure and function are similar in all eutherian mammals.[2,5] In response to either physiological or pharmacological stimuli, the membranes overlying this cap-like structure undergo fusion and vesiculation leading to the subsequent exposure of the acrosomal contents to the extracellular environment. Fusion and vesiculation of the outer acrosomal membrane with the plasma membrane constitutes the *acrosome reaction*, which can be considered an exocytotic process. It is clear in all species studied to date that the successful completion of the acrosome reaction is an absolute prerequisite to fertilization. Individuals whose sperm display mutations of the acrosome are infertile.[6,7] Functionally, the acrosome and the acrosome reaction directly impact on the fertilizing potential of sperm. Therefore, an understanding of acrosomal biosynthesis, as well as the acrosome reaction, will contribute greatly to our general understanding of the biology and chemistry of mammalian fertilization.

The purpose of this review is to focus on some of the current issues pertaining to the regulation of both acrosomal biosynthesis and acrosomal exocytosis with the aim of understanding the fertilization process. This will not be an exhaustive treatise, as there are already a number of excellent reviews of these subjects.[5,8-14] Rather, this review will address key observations and current concepts, and is intended to supplement current knowledge of these areas. We also wish to point to future directions of research in these areas, which would utilize the new technologies of molecular biology and genetics that are now available, and which should enable us to make important progress in these particular areas of reproductive biology. Although we will attempt to provide a complete examination of this subject using the information gathered from different animal models, frequent reference to experiments in the mouse will be cited because of the depth of knowledge in this particular species.

II. ACROSOME BIOGENESIS AND MATURATION

The biogenesis of the mammalian acrosome begins during spermatogenesis and has been the subject of several papers, mostly dealing with the ultrastructural aspects of the formation of this important organelle. In recent years, with the advent of modern techniques of cell and molecular biology, several questions concerning the formation and maturation of the acrosome have begun to emerge. These include:

1. When during spermatogenesis does acrosome formation actually begin?
2. Is the synthesis of all acrosomal components regulated coordinately, or are some components synthesized early in the development of the acrosome and others later?
3. How are acrosomal components targeted to the acrosome?
4. Once packaged into the acrosome, are these components processed further either in the testis or later in the epididymis?

A. NATURE OF THE ACROSOME AS AN ORGANELLE

Before addressing these questions, the nature of the acrosome needs to be considered. In many reports, the acrosome has been referred to as a modified lysosome. This view was popularized by Allison and Hartree[15] because, like a lysosome, the acrosomal interior has an acidic pH and contains a variety of acid hydrolases such as hyaluronidase, β-*N*-acetylglucosaminidase, and acid phosphatase, in addition to the neutral pH protease zymogen proacrosin.[10] However, the acrosome is not involved in degradative processes within the sperm cell and, as discussed later in this review, is an exocytotic vesicle that releases its contents in response to a specific stimulus. Thus, since other secretory vesicles contain acid hydrolases and maintain an internal acidic pH, it is perhaps more appropriate to refer to the acrosome as a secretory (zymogen) granule[16] and consider the synthesis and processing of acrosomal components in this context.

B. ACROSOME DEVELOPMENT DURING SPERMIOGENESIS

The formation of the mammalian acrosome is a prominent feature of spermiogenesis, the last phase of spermatogenesis. Indeed, its characteristic staining by histochemical stains such as periodic acid-Schiff's has been used as an aid in the staging of spermatids. Based principally upon acrosomal staining, nuclear morphology, and the arrangement of other spermatid organelles, spermiogenesis has been classified into four phases: Golgi, cap, acrosome, and maturation phases.[1] During the Golgi and cap phases of spermiogenesis, the acrosome develops in close association with the prominent Golgi apparatus of round spermatids. In very early spermatids, the acrosome is but a collection of granules located in a position near the nucleus, Golgi apparatus, and chromatoid body. As development proceeds, these granules coalesce into a single acrosomal vesicle that attaches to the nucleus and gradually enlarges. With time the acrosome begins to spread across the surface of the spherical nucleus, forming a cap over one pole of the nucleus. During the acrosome phase of spermiogenesis, the spermatid, with its centrally localized nucleus, transforms into a polarized cell by the gradual elongation of the nucleus, morphogenesis of the acrosome, and maturation of the flagellum. During this phase, the acrosome severs its intimate association with the Golgi apparatus with the consequence that the acrosome reorients away from the luminal compartment of the seminiferous tubule. During the latter stages (maturation phase) of spermiogenesis, the nucleus starts to elongate and the chromatin begins to condense. Coordinate with these changes, the acrosome is also remodeled and gradually assumes a shape similar to that found in mature sperm, although additional morphological changes may continue to take place during epididymal maturation (see below). The remodeling of the head during late spermiogenesis and epididymal maturation creates a shape that is characteristic of each species.

C. ROLE OF THE GOLGI APPARATUS IN ACROSOME BIOGENESIS

During the Golgi and cap phases of spermiogenesis, the Golgi is oriented such that the *trans* cisternae are nearest to the developing acrosome. Clathrin-coated vesicles are involved in the transport of acrosomal components from the Golgi to the developing acrosome.[17-19] A recent cytochemical study demonstrated that the medial stacks of the Golgi are positive for nicotinamide adenine dinucleotide phosphatase (NADPase) but negative for cytidine monophosphatase (CMPase) activity.[20] On the other hand, the *trans* stacks are negative for NADPase and postive for CMPase. The acrosome appears, therefore, to derive from vesicles of both the *trans* and medial cisternae because it is positive for both CMPase and NADPase. Interestingly, the granular component of the acrosomal contents and not the outer fringes (head cap) are positive for NADPase, while the total acrosomal contents are positive for CMPase.[20] This result suggests that acrosomal components may be compartmentalized early in development.

The role of the Golgi has also been studied by following the fate of radioactive fucose by autoradiography using electron microscopy.[21] These experiments demonstrated the active Golgi processing of acrosomal components during the Golgi and cap phases of spermiogenesis (steps 1 to 7 in the rat). However, during step 8 the Golgi apparatus dissociated from the acrosome and no longer processed components destined for the acrosome, although it still remained active in processing glycoconjugates destined for other organelles of the cell.[21]

The importance of the Golgi apparatus in the development of spermatids has recently been emphasized by the recent experiments of Willison et al.[22] These investigators have produced monoclonal antibodies to TCP1, a protein coded by the *t*-complex region of mouse chromosome 17. Among other properties, the *t*-complex region alters mouse fertility and the transmission ratio of genetic traits through its effects on sperm. TCP1 was originally identified by two-dimensional gel analysis of testicular polypeptides from inbred mouse strains and those from *t*-haplotype mice.[23] TCP1 present in *t*-haplotype mice is found as a more acidic form when compared to that present in wild-type mice. Curiously, characterization with monoclonal antibodies has demonstrated that TCP1 is a ubiquitous peripheral membrane protein found associated with the cytoplasmic aspect of the *trans*-Golgi network of a wide variety of vertebrate species and tissues.[22] Further characterization of this protein will undoubtedly yield important new information concerning the processing of acrosomal components in spermatids and, more generally, the transport of secretory proteins in all cells.

D. OTHER PROPOSED PATHWAYS FOR THE INCORPORATION OF ACROSOMAL COMPONENTS

Others have reported possible alternative methods for incorporating components into the acrosome. Mollenhauer and Morré provided morphological evidence for the association of polyribosomes with the membrane of the head cap region of the developing acrosome of guinea pig spermatids.[24] Since that time, no corroboration of this report has appeared and we are unaware of similar observations for immature secretory granules of other cells. Earlier freeze-fracture investigations suggested that the endoplasmic reticulum may contribute vesicles to the acrosome during the latter steps of spermiogenesis.[2] On the other hand, Pelletier and Friend have suggested that some of the vesicles that are seen in association with the acrosome at this stage may actually be removing membrane from the acrosome.[25] This is supported by the morphometric studies of Courtens who demonstrated a general decrease in the amount of acrosomal membrane in early elongating spermatids.[26] More recent studies using transgenic mice expressing human growth hormone (hGH) during late spermiogenesis support the conclusion that secretory products are no longer transported to the acrosome in these late stages. In contrast to transgenic mice where an hGH construct is translated and transported to the developing acrosome in round spermatids, when a second hGH transgene is constructed such that the protein is translated during late spermiogenesis, hGH remains in the cytoplasm, confined to a membrane-bound compartment which has yet to be identified.[27]

E. TARGETING OF ACROSOMAL COMPONENTS

The mechanisms involved in targeting acrosomal components to the developing acrosome have not been elucidated. As mentioned earlier, the acrosome has been compared to the lysosome and might utilize pathways such as the mannose-6-phosphate ligand-receptor mechanism for transport of glycoproteins to the acrosome.[28] No evidence exists to support this hypothesis for acrosomal targeting, but it is known that spermatogenic cells possess functional mannose-6-phosphate/insulin-like growth factor II receptors.[29] Alternatively, components could be targeted to the developing acrosome by the ''default'' pathway utilized by proteins of other secretory systems.[30] Support for this pathway can be found in the recent report of Braun et al.[31] These investigators generated a transgenic mouse containing the gene for hGH interspersed between the mouse protamine 1 promoter and 3′-untranslated region of the hGH gene. Expression of the transgene was specific for spermatogenic cells and, although the transcription occurred during early spermiogenesis, the translation of the mRNA also occurred at this time (in constrast to the expression of endogenous protamine). The hGH transgene product was synthesized and incorporated into the acrosome, indicating that a ''regulated'' secretory protein could become incorporated into the acrosome.[30] It should also be noted that the founder mouse for these studies was hemizygous for the transgene, meaning that only half of the spermatids would carry the transgene. However, the hGH product was distributed to essentially all of the spermatids, indicating that these haploid cells can be considered to be phenotypically diploid. These results will undoubtedly influence our concepts regarding the genetic control of the formation of the acrosome and other sperm organelles during spermiogenesis.

F. INITIATION OF ACROSOME BIOGENESIS

Since the acrosome is first clearly distinguished in the earliest round spermatids, it has generally been stated that acrosome biogenesis begins immediately following the completion of meiosis. However, evidence is beginning to accumulate that precursors of the acrosome are present during meiosis. This was suggested by earlier ultrastructural studies showing minute ''proacrosomal'' vesicles present in the Golgi region of secondary spermatocytes.[32,33] Investigations using immunofluorescence and immunoblotting of purified cell populations with specific antibodies have indicated that specific acrosomal components are present in spermatocytes.[34-37] Similar to the ultrastructural observations mentioned above, immunofluorescence in spermatocytes for each of these cases was restricted to a single cluster of very small granules in a region near where the Golgi would be expected to be found.[32,33] Localization of acrosomal antigens should be performed at the ultrastructural level to determine if the granules detected by immunofluorescence colocalize with the electron dense ''proacrosomal'' granules observed by electron microscopy. If these granules prove to be precursors of the acrosome, then some mechanism may exist to partition the proacrosomal vesicles equally among daughter spermatids. Such a mechanism may be similar to that used by somatic cells to apportion the Golgi apparatus evenly during mitosis.[38] As mentioned above, clathrin-coated vesicles are used to transport acrosomal components from the Golgi to the developing acrosome. Using an antibody to clathrin, immunofluorescent staining was detected in the Golgi region of spermatocytes and the Golgi and acrosomal regions of spermatids.[19] Recent studies of mouse spermatogenic cells have also examined the localization of the α-subunit of G_i, a guanine nucleotide-binding regulatory protein that is concentrated in the acrosome of mature sperm.[39,40] By indirect immunofluorescence, $G_{i\alpha}$ is associated with the developing acrosome of spermatids and is also found in a cluster of immunoreactive granules near to, but not coincident with, the Golgi apparatus of primary spermatocytes.[41] Whether $G_{i\alpha}$, or an immunologically related protein, displays this localization as the result of a general assembly of acrosomal components or is actively involved in some sort of regulatory role, such as modulating vesicular traffic or fusion in this region, needs to be determined.[42,43] However, these results clearly indicate that, at least in certain species, the synthesis and assembly of some of the acrosomal components begin during meiosis.

G. CONTROL OF ACROSOMAL PROTEIN BIOSYNTHESIS

The regulation of acrosomal component synthesis has received little attention. This is partly due to the lack of appropriate tools and approaches to study the synthesis of specific molecules. With advances in immunology, molecular biology, and the isolation and culture of mammalian germ cells, it should be possible to determine whether the synthesis of acrosomal proteins is regulated as a cohort or whether specific components are synthesized at different stages of acrosome development. Initial results suggest the latter possibility. For example, the important sperm protease zymogen, proacrosin, appears to be principally synthesized during the first half of spermiogenesis, although an occasional immunoreactive spermatocyte can be detected by immunofluorescence.[44-47] On the other hand, acrogranin, a newly described acrosomal component of guinea pig sperm acrosomes, is synthesized by both pachytene spermatocytes and early spermatids.[37] An interesting question is whether both genes are transcribed during meiosis but translated at different times or whether the proacrosin mRNA is principally the product of haploid gene transcription. Both types of translational regulation are known to occur during mammalian spermatogenesis.[48]

H. MATURATION OF THE ACROSOME AND ITS CONSTITUENTS

Once packaged into the acrosome, some of the components may become subsequently modified either during testicular or epididymal maturation. The actual physiological significance of the modification is unclear at this time. One example of an acrosomal component that is altered following its incorporation into the acrosome is the antigen recognized by monoclonal antibody 1D4 in the mouse.[34] This high molecular weight glycoconjugate is present in the acrosome throughout steps 1 to 14 of mouse spermiogenesis, but becomes undetectable in steps 15 and 16 of this process of differentiation. In the guinea pig, it is present throughout spermiogenesis and persists in epididymal sperm.[15,35] Not only is the pattern of expression different between the mouse and guinea pig, the molecular weights of the antigens recognized in the two species are also distinct. However, although 1D4 antigens persist throughout spermiogenesis and epididymal maturation in the guinea pig, there are changes in the molecular weights of the antigens, indicating that these components are modified subsequent to their assembly into the acrosome.

Proacrosin is another example of an acrosomal component that is modified following its insertion into the acrosome. In most species, proacrosin is first detectable by immunocytochemistry with polyclonal antibodies in very early round spermatids and is present throughout spermiogenesis.[44-46] Occasionally, it can be detected in spermatocytes, probably secondary spermatocytes, based upon the low frequency of observation in suspensions of cells from seminiferous tubules.[46] These results are contradicted by another report demonstrating that a monoclonal antibody against human acrosin, C11H, reacts only with mouse spermatids at step 9 of spermiogenesis and greater.[49] This is coincident with the onset of nuclear elongation and the cessation of RNA transcription, and corresponds approximately to the time when the Golgi apparatus appears to no longer contribute material to the acrosome.[21] Most likely, the discrepancy between this study and those using polyclonal antibodies to proacrosin or acrosin is due to the extreme specificity of the monoclonal antibody. Although the epitope recognized by C11H has not been characterized, it is possible that an oligosaccharide side chain or some other determinant of proacrosin may be modified around step 9 of spermiogenesis, exposing the antibody-binding site. In the guinea pig, proacrosin is also modified following its insertion into the acrosome. It has a $M_r = 55{,}000$ in round spermatids and a $M_r = 50{,}000$ in cauda epididymal sperm.[46] This change occurs as the sperm mature in the epididymis and is due to processing of the carbohydrate side chains of proacrosin.[50]

Modifications of other acrosome-associated proteins have been reported. Usually, these proteins are associated with the plasma membrane overlying the outer acrosomal membrane. PH-20 is a component of guinea pig sperm that has been shown to be involved in sperm-egg recognition.[51] It is found on the cell surface and within the acrosome and is attached to the membrane via a phosphatidylinositol linkage.[52] During epididymal maturation, the distribution of cell surface PH-20 changes from a uniform localization over the sperm head to a restricted localization in the posterior head region; this altered distribution is associated with a change in the molecular weight of PH-20.[53] Preliminary results have shown that the change in PH-20 distribution and molecular weight can be mimicked *in vitro* by trypsin, suggesting that the modification of PH-20 may be due to the action of an epididymal trypsin-like protease.[54] In the mouse, a pair of proteins (M_r = 185,000 and 200,000) are recognized on cauda epididymal sperm by monoclonal antibody M42 and have been implicated in sperm-egg interactions.[55] These cell surface proteins and others that also overlie the acrosomal region of mouse sperm are processed to lower molecular weights as a consequence of epididymal maturation.[56] These changes occur at about the same point in the epididymis where mouse sperm gain the ability to fertilize eggs.[57]

The role of the epididymis in the posttesticular modification of the acrosome is potentially very imortant, since it is generally accepted that mammalian spermatozoa require a period of maturation within the epididymis to become fertile.[58] The case of proacrosin could be physiologically significant. In some species, the acrosins resulting from the activation of the testicular forms of proacrosin are less active than those from epididymal sources and the molecular weights derived from the testicular form differ slightly from the epididymal forms.[46,59,60] Cell surface components with proposed roles in sperm-egg interaction are modified during epididymal transit and these changes may have important functional consequences.[53,57]

Besides reorganizing the distribution of specific components of the sperm surface, alterations in the shape of acrosomes and the distribution of particles in the membranes over the acrosomes occur in many species during epididymal maturation.[61-64] Guinea pig sperm probably exhibit the most dramatic acrosomal shape changes. Fawcett and Hollenberg demonstrated that a sperm taken from the proximal (caput) region of the guinea pig epididymis has a fairly flat head, similar in appearance to that of a ping-pong paddle; after migration through the epididymis, the tip of the acrosome becomes pointed, the acrosome curves, and a dorsal bulge appears in the acrosome so that a longitudinal section through the sperm head looks like a pruning hook.[61] Two zones of differing electron density become apparent in the acrosome during epididymal maturation. This regionalization of electron densities may reflect compartmentalization of the acrosome since it has been demonstrated that the acrosomal enzyme dipeptidyl peptidase is restricted to the dorsal bulge of guinea pig sperm.[65] The epididymis-associated alteration in morphology is under hormonal control since castration of guinea pigs decreases the percentage of cauda sperm that undergo morphological maturation of the acrosome.[66] The conversion to a normal caudal sperm structure can be partially rescued in castrated animals by replacement therapy with testosterone propionate.

The evidence presented from several species suggests that the modifications of the acrosome in the epididymis may be important for the complete maturation of sperm. The changes observed in the morphology of the acrosome, alterations of the components of the plasma membrane overlying the acrosome, redistribution of membrane particles, and modifications in internal acrosomal component such as proacrosin may be related. In cases where it has been examined, the observed changes in the acrosome correlate with the site where sperm become fertile.[57,61] The observed changes also do not occur as a function of residence time within the epididymis but are dependent upon sperm passing through particular regions of the epididymis.[56,58] These changes presumably result from the action of some external factor working upon the sperm surface, perhaps through a paracrine mechanism, to induce internal changes that lead to the restructuring of acrosome shape, modifications in internal acrosomal components, and, perhaps altering the sperm such that they become fertile. Most likely, an epididymal fluid component(s) acts on the sperm to cause a reorganization of cytoskeletal elements between the plasma and outer acrosomal membranes, leading to the observed shape changes. A cytoskeletal structure has been observed in association with the hamster sperm acrosome and proteins such as spectrin have been detected in the region overlying the human acrosome.[67,68] A signal transduction system, possibly involving G-proteins, could mediate the observed effects. In response to an intracellular signal(s), certain enzymes within the acrosomal lumen may become activated and subsequently modify other acrosomal components. A possible model system for this type of change in a secretory granule would be pancreatic islets. As the vesicles mature, the vesicles gradually acidify, leading to the activation of proteases with the vesicle that then convert proinsulin to insulin.[69,70] In a similar manner, changes in the milieu (e.g., pH, ions) of the sperm cytoplasm and/or acrosomal lumen could be induced by external substances acting at the sperm surface. Such changes could cause the restructuring of cytoskeletal elements and the activation of certain acrosomal enzymes, leading to morphological changes in the acrosomes and alterations in the properties of acrosomal components.

III. BIOLOGICAL SIGNIFICANCE OF THE ACROSOME REACTION

In the mammal, sperm-female reproductive tract interaction and sperm-egg interaction prior to sperm-egg fusion occur at a number of different levels *in vivo* and are the culmination of a number of integrated processes designed to deliver the most viable, functionally active sperm to the site of fertiliziation, e.g., the ampullary region of the oviduct. As sperm traverse the female reproductive tract, there are a number of species-specific selection processes that limit the delivery of these cells to the site of fertilization. These selection processes can be considered anatomical, as well as environmental, and may include cervical- and/or uterotubal junction-sperm interactions, cervical mucus-sperm interactions, reproductive tract pH, circulating antisperm antibodies, and the ability of the sperm to undergo capacitation. Sperm arriving at the ampullary region of the oviduct encounter additional selection processes in the form of cellular investments (cells comprising the cumulus oophorus and corona radiata) and extracellular matrices (glycosaminoglycans comprising the cumulus cell matrix; the zona pellucida) associated with the ovulated egg. Although the presence of the cumulus oophorus throughout the fertilizable life span of an ovulated egg *in vivo* is species-dependent, the presence of the zona pellucida during this time period is universal. The aforementioned selection processes may serve, therefore, to ensure that sperm-egg interaction occurs between gametes that are most competent to fuse in a successful manner. The reader is encouraged to examine the excellent reviews by Talbot[71] and Katz et al.[72] which address these processes in greater depth.

Selection processes present within the female reproductive tract to ensure the delivery of an optimal population of sperm to the site of fertilization are not unique and are an important general biological phenomenon. Cellular selection, such as is found in the immune system, is important for singling out the most optimal cell for a particular function. Sperm represent a heterogenous cell population with regard to cellular age, morphology, motility characteristics, and ability to undergo capacitation. In addition, there is certain to be heterogeneity at the biochemical level, possibly reflected in differences in membrane characteristics and ability to respond to external stimuli.

The acrosome reaction, itself, may be considered a selection process designed to ensure the delivery of optimal sperm to the initial site of sperm-egg interaction proper, i.e., the zona pellucida, which is a secretory product of the growing oocyte.[73,74] In many species, sperm that undergo the acrosome reaction prematurely generally display reduced fertilizability and may fail to penetrate the cumulus oophorus mass due to the fact that they adhere to the outer portions of the extracellular matrix surrounding these cells.[75-81] In addition, sperm that undergo the acrosome reaction while within the cumulus oophorus matrix have a reduced ability to penetrate this matrix further since the sperm adhere to the cumulus cells.[71] Finally, sperm that undergo the acrosome reaction after penetrating the cumulus oophorus matrix may be precluded from binding to the zona pellucida since, in many species studied (e.g., mouse,[82] boar,[83-85] hamster,[78,86,87] sheep,[88-90] bull,[91-93] human,[94] rabbit,[95] guinea pig,[96] and rat[97]), acrosome-intact sperm can participate in binding to the zona pellucida. Although this appears to be a general phenomenon, there do appear to be certain species (e.g., guinea pig, rabbit, human) in which both acrosome-intact and acrosome reacted sperm bind to the zona pellucida.[96,98-100] The primary function of the acrosome reaction at this stage is to permit the penetration of the zona pellucida by sperm so that these cells can gain access to the perivitelline space. A consequence of zona pellucida penetration is the development of a sharply defined, thin penetration slit in this extracellular matrix which is the presumed result of limited digestion of the matrix by acrosomal-associated proteases. The relative contribution of mechanical forces generated by sperm motility and enzymatic digestion of the zona pellucida by these proteases in mediating sperm passage through the zona pellucida, however, is not clear.[8,72] The acrosome-reacted sperm, once it has penetrated the zona pellucida and has arrived in the perivitelline space, has the ability to interact and fuse with the plasma membrane of the egg. Since acrosome-intact sperm do not possess this ability,[5,101] acrosomal exocytosis with the resultant exposure of the inner acrosomal membrane and equatorial segment plays a key role in sperm-egg fusion. The nature and mechanisms of those component(s) associated with the inner acrosomal membranes and/or the equatorial segment involved in the fusion process are not clear at this time.

The biological significance of the acrosome reaction, therefore, is two-fold. First, in most species studied, it serves as a selection process to ensure the delivery of viable, acrosome-intact sperm to the initial site of sperm-egg interaction proper (i.e., the zona pellucida). Second, the zona pellucida-induced acrosome reaction ensures that acrosome-reacted sperm capable of fusion with the egg plasma membrane are delivered to the perivitelline space.

IV. ASSAYS TO MONITOR THE ACROSOME REACTION

Analysis of the molecular mechanisms of the mammalian sperm acrosome reaction, as well as the assessment of this exocytotic event under *in vitro* and *in vivo* conditions, requires an unequivocal method(s) to assess acrosomal status of sperm from each species under study. The accurate assessment of this exocytotic event is complicated by the fact that the size of the acrosome is species-specific, making it difficult to apply one particular assay technique to multiple species. In addition, moribund or dead sperm may undergo degenerative changes of the acrosome that resemble the acrosome reaction morphologically, but are obviously nonspecific when compared to an induced acrosome reaction in viable, motile sperm. An ideal assay for this exocytotic event would delineate between acrosomal modifications induced nonspecifically in nonviable cells and induced acrosome reactions in viable cells.

A number of approaches have evolved to monitor sperm acrosomal status. All of these methods utilize either light or electron microscopic evaluation and rely on either direct visualization or the use of specific probes such as histological stains, antibodies, lectins, or membrane-permeable or intercalating dyes. Generally speaking, there is no established universal assay to assess for acrosomal status since each of these techniques has distinct advantages and disadvantages and may only be applicable for a limited number of species.

To date, the most unequivocal method to assess acrosomal status is electron microscopy. Although it may be argued that the fragility of the acrosomal structure may make such an approach prone to fixation artifact, it is clear that this technique is the only way to assess minor modifications of acrosomal or membrane structure that may constitute the various stages of this exocytotic event.[102] The amount of time and expense required to prepare and score the samples generally limits the use of this technique on a daily basis. However, it must be emphasized that the other methods developed to assess acrosomal status must be calibrated against electron microscopic observations to ensure the validity of the assay in question. One disadvantage that all of the other techniques have in common is that they do not possess the degree of resolution that electron microscopy offers for observing potential stages that comprise this exocytotic event.

Direct observation of acrosomal status of sperm by light microscopy is possible in some species possessing large acrosomes (e.g., guinea pig, hamster), but is difficult in species with small acrosomes (e.g., human). Recently, direct light microscopic observation using Nomarski differential interference contrast optics has been utilized to assess acrosomal status in mouse spermatozoa.[103] Rabbit sperm acrosomal status has also been evaluated by light microscopy.[104]

Histological stains are used in some species (hamster, bull, rabbit, boar, ram, stallion) and have the advantage that they can distinguish between acrosome-reacted live and acrosome-reacted dead sperm.[105-107] However, the quality of some of the histological stains can be variable from different suppliers and the techniques used for fixing and dehydrating the sperm can often give rise to samples that are difficult to score.

Antibodies raised against components associated with the membrane(s) overlying the acrosome have been used in conjunction with fluorescently labeled second antibodies by a number of investigators.[108,109] The advantages of such approaches are that samples can be unequivocally scored with the use of a fluorescent microscope once the conditions for sperm incubation with the primary and secondary antibodies are established. Preferably, the specificity of the antibody and the localization of the immunoreactive component(s) should be established by immunoblotting and immunolocalization at the electron microscopic level.

Investigators have also taken advantage of the fact that sperm possess lectin-binding glycoconjugates, which in some cases are associated with the acrosomal region of the cell.[110] Assays for acrosomal status have been developed utilizing direct binding of fluoresceinated lectins.[111] These assays, coupled with a DNA-specific stain,[112] have been used to assess acrosomal status in both viable and nonviable cells. The advantage of this technique is that it is relatively quick, inexpensive, easy to score, and easy to delineate viable from nonviable cells. As with assays based on antibodies, the acrosomal localization of the lectin binding component(s) should be corroborated by a second independent method.

Acrosomal status has also been determined using specific fluorescent probes. In the mouse, the fluorescent probe chlortetracycline has been used to assess acrosomal status of sperm either bound to structurally intact zonae pellucidae or incubated with either soluble zona pellucida or purified ZP3.[113-116] Three major patterns of chlortetracycline fluorescence have been described and are characterized by specific distributions of the dye on the head region of the sperm. This assay has been calibrated using both electron microscopy[82,117,118] and the HS-19 monoclonal antibody.[118,119] The time-dependent changes in the distribution pattern of this fluorescent probe have been correlated with ionic changes occurring in the cell, so there appears to be a biochemical basis for the differential staining of the cells under various incubation conditions.[115,116,118,120] At this time the basis of the chlortetracycline staining patterns is not known. Since the fluorescence of the chlortetracycline-Ca^{2+} complex is greatly enhanced on binding to membranes,[121] the observed patterns on sperm may reflect regional binding of the complex. An advantage to this assay is that it is quick, inexpensive, and can distinguish multiple stages comprising the acrosome reaction. Recently this fluorescent probe has been used to assess acrosomal status in human sperm.[122]

V. SITE OF THE ACROSOME REACTION *IN VIVO*

A major problem in determining the site of the acrosome reaction *in vivo* is that many studies of this exocytotic event have been conducted under *in vitro* conditions. These *in vitro* studies, while invaluable to our understanding of some of the basic mechanisms of acrosomal exocytosis, do not reflect the actual conditions that the sperm encounter at the different levels of the male and female reproductive tracts. For example, many investigators utilize epididymal sperm in their studies, since it is easy to obtain and it is possible to capacitate, to stimulate acrosome reactions, and to fertilize eggs *in vitro* with these cells. However, such cells are not exposed to the environmental selection processes that normally occur *in vivo* in the reproductive tracts (see Section III above). Such differences may have substantial effects on the numbers and kinetics of sperm undergoing capacitation which may ultimately affect their ability to undergo the acrosome reaction. Furthermore, typical *in vitro* fertilization procedures utilize sperm/egg ratios on the order of 1000:1 to 1,000,000:1 when, in fact, the number of sperm at the site of fertilization *in vivo* may be only on the order of 1:1 to 100:1.[123-126] Caution, therefore, must be exercised when interpreting the results of *in vitro* studies and extrapolating these results to the *in vivo* situation.

As discussed earlier (Section III) acrosome reactions may occur at different regions of the female reproductive tract, and these reactions have no functional significance in penetration of the zona pellucida per se, but may be important as a selection process to ensure that only the most viable, acrosome-intact sperm with optimal fertilizing potential arrive at the zona pellucida. In studies where sperm are recovered from naturally mated or artificially inseminated females, a majority of the sperm collected from the ampullary region of the oviduct that are swimming freely and are not associated with the cumulus-egg complex have not undergone the acrosome reaction.[75,104,126-128] Sperm associated with the cumulus oophorus complex, in contrast, are either unreacted or are undergoing the acrosome reaction.[126,127,129,130] Sperm found to be associated with the zona pellucida have either undergone acrosomal changes (perhaps initial stages of the acrosome reaction)[97] or have completed the acrosome reaction.[126] Similar results have been observed under conditions of *in vitro* fertilization.[77,97,131] Since binding to the zona pellucida appears to be initiated by acrosome intact sperm in many species,[78,82-96] it is likely that the site of the physiologically relevant acrosome reaction is at the level of this egg-associated extracellular matrix. This is supported by the fact that the zona pellucida from different species has been demonstrated to induce the acrosome reaction of homologous sperm.[78,88,92,93,95,117,132-134] In species where both acrosome-intact and acrosome-reacted sperm bind to the zona pellucida (e.g., guinea pig, rabbit, human) it is unclear whether the physiologically relevant acrosome reaction is induced by component(s) associated with the cumulus oophorus and/or the zona pellucida. However, there is no unequivocal evidence to support the idea that specific factors associated with the cumulus oophorus induce the acrosome reaction under physiological conditions. Clearly, additional studies will be required to answer this question.

VI. NATURE OF THE PHYSIOLOGICAL SPERM ACROSOME REACTION-INDUCING COMPONENT(S)

Although there is little debate as to the role of the acrosome reaction in mammalian fertilization, there is considerable disagreement regarding the identity of the component(s) associated with the female reproductive tract that induces the physiologically relevant acrosome reaction *in vivo* (e.g., the acrosome reaction required for successful penetration of the zona pellucida). Conflicting observations using the same species have been attributed to differences in technique or to the fact that *in vitro* studies may not be comparable to *in vivo* studies. Distinctions observed between species have been attributed to unique characteristics of the different species.

Although sperm-egg interaction and fertilization constitute an essential biological process for species preservation, it is first and foremost an elegant example of cell-cell interaction and cellular activation. As a consequence of sperm-egg interaction two different cellular activation events occur. The first event is an egg-induced sperm activation (i.e., the acrosome reaction) which appears to be mediated in many species by an extracellular matrix, the zona pellucida. The second event is a sperm-induced egg activation (i.e., activation of the egg and the establishment of the block to polyspermy) which is initiated by fusion of the sperm and egg plasma membranes. Generally speaking, regardless of whether intercellular communication is mediated at a distance (by the elaboration of circulating humoral factors) or locally (by autocrine or paracrine mechanisms), it appears to occur through the interaction of specific ligand-receptor-signal transduction mechanisms. At first glance, the induction of the mammalian sperm acrosome reaction might seem to represent an exception to this general rule, since it has been demonstrated that many biological factors, both specifically and nonspecifically associated with the female reproductive tract, induce the acrosome reaction under *in vitro* conditions.[9,135] Many of these factors would seemingly lack the specificity that one would expect if the regulation of this exocytotic process were a highly coordinated event. How might one reconcile this apparent lack of specificity, in light of the fact that other aspects of sperm-egg interaction and fertilization appear to be so highly coordinated and regulated? One answer to this question might derive from the fact that conclusions made regarding the nature of the *in vivo* acrosome reaction-inducing component(s) are based on *in vitro* experiments, and thus may have little physiological relevance. Alternatively, it is possible that there is nonspecificity, as well as specificity, when it comes to the nature of the *in vivo* acrosome reaction-inducing component, such that the interaction between the egg and an optimally selected sperm is assured.

As discussed earlier, biological selection is a way to overcome problems with inherent heterogeneity in cell populations. If the acrosome reaction can serve as a selection mechanism then it is not surprising that nonspecific as well as highly specific acrosome reaction-inducing components exist within the female reproductive tract which, acting together, ensure that fertilization is accomplished by the most optimal sperm population. Sperm display considerable heterogeneity with regard to their ability to undergo an acrosome reaction. This is supported by observations made in several species. First, there always appears to be a certain low level percentage of spontaneous acrosome reactions which occur nonspecifically in populations of sperm incubated under capacitating conditions. Second, inter- and intraspecies conditions for capacitation appear to be quite variable. Since sperm must be capacitated to undergo an induced acrosome reaction, capacitation conditions impact indirectly on the heterogeneity of the sperm population with regard to their ability to undergo this exocytotic event. Third, since acrosomal exocytosis may consist of discrete steps (see Section VII), a situation observed in many other exocytotic processes, the completion of the acrosome reaction may be variable since the reactions governing these different steps may have different rate constants. All of these factors probably contribute to the observations that the acrosome reaction (both spontaneous and induced) in populations of sperm can be described as a stochastic process. This is supported by the fact that the progression of this exocytotic event, under a variety of incubation conditions, can be described in kinetic terms as first-order reaction and not by an all-or-none event. Since the molecular events comprising capacitation and the acrosome reaction are not fully understood, one must also consider the possibility that many of the observed effects of biological factors on the acrosome reaction may, in fact, be due to effects on later stages of the capacitation process.

If one considers the biological significance of the acrosome reaction as (1) a means to select a population of sperm optimally suited for zona pellucida binding, and (2) a prerequisite for zona pellucida penetration required for successful fertilization, some predictions may be made with regard to the identity and mode of action of acrosome reaction-inducing component(s) associated with the female reproductive tract. If one accepts this premise, it is reasonable to propose that multiple inducers of the acrosome reaction exist with extremely different properties and modes of action. The existence of multiple inducers may reflect the different functions that these factors have in regulating this exocytotic process in subpopulations of sperm. For example, in some species, sperm that are suboptimal because they display morphological and/or motility defects are probably prevented from arriving at the site of fertilization because of anatomical and environmental considerations (e.g., cervical mucus interactions; inability to progress to the various levels of the reproductive tract). In addition, sperm that appear normal by gross criteria of morphology and motility may not necessarily be optimal with regard to membrane structure, and these cells may have a greater tendency to undergo acrosome reactions either spontaneously or in response to a wide range of components present in the female reproductive tract. As stated previously, the induction of the acrosome reaction of such sperm prior to their interaction with the cumulus oophorus would preclude them from entering the cumulus matrix. Furthermore, acrosome reactions of these sperm, while within the cumulus oophorus, would prevent them (in many species) from binding to the zona pellucida, an absolute prerequisite to egg penetration. Acrosome reactions at these points would represent a very effective means to select out suboptimal sperm populations. On the other hand, those few sperm that arrive at the zona pellucida with their acrosomes intact and initiate binding to this extracellular matrix would represent optimally selected cells destined to participate in the final stages of sperm-egg interaction leading to fertilization. In order for successful fertilization to occur these sperm would have to interact with specific egg-associated ligands that would mediate highly specific sperm-egg interactions (i.e., species-specific sperm binding) and sperm activation events (i.e., induction of the acrosome reaction). Presumably, cellular recognition, interaction, and activation at this level would involve unique, biologically potent molecules associated with both the sperm and egg. Furthermore, the egg-associated ligand(s) would be expected to act at a short range. These aforementioned criteria appear to be fulfilled by the zona pellucida. It is reasonable to assume, therefore, that both nonspecific and highly specific acrosome reaction-inducing factors exist in the female reproductive tract that play physiological roles in modulating sperm acrosomal exocytosis.

Three possible sites of the acrosome reaction *in vivo* have, in fact, been postulated as a result of studies in which sperm have been removed from the female productive tract and assessed for acrosomal status as well as from studies *in vitro*. These sites are the oviduct itself, the cumulus oophorus matrix, and the zona pellucida. There is disagreement regarding the extent of acrosome reactions at two of these sites, the oviduct and the cumulus oophorus. Although freely swimming sperm recovered from the oviduct have been found to have undergone acrosomal modification in some studies,[136] other reports demonstrated that a majority of these cells were acrosome intact.[75,104,126-128,131] Likewise, sperm within the cumulus oophorus matrix have been observed to be either acrosome intact or acrosome reacted.[71,78,126,127,129-131] Sperm, once bound to the zona pellucida with their acrosomes intact,[77,113,134] have been observed to undergo the acrosome reaction.[126,131,134]

Factors associated with these three sites have been found to induce acrosome reactions *in vitro*, although in many cases the nature of the component(s) has not been identified.[8,9,135,137] The identities of some of these factors, however, have been determined and are briefly considered below.

A. SERUM COMPONENTS

Serum albumin, which has been demonstrated to be a component of both oviductal and follicular fluids, has been postulated to stimulate capacitation and initiate the acrosome reaction.[137-139] There is no doubt as to the importance of a protein supplement (in most cases provided by serum albumin) in *in vitro* culture media to support sperm capacitation.[140] The ability of this serum protein to promote capacitation is thought to be due, in part, to its ability to bind a number of biologically active molecules. Prostaglandins, fatty acids, steroids, and biogenic amines have all been suggested to play a role in the induction of the acrosome reaction, and these molecules have been demonstrated to bind to albumin.[141-145] The possibility that albumin may be exerting its effects on the acrosome reaction in an indirect manner by affecting capacitation through its ability to bind cholesterol and fatty acids has also been proposed,[9] since the removal of this sterol from the sperm plasma membrane has been postulated to play an important role in capacitation.[146-151] Albumin has also been proposed to promote the acrosome reaction by its ability to serve as a metal chelator.[9,145,152] The mechanism by which this ubiquitous serum protein affects acrosomal exoctyosis is not clear, and it is difficult to conceive how this protein could elicit its effects on acrosomal exocytosis in a specific, receptor-mediated manner.

B. CIRCULATING BIOACTIVE SUBSTANCES

As alluded to above, a variety of hormones, neurotransmitters, and products of fatty acid metabolism have been suggested as possible acrosome reaction-inducing components of the female reproductive tract. Catecholamines, prostaglandins, steroids, serotonin, and *cis*-unsaturated fatty acids exist within the female reproductive tract and have been shown to stimulate acrosome reactions *in vitro*.[141-143,153]

Recently, it has been reported that the acrosome reaction-inducing activity of human follicular fluid appears to be progesterone and 17α-hydroxyprogesterone.[154,155] There have been conflicting reports, however, as to the potency of the acrosome reaction-inducing effects of unfractionated and fractionated human follicular fluids and these differences may be attributed to a variety of factors, including techniques of sperm and follicular fluid preparation and incubation, assessment of acrosomal status, and the endocrine status of the follicular fluid donors.[154,156-159]

There are a few factors to consider when evaluating the physiological mechanisms by which these compounds are exerting their effects on this exocytotic process. In some cases the concentration of these bioactive substances in the different parts of the female reproductive tract of the animal model being studied was not known, and high concentrations (relative to the normal physiological concentrations required to elicit physiological responses in somatic cells) were required to induce the acrosome reaction. Furthermore, in some of these studies one cannot distinguish between the effects of these compounds on capacitation and effects on the acrosome reaction.[141,144] Finally, there is little evidence to support the idea that these compounds are acting on sperm in a manner analogous to the way they act on somatic cells. For example, there is no evidence for the existence of sperm-associated catecholamine, serotonin, or steroid receptors, and, in some studies, conclusions of receptor-mediated events based on the use of specific agonists and antagonists are difficult to interpret.[144,153,160]

C. GLYCOSAMINOGLYCANS

Large molecular weight glycoconjugates which are present within the follicular fluid[161] as well as within the cumulus oophorus matrix complex (as hyaluronic acid),[162,163] have been demonstrated to have positive effects on capacitation and/or the acrosome reaction.[161,164-168] It has been demonstrated that prolonged incubation of sperm with these molecules induces significant numbers of acrosome reactions over respective controls, and the order of potency of a variety of glycosaminoglycans is heparin > chondroitin sulfates > hyaluronic acid.[161] Further analysis demonstrated that the acrosome reaction-inducing activity of the chondroitin sulfates appears to reside with both the sulfate and amino-sugar complex moieties.[164] More recent data, however, suggest that glycosaminoglycans may be acting on sperm to stimulate capacitation, since in many of the previous studies the sperm were incubated with these compounds throughout the entire capacitation period.[168] Although there appears to be evidence for the binding of heparin to the sperm surface, the specificity of this binding and the nature of the sperm-associated binding moiety(s) are unknown.[169,170] It is also unclear how glycosaminoglycans might be acting to stimulate capacitation and/or the acrosome reaction. The highly anionic nature of these molecules might have significant effects on membrane stability, which may lead to their observed effects on these aforementioned parameters.

D. ZONA PELLUCIDA

The zona pellucida is an oocyte-derived acellular coat that surrounds the egg and can be considered an extracellular matrix. This structure plays an important role both prior to and after fertilization, as well as during subsequent preimplantation development. The zonae pellucidae from all mammals studied to date appear to be composed of two to four glycoproteins, the number of which appears to be species dependent.[171-179] Considerable charge heterogeneity of the zona pellucida glycoproteins exists in different species, and such heterogeneity is probably reflected in the degree of glycosylation of the individual polypeptide chains.

The observations that acrosome-intact sperm bind to the zona pellucida[77,78,82-96,113,134] and then undergo the acrosome reaction[113,117,126,131,134] strongly suggest that a component(s) of the zona pellucida mediates this exocytotic event. Subsequent studies have demonstrated that either structurally intact or solubilized zonae pellucidae from different species can induce acrosome reactions in homologous sperm.[78,92,95,117,132,134]

Presently, only the zona pellucida of the mouse egg has been examined in sufficient detail to permit an analysis of the component(s) that possesses acrosome reaction-inducing activity. The reader is encouraged to examine the chapter by Bleil (Chapter 6) in this volume, as well as an earlier review by Wassarman,[14] for an in-depth analysis of the structure and function of the mouse zona pellucida.

Briefly, the murine zona pellucida is composed of three major sulfated glycoproteins designated as ZP1 (M_r = 200,000), ZP2 (M_r = 120,000), and ZP3 (M_r = 83,000), which are present in molar ratios approximately 1:10:10, respectively.[14,171,180] These glycoproteins are truly egg-associated products, as they are the major biosynthetic products of the growing oocyte and are secreted throughout the period of oocyte growth.[14,73] ZP1 is a dimer connected by intermolecular disulfide bonds and may function to maintain the three-dimensional structure of the zona pellucida by cross-linking filaments that are composed of repeating structures of ZP2/ZP3 heterodimers.[181]

ZP2 (M_r = 120,000 under nonreducing and reducing conditions[171]) may mediate the binding of acrosome-reacted sperm to the zona pellucida.[103,182] Upon fertilization, the egg effects a modification of ZP2 to a form called $ZP2_f$.[183] This modification is brought about by the action of a protease which is most likely secreted from the egg as a consequence of cortical granule exocytosis.[184] $ZP2_f$ has a M_r = 120,000 under nonreducing conditions, which shifts under reducing conditions to M_r = 90,000, suggesting that the proteolysis of the ZP2 molecule results in the generation of fragments that are held together by disulfide bonds. The biological consequence of the ZP2-to-$ZP2_f$ shift is that $ZP2_f$ no longer will bind to acrosome-reacted sperm.[103,182]

ZP3 (M_r = 83,000) accounts for both the sperm binding and the acrosome reaction-inducing activities of the ZP of unfertilized eggs.[132,185] The sperm binding activity of ZP3 appears to be conferred by O-linked carbohydrate moieties and not by its polypeptide chain.[108,186] Alpha-linked terminal galactose residues at the nonreducing termini of these O-linked oligosaccharide chains play a critical role in this binding activity.[187] On the other hand, the acrosome reaction-inducing activity of ZP3 may be conferred by both the carbohydrate and protein portions of the molecule,[108,188] although the exact nature of the interaction between the protein and carbohydrate required for biological activity is not clear at this time. Associated with fertilization is a loss of both the sperm binding and acrosome reaction-inducing activities of the ZP3 molecule.[132,185] The loss of these two important biological activities is associated with a relatively minor biochemical modification of the

ZP3 molecule, since the electrophoretic mobility of ZP3 from fertilized eggs is similar to that of ZP3 from unfertilized eggs.

In addition to the general biological properties described above, ZP3 possesses additional properties which make it ideally suited as a candidate to mediate the initial steps of sperm-egg interaction proper (e.g., sperm binding) and subsequent sperm activation (e.g., induction of the acrosome reaction). Although conserved at the genomic level in a variety of species, ZP3 subserves very specific functions as a component of the egg-associated extracellular matrix.[189]

1. ZP3 is synthesized only by the growing oocyte.[73,74,180]
2. There is little apparent amino acid sequence homology between ZP3 and any other known proteins or glycoproteins thus far examined.[189]
3. The cross-linking of ZP3/ZP2 heterodimers by ZP1 in a nonrandom process[14,181] gives rise to structural domains that ensure that the ZP3 ligand is immobilized and can thus only act at short distances.
4. Both the sperm binding and acrosome reaction-inducing activities of ZP3 are observed in the nanomolar range.[103,186]
5. Mouse sperm appear to possess complementary binding sites (receptors?) for ZP3 that are localized over the acrosomal cap region and are present in numbers (10,000 to 50,000 binding sites/cell)[103] similar to that observed for receptor numbers in many hormonally responsive cells.
6. ZP3 covalently bound to silica beads will specifically bind acrosome intact sperm via the sperm head,[190] further supporting the idea that the sperm-associated ZP3 binding moiety(s) is associated with the plasma membrane overlying the acrosomal region.

The unique structure of ZP3, its biological potency, its association with ZP1 and ZP2 in the zona pellucida to form an immobilized ligand in a highly ordered three-dimensional matrix capable of acting at limited distances, and the probable existence on the plasma membrane of acrosome intact sperm of a specific ZP3 receptor(s) satisfy a number of criteria that would have to be met if one were to construct a macromolecule that would control specific cell-cell recognition events in a receptor-mediated fashion. Acrosomal exocytosis, which would occur subsequent to ZP3-sperm interaction, might then be expected to have characteristics similar to receptor-mediated signal transduction processes observed in somatic cells.

Presently, there are a number of questions to be asked regarding the biochemical mechanisms by which ZP3 binds to sperm and stimulates the acrosome reaction. First, what is the molecular nature of the ZP3 functional group(s) that induces the acrosome reaction? Second, what is the mode of ZP3 binding to the sperm surface? Third, what is the molecular identity of the sperm-associated "receptor" for ZP3? Fourth, what is the nature of the transmembrane signaling event(s) distal to the binding of the ZP3 ligand by its putative sperm-associated receptor? Finally, what is the nature of the intracellular second messenger signaling system(s) which ultimately regulates acrosomal exocytosis? Some of these questions will be addressed in greater detail in Section VII.

In conclusion, the existence and operation of both nonspecific and specific acrosome reaction-inducing substances associated with the female reproductive tract can be postulated based on the premise that the induction of the acrosome reaction represents a mechanism to limit the accessibility of suboptimal sperm to the site of fertilization and the fact that this exocytotic event is an absolute prerequisite to successful sperm-egg fusion. It would be expected that suboptimal sperm could respond to a variety of substances with induction of acrosomal exocytosis. In contrast, physiologically relevant acrosome reactions (i.e., those required for zona pellucida penetration) would occur in an optimally selected sperm population through a ligand-receptor-mediated process. In many species that ligand appears to be a component of the zona pellucida. Section VII will consider the possible biochemical mechanisms which have been proposed to mediate acrosomal exocytosis.

VII. BIOCHEMICAL MECHANISMS CONTROLLING THE MAMMALIAN SPERM ACROSOME REACTION

As described in Section VI, the acrosome reaction of mammalian sperm has many characteristics of a true exocytotic reaction in that the outer acrosomal membrane fuses with the plasma membrane overlying the acrosome, thus producing hybrid membranes which vesiculate and are eventually sloughed from the sperm head. Exocytosis in somatic cells is regulated in a receptor-mediated fashion by specific signal transduction-intracellular second messenger processes,[191] so it is reasonable to expect that conserved inter- and intracellular signaling processes are operating in sperm to mediate sperm-egg recognition and sperm acrosomal exocytosis. This should not be surprising when one considers the fact that reproduction is a highly ordered process at multiple levels, and that species-specific gamete recognition, interaction, and activation represent an absolute prerequisite to the successful propagation of the species.

A. EVIDENCE FOR RECEPTOR-MEDIATED MODULATION OF SPERM FUNCTION

Presently, sea urchin sperm remain the experimental prototype in which receptor-mediated sperm signal transduction has been unequivocally demonstrated. In this system peptides associated with the jelly coat of the sea urchin egg (resact, speract)[192-194] interact with specific cell surface receptors in a species-specific manner[195-198] to elevate sperm cGMP concentrations.[194,198,199] The receptor for these peptides appears to be identical with or closely associated with the membrane-bound enzyme guanylate cyclase.[196,200] Peptide-receptor interaction has been demonstrated to result in sperm chemotaxis,[201] as well as changes in motility and metabolism.[192,202,203] Sperm, therefore, respond to external stimuli in their environment through receptor-mediated processes coupled to intracellular signaling systems. It is of interest to note that the membrane-bound form of guanylate cyclase also appears to be the cell surface receptor for atrial natriuretic factor (ANF) in specific mammalian tissues.[204,205] The conservation of receptor function of an enzyme from both an invertebrate germ cell and mammalian somatic cells is notable, and is even more interesting in light of recent observations that both of these enzymes have considerable sequence homology to one another[206] and also belong to the superfamily of protein kinases,[204,206,207] members of which include cell surface receptors for growth factors.[208] Such evolutionary conservation should be kept in mind when assessing the molecular mechanisms by which mammalian sperm function is modulated.

Presently, evidence for the existence of specific receptor-mediated events controlling mammalian sperm functions such as the acrosome reaction is only starting to evolve. As described in the previous section the mouse zona pellucida glycoprotein, ZP3, has a number of properties that make it an ideal candidate as a physiological regulator of sperm function. It is important to determine whether molecules with properties similar to ZP3 are responsible for the sperm binding and/or the acrosome reaction-inducing activities of the zona pellucida of other species.[78,92,95,134]

B. NATURE OF SPERM-ASSOCIATED COMPONENT(S) INVOLVED IN THE BINDING OF ACROSOME REACTION-INDUCING LIGANDS

1. General Requirements for Ligand Binding to the Sperm Surface

Although a number of biological substances have been shown to induce the sperm acrosome reaction, with the exception of the zona pellucida little is known about the minimal requirements for the binding of the active substance(s) to the sperm surface. The reason for this is that in most cases specific binding to the sperm surface has not been demonstrated. One must also consider the nature of the component that interacts with the sperm surface and modulates sperm function. For example, mouse sperm interaction with the component(s) of the zona pellucida which mediates sperm binding is probably different from that component(s) which mediates acrosomal exocytosis. Ligand binding to the sperm surface and sperm activation may, therefore, represent intimately associated but distinct processes, which have separate requirements.

Although capacitation and the presence of Ca^{2+} appear to be two absolute prerequisites to the successful induction of the acrosome reaction in all mammalian species, it is not clear whether these two requirements are necessary for the binding of the acrosome reaction-inducing ligand to the sperm surface. Handrow et al. demonstrated that the binding of ^3H-heparin to the bovine sperm surface is Ca^{2+}, pH, and temperature dependent.[170] Mouse sperm binding to the zona pellucida requires Ca^{2+},[209-211] is temperature dependent,[212] but does not require agents such as serum albumin or metabolic substrates.[114,209,213,214] The Ca^{2+} requirement appears to be important for initial stages of sperm-zona pellucida binding.[209,211] Sperm binding to the zona pellucida can occur under noncapacitating conditions,[82,117,213] although the subsequent acrosome reaction will not occur.[82,114,215] It has not been demonstrated whether these requirements must be satisfied to permit sperm binding to purified ZP3. It is clear that binding studies using purified ZP3 and either intact mouse sperm or mouse sperm membrane preparations will be needed to construct a detailed analysis of the binding requirements of this glycoprotein to the sperm surface. Similar approaches must be taken in other species.

2. Identity of Sperm Surface-Associated Binding Component(s) for Acrosome Reaction-Inducing Ligands

Any discussion dealing with the identity of a sperm-associated binding component for an acrosome reaction-inducing ligand must consider the relationship between the binding of that ligand to the sperm surface and the ability of that ligand to bring about acrosomal exocytosis. For example, in the mouse a single zona pellucida glycoprotein, ZP3, mediates *both* sperm binding to the zona pellucida and the induction of the acrosome reaction. Since it is possible to uncouple sperm-zona pellucida binding from acrosomal exocytosis under a variety of experimental conditions, any discussion of sperm-associated zona pellucida binding components must consider the possibility that the sperm-associated component(s) that is involved in sperm-ZP3 binding does not necessary have to be identical to the sperm-associated component(s) involved in the ZP3-induced acrosome reaction, even though ZP3, itself, possesses ligands which mediate both sperm binding and the acrosome reaction. Although a detailed analysis of mechanisms involved in sperm-egg binding is addressed elsewhere in this volume (see Chapter 6) a discussion of this topic is necessary at this juncture, since the nature and identity of sperm-associated components involved in such binding may have a direct bearing on the biochemical mechanisms controlling acrosomal exocytosis. The reader is also referred to the review article by Saling[216] dealing with mammalian sperm-extracellular matrix interactions.

Presently, the identity of the sperm-associated binding component(s) for those ligands that induce the acrosome reaction is not clear. As mentioned in Section VI, there is no evidence for the presence of specific receptors on the sperm surface for steroids, biogenic amines, prostaglandins, or glycosaminoglycans. However, there is preliminary evidence to suggest that there might be complementary binding sites for ZP3 associated with the mouse sperm.[103] These workers utilized whole mount autoradiography to examine the binding of ^{125}I-ZP3 to mouse sperm under various conditions. Binding appeared to be solely associated with the plasma membrane overlying the acrosomal region of mouse sperm and was not observed on acrosome-reacted sperm. Binding of the radiolabeled ligand could be competed by unlabeled ZP3 but not ZP2, and ZP3 binding did not occur on somatic cells. Although the use of the whole mount autoradiography has definite limitations with regard to characterization of the moiety(s) involved in ZP3 binding, these results are suggestive of a specific sperm-associated binding site(s) for this ligand.

Similar approaches have been taken in other species to identify specific sperm-associated zona pellucida binding components. Using fluorescein isothiocyanate-conjugated solubilized zonae pellucidae, acrosome-intact rabbit, boar, and hamster sperm have been demonstrated to bind the fluorescently tagged marker over the acrosomal region.[95,217] Acrosome-reacted sperm, on the other hand, exhibited fluorescence in the postacrosomal and equatorial regions. Characterization of the binding moiety(s) at the molecular level was not possible in these studies, since (1) the authors used whole zonae pellucidae preparations and not individual zona pellucida glycoproteins, and (2) the use of visualization at the light microscopic level is qualitative at best. However, these studies do suggest that a zona pellucida binding component(s) exists on the acrosomal region of acrosome-intact sperm of other species.

Direct binding of the zona pellucida to sperm plasma membranes from broken cells has also been observed. Three proteins of M_r = 30,000, 45,000, and 70,000 have been found to bind to immobilized pig zonae pellucidae when a partially purified boar sperm plasma membrane fraction was applied to the affinity matrix.[218,219] More recent studies by these workers suggest that an integral membrane protein of M_r = 55,000 isolated from boar sperm is the only protein that participates in sperm adhesion of boar sperm to the pig zona pellucida;[220] this protein does not appear to be proacrosin. Specific binding of radiolabeled zona pellucida glycoproteins to sperm proteins that have been first resolved by electrophoresis using SDS and then blotted onto nitrocellulose has also been attempted. O'Rand et al. demonstrated that rabbit, pig, human, and mouse sperm possessed both low and high molecular weight zona pellucida binding proteins,[221] but it is difficult to interpret these results since the number and type of binding proteins were quite variable and both homologous and heterologous sperm-zona pellucida combinations were used. Hamster sperm have been demonstrated to possess a zona pellucida binding protein of M_r = 26,000 that also displays Concanavalin A binding activity.[174] Boar sperm contain a protein M_r = 53,000 that binds to solubilized pig zonae pellucidae[222] and was subsequently demonstrated to be proacrosin.[223] It is difficult to assess the specificity of the zona pellucida-sperm protein interactions in these particular studies, since they were carried out with whole sperm extracts instead of sperm preparations enriched in plasma membranes which would contain the presumed zona pellucida binding activity. An additional problem with many of the aforementioned studies is that the localization of the zona pellucida binding activity was not determined.

Recently, Leyton and Saling demonstrated that polyclonal antiphosphotyrosine antibodies reacted with mouse sperm plasma membrane proteins of M_r = 52,000, 75,000, and 95,000 and this immunoreactivity appears to be localized to the acrosomal region of the sperm head.[224] ZP3 appears to bind, on nitrocellulose blots, to the M_r = 95,000 protein and these workers provided evidence to indicate that the M_r = 95,000 protein which bound ZP3 also reacted with the antiphosphotyrosine antibodies. These investigators suggest that the M_r = 95,000 protein may be a receptor for ZP3 which acts either as a substrate for tyrosine phosphorylation or possesses tyrosine kinase activity itself. These results are extremely interesting in light of the fact that somatic cell receptors possess intrinsic protein kinase activity.[208]

In addition to those studies directed at determining the complementary sperm binding site(s) for the zona pellucida by the approaches described above, a number of investigators have also attempted to characterize such sperm binding site(s) by examining whether specific agents interfere with sperm-zona pellucida interaction.

a. Proteases or Protease Inhibitor-Sensitive Sites

Sperm-associated proteinases have been implicated in sperm-zona pellucida binding. These conclusions were initially made based on the observations that a variety of compounds known to inhibit serine proteinases (e.g., trypsin, acrosin) also inhibit sperm-zona pellucida interaction *in vitro* in both hamsters and mice.[225-227] The mode of inhibition appears to be at the level of sperm-zona pellucida interaction, since sperm penetration through the zona pellucida and sperm-egg fusion does not appear to be affected.[226] The proteinase inhibitor-sensitive site in mouse sperm has been characterized in more detail by Benau and Storey who demonstrated that 4-methylumbelliferyl-*p*-guanidinobenzoate (MUGB), a serine proteinase active site titrant, is hydrolyzed by sperm and inhibits the binding of sperm to the zona pellucida.[228] Hydrolysis of MUGB is inhibited by both large (soybean trypsin inhibitor) and small (*p*-nitrophenyl guanidinobenzoate) molecular weight serine proteinase inhibitors, as well as by solubilized zonae pellucidae. In contrast, lysyl and arginyl trypsin substrates are not hydrolyzed by mouse sperm and do not inhibit sperm binding to the zona pellucida. These investigators concluded that the sperm-associated trypsin inhibitor-sensitive site has unique properties in that it interacts with inhibitors but does not turnover trypsin substrates. The molecular identity of this trypsin inhibitor-sensitive site is not presently known, and it is also unclear what component of the zona pellucida is interacting with this site. However, studies by other investigators have demonstrated the existence of a $M_r = 15,000$ membrane protein from mouse sperm which binds serine protease inhibitors and substrates, but does not hydrolyze the substrates.[227,229] Furthermore, addition of preparations enriched in this protein to zonae pellucidae inhibit subsequent sperm binding to this extracellular matrix.[229] It will be of interest to determine whether this protein and the trypsin inhibitor-sensitive site described by Benau and Storey[228] are related. More recent studies suggest that the galactosyltransferase site (see below) and the trypsin inhibitor-sensitive site on the sperm surface are independent with regard to their presumed interaction with the zona pellucida, and that inhibitors of one or the other site are sufficient to inhibit sperm-zona pellucida binding.[230] These two sites, although independent, are thought to be very closely apposed in functional terms since the inhibition of one of these sites results in an inhibition of the other site.[230] The potential interrelationship between these two sites in modulating sperm binding to the zona pellucida and induction of the acrosome reaction is not known, and should be investigated further. The interaction of the zona pellucida with multiple sites on the sperm surface is an interesting possibility in light of some recent studies directed at understanding the mode of interaction of ZP3 with the sperm surface to mediate both sperm-ZP3 binding and induction of the acrosome reaction. These studies will be addressed in greater detail below.

The serine proteinase, acrosin, has also been postulated to serve as a zona pellucida binding protein in porcine sperm that are either in the process of undergoing the acrosome reaction or have completed this exocytotic process.[222,223,231-233] Acrosin has been demonstrated to be a fucose binding protein,[232,234] and has been postulated to work in concert with a $M_r = 17,000$ fucose-binding protein to mediate both the primary binding of acrosome intact sperm and the secondary binding of acrosome-reacted sperm,[235] although the exact mechanism by which this might occur is unclear. It is difficult to reconcile the potential role of this protease in mediating the binding of acrosome intact sperm to the zona pellucida, since proacrosin/acrosin appears to reside within the acrosome and would not be accessible to the external environment until the acrosome reaction occurred. Perhaps a role for this enzyme in the subsequent binding of acrosome-reacted sperm to the zona pellucida might be considered in light of the observations made by Bleil et al. that trypsin inhibitors may block the secondary binding of acrosome-reacted sperm to this extracellular matrix.[182] Such a possibility has recently been proposed by Hedrick et al.[236]

b. Galactosyltransferase

Since cell surface-associated glycosyltransferases have been implicated in mediating cell-cell interactions in a variety of systems,[237] a number of investigators have examined the potential role of this family of enzymes in mediating sperm-zona pellucida interaction. The possibility that sperm-associated glycosyltransferases are involved in fertilization is reasonable when one considers the fact that the sperm binding activity of the ZP3 molecule appears to reside with the carbohydrate moieties of the glycoprotein,[108,186,187] and that a number of other investigators have demonstrated the potential importance of gamete-associated carbohydrate determinants in fertilization.[238] Shur and co-workers have provided evidence in the mouse that a sperm surface-associated galactosyltransferase (GalTase) activity may serve as the primary receptor for the zona pellucida.[239-244] This enzyme catalyzes the transfer of galactose from uridine 5′-diphosphate-galactose to terminal *N*-acetylglucosamine residues to form *N*-acetyllactosamine. Presumably, *N*-acetylglucosamine residues of ZP3 serve as the substrate acceptors for this enzyme and are subsequently modified to *N*-acetyllactosamine residues if UDP-galactose is available. Evidence that this enzyme participates in sperm-zona pellucida binding is based on the observations that (1) inhibitors of the enzyme (α-lactalbumin, UDP-dialdehyde) and the donor substrate UDP-galactose[230] also inhibit sperm binding to the zona pellucida,[240] (2) both bovine milk-soluble GalTase[241] and purified sperm surface GalTase[244] inhibit mouse sperm binding to the zona pellucida and these enzymes have been shown to galactosylate this extracellular matrix,[240,241] and (3) antibodies raised against the bovine milk-soluble GalTase inhibit the sperm GalTase activity, inhibit sperm-zona pellucida binding, and have been used to immunolocalize the sperm enzyme to the plasma membrane overlying the acrosomal crest region of sperm.[241] GalTase appears to localize to the acrosome during spermatogenesis.[245] Since ZP3 mediates sperm binding, it is presumed that ZP3 contains the sites for enzyme-mediated galactosylation. Presently, it is not known whether ZP3 possesses these sites, particularly since the acid-solubilized zona pellucida has proven to be a poor substrate for the transferase reaction. Furthermore, the oligosaccharide(s) that were galactosylated in the aforementioned studies has not been identified and characterized. The role of the enzymatic activity of GalTase in mediating sperm binding is also unclear, since conditions that favor sperm-zona pellucida binding do not necessarily favor enzymatic activity.[246] It is possible that GalTase may play a role in secondary binding of acrosome-reacted sperm to the zona pellucida, since the acrosome reaction is accompanied by a redistribution of enzyme activity to the inner acrosomal membrane.[243]

In addition to GalTase, possible roles for mouse sperm-associated fucosyltransferase[247] and α-D-mannosidase[248] activities in binding to the zona pellucida have recently been proposed. Their relationship to GalTase and/or the protease inhibitor-sensitive sites is not known.

c. Guinea Pig Sperm PH-20 Antigen

This M_r − 60,000 sperm membrane protein is associated with both the posterior head plasma membrane and the inner acrosomal membrane of mature guinea pig sperm,[53,249] and undergoes alterations in its surface distribution upon induction of the acrosome reaction which is consistent with antigen movement in the membrane.[249,250] These changes in membrane lateral diffusion may occur because the PH-20 antigen is a member of a class of membrane proteins anchored in the membrane by attachment to phosphatidylinositol.[52] Monoclonal antibodies generated against this antigen specifically block sperm binding to the zona pellucida,[251] and additional studies suggest that this antigen may play an important role in the binding of acrosome-reacted sperm to the zona pellucida.[96] Subsequent purification of the PH-20 antigen revealed multiple molecular weight forms that are structurally related and may undergo site-specific endoproteolytic cleavage during the acrosome reaction.[252] These forms have been postulated to be confined to specific domains of the sperm surface. The nature of the zona pellucida component(s) that the PH-20 antigen (or its different forms) interacts with and its role in sperm-zona pellucida binding are still unclear at this time.

d. Rabbit Sperm Autoantigen

Rabbit sperm autoantigens (RSA) have been postulated to function as lectins in mediating sperm-zona pellucida binding in the rabbit.[253] These low molecular weight proteins (M_r = 14,000 to 19,000) are associated with the post acrosomal region of rabbit sperm, and antibodies against these proteins block the interaction of rabbit sperm with the zona pellucida under *in vitro* conditions.[11,254] Binding of solubilized zonae pellucidae to proteins with molecular weights similar to RSA have been observed, but it is unclear whether these proteins are one and the same.[221] The fact that certain oligosaccharides can interfere with the interaction between RSA and the zona pellucida suggests that RSA may have lectin-like properties,[133] but the complete specificity of these interactions has not been determined.

In summary, a number of sperm-associated proteins or "sites" (based on functional activity) that may function in sperm-zona pellucida binding have been described in different species. Presently, however, it is unclear as to how these sites function on a molecular basis. For example, there is no proof at this time that these sites have the characteristics common to classic cell-surface receptors known to exist in somatic cells. It is also unclear whether or how these zona pellucida binding protein(s) are coupled to events leading to the induction of the acrosome reaction. Research in this area will continue to yield important and exciting information regarding the molecular nature of sperm-egg interaction specifically, and may provide insights into mechanisms of cell-extracellular matrix interactions in general.

C. NATURE OF THE INTERACTION BETWEEN THE ACROSOME REACTION-INDUCING LIGAND AND THE SPERM SURFACE

Presently, little is known about the molecular nature of the physiologically relevant acrosome reaction-inducing component(s) and even less about the sperm-associated site(s) for the binding of that component(s). It would seem premature, therefore, to discuss the nature of the interaction between these two moieties. However, recent data have emerged from a few laboratories that might reveal information as to the type of interaction that may occur to bring about this important exocytotic event. These data have emerged from studies of sperm-zona pellucida binding and the zona pellucida-induced acrosome reaction in the mouse, and have evolved from observations that the zona pellucida glycoprotein, ZP3, is responsible for both of these events.

Experiments by a number of investigators have demonstrated that the sperm binding activity and the acrosome reaction-inducing activity of ZP3 can be either partially or completely dissociated from one another by agents that act either at the level of the sperm or at the level of ZP3. In cases where acrosomal exocytosis is blocked, the inhibition by these agents appears to occur at a specific stage of the zona pellucida-induced acrosome reaction. In these experiments the acrosomal status of mouse sperm was monitored using the fluorescent probe chlortetracycline (CTC).[113-115] CTC is fluorescent when present in a hydrophobic environment and complexed with calcium. Mouse sperm treated with this probe exhibit characteristic patterns of fluorescence that are correlated with different stages leading to the completion of the acrosome reaction. Capacitated, acrosome-intact sperm display a characteristic "B pattern", which is characterizied by bright fluorescence over the anterior portion of the head and midpiece; the postacrosomal region displays no fluorescence. An "S pattern", which represents a true intermediate stage that appears prior to the completion of the acrosome reaction, is characterized by bright fluorescence over the midpiece and punctate (or spotty) fluorescence over the anterior portion of the head. The appearance of this pattern correlates with the loss of a transmembrane pH gradient.[115,120] An "AR pattern" is characteristic of sperm that have completed the acrosome reaction and is characterized by a lack of fluorescence over the anterior portion of the sperm head. Both the B and AR patterns have been characterized as acrosome intact and acrosome reacted, respectively, by electron microscopy.[82,117] The zona pellucida-induced acrosome reaction in mouse sperm as monitored by the CTC assay, therefore, consists of B to S and S to AR transitions which may be independently regulated.[255]

3-Quinuclidinyl benzilate (QNB), a classical muscarinic-cholinergic antagonist, has been shown to bind to a single class of high-affinity sites on mouse sperm[256] and to inhibit the zona pellucida-induced acrosome reaction without affecting the ability of the sperm to bind to this extracellular matrix.[117,256,257] Sperm incubated with QNB are inhibited from undergoing a B to S transition in response to either solubilized zonae pellucidae or purified ZP3.[115,116] Although specific binding of this antagonist to sperm can be demonstrated, the characteristics of this binding are different from the binding characteristics of a classical muscarinic-cholinergic receptor, and the molecular nature of this sperm-associated site has yet to be determined.[256] M-42, a monoclonal antibody against a M_r = 200,000/220,000 sperm protein, represents a second reagent that has been demonstrated to inhibit the zona pellucida-induced acrosome reaction without affecting sperm binding to this extracellular matrix.[258,259] The inhibitory effect of this antibody is at the B to S transition, does not block more distal events leading to acrosomal exocytosis (e.g., S to AR transition), and is observed with solubilized zonae pellucidae as well as purified ZP3.[260] Pertussis toxin (PT) represents a third agent that, when added to capacitated mouse sperm, inhibits specifically the B to S transition induced by structurally intact zonae pellucidae, solubilized zonae pellucidae, or purified ZP3[116,261] without affecting sperm binding to this extracellular matrix.[261] This bacterial toxin has been demonstrated to functionally inactivate a number of guanine nucleotide binding regulatory proteins (G-proteins) by its ability to catalyze the ADP-ribosylation of the α subunit of these proteins[262] and is thought to inactivate the G_i-like protein present in the mouse sperm.[263] All of these agents, when added to sperm, inhibit a specific stage(s) of the acrosome reaction induced solely by the zona pellucida (or ZP3) without affecting reactions involved in sperm-zona pellucida binding (mediated by ZP3). Furthermore, incubation of mouse sperm with either biologically active phorbol diesters or diacylglycerols, which are activators of the Ca^{2+} and phospholipid-dependent protein kinase (protein kinase C; PK-C), in the presence of zonae pellucidae accelerates the B to S transition, but inhibits the S to AR transition, suggesting that these two transitions can be independently regulated.[255] These PK-C activators do not affect the ability of sperm to bind to the zona pellucida and, therefore, can affect the different stages of the acrosome reaction without affecting reactions involved in zona pellucida binding.

Biologically active phorbol diesters and diacylglycerols also effect egg-induced dissociations of the two biological activities of the zona pellucida. These agents, when added to zona pellucida-intact mouse eggs, cause egg-induced modifications of ZP2 to $ZP2_f$ and a modification of ZP3, such that the sperm binding activity of ZP3 is completely retained and the acrosome reaction-inducing component of ZP3 is partially modified.[264,265] Sperm incubated with either solubilized zonae pellucidae or purified ZP3 from these treated eggs bind normally and undergo the B to S transition with normal kinetics, but are inhibited from completing the acrosome reaction as a consequence of their inability to undergo an S to AR transition.[118,264,265] Sperm suspended in intermediate S are not irreversibly stuck in this pattern since solubilized zonae pellucidae from untreated eggs (but not treated or fertilized eggs) can bring about an S to AR transition of those S pattern sperm.[118,119,260]

These observations, taken together, demonstrate that (1) sperm binding to the zona pellucida and the zona pellucida-induced acrosome reaction are two independent processes, and (2) the zona pellucida-induced acrosome reaction in the mouse may consist of discrete, independently regulated events, as monitored by the CTC assay. The reaction(s) comprising these different events (i.e., comprising the B to S and S to AR transitions) can be inhibited by a number of agents in a very specific manner. The concept that this exocytotic process proceeds in discrete steps is consistent with known regulatory events comprising exocytotic reactions in other cells.[191]

Insight into the nature of the interaction of ZP3 with the sperm surface to modulate both sperm binding and the induction of a complete acrosome reaction can be inferred when attempting to explain the effects of PK-C activators on egg-induced modifications of ZP3. Since preliminary experiments indicate that treatment of eggs with PK-C activators result in a partial reduction of cortical granules,[266] it is reasonable to assume that these agents are acting to stimulate cortical granule exocytosis. A model has been hypothesized by Kopf et al.[267] to explain these results, and takes into account the properties of sperm binding and the induction of the acrosome reaction observed with varying concentrations of solubilized zonae pellucidae or purified ZP3.[132] This model is based on the fact that multiple interactions between ZP3 molecules and sperm surface sites may be required *in toto* for binding and the induction of the acrosome reaction. This model also proposes that there are differences in the concentration dependence of ZP3 to express sperm binding activity and acrosome reaction-inducing activity. Specifically, the concentration response curve for ZP3 acrosome reaction-induced activity is shifted to the right of the concentration response curve for ZP3 sperm binding activity. Finally, this model predicts that ZP3 is composed of multiple "functional ligands", and that the interaction of these ligands with the sperm surface is responsible for both the sperm binding activity and the ability to induce a complete acrosome reaction. These functional ligands do not necessarily have to be identical. Successful sperm binding to ZP3 requires the interaction with a single ligand of the ZP3 molecule, and this accounts for the concentration-dependence relationship of ZP3 to effect sperm binding that is described by a hyperbolic function.[132] Such a hyperbolic function is suggestive of single-site interactions between the sperm surface and ZP3 to mediate sperm binding. Once this interaction to mediate binding has occurred interactions of the sperm with multiple ligands of the ZP3 molecule are required to induce a complete acrosome reaction. This is consistent with the observation that the concentration dependence of ZP3 to induce acrosomal exocytosis is described by a sigmoidal function.[132] Such a function is suggestive of cooperative or multisite interactions between the sperm surface and ZP3 to mediate the complete acrosome reaction. Fertilization normally results in a modification or inactivation of a majority (or all) of the ligands associated with the ZP3 molecule and hence the loss of both biological activities. In contrast, the partial release of cortical granules that is stimulated by PK-C activators results in a submaximal modification of these ligands. Consequently, ZP3 still possesses sperm binding activity, but the suboptimal number of active ligands remaining are not sufficient to permit the multiple interactions with the sperm surface necessary to induce a complete acrosome reaction. As a result, sperm initiate (B to S transition) but do not complete the acrosome reaction because they remain arrested in the S pattern. This partial inhibition can be relieved when zonae pellucidae (or ZP3) which contain a sufficient number of active ligands (e.g., from unfertilized eggs) are now added to the sperm arrested at the S pattern.[118,119,260] Induction of the complete acrosome reaction, therefore, may involve multiple interactions between sperm-associated ZP3 binding components and the ZP3 molecule.

Such interactions have also been recently proposed by Leyton and Saling using a different approach.[268] Using monospecific polyclonal antibodies to ZP2 and ZP3 these investigators have provided evidence that the aggregation of sperm-associated receptors for ZP3 may play a role in the ZP3-induced acrosome reaction. Briefly, ZP3 glycopeptides, which possess sperm binding activity but do not possess the ability to induce an acrosome reaction when added to sperm alone, bring about an acrosome reaction when incubated with sperm in the presence of anti-ZP3 IgG. This effect is not observed with anti-ZP2 IgG or with monovalent anti-ZP3 Fab fragments. However, if sperm are incubated with ZP3 glycopeptides and anti-ZP3 Fab fragments in the presence of an anti-IgG, the ability of the sperm to undergo acrosome reactions is restored. Although the addition of anti-ZP3 IgG to capacitated B pattern sperm does not induce an acrosome reaction, the addition of this antibody to sperm that are suspended in the S pattern (obtained by incubation of sperm with zonae pellucidae from phorbol diester-treated eggs) brings about this exocytotic event. These investigators concluded that the aggregation of sperm-associated receptors for ZP3, which were recognized by the ZP3 glycopeptides or by zonae pellucidae from phorbol diester-treated eggs and then cross-linked by the addition of the anti-ZP3 IgG fraction, is required for the induction of the complete acrosome reaction. This concept is further elaborated by Saling.[216]

Figure 1 summarizes these concepts in a diagrammatic form. In this cartoon the ZP3 ligand involved in sperm binding is shown as being different from those ligands participating

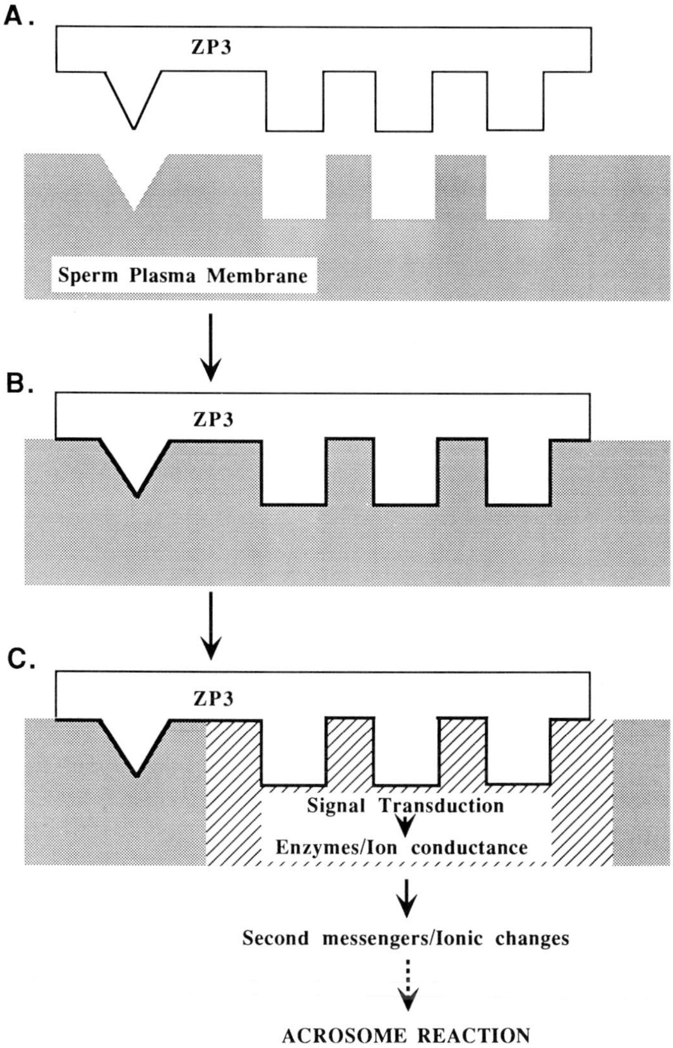

FIGURE 1. Model for the interaction of the zona pellucida glycoprotein, ZP3, with the plasma membrane overlying the acrosome of mouse spermatozoa to mediate sperm binding and acrosomal exocytosis. (A) In this model, the ZP3 molecule is composed of multiple "functional ligands" (∨; ⊔) which interact with complementary cell surface receptors present in the sperm plasma membrane. These ligands are shown as being different from one another in this cartoon, but this does not necessarily have to be the case. (B) Successful sperm binding to ZP3 requires the interaction with a single ligand (∨), thus accounting for the concentration dependence of ZP3 to mediate sperm binding that is described by a hyperbolic function. Once this interaction has occurred multiple interactions of the sperm with additional ligands (⊔) associated with the ZP3 molecule are then required to induce a complete acrosome reaction; this accounts for a concentration dependence of ZP3 to induce acrosomal exocytosis that is described by a sigmoidal function. The interaction of sperm with ZP3 to promote binding presumably allows for the proper orientation of the ZP3-associated functional ligands and complementary sperm cell surface receptors involved in mediating the acrosome reaction. In this cartoon three ZP3-associated ligands are shown which are involved in inducing the acrosome reaction. The actual number of ligands involved are not known and are shown in this fashion for illustrative purposes only. (C) Once the proper interaction of these ZP3 ligands with their complementary receptors has been accomplished, signal transduction and second messenger/ion conductance changes occur which ultimately lead to the acrosome reaction (shown in the figure by a change in the shading of the sperm plasma membrane). The interaction of multiple ZP3 ligands to bring about this exocytotic event may be accomplished by some type of functional coupling of the receptors for these different ligands, i.e., perhaps by receptor aggregation.

in acrosomal exocytosis. This does not necessarily have to be the case but is presented in this manner for clarity. The relationship between the sperm-associated galactosyltransferase activity, protease inhibitor-sensitive site, fucosyltransferase, and α-D-mannosidase in mediating sperm binding to the zona pellucida is not clear at this time.

The idea that receptor aggregation is required to elicit ligand-dependent signal transduction and a cellular response is not unique.[269-273] Since it has been recently demonstrated that extracellular matrices from a variety of tissues may contain domains that aggregate acetylcholine receptors,[274] it is certainly possible that the zona pellucida, itself an extracellular matrix, also possesses this property. The mechanisms of sperm-ZP3 interaction and cellular activation, therefore, has implications to the field of cell-extracellular matrix interactions in general, and will be of great interest for future studies.

D. NATURE OF SPERM SIGNAL TRANSDUCTION MECHANISMS MODULATING THE ACROSOME REACTION

Results from experiments emanating from a number of laboratories suggest that sperm-egg interaction in mammals, as well as in invertebrates, has elements of intercellular communication systems common to many hormonally responsive somatic cells. That is, biologically relevant ligands which modulate sperm function appear to bind to sites (receptors?) on the sperm surface and to transmit their information intracellularly through specific second messenger systems or changes in ionic permeability (see Section VII.E).[8,135] Although the components of these systems are only starting to be recognized and characterized, an additional question comes to mind. That question pertains to the mechanism by which an extracellular stimulus effects information flow across the sperm plasma membrane to elicit the appropriate cellular response.

The guanine nucleotide-binding regulatory proteins (G-proteins), which include G_s, G_i, G_o, and transducin, occupy critical roles as signal transducing elements in coupling many ligand-receptor interactions with the generation of intracellular second messengers.[262] These heterotrimeric plasma membrane-associated proteins are composed of distinct α-subunits, which contain a GTP-binding domain, and more highly conserved β- and γ-subunits. The α-subunits can be covalently modified by ADP-ribosylation in the presence NAD^+ by the action of a variety of bacterial toxins, including cholera toxin and PT. The ability to undergo toxin-catalyzed ADP-ribosylation has been used by many investigators to identify general classes of G-proteins. Activation of G-proteins occurs upon GTP binding to the α-subunit, causing dissociation of α-GTP from the βγ dimer. These dissociated subunits then exert their regulatory actions on respective targets, ultimately resulting in ligand coupling to intracellular signaling systems. G-proteins regulate the hormonally responsive adenylyl cyclase of somatic cells,[275] the cyclic GMP phosphodiesterase of retinal rod outer segments,[276] and atrial K^+ channels,[277] and have been implicated in regulating phospholipase C,[278] phospholipase A_2,[279,280] phospholipase D,[281,282] and Ca^{2+} channels.[283,284] Since many of these intracellular signaling mechanisms have been reported to be important in the control of sperm function,[135,285] it is possible that sperm utilize G-proteins as signal transducing elements in mediating cellular function in response to the extracellular environment.

Until a few years ago, sperm were thought to be "unique" because it was thought they lacked G-proteins. These conclusions were based largely on the properties of the adenylyl cyclase, which in somatic cells is regulated by guanine nucleotides, a stimulatory G-protein (G_s), and an inhibitory G-protein (G_i).[262] The sperm enzyme is particulate and the activity of the enzyme does not appear to be affected by fluoride ions, cholera toxin plus NAD^+, or forskolin, agents which are known to stimulate G-protein-regulated adenylyl cyclases in other cells.[135,286] Catalytic activity, for the most part, is poorly supported by Mg^{2+}, which is similar to the adenylyl cyclase activity from the S-49 lymphoma cell line which is devoid of G_s-regulated enzymatic activity. These properties are consistent with an enzyme either devoid of, or uncoupled from, G-proteins.[135,286] This was further confirmed by the work of Hildebrandt et al., who provided evidence that mammalian sperm lack both G_s and G_i, and that the sperm adenylyl cyclase is not able to interact with G_s from exogenous sources.[287] Although guanine nucleotides have been reported to stimulate sperm adenylyl cyclases, the nature of these stimulatory effects remains unclear since there is no apparent effect of cholera toxin.[288-291] The activity of the sperm enzyme has, however, been demonstrated to be regulated by extracellular effectors of sperm function, including egg-associated components and Ca^{2+} and calmodulin.[292-295] Sperm adenylyl cyclase activity, therefore, can be regulated by biologically relevant factors, but the mode of this regulation is presently unclear.

A reexamination of the question of whether sperm contain G-proteins was carried out independently by Kopf et al. and Bentley et al.[263,296] It was demonstrated that all mammalian sperm studied (mouse, bull, boar, human, guinea pig) possess a single $M_r = 41{,}000$ substrate for PT-catalyzed ADP-ribosylation. Peptide mapping of these substrates in human and mouse sperm revealed a similarity to the α-subunit of somatic cell G_i. Subsequent observations by Jones et al. have revealed that mouse sperm contain two closely migrating PT substrates which can only be resolved when sodium dodecyl sulfate containing a substantial portion of C_{14} and C_{16} alkyl sulfates is used for polyacrylamide gel electrophoresis.[297] The nature of these multiple forms is not presently known. Although sperm from invertebrates such as abalone also possess a single $M_r = 41{,}000$ PT substrate, the substrate observed in sea urchin sperm is $M_r = 39{,}000$, and has been demonstrated to bind $GTP\gamma^{35}S$. The β-subunit of the sperm G_i-like protein was shown to be $M_r = 35{,}000/36{,}000$ by immunoblotting using antisera raised against purified somatic cell β-subunit. These data suggest that mammalian sperm contain G-proteins with properties similar to that of somatic cell G_i. It will be interesting to determine the nature of the G_i subtype(s) (i.e,. $G_{i\alpha1}$, $G_{i\alpha2}$, $G_{i\alpha3}$) present in sperm, since antisera specific to the different forms are now becoming available. Attempts to determine whether sperm contain G_s-like proteins have been unsuccessful to date, consistent with the observations of Hildebrandt et al.[287] It is possible that G_s-like proteins are present in sperm but the current methods used to probe for their existence may be inadequate at this time.

Experiments have been carried out to determine whether the sperm G_i-like protein plays a physiological role in sperm function. Endo et al. examined the role of the mouse sperm G_i-like protein in acrosomal exocytosis induced by the zona pellucida.[116,261] In these experiments, intact sperm were incubated with PT under conditions conducive to capacitation and it was demonstrated that the toxin could enter the cell and ADP-ribosylate the sperm G_i-like protein The ability of PT to enter intact cells and ADP-ribosylate G-proteins has been established in a number of systems and it has been demonstrated that PT-catalyzed ADP ribosylation functionally inactivates G_i.[298] Under these incubation conditions the ability of the sperm to bind to structurally intact zonae pellucidae is not affected, but the ability of the sperm to undergo the acrosome reaction is inhibited in a concentration-dependent manner by the toxin. The PT-induced inhibition of acrosomal exocytosis appears to be specifically mediated by its effects on the sperm G_i-like protein, since conditions conducive to G_i-protein activation, which preclude the ability of the toxin to functionally inactivate G_i, abolish the ability of the toxin to inhibit the acrosome reaction. Furthermore, the level at which the toxin inhibits the acrosome reaction appears to be very specific and is confined to the B to S transition. Additional experiments by Endo et al. further defined the nature of the inhibitory effect of PT on the sperm acrosome reaction.[116] They demonstrated that the ZP3-induced acrosome reaction, and not acrosome reactions induced either spontaneously or nonspecifically (with A23187), is inhibited by PT. These data are consistent with the idea that the PT-sensitive site (G_i-like protein) in mouse sperm plays an important intermediary role in the acrosomal exocytosis induced specifically by ZP3. Similar observations in the mouse have been recently reported by other investigators[190] and Florman et al.[299] have recently demonstrated that PT inhibits the bovine sperm acrosome reaction induced by solubilized bovine zonae pellucidae.

Localization of G-proteins in bovine spermatozoa by indirect immunofluorescence using antisera directed against a variety of purified G-proteins (holotransducin, βγ-transducin, α-transducin, and $G_{o\alpha}$) has demonstrated domains of immunoreactivity in both the head (acrosomal and postacrosomal) and tail (midpiece) regions, depending on the antiserum used.[300] It was concluded that the differential localization of immunoreactive material may reflect the association of the G-proteins with various effector systems. Similar studies carried out in mouse and guinea pig sperm have concluded that the positive immunoreactive material associated with the acrosome of these cells using antisera generated against a peptide sequence of a conserved region of $G_{i\alpha 1}$, $G_{i\alpha 2}$, $G_{i\alpha 3}$, and $G_{z\alpha}$ appears confined to either the plasma and/or outer acrosomal membranes, since the immunoreactive material is lost when these cells undergo either a spontaneous or zona pellucida-induced acrosome reaction.[40] The effects of PT and the acrosomal localization of G-proteins by indirect immunofluorescence support the idea that sperm-associated G-proteins may occupy critical signal transducing functions in a variety of mammalian species to mediate the acrosome reaction in response to the zona pellucida.

Although G-proteins have been implicated in receptor-mediated exocytosis in other cells,[191] the mechanism by which sperm G-proteins mediate acrosomal exocytosis is unknown. If one considers the zona pellucida-induced acrosome reaction as an example of stimulus-secretion coupling which occurs in a receptor-mediated fashion, it is reasonable to propose that receptor G-protein interaction subsequently leads to the generation of intracellular second messengers and/or the modulation of ionic changes within the sperm. Two criteria would have to be met to establish the signal transducing function of this GTP binding protein in such a system. First, the occupation of a putative receptor for the active component(s) of the zona pellucida responsible for the induction of the acrosome reaction (e.g., ZP3 in the mouse) should result in G-protein activation in a manner described for other ligand receptor G-protein interactions. Second, the resultant G-protein activation should then modulate second messenger systems/ionic changes distal to receptor occupancy. Presently, there is no evidence to support the first criterion and little evidence in support of the second criterion. Endo et al. demonstrated that the discharge of the transacrosomal pH gradient in mouse sperm that normally accompanies the zona pellucida-mediated B to S transition is inhibited by PT. These data suggest that ionic changes leading to the discharge of this transmembrane pH gradient may be regulated by a PT-sensitive G-protein. Florman et al. have recently demonstrated that the elevations of internal Ca^{2+} and pH which accompany the bovine sperm acrosome reaction induced by the addition of solubilized zonae pellucidae are reduced by PT under conditions where the toxin inhibits exocytosis.[299] Although the techniques used in these studies to monitor pH and Ca^{2+} changes were superior to the fluorescent probe methods utilized by Endo et al.,[116] results from both laboratories suggest that one possible regulatory event (e.g., zona pellucida-induced pH changes) may be regulated either directly or indirectly through a G-protein. The observation that internal GTP can stimulate speract receptor-mediated voltage changes in membrane vesicles from sea urchin sperm points to the possibility that G-proteins may regulate receptor-mediated ionic events in sperm from a variety of species.[301]

The role of G-proteins in regulating intracellular second messenger cascade systems in mammalian sperm has not been investigated because we are just starting to learn about what signaling systems might be activated in mammalian sperm in response to biologically relevant molecules that induce the acrosome reaction. As stated above, adenylyl cyclase, cyclic nucleotide phosphodiesterase, phospholipase C, phospholipase A_2, and phospholipase D have been either demonstrated or proposed to be regulated by G-proteins in a receptor-mediated fashion in various somatic cells. Intracellular signaling systems involving all of these enzymes have been proposed by various investigators to play a role in the mammalian sperm acrosome reaction, and will be considered in the following section of this chapter.

It will be necessary to evaluate systematically the role of each of these different signaling systems in sperm in order to determine (1) whether they operate in response to physiologically relevant ligands in a receptor-mediated fashion and (2) whether they are regulated by G-proteins.

E. IONIC EVENTS AND INTRACELLULAR SECOND MESSENGERS MODULATING THE ACROSOME REACTION

Since the acrosome reaction is an exocytotic event, the nature of the intracellular signals that are generated in response to a biological stimulus that ultimately result in membrane fusion and vesiculation will most probably be similar to those signals that have been proposed to regulate exocytotic events in other cells. Such signals include changes in ionic conductance, changes in cyclic nucleotide metabolism, and changes in phospholipid metabolism. Similar to observations in other cells, one would expect these sperm-signaling processes to be regulated in a receptor-mediated fashion. Although there have been innumerable reports describing the ions and/or second messenger systems proposed to play a role in the mammalian sperm acrosome reaction (see review by Yanagimachi[8]), there is surprisingly little known about what system(s) may be operating physiologically, since there has been little attention paid to the effects of biologically relevant acrosome reaction-inducing components on sperm signaling systems. In the next few paragraphs an attempt will be made to summarize what is currently known with regard to the nature of sperm intracellular signaling during the acrosome reaction, focusing specifically on those cellular events which occur in response to biologically relevant molecules which have been demonstrated to induce acrosomal exocytosis. It is not the intent of this review to provide a comprehensive overview of every known and suspected intracellular mediator of the acrosome reaction, since this has been covered in earlier reviews of this subject.[5,8,9,135] It is obvious, however, that the interrelationships between ionic events and intracellular signaling systems in sperm are far from clarified.

1. Ions

Of all the ions that have been studied in relationship to the acrosome reaction, Ca^{2+} appears to be an absolute prerequisite for the successful completion of this exocytotic event in every species studied.[8,135,302] Incubation of mammalian sperm in a Ca^{2+}-deficient medium prevents the acrosome reaction from occurring under a variety of conditions. Likewise, incubation of sperm with cation ionophores that catalyze $Ca^{2+}/2H^{+}$ exchange[303-305] or incubation of sperm under other conditions which promote Ca^{2+} entry[306,307] will induce the acrosome reaction. This is presumed to occur as a consequence of increased internal Ca^{2+} and an increase in internal pH. In the sea urchin, the component of egg jelly that has been demonstrated to induce the acrosome reaction will effect an increase in sperm Ca^{2+} uptake that can be inhibited by the specific calcium transport antagonists D-600 and verapamil,[292] indicating that Ca^{2+} channels may play a role in this exocytotic event.[308] Only recently, however, have changes in mammalian sperm Ca^{2+} concentrations been observed to occur in response to reproductive tract or egg-associated components that induce the acrosome reaction. Using the intracellular fluorescent Ca^{2+} probe, Fura-2, Thomas and Meizel have demonstrated that human follicular fluid extracts that induce acrosome reactions in human sperm also increase intracellular sperm Ca^{2+} concentrations.[309] Florman and co-workers have recently demonstrated that solubilized bovine zonae pellucidae promote elevations of both intracelluar Ca^{2+} (using Fura-2) and pH (using carboxy fluorescein) under conditions that support acrosomal exocytosis.[299] These observations are significant because they are the first quantitative studies to conclusively demonstrate that a zona pellucida glycoprotein(s) can alter sperm ionic permeability under conditions conducive to acrosomal exocytosis. The recent observations of Lee and Storey in the mouse also support the idea that the zona pellucida can bring about alterations in sperm Ca^{2+} permeability.[120] At this time the mechanism by which follicular fluid and the zona pellucida mediate these Ca^{2+} permeability changes is not known. Additional studies will be required to determine whether these agents act at the level of ligand-gated Ca^{2+} channels, voltage-gated Ca^{2+} channels, ion exchange systems, or ion cotransporters. Evidence against the existence of Ca^{2+} channels in the regulation of the acrosome reaction in guinea pig and hamster sperm has been provided by investigators who have reported that Ca^{2+} channel antagonists (i.e., D600 and verapamil) do not inhibit exocytosis[307,310,311] and, in some cases, accelerate the onset of this exocytotic event[307,311] and stimulate sperm Ca^{2+} uptake.[307] However, the effects of these antagonists on acrosomal exocytosis induced by biological effectors of the acrosome reaction has yet to be determined.

Changes in internal pH also appear to be important in mediating the mammalian sperm acrosome reaction.[312,313] Elevated external pH (> pH 7.8) may induce the acrosome reaction independently of the presence of a particular stimulus and, in one study, high media pH precluded the need for serum for the acrosome reaction.[314] The zona pellucida of murine[115] and bovine[299] eggs induces an alkalinization of acrosomal pH under conditions conducive to the induction of the acrosome reaction. Although the zona pellucida-induced pH changes in the mouse at first appeared to be correlated with the initial stages of the acrosome reaction (i.e., the B to S transition using the chlortetracycline assay),[115] it is clear from more recent work that this pH change might be associated with later stages (i.e., intermediate S pattern[s] or the S to AR transition) of this exocytotic event.[118-120] It is not currently known what ions are supporting these induced pH changes. Although changes in both intracellular Ca^{2+} and pH occur when bull[299] and mouse[120] sperm are incubated with their respective zonae pellucidae, it is not clear how these two events are linked temporally and spatially. Lee and Storey have demonstrated that the Ca^{2+} and pH effects can be dissociated by the ionophore nigericin (i.e., nigericin causes pH changes in the apparent absence of Ca^{2+} changes) suggesting that pH changes may not lead to increased Ca^{2+} permeability.[115,120] However, it is unclear whether the nigericin-induced pH changes can be equated with zonae pellucidae-induced pH changes. Babcock and Pfeiffer have demonstrated that the elevation of cytosolic Ca^{2+} and internal pH in bovine sperm can occur independently through distinct mechanisms,[315] but this has not been examined in sperm challenged with zonae pellucidae.

Other ions (e.g., Na^+, K^+ and HCO_3^-) have been implicated in the acrosome reaction.[5,8] A sodium requirement for the acrosome reaction in guinea pig sperm has been demonstrated.[310,316] In these studies the acrosome reaction was enhanced by monensin, an ionophore which catalyzes an electroneutral Na^+/H^+ exchange, and was inhibited by the Na^+ channel blocker, amiloride.[316] This Na^+ requirement has been postulated to be involved in the regulation of internal Ca^{2+} concentrations through the operation of a Na^+/Ca^{2+} exchanger[310,317] which exists in the sperm plasma membrane.[318] K^+ ion influx has been proposed to regulate the acrosome reaction of hamster sperm through the action of a Na^+, K^+-ATPase.[319] The operational relationship between such a Na^+/Ca^{2+} exchanger and Na^+, K^+-ATPase activity is not presently known. It is also not known whether these transport systems are modulated in response to biological effectors of the acrosome reaction.

HCO_3^- has been demonstrated to be required for capacitation,[320] as well as the acrosome reaction induced either by monensin[310] or the zona pellucida.[321] Whether the HCO_3^- requirement for this exocytotic event is a direct effect or coupled to other ionic events such as Na^+-independent or -dependent-HCO_3^-/Cl^- antiporters[322] or HCO_3^-/Na^+ cotransporters[323] is not clear at this time. Of particular note, however, is the apparent relationship between HCO_3^-, Ca^{2+}, and cAMP metabolism in sperm, which will be discussed at greater length in the following section. It has been demonstrated that the Ca^{2+}-dependent elevation of guinea pig sperm cAMP concentrations is dependent on the presence of HCO_3^-.[324a] This anion is thought to be required for the uptake or exchange of Ca^{2+} which in some manner is responsible for the elevation of sperm cAMP.

Although the regulation of ionic conductance in sperm during acrosomal exocytosis is poorly understood, it is probable that the concerted action of enzymatic transport systems, as well as ion channels, play a role in modulating ionic changes needed for the successful completion of this important physiological event. The problem in currently dissecting those transport systems that operate during acrosomal exocytosis is the fact that ion transport processes are active continuously in sperm to maintain cellular homeostasis. As with most other cell types, the pertinent question is to elucidate those changes in steady-state ionic fluxes which occur upon ligand-induced cellular activation. The demonstration of sperm G-proteins and their regulatory role in the zona pellucida-mediated acrosome reaction adds an extra dimension regarding the regulation of ionic events in sperm, since these proteins have been shown to mediate ligand-stimulated ionic changes in somatic cells. The inhibitory effects of PT observed on zona pellucida-induced changes in intracellular pH and Ca^{2+} suggest that G-proteins may be involved in these important ionic events, either directly or indirectly.[116,299]

2. Cyclic Nucleotides

Although cyclic nucleotide metabolism has been implicated in the regulation of sperm function in both nonmammalian and mammalian species, little is known regarding the role of this intracellular second messenger system in the regulation of the mammalian sperm acrosome reaction.[135,285] Both nonmammalian and mammalian sperm contain the enzymes involved in both the synthesis and degradation of these important intracellular second messengers.[135]

In the sea urchin, the egg jelly component that induces the acrosome reaction causes a Ca^{2+}-dependent activation of adenylyl cyclase and the resulting elevation of cAMP is correlated with the induction of the acrosome reaction.[292] No effects on sperm cGMP concentrations are observed. Concomitant with the cAMP elevation is an increase in the activity of the cAMP-dependent protein kinase, the only known target for cAMP action.[324b] Although the elevation of cAMP precedes the induction of the acrosome reaction it appears as if Ca^{2+}, and not cAMP, is the primary mediator of this exocytotic event and that cAMP metabolism is regulated in some manner by this divalent cation.[292] Although elevations of cAMP and the acrosome reaction are closely correlated, the exact relationship between these events is still unclear. However, it appears that cAMP is a strong candidate to be at least one intracellular mediator of this exocytotic event.

The direct addition of cyclic nucleotide analogues to either mammalian or nonmammalian sperm has been reported to have either no effects, inhibitory effects, or stimulatory effects on the acrosome reaction.[8,135] These differences may be due, in part, to the nucleotide used, as well as the species tested. The lack of effect of endogenously applied cyclic nucleotides is not surprising in light of previous observations that cAMP appears to play a required, albeit secondary role to Ca^{2+} in the sea urchin sperm acrosome reaction and has also been demonstrated to play a similar role in guinea pig sperm.[306]

There have been very few reports in which the endogenous levels of mammalian sperm cyclic nucleotide concentrations have been measured under conditions conducive to capacitation and/or the acrosome reaction. Stein and Fraser reported that the adenylyl cyclase activity of mouse epididymal sperm increases during *in vitro* capacitation, but that the endogenous steady-state level of cAMP decreases during this period of time.[325] This was accompanied by a decrease in the activity of the cyclic nucleotide phosphodiesterase. These investigators also reported that the cGMP/cAMP ratio increases during capacitation as a consequence of decreasing endogenous cAMP concentrations in the absence of cGMP changes and concluded that this increase cGMP/cAMP ratio may play a role in capacitation. Experiments in our laboratory have demonstrated that the cAMP concentrations of mouse epididymal sperm do not change under conditions conducive to capacitation[326] and that the cAMP concentrations in sperm incubated under these conditions is 6- to 35-fold lower than that reported by Stein and Fraser.[325] Furthermore, cGMP concentrations are undetectable in our system.[326] It is difficult, therefore, to interpret the results of these investigations regarding the role of cyclic nucleotides in capacitation.

Hyne and Garbers reported that the addition of guinea pig sperm to capacitation media containing Ca^{2+} causes rapid increases (within 30 s) of cAMP up to 30-fold over baseline.[306] These cAMP elevations are dramatically reduced when the same experiment is performed in media devoid of added Ca^{2+}, and these Ca^{2+}-dependent cAMP elevations are blocked by the Ca^{2+}-transport antagonist D-600. The Ca^{2+}-dependent elevations of cAMP are associated temporally with the induction of the acrosome reaction, and a synergistic elevation of cAMP is observed when sperm are incubated in the presence of Ca^{2+} plus 1-methyl-3-isobutylxanthine (MIX), a cyclic nucleotide phosphodiesterase inhibitor. The addition of either MIX or 8-bromo-cAMP to capacitated sperm does not induce an acrosome reaction in the absence of Ca^{2+}, but significantly reduces the time period required for maximal expression of the acrosome reaction in the presence of Ca^{2+}. These data suggest that the acrosome reaction in guinea pig sperm is associated with both a primary transport of Ca^{2+} and a Ca^{2+}-dependent elevation of sperm cAMP concentrations. The absence of an effect of added 8-bromo-cAMP on the acrosome reaction of sperm incubated in the absence of Ca^{2+} is consistent with the notion that cAMP, although perhaps an important intracellular mediator of the acrosome reaction, does not appear to be the primary mediator of this exocytotic event.

Recently, Noland et al. examined the effects of solubilized mouse egg zonae pellucidae on the cAMP concentrations and acrosome reaction of capacitated mouse sperm.[327] The zona pellucida causes rapid and transient four- to six-fold increases in sperm cAMP concentrations over untreated controls, which occur prior to the onset of the acrosome reaction. Both the degree of sperm cAMP elevation and the percentage of acrosome reactions are dependent on the concentration of solubilized zonae pellucidae glycoproteins added to the assay. When a saturating concentration of zonae pellucidae is added to capacitated sperm in the presence of 2 m*M* EGTA, a concentration great enough to chelate the extracellular Ca^{2+} in the medium, both the zona pellucida-induced cAMP elevation and induction of the acrosome reaction are inhibited. It appears, therefore, that one of the responses of sperm to zonae pellucidae is a Ca^{2+}-dependent cAMP elevation which is correlated temporally with the induction of acrosomal exocytosis. These data represent the first report on an intracellular second messenger system activated in mammalian sperm in response to an egg-associated component and are consistent with the proposed role of cAMP as a mediator of acrosomal exocytosis.

Additional work will be required to determine the interrelationship between cAMP elevation and the induction of the acrosome reaction induced by the zona pellucida. For instance, it is not known whether zona pellucida-induced cAMP elevations occur in non-capacitated sperm. It is also not known which zona pellucida glycoprotein (ZP1, ZP2, ZP3) is responsible for the cAMP elevations; it is presumed that ZP3 is the active moiety but this must be tested. Furthermore, it remains to be demonstrated whether cAMP-dependent changes in the phosphorylation state of specific proteins (through the activation of a cAMP-dependent protein kinase) occurs in response to the zona pellucida. The relationship between extracellular Ca^{2+} and sperm cAMP metabolism has also yet to be established in this system. Since sperm adenylyl cyclases and cyclic nucleotide phosphodiesterases have been demonstrated to be regulated by Ca^{2+} [292,294,328,329] and/or Ca^{2+}-calmodulin interactions,[293,295,330] it is possible that calmodulin, which is present in sperm,[331-333] plays an important regulatory role in modulating the effects of Ca^{2+} on the cyclic nucleotide metabolism.

As previously noted, HCO_3^- appears to play an important role both in the acrosome reaction[310,321] and the regulation of sperm cAMP concentrations by Ca^{2+}.[324a] Although

Garbers et al.[324a] postulated that the effect of HCO_3^- on the Ca^{2+}-induced cAMP elevation in guinea pig sperm was due to the effects of this anion on the uptake (exchange) or membrane sequestration of small amounts of physiologically active Ca^{2+}, more recent work indicates that this anion might have direct effects on the activity of the sperm adenylyl cyclase. Okamura and co-workers have demonstrated that the low molecular weight factor from boar seminal plasma which stimulates sperm adenylyl cyclase is HCO_3^-.[334,335] The effects of this anion on sperm adenylyl cyclase activity was confirmed by Garty and Salomon, who demonstrated that HCO_3^- was probably interacting directly with the enzyme.[300] It will be important to determine the mechansims by which HCO_3^- affects the activity of the sperm adenylyl cyclase, since little is known about its regulatory properties.

In light of the findings that sperm contain G_i-like proteins which mediate the ZP3-induced acrosome reaction, it will also be of interest to determine whether these signal transducing proteins play either a direct or indirect role in regulating the zona pellucida-induced changes in sperm cAMP concentrations. Presently, it is difficult to reconcile a direct regulatory effect of G_i on adenylyl cyclase. However, since sperm adenylyl cyclase activity does appear to be regulated by biological factors associated with the egg in invertebrates, it is possible that sperm adenylyl cyclases, in general, possess unique regulatory properties. Effects of G-proteins on sperm cyclic nucleotide phosphodiesterase activity must also be considered, since particular forms of this enzyme have been demonstrated to be regulated by these signal-transducing proteins.[276] This promises to be a very exciting area of research in the future.

3. Lipid and Phospholipid Turnover

Changes in sperm lipid and phospholipid metabolism have been proposed by a number of investigators as playing a role in mammalian sperm capacitation, as well as the acrosome reaction.[150] Capacitation has been postulated to be accompanied by an increase in sperm plasma membrane fluidity[336] due, in part, to alterations in the fatty acid, phospholipid, and sterol content of the membrane.[150] These changes may be accomplished by reductions in the sperm membrane cholesterol/phospholipid ratio.[146-149,151] Membrane composition and fluidity changes could directly increase the permeability of the sperm membrane to Ca^{2+} and/or other ions. Since one end point of capacitation is the ability of the sperm to undergo acrosomal exocytosis in response to a physiologically relevant stimulus, such changes in membrane composition may make the membranes overlying the acrosome more fusigenic and therefore more prone to undergo membrane fusion and vesiculation in response to the appropriate stimulus. The relationship between sperm membrane lipid and phospholipid turnover during the processes of capacitation and the acrosome reaction are, therefore, probably tightly coupled to one another and must be considered when evaluating the role of lipids and phospholipids in acrosomal exocytosis.

Sperm membrane phospholipid changes thought to be involved in the induction of the acrosome reaction presumably occur through the action of specific phospholipases. Phospholipase A_2, which hydrolyzes phospholipids to their respective lysophospholipids, exists in the sperm of a variety of mammalian species.[150] It is thought that the generation of lysophospholipids, which possess fusigenic properties,[337] may promote the fusion and subsequent vesiculation of the plasma and outer acrosomal membranes.[75,150,338,339] Inhibitors of phospholipase A_2 activity have been demonstrated to block the acrosome reaction *in vitro*[139] and the addition of lysophospholipids or unsaturated fatty acids (products of phospholipase A_2 activity) to sperm can stimulate this exocytotic event.[142,338,340] It must be noted, however, that the phospholipase A_2 inhibitors that were used in these studies are not necessarily specific and the addition of exogenous lysophospholipids to any cell may have drastic effects on its membrane structure and permeability.

Although lysophospholipids have been reported to accumulate during sperm incubation *in vitro* and exogenously added phospholipids have been demonstrated to be degraded by sperm,[150] there have been no reports of a modulation of sperm-associated phospholipase A_2 activity in response to biologically relevant molecules which induce the acrosome reaction. Future experiments should address this issue since a ligand-induced effect on this enzyme might provide more convincing evidence of the potential role of this enzyme in acrosomal exocytosis. Regulation of this enzyme by a G-protein might also be considered.[279,280]

Receptor-mediated membrane polyphosphoinositide turnover, which generates multiple second messengers such as inositol, 1,3,5-trisphosphate (IP_3), *sn*-1,2,-diacylglycerol (DAG), arachidonic acid, and phosphatidic acid, represents an additional mechanism by which cells respond to external stimuli to undergo changes in physiological function.[341] The generation of these second messengers occurs as a consequence of the hydrolysis of phosphatidylinositol 4,5,-bisphosphate (PIP_2) by a PIP_2-specific phospholipase C, and the activity of this enzyme has been proposed to be regulated by G-proteins.[278] IP_3 modulates intracellular Ca^{2+} concentrations by causing the release of this cation from intracellular nonmitochondrial stores and DAG functions to activate the Ca^{2+} and phospholipid-dependent protein kinase (PK-C). Both IP_3 and PK-C have been implicated in ligand-induced exocytosis,[191,342-344] so this system seems a logical choice for investigation as a potential regulatory process controlling sperm acrosomal exocytosis. Information regarding the role of this system in sperm function is just starting to emerge.

Domino and Garbers have demonstrated that the fucose-sulfate glycoconjugate component of sea urchin egg jelly which induces the acrosome reaction of sea urchin sperm causes an accumulation of IP_3 up to ten-fold over that of the respective nontreated controls.[345] The glycoconjugate-induced increase in sperm IP_3 concentrations is completely Ca^{2+}-dependent and these Ca^{2+}-dependent increases are blocked by the Ca^{2+} channel blockers, nifedipine and verapamil. The characteristics of the IP_3 increase are similar to the cAMP increases reported earlier in this system.[292] Induction of the acrosome reaction by a variety of other treatments (e.g., elevated pH, nigericin, or gramicidin) is also accompanied by an increase in IP_3. These data provide the first demonstration that ligand-mediated acrosomal exocytosis is accompanied by an alteration of polyphosphoinositide turnover, and suggests that this intracellular second messenger may regulate a process(es) leading to exocytosis. The apparent Ca^{2+} dependency for this effect may suggest that Ca^{2+} dependent events precede and may be responsible for the stimulation of phospholipase C; alternatively, phospholipase C may be regulated by Ca^{2+} (or Ca^{2+} conductance) in some manner. Although the function of IP_3 in the acrosome reaction is not known, a number of possibilities might be proposed. This inositol phosphate might modulate intracellular Ca^{2+} concentrations by causing the release of this cation from internal stores such as the plasma or outer acrosomal membranes.[346,347] Alternatively, IP_3, or its biologically active phosphorylated intermediate inositol-1,3,4,5,-tetrakisphosphate (IP_4), might regulate the opening of plasma membrane Ca^{2+} channels, as there is evidence for such a mechanism of action in a number of cellular systems.[348-350]

Berruti and Franchi have localized phosphoinositides to the outer acrosomal membrane of boar sperm and have demonstrated that extracts from sperm contain phospholipase C-like activity.[347] These investigators did not, however, examine the potential relationship between signaling components and the acrosome reaction. More recently, Lee et al. demonstrated that activators of PK-C (e.g., 12-tetradecanoyl phorbol 13-acetate [TPA], 4-β-phorbol-12,13-didecanoate [4β-PDD], or 1-oleyl-2-acetylglycerol [OAG]), when added to mouse sperm bound to structurally intact zonae pellucidae, influence the kinetics of the different stages of the acrosome reaction but have no effect on the acrosome reaction induced spontaneously in the absence of zonae pellucidae.[255] Using the chlortetracycline fluorescence assay to monitor the different states of the acrosome reaction (B to S and S to AR transitions) it was determined that these phorbol diesters and diacylglycerol accelerate the B to S transition, but that they retard the S to AR transition. Biologically inactive phorbol diesters (4α-phorbol 12,13-didecanoate) have no effect on these transitions. Furthermore, the loss of the transacrosomal pH gradient that normally accompanies the B to S transition is also accelerated in the presence of these PK-C activators and QNB, which inhibits the B to S transition, also inhibits the effects of these PK-C activators on the B to S transition. These data suggest that activators of PK-C influence an event(s) which normally modulates the B to S and S to AR transitions in response to the zona pellucida and support the idea that the B to S and S to AR transitions represent two independently regulated states comprising the acrosome reaction. Although PK-C activity was not demonstrated in these studies, the stereospecificity of the phorbol diester effects strongly points to the existence of this enzyme in sperm. Subsequent experiments have demonstrated the existence of this enzyme in mouse sperm by both immunological[351] and biochemical[352] methods.

A phospholipase C specific for phosphoinositides has been described in human sperm.[353] This enzyme is activated by Ca^{2+}, will hydrolyze PIP_2 to IP_3, and resides in both particulate and soluble portions of the cell. These workers estimated that a majority (80%) of the particulate enzyme activity resides in the sperm head. By monitoring the end products of phospholipase A_2 (arachidonic acid) and phospholipase C (diacylglycerol) activities, this group subsequently demonstrated that the A23187-induced acrosome reaction of human sperm is accompanied by the formation of free arachidonic acid, DAG, and phosphatidic acid, which these workers attributed to the phosphorylation of DAG.[354] The generation of phosphatidic acid is of note, since this compound has been proposed to act as an ionophore.[355] Of additional interest is the observation that phosphatidylcholine hydrolysis is stimulated preferentially over phosphatidylinositol hydrolysis in this system and was proposed to be responsible for the generation of phosphatidic acid. The hydrolysis of the particular phospholipid may be of interest for future studies, since agonist-induced phosphatidylcholine hydrolysis of phosphatidylcholine-specifc phospholipases has been demonstrated to occur in other cell types and may provide an important source of DAG, which can be phosphorylated to phosphatidic acid.[356] Phosphatidylcholine hydrolysis in this study, therefore, might not only result from the action of a phospholipase A_2.

Recently, Roldan and Harrison took a similar approach to study the role of polyphosphoinositides in the A23187-induced acrosome reaction of ram, boar, guinea pig, mouse, and human sperm.[357] Using sperm prelabeled with [^{32}P]Pi, they demonstrated that the addition of Ca^{2+} and A23187 causes a rapid breakdown of PIP_2 and phosphatidylinositol 4-phosphate (PIP) with the formation of phosphatidic acid; these events preceded the acrosome reaction. EGTA inhibits both the hydrolysis of the polyphosphoinositides and the acrosome reaction, suggesting that these two events might be correlated with one another. Neomycin, an inhibitor of polyphosphoinositide hydrolysis, also inhibits the acrosome reaction when added at the time of the A23187. However, the inhibitory effect of this antibiotic on the acrosome reaction is not observed when it was added after polyphosphoinositide breakdown had occurred. These studies argue for the role of polyphosphoinositide breakdown in acrosomal exocytosis, but both this study and the study of Bennet et al.[354] must be interpreted with caution, since the acrosome reaction is induced in these studies by a relatively nonspecific agent (A23187). This ionophore could mask more subtle changes in phospholipid turnover that might be occurring under more physiological conditions, since any putative receptor-mediated changes in phospholipase C activity would be completely overridden.

The mechanism whereby elevations of mammalian sperm phosphatidic acid occur concomitant with phospholipid turnover and acrosomal exocytosis induced by A23187[354,357] should be examined in more detail as a result of the recent studies of Domino et al.[358] In these studies the sea urchin egg jelly fucose-sulfate glycoconjugate that induces the acrosome reaction in sea urchin sperm was demonstrated to bring about Ca^{2+}-dependent increases in sperm 1,2-diacylglycerol and phosphatidic acid concentrations. There was a disparity, however, in the degree of elevation of these two products, 1,2-diacylglycerol being elevated two-fold and phosphatidic acid ten-fold. Formation of phosphatidic acid by either the phosphorylation of 1,2-diacylglycerol or by the synthesis *de novo* from glycerol does not appear to occur under these conditions. Phosphatidylcholine hydrolysis to phosphatidic acid occurs when sperm are incubated with the egg jelly component, and isolated sperm membranes display the ability to catalyze this hydrolysis as well. These, as well as other observations, indicate that phosphatidylcholine is the major precursor for the formation of phosphatidic acid and that this occurs through the action of a phospholipase D. Although phospholipase D may be regulated by a G-protein in other cells,[281,282] it is unclear whether it may play a role in the sea urchin system. It is also unknown whether the formation of 1,2-diacylglycerol occurs through the action of a phospholipase C or through the dephosphorylation of phosphatidic acid by phosphatidate phosphohydrolase. It appears, therefore, that the acrosome reaction in sea urchin sperm, as in mammalian sperm,[354] may be accompanied by the activation of multiple phospholipases resulting in the generation of a number of different intracellular signals.

It appears as if phospholipid turnover may play a key role in the induction of the mammalian sperm acrosome reaction. It is clear from the studies carried out thus far that the mechanisms by which this occurs may be complex and that studies utilizing biologically relevant inducers of acrosomal exocytosis will be necessary to sort out these different pathways.

F. HYPOTHETICAL MODEL FOR LIGAND-INDUCED ACROSOMAL EXOCYTOSIS IN MAMMALIAN SPERM

Figure 2 illustrates some of the potential pathways by which biological effectors of acrosomal exocytosis may mediate their effects on mammalian spermatozoa. This diagram is in no way meant to presume that all of the pathways shown function to bring about acrosomal exocytosis, since information regarding signal transduction and intracellular signaling systems activated in response to physiological ligands is only starting to become available. Rather it is meant to stimulate future investigation into this particular area of cell biology. There are a few points that should be considered when evaluating potential signal transduction and intracellular signaling pathways involved in the acrosome reaction.

In other cells that respond to external stimuli in a receptor-mediated fashion, G-protein activation has been demonstrated to reduce ligand-receptor affinity, to reduce receptor-G-protein affinity, and to alter receptor-G-protein coupling efficiency.[262,359] Consequently, dissociation of ligand-receptor and receptor-G-protein complexes permits receptor recycling and allows for subsequent ligand-receptor-G-protein interactions to occur. These effects are obviously important for dampening signal generation after receptor occupancy so that the cell can prepare itself for subsequent receptor-mediated activation events. The effects of G-proteins to alter ligand-receptor dynamics, therefore, is very important in a cell which normally has to respond to the same stimulus multiple times during its functional life span. This does not apply to the acrosome reaction in sperm, however, since this exocytotic event is designed to occur only once in the fertilizable life span of the sperm cell. There is no need for reversible modulation of zona pellucida ligand-receptor interactions leading to exocytosis, since it would be predicted that once these interactions occurred this event would be driven to completion. Therefore, the involvement of G-proteins to mediate signal transduction distal to the binding of a zona pellucida component(s) to its respective sperm surface-associated receptor(s) might be confined solely to the activation of ionic events and/or second messenger systems involved in acrosomal exocytosis, and not necessarily to the modulation of ligand-receptor dynamics subsequent to initial ligand-receptor interaction. It is also possible that multiple signaling processes could be triggered in response to zona pellucida-sperm interaction to insure the successful completion of the acrosome reaction. Such "signal

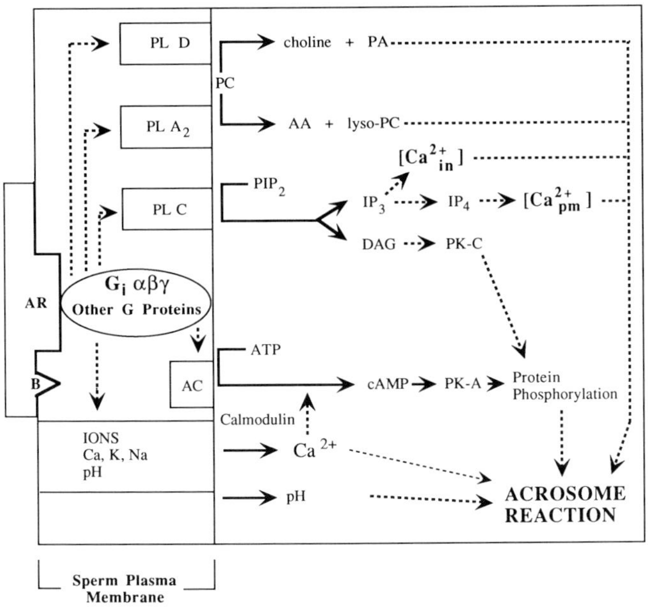

FIGURE 2. Hypothetical pathways by which biological effectors of acrosomal exocytosis mediate their effects on mammalian spermatozoa. Although this model incorporates results obtained from experiments using different species, many of the elements shown in this figure have been taken from observations made using the zona pellucida-induced acrosome reaction of mouse sperm as a model system. The sperm plasma membrane is denoted in this figure. The signaling systems distal to ligand-receptor G-protein interactions may be associated with a number of different regions of the sperm head, including the plasma, outer, and inner acrosomal membranes, as well as the regions confined within these membranes. The zona pellucida glycoprotein ZP3 possesses ligands which mediate both sperm binding (B) and the induction of a complete acrosome reaction (AR). These ligands bind to complementary receptors on the sperm plasma membrane overlying the acrosome. Sperm binding is accomplished by the single site interaction of a ZP3 ligand with a sperm plasma membrane receptor (\/) and subsequent acrosomal exocytosis is accomplished by multiple site interactions of ZP3 ligands with sperm-associated receptors to promote receptor aggregation (˥_Γ) as described in Figure 1. An acrosomal-associated G_i-like protein (G_i) (or other possible presently unidentified G-proteins) functions as a signal transducing element distal to ZP3-ZP3 receptor interactions, and couples receptor occupancy to changes in ionic conductance and/or a variety of intracellular second messenger cascade systems. These second messenger systems may include the polyphosphoinositide-specific phospholipase C (PLC), which hydrolyzes phosphatidylinositol 4,5-bisphosphate (PIP_2) to form *sn* 1,2-diacylglycerol (DAG) and inositol 1,4,5-trisphosphate (IP_3). IP_3 formation may regulate intracellular sperm Ca^{2+} concentrations through effects on intracellular nonmitochondrial Ca^{2+} stores ([Ca^{2+}in]) such as the plasma or outer acrosomal membranes. In addition, formation of inositol 1,3,4,5-tetrakisphosphate (IP_4) by the phosphorylation of IP_3 may ultimately regulate sperm Ca^{2+} uptake through effects on plasma membrane channels ([Ca^{2+}pm]). Liberated DAG activates a Ca^{2+} and phospholipid-dependent protein kinase (PK-C) that then phosphorylates proteins which may ultimately lead to the induction of the acrosome reaction. Hydrolysis of other phospholipids such as phosphatidylcholine (PC) may occur through the action of phospholipase A_2 (PLA_2) or phospholipase D (PLD), resulting in the generation of either lysophosphatidylcholine (lyso-PC) plus arachidonic acid (AA) or phosphatidic acid (PA) plus AA, respectively. Thse enzymes have also been proposed to be regulated by G-proteins. Membrane-associated lysophospholipids may promote acrosomal exocytosis by virtue of their fusigenic properties. PA may promote the acrosome reaction by virtue of its ability to function as an ionophore. PA formation may also occur through the phosphorylation of DAG. An alternative/additional signaling pathway that may play a role in the induction of the acrosome reaction is the adenylyl cyclase (AC) system. This enzyme might be regulated by extracellular Ca^{2+}, through the Ca^{2+} binding protein, calmodulin. Alternatively, AC may be regulated by the G_i-like protein in some manner. Generation of cyclic AMP (cAMP), with resultant stimulation of a cAMP-dependent protein kinase (PK-A), could also lead to protein phosphorylation and acrosomal exocytosis. G-protein-mediated effects on ion conductance might directly or indirectly lead to changes in intracellular pH and/or the regulation of the aforementioned signaling systems. Integration of a number of these signaling systems may be required for the successful induction of acrosomal exocytosis. Solid arrows (⟶) indicate pathways that have already been demonstrated to exist in sperm (invertebrate and/or mammalian) and dashed arrows (····➤) indicate potential interactions that have been proposed to regulate sperm function.

redundancy'' may be an important way to guarantee that a one-time exocytotic event occurs rapidly and at the appropriate time.

ACKNOWLEDGMENTS

The authors would like to acknowledge the contributions made by the various members of our laboratories, as well as our other scientific colleagues whose works were cited in this chapter: A. Abisogun, O. Anakwe, C. Arboleda, V. Dabney, Y. Endo, D. L. Garbers, M. Glassner, I. Kligman, S. Kurasawa, M. A. Lee, S. Newman, T. D. Noland, R. M. Schultz, N. M. Schwab, B. T. Storey, M. Wilde, and M. J. Woolkalis. Work from our laboratories has been supported by NIH (HD 19096, HD 06274 to G.S.K.; HD 22732 to G.S.K. and R.M.S.; HD 22899, HD 06274 to G.L.G.), the Andrew W. Mellon Foundation (G.S.K. and G.L.G.), and the Rockefeller Foundation (to A.A., C.A., Y.E., I.K., and S.K.). The authors wish to thank B. T. Storey for critically reviewing this manuscript and P. M. Saling and H. M. Florman for making available preprints of their work prior to publication.

REFERENCES

1. **Leblond, C. P. and Clermont, Y.,** Spermiogenesis of rat, mouse, hamster and guinea pig as revealed by the ''periodic acid-fuschin sulfurous acid'' technique, *Am. J. Anat.,* 90, 167, 1952.
2. **Fawcett, D. W.,** Morphogenesis of the mammalian sperm acrosome in new perspective, in *The Functional Anatomy of the Spermatozoon,* Afzelius, B. A., Ed., Pergamon Press, Oxford, 1975, 199.
3. **Bellué, A. R.,** The molecular biology of mammalian spermatogenesis, *Oxford Rev. Reprod. Biol.,* 1, 159, 1979.
4. **Setchell, B. P.,** Spermatogenesis and spermatozoa, in *Reproduction in Mammals: 1. Germ Cells and Fertilization,* Austin C. R. and Short, R. V., Eds., Cambridge University Press, Cambridge, 1982, 63.
5. **Yanagimachi, R.,** Mechanisms of fertilization in mammals, in *Fertilization and Embryonic Development in Vitro,* Mastrioanni, L. and Biggers, J. D., Eds., Plenum Press, New York, 1981, 81.
6. **Jeyendran, R. S., Van der Ven, H. H., Kennedy, W. P., Heath, E., Perez-Pelacz, M., Sobrero, A. J., and Zaneveld, L. J. D.,** Acrosomeless sperm: a primary cause of male infertility, *Andrologia,* 17, 31, 1985.
7. **Sotomayor, R. E. and Handel, M. A.,** Failure of acrosome assembly in a male sterile mutant mouse, *Biol. Reprod.,* 34, 171, 1986.
8. **Yanagimachi, R.,** Mammalian fertilization, in *The Physiology of Reproduction,* Knobil, E. and Neill, J. D., et al., Eds., Raven Press, New York, 1988, 135.
9. **Meizel, S.,** Molecules that initiate or help stimulate the acrosome reaction by their interaction with the mammalian sperm surface, *Am. J. Anat.,* 174, 285, 1985.
10. **Eddy, E. M.,** The spermatozoon, in *The Physiology of Reproduction,* Knobil, E. and Neill, J. D., et al., Eds., Raven Press, New York, 1988, 27.
11. **O'Rand, M. G.,** Sperm-egg recognition and barriers to interspecies fertilization, *Gamete Res.,* 19, 315, 1988.
12. **Wassarman, P. M.,** The biology and chemistry of fertilization, *Science,* 235, 553, 1987.
13. **Wassarman, P. M.,** Early events in mammalian fertilization, *Annu. Rev. Cell Biol.,* 3, 109, 1987.
14. **Wassarman, P. M.,** Zona pellucida glycoproteins, *Annu. Rev. Biochem.,* 57, 415, 1988.
15. **Allison, A. C. and Hartree, E. F.,** Lysosomal enzymes in the acrosome and their possible role in fertilization, *J. Reprod. Fertil.,* 21, 501, 1970.
16. **Friend, D. S.,** The organization of the spermatozoal membrane, in *Immunobiology of Gametes,* Edidin, M. and Johnson, M. H., Eds., Cambridge University Press, Cambridge, 1977, 5.
17. **Sandoz, D.,** Evolution des ultrastructures au cours de la formation de l'acrosome du spermatozoide chez la souris, *J. Micros.,* 9, 535, 1970.
18. **Susi, R. F., Leblond, C. P., and Clermont, Y.,** Changes in the Golgi apparatus during spermiogenesis in the rat, *Am. J. Anat.,* 130, 251, 1971.

19. **Griffiths, G., Warren, G., Stuhlfauth, I., and Jockusch, B. M.,** The role of clathrin-coated vesicles in acrosome formation, *Eur. J. Cell Biol.*, 26, 52, 1981.
20. **Thorne-Tjomsland, G., Clermont, Y., and Hermo, L.,** Contribution of the Golgi apparatus components to the formation of the acrosomic system and chromatoid body in rat spermatids, *Anat. Rec.*, 221, 591, 1988.
21. **Clermont, Y. and Tang, X. M.,** Glycoprotein synthesis in the Golgi apparatus of spermatids during spermiogenesis of the rat, *Anat. Rec.*, 213, 33, 1985.
22. **Willison, K., Lewis, V., Zuckerman, K. S., Cordell, J., Dean, C., Miller, K., Lyon, M., and Marsh, M.** The *t* complex polypeptide 1 (TCP-1) is associated with the cytoplasmic aspect of Golgi membranes, *Cell*, 57, 621, 1989.
23. **Silver, L. M., Artzt, K., and Bennett, D.,** A major testicular cell protein specified by a mouse T/t complex gene, *Cell*, 17, 275, 1979.
24. **Mollenhauer, H. H. and Morré, D. J.,** Polyribosomes associated with forming acrosome membranes in guinea pig spermatids, *Science*, 200, 85, 1978.
25. **Pelletier, R. M. and Friend, D. S.,** Development of membrane differentiations in the guinea pig spermatid during spermiogenesis, *Am. J. Anat.*, 167, 119, 1983.
26. **Courtens, J. L.,** Release of carbohydrate containing vesicles by the developing acrosome of ram sperm: relation to morphometric changes of the acrosome, *J. Ultrastruct. Res.*, 65, 182, 1978.
27. **Braun, R. E., Behringer, R. R., Peschon, J. J., Brinster, R. L., and Palmiter, R. D.,** Genetically haploid spermatids are phenotypically diploid, *Nature*, 337, 373, 1989.
28. **Kornfeld, S.,** Trafficking of lysosomal enzymes, *FASEB J.*, 1, 462, 1987.
29. **O'Brien, D. A., Gabel, C. A., Rockett, D. L., and Eddy, E. M.,** Receptor-mediated endocytosis and differential synthesis of mannose 6-phosphate receptors in isolated spermatogenic cells and Sertoli cells, *Endocrinology*, 125, 2973, 1989.
30. **Burgess, T. L. and Kelly, R. B.,** Constitutive and regulated secretion of proteins, *Annu. Rev. Cell Biol.*, 3, 243, 1987.
31. **Braun, R., Peschon, J. J., Behringer, R. R., Brinster, R. L., and Palmiter, R. D.,** Protamine 3′ untranslated sequences regulate temporal translational control and subcellular localization of growth hormone in spermatids of transgenic mice, *Genes Dev.*, 3, 793, 1989.
32. **Bloom, W. and Fawcett, D. W.,** *A Textbook of Histology*, W. B. Saunders, Philadelphia, 1975, 825.
33. **Holstein, A. F. and Roosen-Runge, E. C.,** *Atlas of Human Spermatogenesis*, Grosse-Verlag, Berlin, 1981.
34. **O'Brien, D. A., Gerton, G. L., and Eddy, E. M.,** Acrosomal constituents identified with a monoclonal antibody are modified during late spermiogenesis in the mouse, *Biol. Reprod.*, 38, 955, 1988.
35. **Gerton, G. L., O'Brien, D. A., and Eddy, E. M.,** Antigens recognized by monoclonal antibody to mouse acrosomal components differ in guinea pig spermatogenic cells and sperm, *Biol. Reprod.*, 39, 431, 1988.
36. **Hardy, D. M., Huang, T. T. F., Jr., Driscoll, W. J., Tung, K. S. K., and Wild, G. C.,** Purification and characterization of the primary acrosomal autoantigen of guinea pig epididymal spermatozoa, *Biol. Reprod.*, 38, 423, 1988.
37. **Anakwe, O. O. and Gerton, G. L.,** Acrosome biogenesis begins during meiosis: evidence from the synthesis and distribution of an acrosomal glycoprotein, acrogranin, during guinea pig spermatogenesis, *Biol. Reprod.*, 42, 317, 1990.
38. **Lucocq, J. M. and Warren, G.,** Fragmentation and partitioning of the Golgi apparatus during mitosis in HeLa cells, *EMBO J.*, 6, 3239, 1987.
39. **Garty, N. B., Galiani, D., Aharonheim, A., Ho, Y. K., Phillips, D. M., Dekel, N., and Salomon, Y.,** G-proteins in mammalian gametes: an immunocytochemical study, *J. Cell Sci.*, 91, 21, 1988.
40. **Glassner, M., Abisogun, A. O., Kligman, I., Woolkalis, M. J., Gerton, G. L., and Kopf, G. S.,** Immunocytochemical and biochemical analysis of guanine nucleotide-binding regulatory proteins (G-proteins) in mammalian spermatozoa, *J. Cell Biol.*, 109, 250a, 1989.
41. **Newman, S., Abisogun, A. O., Kopf, G. S., and Gerton, G. L.,** Identification of guanine nucleotide-binding regulatory proteins (G-proteins) in mammaliam spermatogenic cells, *J. Cell Biol.*, 109, 68a, 1989.
42. **Segev, N., Mulholland, J., and Botstein, D.,** The yeast GTP-binding YPT1 protein and a mammalian counterpart associated with the secretion machinery, *Cell*, 52, 915, 1988.
43. **Goud, B., Salminen, A., Walworth, N. C., and Novick, P. J.,** A GTP-binding protein required for secretion rapidly associates with secretory vesicles and the plasma membrane in yeast, *Cell*, 53, 753, 1988.
44. **Flörke, S., Phi-van, L., Müller-Esterl, W., Scheuber, H.-P., and Engel, W.,** Acrosin in the spermiohistogenesis of mammals, *Differentiation*, 24, 250, 1983.
45. **Phi-van, L., Müller-Esterl, W., Flörke, S., Schmid, M., and Engel, W.,** Proacrosin and differentiation of the spermatozoa, *Biol. Reprod.*, 29, 479, 1983.
46. **Arboleda, C. E. and Gerton, G. L.,** Proacrosin/acrosin during guinea pig spermatogenesis, *Dev. Biol.*, 125, 217, 1988.

47. **Anakwe, O. O., Hardy, D. M., and Gerton, G. L.,** Synthesis and post-translational modification of the protease zymogen proacrosin by primary cultures of guinea pig spermatogenic cells, *J. Cell Biol.,* 109, 67a, 1990.
48. **Hecht, N. B.,** Evidence for haploid-specific genes in the mammalian testis, in *Molecular and Cellular Endocrinology of the Testis,* Stefanini, M., Conti, M., Geremia, R., and Ziparo, E., Eds., Elsevier, Amsterdam, 1986, 199.
49. **Kallajoki, M., Parvinen, M., and Suominen, J. J. O.,** Expression of acrosin during mouse spermatogenesis: a biochemical and immunocytochemical analysis by a monoclonal antibody C11H, *Biol. Reprod.,* 35, 157, 1986.
50. **Anakwe, O. O., Hoff, H. B., Hardy, D. M., and Gerton, G. L.,** Guinea pig proacrosin is converted to a lower molecular weight during sperm transit through the epididymis, *Biol. Reprod.,* 40, (Suppl. 1), 230, 1989.
51. **Primakoff, P., Hyatt, H., and Tredick-Kline, J.,** Identification of a sperm surface protein with a potential role in sperm-egg membrane fusion, *J. Cell Biol.,* 104, 141, 1987.
52. **Phelps, B. M., Primakoff, P., Koppel, D. E., Low, M. G., and Myles, D. G.,** Restricted lateral diffusion of PH-20, a PI-anchored sperm membrane protein, *Science,* 240, 1780, 1988.
53. **Phelps, B. M. and Myles, D. G.,** The guinea pig sperm plasma membrane protein, PH-20 reaches the surface via two transport pathways and becomes localized to a domain after an initial uniform distribution, *Dev. Biol.,* 123, 63, 1987.
54. **Phelps, B. M., Koppel, D. E., and Myles, D. G.,** In vitro induction of sperm surface domain formation that normally occurs during sperm maturation, *J. Cell Biol.,* 107, 778a, 1988.
55. **Saling, P. M.,** Mouse sperm antigens that participate in fertilization. IV. A monoclonal antibody prevents zona penetration by inhibition of the acrosome reaction, *Dev. Biol.,* 117, 511, 1986.
56. **Lakoski, K., Williams, C., and Saling, P. M.,** Proteins of the acrosome region in mouse sperm: immunological probes reveal post-testicular modifications, *Gamete Res.,* 23, 21, 1989.
57. **Lakoski, K. A., Carron, C. P., Cabot, C. L., and Saling, P. M.,** Epididymal maturation and the acrosome reaction in mouse sperm: response to zona pellucida develops coincident with modification of M42 antigen, *Biol. Reprod.,* 38, 221, 1988.
58. **Robaire, B. and Hermo, L.,** Efferent ducts, epididymis, and vas deferens: structure, functions, and their regulation, in *The Physiology of Reproduction,* Knobil, E. and Neill, J., Eds., Raven Press, New York, 1988, 999.
59. **Harrison, R. A. P. and Brown, C. R.,** The zymogen form of acrosin in testicular, epididymal, and ejaculated spermatozoa from ram, *Gamete Res.,* 2, 75, 1979.
60. **Bhattacharyya, A. and Bhattacharyya, A. K.,** Biochemical changes in some acrosomal enzymes of spermatozoa during maturation, *Experientia,* 41, 62, 1985.
61. **Fawcett, D. W. and Hollenberg, R. D.,** Changes in the acrosome of guinea pig spermatozoa during passage through the epididymis, *Z. Zellforsch.,* 60, 276, 1963.
62. **Harding, H. R., Carrick, F. M., and Shorey, C. D.,** Ultrastructural changes in spermatozoa of the brush-tailed possum, Trichosurus vulpecula (Marsupialia), during epididymal transit. II. The acrosome, *Cell Tissue Res.,* 171, 61, 1976.
63. **Suzuki, F. and Nagano, T.,** Epididymal maturation of rat spermatozoa studied by thin-sectioning and freeze-fracture, *Biol. Reprod.,* 22, 1219, 1980.
64. **Yanagimachi, R., Kamiguchi, Y., Mikamo, K., Suzuki, F., and Yanagimachi, H.,** Maturation of spermatozoa in the epididymis of the Chinese hamster, *Am. J. Anat.,* 172, 317, 1985.
65. **Talbot, P. and Dicarlantonio, G.,** Cytochemical localization of dipeptidyl peptidase II (DPP-II) in mature guinea pig sperm, *J. Histochem. Cytochem.,* 33, 1169, 1985.
66. **Blaquier, J. A., Cameo, M. S., and Burgos, M. H.,** The role of androgens in the maturation of epididymal spermatozoa in the guinea pig, *Endocrinology,* 90, 839, 1972.
67. **Olson, G. E. and Winfrey, V. P.,** Substructure of cytoskeletal complex associated with hamster sperm acrosome, *J. Ultrastruc. Res.,* 92, 167, 1985.
68. **Virtanen, I., Badley, R. A., Paasivuo, R., and Lehto, V.-P.,** Distinct cytoskeletal domains revealed in sperm cells, *J. Cell Biol.,* 99, 1083, 1984.
69. **Orci, L., Ravazzola, M., Amherdt, M., Masden, O., Perrelet, A., Vassalli, J.-D., and Anderson, R. G. W.,** Conversion of proinsulin to insulin occurs coordinately with the acidification of maturing vesicles, *J. Cell Biol.,* 103, 2273, 1986.
70. **Orci, L., Ravazzola, M., Storch, M.-G., Anderson, R. G. W., Vassalli, J.-D., and Perrelet, A.,** Proteolytic maturation of insulin begins in clathrin-coated secretory vesicles, *Cell,* 49, 865, 1987.
71. **Talbot, P.,** Sperm penetration through oocyte investments in mammals, *Am. J. Anat.,* 174, 331, 1985.
72. **Katz, D. F., Drobnis, E. Z., and Overstreet, J. W.,** Factors regulating mammalian sperm migration through the female reproductive tract and oocyte vestments, *Gamete Res.,* 22, 443, 1989.
73. **Bleil, J. D. and Wassarman, P. M.,** Synthesis of zona pellucida proteins by denuded and follicle-enclosed mouse oocytes during culture, *in vitro, Proc. Natl. Acad. Sci. U.S.A.,* 77, 1029, 1980.

74. **Philpott, C. C., Ringuette, M. R., and Dean, J.,** Oocyte-specific expression and developmental regulation of ZP3, the sperm receptor of the mouse zona pellucida, *Dev. Biol.*, 121, 568, 1987.
75. **Suarez, S. S., Katz, D. F., and Overstreet, J. W.,** Movement characteristics and acrosomal status of rabbit spermatozoa recovered at the site and time of fertilization, *Biol. Reprod.*, 29, 1277, 1983.
76. **Suarez, S. S., Katz, D. F., and Meizel, S.,** Changes in motility that accompany the acrosome reaction in hyperactivated hamster spermatozoa, *Gamete Res.*, 10, 253, 1984.
77. **Storey, B. T., Lee, M. A., Muller, C., Ward, C. R., and Wirtshafter, D. G.,** Binding of mouse spermatozoa to the zonae pellucidae of mouse eggs in cumulus: evidence that the acrosomes remain substantially intact, *Biol. Reprod.*, 31, 119, 1984.
78. **Cherr, G. N., Lambert, H., Meizel, S., and Katz, D. F.,** *In vitro* studies of the golden hamster sperm acrosome reaction: completion on the zona pellucida and induction by homologous soluble zonae pellucidae, *Dev. Biol.*, 114, 119, 1986.
79. **Cummins, J. M. and Yanagimachi, R.,** Development of ability to penetrate the cumulus oophorous by hamster spermatozoa capacitated *in vitro* in relation to the time of the acrosome reaction, *Gamete Res.*, 15, 187, 1986.
80. **Corselli, J. and Talbot, P.,** An *in vitro* technique to study penetration of hamster oocyte-cumulus complexes by using physiological numbers of sperm, *Gamete Res.*, 13, 293, 1986.
81. **Corselli, J. and Talbot, P.,** In vitro penetration of hamster oocyte-cumulus complexes using physiological numbers of sperm, *Dev. Biol.*, 122, 227, 1987.
82. **Saling, P. M., Sowinski, J., and Storey, B. T.,** An ultrastructural study of epididymal mouse spermatozoa binding to zonae pellucidae *in vitro*: sequential relationship to the acrosome reaction, *J. Exp. Zool.*, 209, 229, 1979.
83. **Peterson, R. N., Russell, L. D., Bundman, D., and Freund, M.,** Sperm-egg interaction: evidence for boar sperm plasma membrane receptors for porcine zona pellucida, *Science*, 207, 73, 1980.
84. **Peterson, R. N., Russell, L. D., Bundman, D., Conway, M., and Freund, M.,** The interaction of living boar sperm and sperm plasma membrane vesicles with the porcine zona pellucida, *Dev. Biol.*, 84, 144, 1981.
85. **Peterson, R. N., Bovi, J. M., Dzuik, P. J., and Russell, L. D.,** The effects of antisperm plasma membrane antibodies on sperm-egg binding, penetration, and fertilization in the pig, *J. Exp. Zool.*, 223, 79, 1982.
86. **Shalgi, R. and Phillips, D.,** Mechanics of *in vitro* fertilization in the hamster, *Biol. Reprod.*, 23, 433, 1980.
87. **Katz, D. F., Cherr, G. N., and Lambert, H.,** The evolution of hamster sperm motility during capacitation and interaction with the ovum vestments in vitro, *Gamete Res.*, 14, 333, 1986.
88. **Crozet, N. and Dumont M.,** The site of the acrosome reaction during *in vivo* penetration of the sheep oocyte, *Gamete Res.*, 10, 97, 1984.
89. **Fournier-Delpech, S., Crozet, N., and Courot, M.,** Electron microscopic study of the *in vivo* zona pellucida binding ability of epididymal ram spermatozoa, *Gamete Res.*, 14, 225, 1986.
90. **Crozet N.,** Fine structure of sheep fertilization in vitro, *Gamete Res.*, 19, 291, 1988.
91. **Crozet, N.,** Ultrastructural aspects of *in vivo* fertilization in the cow, *Gamete Res.*, 10, 241, 1984.
92. **Florman, H. M. and First, N. L.,** The regulation of acrosomal exocytosis: sperm capacitation is required for the induction of acrosome reactions by the bovine zona pellucida, *in vitro, Dev. Biol.*, 128, 453, 1988
93. **Florman, H. M. and First, N. L.,** Regulation of acrosomal exocytosis: the zona pellucida-induced acrosome reaction of bovine spermatozoa is controlled by extrinsic positive regulatory elements, *Dev. Biol.*, 128, 464, 1988.
94. **Inger, S. L., Lambert H., Overstreet, J. W., Hanson, F. W., and Yanagimachi, R.,** The kinetics of human sperm binding to the human zona pellucida and zona-free hamster oocyte *in vitro, Gamete Res.*, 12, 29, 1985.
95. **O'Rand, M. G. and Fisher, S. J.,** Localization of zona pellucida binding sites on rabbit spermatozoa and induction of the acrosome reaction by solubilized zonae, *Dev. Biol.*, 119, 551, 1987.
96. **Myles, D. G., Hyatt, H., and Primakoff, P.,** Binding of both acrosome-intact and acrosome-reacted guinea pig sperm to the zona pellucida during *in vitro* fertilization, *Dev. Biol.*, 121, 559, 1987.
97. **Shalgi, R., Phillips, D. M., and Jones, R.,** Status of the rat acrosome during sperm-zona pellucida interactions, *Gamete Res.*, 22, 1, 1989.
98. **Huang, T. T. F., Fleming, A. D., and Yanagimachi, R.,** Only acrosome-reacted spermatozoa can bind to and penetrate the zona pellucida: a study using the guinea pig, *J. Exp. Zool.*, 217, 287, 1981.
99. **Kuzan, F. B., Fleming, A. D., and Seidel, G. E.,** Successful fertilization in vitro of fresh intact oocytes by perivitelline (acrosome reacted) spermatozoa of the rabbit, *Fertil. Steril.*, 41, 766, 1984.
100. **Morales, P., Cross, N. L., Overstreet, J. W., and Hanson, F. W.,** Acrosome-intact and acrosome-reacted human sperm can initiate binding to the zona pellucida, *Dev. Biol.*, 133, 385, 1989.
101. **Yanagimachi, R. and Noda, Y. D.,** Physiological changes in the post-nuclear cap region of mammalian spermatozoa: a necessary preliminary to the membrane fusion between sperm and egg cells, *J. Ultrastruct. Res.*, 31, 486, 1970.

102. **Fawcett, D. W.,** The mammalian spermatozoon, *Dev. Biol.,* 44, 394, 1975.
103. **Bleil, J. D. and Wassarman, P. M.,** Autoradiographic visualization of the mouse egg's sperm receptor bound to sperm, *J. Cell Biol.,* 102, 1363, 1986.
104. **Overstreet, J. W. and Cooper, G. W.,** The time and location of the acrosome during sperm transport in the female rabbit, *J. Exp. Zool.,* 209, 97, 1979.
105. **Talbot, P. and Chacon, R. S.,** A triple stain technique for scoring acrosome reaction of human sperm, *J. Exp. Zool.,* 215, 201, 1981.
106. **Bryan, J. H. D. and Akruk, S. R.,** A naphthol-yellow-S and erythrosin B staining procedure for use in studies of the acrosome reaction of rabbit spermatozoa, *Stain. Technol.,* 52, 47, 1977.
107. **Didion, B. A., Dobrinsky, J. R., Giles, J. R., and Graves, C. N.,** Staining procedure to detect viability and the true acrosome reaction in spermatozoa of various species, *Gamete Res.,* 22, 51, 1989.
108. **Florman, H. M., Bechtol, K. B., and Wassarman, P. M.,** Enzymatic dissection of the functions of the mouse egg's receptor for sperm, *Dev. Biol.,* 106, 243, 1984.
109. **Wolf, D. P., Boldt, J., Byrd, W., and Bechtol, K. B.,** Acrosomal status evaluation in human ejaculated sperm with monoclonal antibodies, *Biol. Reprod.,* 32, 1157, 1985.
110. **Koehler, J. K.,** Lectins as probes of the spermatozoa surface, *Arch. Androl.,* 6, 197, 1981.
111. **Talbot, P. and Chacon, R. S.,** A new procedure for rapidly scoring acrosome reaction of human sperm, *Gamete Res.,* 3, 211, 1980.
112. **Cross, N. L., Morales, P., Overstreet, J. W., and Hanson, F. W.,** Two simple methods for detecting human sperm acrosome reactions, *Gamete Res.,* 15, 213, 1986.
113. **Saling, P. M. and Storey, B. T.,** Mouse gamete interactions during fertilization *in vitro*: chlortetracycline as fluorescent probe for the mouse sperm acrosome reaction, *J. Cell Biol.,* 83, 544, 1979.
114. **Ward, C. R. and Storey, B. T.,** Determination of the time course of capacitation in mouse spermatozoa using a chlortetracycline fluorescence assay, *Dev. Biol.,* 104, 287, 1984.
115. **Lee, M. A. and Storey, B. T.,** Evidence for plasma membrane impermeability to small ions in acrosome-intact mouse spermatozoa bound to mouse *zonae pellucidae*, using an aminoacridine fluorescent probe: time course of the *zona*-induced acrosome reaction monitored by both chlortetracycline and pH probe fluorescence, *Biol. Reprod.,* 33, 235, 1985.
116. **Endo, Y., Lee, M. A., and Kopf, G. S.,** Characterization of an islet activating protein-sensitive site in mouse sperm that is involved in the *zona pellucida*-induced acrosome reaction, *Dev. Biol.,* 129, 12, 1988.
117. **Florman, H. M. and Storey, B. T.,** Mouse gamete interactions: the *zona pellucida* is the site of the acrosome reaction leading to fertilization *in vitro, Dev. Biol.,* 91, 121, 1982.
118. **Kligman, I., Storey, B. T., and Kopf, G. S.,** The *zona pellucida*-mediated acrosome reaction of mouse sperm: characterization of an intermediate stage prior to the completion of the acrosome reaction, in 10th Testis Workshop, NICHD, Baltimore, MD, December 8 to 11, 1988.
119. **Kligman, I., Glassner, M., Storey, B. T., and Kopf, G. S.,** Zona pellucida-mediated acrosomal exocytosis in mouse spermatozoa: characterization of an intermediate stage prior to the completion of the acrosome reaction, submitted.
120. **Lee, M. A. and Storey, B. T.,** Endpoint of first stage of zona pellucida-induced acrosome reaction in mouse spermatozoa characterized by acrosomal H^+ and Ca^{2+} permeability: population and single cell kinetics, *Gamete Res.,* 24, 303, 1989.
121. **Caswell, A. H. and Hutchison, J. D.,** Selectivity of cation chelation to tetracyclines: evidence for special conformation of calcium chelate, *Biochem. Biophys. Res. Commun.,* 43, 625, 1971.
122. **Lee, M. A., Trucco, G. S., Bechtol, K. B., Wummer, N., Kopf, G. S., Blasco, L., and Storey, B. T.,** Capacitation and acrosome reactions in human spermatozoa monitored by a chlortetracycline fluorescence assay, *Fertil. Steril.,* 48, 649, 1987.
123. **Moricard, R. and Bossu, J.,** Arrival of fertilizing sperm at the follicular cell of the secondary oocyte: a study of the rat, *Fertil. Steril.,* 2, 260, 1951.
124. **Zamboni, L.,** Fertilization in the mouse, in *Biology of Mammalian Fertilization and Implantation,* Moghissi, K. S. and Hafez, E. S. E., Eds., Charles C Thomas, Springfield, IL, 1972, 213.
125. **Shalgi, R. and Kraicer, P. F.,** Timing of sperm transport, sperm penetration, and cleavage in the rat, *J. Exp. Zool.,* 204, 353, 1978.
126. **Cummins, J. M. and Yanagimachi, R.,** Sperm-egg ratio and the site of the acrosome reaction during in vivo fertilization in the hamster, *Gamete Res.,* 5, 239, 1982.
127. **Austin, C. R. and Bishop, M. W. H.,** Some features of the acrosome and perforatorium in mammalian spermatozoa, *Proc. R. Soc. London (Biol.),* 149, 234, 1958a.
128. **Bryan, J. H. D.,** Capacitation in the mouse: the response of murine acrosomes to the environment of the female genital tract, *Biol. Reprod.,* 10, 414, 1974.
129. **Bedford, J. M.,** An electron microscopic study of sperm penetration into rabbit eggs after natural mating, *Am. J. Anat.,* 133, 213, 1972.
130. **Yanagimachi, R. and Noda, Y. D.,** Ultrastructural changes in the hamster sperm head during fertilization, *J. Ultrastruct. Res.,* 31, 465, 1970.

131. **Yanagimachi, R. and Phillips, D. M.,** The status of acrosomal caps of hamster spermatozoa immediately before fertilization *in vitro, Gamete Res.,* 9, 1, 1984.
132. **Bleil, J. D. and Wassarman, P. M.,** Sperm-egg interactions in the mouse: sequence of events and induction of the acrosome reaction by a *zona pellucida* glycoprotein, *Dev. Biol.,* 95, 317, 1983.
133. **O'Rand, M. G., Welch, J. E., and Fisher, S. J.,** Sperm membrane and zona pellucida interactions during fertilization, in *Molecular and Cellular Aspects of Reproduction,* Dhindsa, D. S. and Bahl, O. P., Eds., Plenum Press, New York, 1986, 131.
134. **Cross, N. L., Morales, P., Overstreet, J. W., and Hanson, F. W.,** Induction of acrosome reactions by the human zona pellucida, *Biol. Reprod.,* 38, 235, 1988.
135. **Garbers, D. L. and Kopf, G. S.,** The regulation of spermatozoa by calcium and cyclic nucleotides, *Adv. Cyclic Nucleotide Res.,* 13, 251, 1980.
136. **Cummins, J. M.,** Hyperactivated motility patterns of ram spermatozoa recovered from the oviducts of mated ewes, *Gamete Res.,* 6, 53, 1982.
137. **Meizel, S.,** The mammalian sperm acrosome reaction: a biochemical approach, in *Development in Mammals,* Johnson, M. H., Ed., North-Holland, Amsterdam, 1978, 1.
138. **Lui, C. W. and Meizel, S.,** Biochemical studies of the in vitro acrosome reaction inducing activity of bovine serum albumin, *Differentiation,* 9, 59, 1977.
139. **Lui, C. W. and Meizel, S.,** Further evidence in support of a role for hamster sperm hydrolytic enzymes in the acrosome reaction, *J. Exp. Zool.,* 207, 173, 1979.
140. **Rogers, B. J.,** Mammalian sperm capacitation and fertilization in vitro: a critique of methodology, *Gamete Res.,* 1, 165, 1978.
141. **Meizel, S. and Turner, K. O.,** Serotonin or its agonist 5-methoxytryptamine can stimulate hamster sperm acrosome reactions in a more direct manner than catecholamines, *J. Exp. Zool.,* 226, 171, 1983.
142. **Meizel, S. and Turner, K. O.,** Stimulation of an exocytotic event, the hamster sperm acrosome reaction by cis-unsaturated fatty acids, *FEBS Lett.,* 161, 315, 1983.
143. **Meizel, S. and Turner, K. O.,** The effects of products and inhibitors of arachidonic acid metabolism on the hamster sperm acrosome reaction, *J. Exp. Zool.,* 231, 283, 1984.
144. **Cornett, L. E. and Meizel, S.,** Stimulation of in vitro activation and the acrosome reaction of hamster spermatozoa by catecholamines, *Proc. Natl. Acad. Sci. U.S.A.,* 75, 4954, 1978.
145. **Meizel, S. and Working, P. K.,** Further evidence suggesting the hormonal stimulation of hamster sperm acrosome reactions by catecholamines in vitro, *Biol. Reprod.,* 22, 211, 1980.
146. **Davis, B. K., Byrne, R., and Jungund, B.,** Studies on the mechanisms of capacitation in vitro. II. Evidence for lipid transfer between plasma membrane of rat sperm and serum albumin during capacitation in vitro, *Biochim. Biophys. Acta,* 558, 257, 1979.
147. **Davis, B. K., Byrne, R., and Bedigian, K.,** Studies on the mechanism of capactiation: albumin-mediated changes in plasma membrane lipids during in vitro incubation of rat sperm cells, *Proc. Natl. Acad. Sci. U.S.A.,* 77, 1546, 1980.
148. **Go, K. J. and Wolf, D. P.,** The role of sterols in sperm capacitation, *Adv. Lipids Res.,* 20, 317, 1983.
149. **Go, K. J. and Wolf, D. P.,** Albumin-mediated changes in sperm sterol content during capacitation, *Biol. Reprod.,* 32, 145, 1985.
150. **Langlais, J. and Roberts, K. D.,** A molecular membrane model of sperm capacitation and the acrosome reaction of mammalian spermatozoa, *Gamete Res.,* 12, 183, 1985.
151. **Langlais, J., Kan, F. W. K., Granger, L., Raymond, L., Bleau, G., and Roberts, K. D.,** Identification of sterol acceptors that stimulate cholesterol efflux from human spermatozoa during in vitro capacitation, *Gamete Res.,* 20, 185, 1988.
152. **Aonuma, S., Okabe, M., Kishi, Y., Kawaguchi, M., and Yamada, H.,** Capacitation and inducing activity of serum albumin in fertilization of mouse ova in vitro, *J. Pharm. Dyn.,* 5, 980, 1982.
153. **Meizel, S., Turner, K. O., and Thomas, P.,** The stimulation of hamster sperm capacitation and acrosome reactions by biogenic amines, in *Catecholamines as Hormone Regulators,* Ben-Jonathan, N., Bahr, J. M., and Weiner, R. I., Eds., Raven Press, New York, 1985, 329.
154. **Suarez, S. S., Wolf, D. P., and Meizel, S.,** Induction of the acrosome reaction in human spermatozoa by a fraction of human follicular fluid, *Gamete Res.,* 14, 107, 1986.
155. **Osman, R. A., Andria, J. L., Jones, A. D., and Meizel, S.,** Steroid induced exocytosis: the human sperm acrosome reaction, *Biochem. Biophys. Res. Commun.,* 160, 828, 1989.
156. **Yudin, A. I., Gottlieb, W., and Meizel, S.,** Ultrastructural studies of the early events of the human sperm acrosome reaction as initiated by human follicular fluid, *Gamete Res.,* 20, 11, 1988.
157. **Siiteri, J. E., Gottlieb, W., and Meizel, S.,** Partial characterization of a fraction from human follicular fluid that initiates the human sperm acrosome reaction in vitro, *Gamete Res.,* 20, 25, 1988.
158. **Mortimer, D. and Camenzind, A. R.,** The role of follicular fluid in inducing the acrosome reaction in human spermatozoa incubated in vitro, *Hum. Reprod.,* 4, 169, 1989.

159. **Stock, C. E., Bates, R., Lindsay, K. S., Edmonds, D. K., and Fraser, L. R.,** Extended exposure to follicular fluid is required for significant stimulation of the acrosome reaction in human spermatozoa, *J. Reprod. Fertil.*, 86, 401, 1989.
160. **Cornett, L. E., Bavister, B. D., and Meizel, S.,** Adrenergic stimulation of fertilizing ability in hamster spermatozoa, *Biol. Reprod.*, 20, 925, 1979.
161. **Lenz, R. W., Ax, R. L., Grimek, H. J., and First, N. L.,** Proteoglycan from bovine follicular fluid enhances an acrosome reaction in bovine spermatozoa, *Biochem. Biophys. Res. Commun.*, 106, 1092, 1982.
162. **Eppig, J. J.,** FSH stimulates hyaluronic acid synthesis by oocyte-cumulus cell complexes from mouse preovulatory follicles, *Nature (London)*, 281, 483, 1979.
163. **Ball, G. D., Bellin, M. E., Ax, R. L., and First, N. L.,** Glycosaminoglycans in bovine cumulus-oocyte complexes: morphology and chemistry, *Mol. Cell Endocrinol.*, 28, 113, 1982.
164. **Handrow, R. R., Lenz, R. W., and Ax, R. L.,** Structural comparisons among glycosaminoglycans to promote an acrosome reaction in bovine spermatozoa, *Biochem. Biophys. Res. Commun.*, 107, 1326, 1982.
165. **Lenz, R. W., Ball, G. D., Lohse, J. K., First, N. L., and Ax, R. L.,** Chondroitin sulfate facilitates an acrosome reaction in bovine spermatozoa as evidenced by light microscopy, electron microscopy and in vitro fertilization, *Biol. Reprod.*, 28, 683, 1983.
166. **Lenz, R. W., Bellin, M. E., and Ax, R. L.,** Rabbit spermatozoa undergo an acrosome reaction in the presence of glycosaminoglycans, *Gamete Res.*, 8, 11, 1983.
167. **Parrish, J. J., Susko-Parrish, J. L., and First, N. L.,** Effect of heparin and chondroitin sulfate on the acrosome reaction and fertility of bovine sperm in vitro, *Theriogenology*, 24, 537, 1985.
168. **Parrish, J. J., Susko-Paris, J., Winer, M. A., and First, N. L.,** Capacitation of bovine sperm by heparin, *Biol. Reprod.*, 38, 1171, 1988.
169. **Delgado, N. M., Reyes, R., Huacuja, L., Merchant, H., and Rasado, A.,** Heparin binding sites in the human spermatozoal membrane, *Arch. Androl.*, 8, 87, 1982.
170. **Hendrow, R. R., Boehm, S. K., Lenz, R. W., Robinson, J. A., and Ax, R. L.,** Specific binding of the glycosaminoglycan ^{3}H-heparin to bull, monkey, and rabbit spermatozoa in vitro, *J. Androl.*, 5, 551, 1984.
171. **Bleil, J. D. and Wassarman, P. M.,** Structure and function of the zona pellucida: identification and characterization of the proteins of the mouse oocyte's zona pellucida, *Dev. Biol.*, 76, 185, 1980.
172. **Dunbar, B. S., Wardrip, N. J., and Hedrick, J. L.,** Isolation, physiochemical properties, and macromolecular composition of zona pellucida from porcine oocytes, *Biochemistry*, 19, 356, 1980.
173. **Dunbar, B. S., Liu, C., and Sammons, D. W.,** Identification of the three major proteins of porcine and rabbit zonae pellucidae by high resolution two-dimensional gel electrophoresis: comparsion with serum, follicular fluid, and ovarian cell proteins, *Biol. Reprod.*, 24, 1111, 1981.
174. **Sullivan, R. and Bleau, G.,** Interaction of isolated components from mammalian sperm and egg, *Gamete Res.*, 12, 101, 1985.
175. **Brown, C. R.,** The morphological and molecular susceptibility of sheep and mouse zona pellucida to acrosin, *J. Reprod. Fertil.*, 77, 411, 1986.
176. **Brown, C. R. and Cheng, W. K. T.,** Changes in composition of the porcine zona pellucida during development of the oocyte to the 2- to 4-cell embryo, *J. Embryol. Exp. Morphol.*, 92, 183, 1986.
177. **Oikawa, T., Sendai, Y., Kurata, S., and Yanagimachi, R.,** A glycoprotein of oviductal origin alters biochemical properties of the zona pellucida of hamster egg, *Gamete Res.*, 19, 113, 1987.
178. **Yurewicz, E. C., Sacco, A. G., and Subramanian, M. G.,** Structural characterization of the $M_r = 55{,}000$ antigen (ZP3) of porcine oocyte zona pellucida. Purification and characterization of alpha and beta-glycoproteins following digestion of lactosaminoglycan with endo-beta-galactosidase, *J. Biol. Chem.*, 262, 564, 1987.
179. **Shabanowitz, R. B. and O'Rand, M. G.,** Characterization of the human zona pellucida from fertilized and unfertilized eggs, *J. Reprod. Fertil.*, 82, 151, 1988.
180. **Shimizu, S., Tsuji, M., and Dean, J.,** In vitro biosynthesis of three sulfated glycoproteins of murine zonae pellucidae by oocytes grown in follicle culture, *J. Biol. Chem.*, 258, 5858, 1983.
181. **Greve, J. M. and Wassarman, P. M.,** Mouse egg extracellular coat is a matrix of interconnected filaments possessing a structural repeat, *J. Mol. Biol.*, 181, 253, 1985.
182. **Bleil, J. D., Greve, J. M., and Wassarman, P. M.,** Identification of a secondary sperm receptor in the mouse egg zona pellucida: role in maintenance of binding of acrosome-reacted sperm to eggs, *Dev. Biol.*, 128, 376, 1988.
183. **Bleil, J. D. and Wassarman, P. M.,** Mammalian sperm-egg interaction: fertilization of mouse eggs triggers modification of the major zona pellucida glycoprotein, ZP2, *Dev. Biol.*, 86, 189, 1981.
184. **Moller, C. C. and Wassarman, P. M.,** Characterization of a proteinase that cleaves zona pellucida glycoprotein ZP2 folllowing activation of mouse eggs, *Dev. Biol.*, 132, 103, 1989.
185. **Bleil, J. D. and Wassarman, P. M.,** Mammalian sperm-egg interaction: identification of a glycoprotein in mouse egg zonae pellucidae possessing receptor activity for sperm, *Cell*, 20, 873, 1980.

186. **Florman, H. M. and Wassarman, P. M.,** O-linked oligosaccharides of mouse egg ZP3 account for its sperm receptor activity, *Cell,* 41, 313, 1985.
187. **Bleil, J. D. and Wassarman, P. M.,** Galactose at the non-reducing terminus of O-linked oligosaccharides of mouse egg zona pellucida glycoprotein ZP3 is essential for the glycoprotein's sperm receptor activity, *Proc. Natl. Acad. Sci. U.S.A.,* 85, 6778, 1988.
188. **Wassarman, P. M., Bleil, J. D., Florman, H. M., Greve, J. M., and Roller, R. J.,** The mouse egg's receptor for sperm: what is it and how does it work?, *Cold Spring Harbor Symp. Quant. Biol.,* 50, 11, 1985.
189. **Ringuette, M. J., Chamberlin, M. E., Baur, A. W., Sobieski, D. A., and Dean, J.,** Molecular analysis of cDNA coding for ZP3, a sperm binding protein of the mouse zona pellucida, *Dev. Biol.,* 127, 287, 1988.
190. **Vazquez, M. H., Phillips, D. M., and Wassarman, P. M.,** Interaction of mouse sperm with purified sperm receptors covalently linked to silica beads, *J. Cell Sci.,* 92, 713, 1989.
191. **Gomperts, B. D. and Tatham, P. E. R.,** GTP-binding proteins in the control of exocytosis, *Cold Spring Harbor Symp. Quant. Biol.,* 53, 983, 1988.
192. **Hansbrough, J. R. and Garbers, D. L.,** Speract; purification and characterization of a peptide associated with eggs that activates spermatozoa, *J. Biol. Chem.,* 256, 2235, 1981.
193. **Suzuki, N., Nomura, K., Ohatake, H., and Isaka, S.,** Purification and the primary structure of sperm-activating peptides from the jelly coat of sea urchin eggs, *Biochem. Biophys. Res. Commun.,* 99, 1238, 1981.
194. **Garbers, D. L., Watkins, H. D., Hansbrough, J. R., Smith, A., and Misono, K. S.,** The amino acid sequence and chemical synthesis of speract analogues, *J. Biol. Chem.,* 257, 2734, 1982.
195. **Dangott, L. J., and Garbers, D. L.,** Identification and partial characterization of the receptor for speract, *J. Biol. Chem.,* 259, 13712, 1984.
196. **Shimomura, H., Dangott, L. J., and Garbers, D. L.,** Covalent coupling of a resact analogue to guanylate cyclase, *J. Biol. Chem.,* 261, 15778, 1986.
197. **Bentley, J. K. and Garbers, D. L.,** Retention of the speract receptor by isolated plasma membranes of sea urchin spermatozoa, *Biol. Reprod.,* 34, 413, 1986.
198. **Bentley, J. K., Shimomura, H., and Gabers, D. L.,** Retention of a functional resact receptor in isolated sperm plasma membranes, *Cell,* 45, 281, 1986.
199. **Bentley, J. K., Tubb, D. J., and Garbers, D. L.,** Receptor-mediated activation of spermatozoan guanylate cyclase, *J. Biol. Chem.,* 261, 14859, 1986.
200. **Bentley, J. K., Khatra, A. S., and Garbers, D. L.,** Receptor-mediated activation of detergent-solubilized guanylate cyclase, *Biol. Reprod.,* 39, 639, 1988.
201. **Ward, G. E., Brokaw, C. J., Garbers, D. L., and Vacquier, V. D.,** Chemotaxis of *Arbacia punctulata* spermatozoa to resact, a peptide from the egg jelly layer, *J. Cell Biol.,* 101, 2324, 1985.
202. **Kopf, G. S., Tubb, D. J., and Garbers, D. L.,** Activation of sperm respiration by a low molecular weight egg factor and by 8-bromoguanosine 3′,5′-monophosphate, *J. Biol. Chem.,* 254, 8554, 1979.
203. **Hansbrough, J. R. and Garbers, D. L.,** Sodium dependent activation of sea urchin spermatozoa by speract and monensin, *J. Biol. Chem.,* 256, 2235, 1981b.
204. **Garbers, D. L., Lowe, D. G., Dangott, L. J., Chinkers, M., Thorpe, D. S, Bentley, J. K., Ramarao, C. S., Goeddel, D. V., and Singh, S.,** The membrane form of guanylate cyclase, *Cold Spring Harbor Symp. Quant. Biol.,* 53, 993, 1988.
205. **Chinkers, M., Garbers, D. L., Chang, M. S., Lowe, D. G., Chin, H., Goeddel, D. V., and Schulz, S.,** A membrane form of guanylate cyclase is an atrial natriuretic peptide receptor, *Nature (London),* 338, 78, 1989.
206. **Thorpe, D. S. and Garbers, D. L.,** The membrane form of guanylate cyclase; homology with subunit of the cytoplasmic form of the enzyme, *J. Biol. Chem.,* 264, 6545, 1989.
207. **Singh, S., Lowe, D. G., Thorpe, D. S., Rodriguez, H., Juang, W. J., Dangott, L. J., Chinkers, M., Goeddel, D. V., and Garbers, D. L.,** Membrane guanylate cyclase is a cell-surface receptor with homology to protein kinases, *Nature (London),* 334, 708, 1988.
208. **Yarden, Y. and Ullrich, A.,** Growth factor receptory tyrosine kinases, *Annu. Rev. Biochem.,* 57, 443, 1988.
209. **Saling, P. M., Wolf, D. P., and Storey, B. T.,** Calcium-dependent binding of mouse epididymal spermatozoa to the zona pellucida, *Dev. Biol.,* 65, 515, 1978.
210. **Heffner, L. J., Saling, P. M., and Storey, B. T.,** Separation of calcium effects on motility and zona binding ability in mouse spermatozoa, *J. Exp. Zool.,* 212, 53, 1980.
211. **Florman, H. M., Saling, P. M., and Storey, B. T.,** Fertilization of mouse eggs in vitro: time resolution of the reactions preceding penetration of the zona pellucida, *J. Androl.,* 3, 373, 1982.
212. **Heffner, L. J. and Storey, B. T.,** Cold lability of mouse sperm binding to zona pellucida, *J. Exp. Zool.,* 219, 155, 1982.

213. **Saling, P. M.,** Development of the ability to bind to zonae pellucidae during epididymal maturation: reversible immobilization of mouse spermatozoa by lanthanum, *Biol. Reprod.,* 26, 429, 1982.
214. **Neill, J. M. and Olds-Clarke, P.,** Incubation of mouse sperm with lactate delays capacitation and hyperactivation and lowers fertilization levels in vitro, *Gamete Res.,* 20, 459, 1988.
215. **Soldani, P. and Rosati, F.,** Sperm-egg interaction in the mouse using live and glutaraldehyde-fixed eggs, *Gamete Res.,* 18, 225, 1987.
216. **Saling, P. M.,** Mammalian sperm interaction with extracellular matrices of the egg, in *Oxford Reviews of Reproductive Biology,* Milligan, S. R., Ed., Oxford University Press, Oxford, 11, 339, 1989.
217. **Gwatkin, R. B. L.,** Zona binding sites of the spermatozoa, in *The Role of the Zona Pellucida in Mammalian Reproduction,* Dietl, J., Ed., Springer-Verlag, Berlin, 1988.
218. **Peterson, R. N., Henry, L., Hunt, W., Saxena, N., and Russell, L. D.,** Further characterization of boar sperm plasma membrane proteins with affinity for the porcine zona pellucida, *Gamete Res.,* 12, 91, 1985.
219. **Peterson, R. N. and Hunt, W. P.,** Purification and further characterization of the boar sperm plasma membrane (PM) protein complex involved in sperm-egg recognition, *J. Cell Biol.,* 105, 107a, 1987.
220. **Peterson, R. N. and Hunt, W. P.,** Identification, isolation, and properties of a plasma membrane protein involved in the adhesion of boar sperm to the porcine zona pellucida, *Gamete Res.,* 23, 103, 1989.
221. **O'Rand, M. D., Matthews, J. E., Welch, J. E., and Fisher, S. J.,** Identification of zona binding proteins of rabbit, pig, human and mouse spermatozoa on nitrocellulose blots, *J. Exp. Zool.,* 235, 423, 1985.
222. **Brown, C. R. and Jones, R.,** Binding of zona pellucida proteins to a boar sperm polypeptide of M_r53,000 and identification of zona moieties involved, *Development,* 99, 333, 1987.
223. **Jones, R., Brown, C. R., and Lancaster, R. T.,** Carbohydrate-binding properties of boar sperm proacrosin and assessment of its role in sperm-egg recognition and adhesion during fertilization, *Development,* 102, 781, 1988.
224. **Leyton, L. and Saling, P.,** 95 kD sperm proteins bind ZP3 and serve as tyrosine kinase substrates in response to zona binding, *Cell,* 57, 123, 1989.
225. **Hartmann, J. F. and Hutchinson, C. F.,** Nature of the prepenetration contact interactions between hamster gametes in vitro, *J. Reprod. Fertil.,* 36, 49, 1974.
226. **Saling, P. M.,** Involvement of trypsin-like activity in binding of mouse spermatozoa to zonae pellucidae, *Proc. Natl. Acad. Sci. U.S.A.,* 78, 6231, 1981.
227. **Aarons, D., Speake, J. L., and Poirier, G. R.,** Evidence for a proteinase inhibitor binding component associated with murine spermatozoa, *Biol. Reprod.,* 31, 811, 1984.
228. **Benau, D. A. and Storey, B. T.,** Characterization of the mouse sperm plasma membrane zona-binding site sensitive to trypsin inhibitors, *Biol. Reprod.,* 36, 282, 1987.
229. **Poirier, G. R., Robinson, R., Richardson, R., Hinds, K., and Clayton, D.,** Evidence for a binding site of the sperm plasma membrane which recognizes the murine zona pellucida: a binding site on the sperm plasma membrane, *Gamete Res.,* 14, 235, 1986.
230. **Benau, D. A. and Storey, B. T.,** Relationship between two types of mouse sperm surface sites that mediate binding of sperm to the zona pellucida, *Biol. Reprod.,* 39, 235, 1988.
231. **Jones, R. and Brown, C. R.,** Identification of a zona-binding protein from boar spermatozoa as proacrosin, *Exp. Cell Res.,* 171, 503, 1987.
232. **Topfer-Peterson, E. and Henschen, A.,** Acrosin shows zona and fucose binding; novel properties for a serine proteinase, *FEBS Lett.,* 226, 38, 1987.
233. **Topfer-Peterson, E. and Henschen, A.,** Zona pellucida-binding and fucose-binding of boar sperm acrosin is not correlated with proteolytic activity, *Biol. Chem. Hoppe Seyler,* 369, 69, 1988.
234. **Jones, R.,** Evidence for boar sperm proacrosin as a carbohydrate binding protein, *Cell Biol. Int. Rep.,* 11, 833, 1987.
235. **Topfer-Peterson, E., Freiss, A. E., Nguyen, H., and Schill, W. B.,** Is a fucose-binding protein involved in sperm-egg interaction in the pig?, in *New Horizons in Sperm Cell Research,* Mohri, H., Ed., Japan Scientific Society Press, Tokyo, 1987, 287.
236. **Hedrick, J. L., Urch, U. A., and Hardy, D. M.,** Structure-function properties of the sperm enzyme acrosin, in *Biocatalysis in Agricultural Biotechnology,* Whitaker, J. R. and Sonnet, P. E., Eds., American Chemical Society, Washington, D.C., 1989, 215.
237. **Pierce, M., Turley, E. A., and Roth, S.,** Cell surface glycosyltransferase activities, *Int. Rev. Cytol.,* 65, 1, 1980.
238. **Ahuja, K. K.,** Carbohydrate determinants involved in mammalian fertilization, *Am. J. Anat.,* 174, 1, 1985.
239. **Shur, B. D. and Hall, N. G.,** Sperm surface galactosyltransferase activities during in vitro capacitation, *J. Cell. Biol.,* 95, 567, 1982.
240. **Shur, B. D. and Hall, N. G.,** A role for mouse sperm surface galactosyltransferase in sperm binding to the egg zona pellucida, *J. Cell Biol.,* 95, 574, 1982.
241. **Lopez, L. C., Bayna, E. M., Litoff, D., Shaper, N. L., Shaper, J. H., and Shur, B. D.,** Receptor function of mouse sperm surface galactosyltransferase during fertilization, *J. Cell Biol.,* 101, 1501, 1985.

242. **McLaughlin, J. D. and Shur, B. D.,** Binding of caput epididymal mouse sperm to the zona pellucida, *Dev. Biol.,* 124, 557, 1987.
243. **Lopez, L. C. and Shur, B. D.,** Redistribution of mouse sperm surface galactosyltransferase after the acrosome reaction, *J. Cell Biol.,* 105, 1663, 1987.
244. **Shur, B. D. and Neely, C. A.,** Plasma membrane association, purification, and partial characterization of mouse sperm β 1,4-galactosyltransferase, *J. Biol. Chem.,* 263, 17706, 1988.
245. **Scully, N. F., Shaper, J. H., and Shur, B. D.,** Spatial and temporal expression of cell surface galactosyltransferase during mouse spermatogenesis and epididymal maturation, *Dev. Biol.,* 124, 111, 1987.
246. **Benau, D. A., McGuire, E. J., and Storey, B. T.,** A further characterization of the mouse sperm zona-binding site with galactosyltransferase activity, *Biol. Reprod.,* 38, 142, 1988.
247. **Apter, F. M., Baltz, J. M., and Millette, C.F.,** A possible role for cell surface fucosyl transferase (FT) activity during sperm-zona pellucida binding in the mouse, *J. Cell Biol.,* 107, 175a, 1988.
248. **Cornwall, G. A., Tulsiani, D. R. P., and Orgebin-Crist, M. C.,** Inhibition of mouse sperm surface α-mannosidase inhibits sperm-egg binding, *Am. Soc. Androl. 14th Annu. Meet.,* p. 45, 1989.
249. **Cowan, A. E., Primakoff, P., and Myles, D. G.,** Sperm exocytosis increases the amount of PH-20 antigen on the surface of guinea pig sperm, *J. Cell Biol.,* 103, 1289, 1986.
250. **Primakoff, P. and Myles, D. G.,** Localized surface antigens of guinea pig sperm migrate to new regions prior to fertilization, *J. Cell Biol.,* 99, 1634, 1984.
251. **Primakoff, P., Hyatt, H., and Myles, D. G.,** A role for the migrating sperm surface antigen PH-20 in guinea pig sperm binding to the egg zona pellucida, *J. Cell Biol.,* 101, 2239, 1985.
252. **Primakoff, P., Cowan, A., Hyatt, H., Tredick-Kline, J., and Myles, D. G.,** Purification of the guinea pig sperm PH-20 antigen and detection of a site-specific endoproteolytic activity in sperm preparations that cleaves PH-20 into two disulfide-linked fragments, *Biol. Reprod.,* 38, 921, 1988.
253. **O'Rand, M. G., Widgren, E. E., and Fisher, S. J.,** Characterization of the rabbit sperm membrane autoantigen, RSA, as a lectin-like zona-binding protein, *Dev. Biol.,* 129, 231, 1988.
254. **O'Rand, M. G., Irons, G. P., and Porter, J. P.,** Monoclonal antibodies to rabbit sperm autoantigens. I. Inhibition of in vitro fertilization and localization on the egg, *Bio. Reprod.,* 30, 721, 1984.
255. **Lee, M. A., Kopf, G. S., and Storey, B. T.,** Effects of phorbol esters and a diacylglcyerol on the mouse sperm acrosome reaction induced by the zona pellucida, *Biol. Reprod.,* 36, 617, 1987.
256. **Florman, H. M. and Storey, B. T.,** Characterization of cholinomimetic agents that inhibit in vitro fertilization in the mouse, *J. Androl.,* 3, 157, 1982.
257. **Florman, H. M. and Storey, B. T.,** Inhibition of in vitro fertilization of mouse eggs: 3-quinuclidinyl benzilate specifically blocks penetration of zonae pellucidae by mouse spermatozoa, *J. Exp. Zool.,* 216, 159, 1981.
258. **Saling, P. M. and Lakoski, K. A.,** Mouse sperm antigens that participate in fertilization. II. Inhibition of sperm penetration through the zona pellucida using monoclonal antibodies, *Biol. Reprod.,* 33, 527, 1985.
259. **Saling, P. M. and Lakoski, K. A.,** Mouse sperm antigens that participate in fertilization. II. Inhibition of sperm penetration through the zona pellucida using monoclonal antibodies, *Biol. Reprod.,* 33, 527, 1985.
260. **Leyton, L., Robinson, A., and Saling, P. M.,** Relationship between the M42 antigen of mouse sperm and the acrosome reaction induced by ZP3, *Dev. Biol.,* 132, 174, 1989.
261. **Endo, Y., Lee, M. A., and Kopf, G. S.,** Evidence for the role of a guanine nucleotide-binding regulatory protein in the zona pellucida-induced mouse sperm acrosome reaction, *Dev. Biol.,* 119, 210, 1987.
262. **Casey, P. J. and Gilman, A. G.,** G protein involvement in receptor-effector coupling, *J. Biol. Chem.,* 263, 2577, 1988.
263. **Kopf, G. S., Woolkalis, M. J., and Gerton, G. L.,** Evidence for a guanine nucleotide-binding regulatory protein in invertebrate and mammalian sperm: identification by islet-activating protein-catalyzed ADP-ribosylation and immunochemical methods, *J. Biol. Chem.,* 261, 7327, 1986.
264. **Endo, Y., Schultz, R. M., and Kopf, G. S.,** Effects of phorbol esters and a diacylglycerol on mouse eggs: inhibition of fertilization and modification of the zona pellucida, *Dev. Biol.,* 119, 199, 1987.
265. **Endo, Y., Mattei, P., Kopf, G. S., and Schultz, R. M.,** Effects of a phorbol ester on mouse eggs: dissociation of sperm receptor activity from acrosome reaction-inducing activity of the mouse pellucida protein, ZP3, *Dev. Biol.,* 123, 574, 1987.
266. **Ducibella, T., Kopf, G. S., and Schultz, R. M.,** unpublished data.
267. **Kopf, G. S., Endo, Y., Mattei, P., Kurasawa, S., and Schultz, R. M.,** Egg-induced modifications of the murine *zona pellucida,* in *Mechanisms of Egg Activation,* Nuccitelli, R. L., Clark, W. H., and Cherr, G. N., Eds., Plenum Press, New York, 1989, 249.
268. **Leyton, L. and Saling, P.,** Evidence that aggregation of mouse sperm receptors by ZP3 triggers the acrosome reaction, *J. Cell Biol.,* 108, 2163, 1989.
269. **Chen, J. J., Kowsower, N. S., Petryshyn, R., and London, I. M.,** The effects of N-ethylmaleimide on the phosphorylation and aggregation of insulin receptors in the isolated plasma membranes of 3T3-F442A adipocytes, *J. Biol. Chem.,* 261, 902, 1986.

270. **Heffetz, D. and Zick, Y.,** Receptor aggregation is necessary for activation of the soluble insulin receptor kinase, *J. Biol. Chem.,* 261, 889, 1986.
271. **Ledbetter, J. A., June, C. H., Grosmaire, L. S., and Rabinovitz, P. S.,** Crosslinking of surface antigens causes mobilization of intracellular calcium in T lymphocytes, *Proc. Natl. Acad. Sci. U.S.A.,* 84, 1384, 1987.
272. **O'Brien, R. M., Soos, M. A., and Siddle, K.,** Monoclonal antibodies to the insulin receptor stimulate the intrinsic protein kinase activity by cross-linking receptor molecules, *EMBO J.,* 6, 4003, 1987.
273. **Yarden, Y. and Schlessinger, J.,** Epidermal growth factor induces rapid, reversible aggregation of the purified epidermal growth factor receptor, *Biochemistry,* 26, 1443, 1987.
274. **Godfrey, E. W., Dietz, M. E., Morstad, A. L., Wallskog, P. A., and Yorde, D. E.,** Acetylcholine receptor-aggregating proteins are associated with the extracellular matrix of many tissues in Torpedo, *J. Cell Biol.,* 106, 1263, 1988.
275. **Gilman, A. G.,** G proteins: transducers of receptor-generated signals, *Annu. Rev. Biochem.,* 56, 615, 1987.
276. **Stryer, L.,** Cyclic GMP cascade of vision, *Annu. Rev. Neurosci.,* 9, 87, 1986.
277. **Mattera, R., Yatani, A., Kirsch, G. E., Graf, R., Okabe, K., Olate, J., Codina, J., Brown, A. K., and Birnbaumer, L.,** Recombinant α_{i3} subunit of G protein activates G_k-gated K channels, *J. Biol. Chem.,* 264, 465, 1989.
278. **Cockcroft, S.,** Polyphosphoinositide phosphodiesterase: regulation by a novel guanine nucleotide binding protein, G_p, *Trends Biochem. Sci.,* 12, 75, 1987.
279. **Jelsema, C. L. and Axelrod, J.,** Simulation of phospholipase A_2 activity in bovine rod outer segments by the beta gamma subunits of transducin and its inhibition by the alpha subunit, *Proc. Natl. Acad. Sci. U.S.A.,* 84, 3623, 1987.
280. **Kim, D., Lewis, D. L., Graziadei, L., Neer, E. J., Bar-Sagi, D., and Clapham, D. E.,** G-protein $\beta\gamma$-subunits activate the cardiac muscarinic K^+-channel via phospholipase A_2, *Nature (London),* 337, 557, 1987.
281. **Bocckino, S. B., Blackmore, P. F., Wilson, P. B., and Exton, J. H.,** Phosphatidate accumulation in hormone treated hepatocytes via a phospholipase D mechanism, *J. Biol. Chem.,* 262, 15309, 1987.
282. **Boccino, S. B., Wilson, P. B., and Exton, J. H.,** Ca^{2+} mobilizing hormones elicit phosphatidylethanol accumulation via phospholipase D activation, *FEBS Lett.,* 225, 201, 1987.
283. **Hescheler, J., Rosenthal, W. Trautwein, W., and Schultz, G.,** The GTP binding protein, Go, regulates neuronal calcium channels, *Nature (London),* 325, 445, 1987.
284. **Mattera, R., Graziano, M. P., Yatani, A., Zhou, Z., Graf, R., Codina, J., Birnbaumer, L., Gilman, A. G., and Brown, A. K.,** Splice variants of the α-subunit of the G protein G_s activate both adenylate cyclase and calcium channels, *Science,* 243, 804, 1989.
285. **Tash, J. S. and Means, A. R.,** Cyclic adenosine 3′,5′-monophosphate, calcium and protein phosphorylation in flagellar motility, *Biol. Reprod.,* 28, 75, 1983.
286. **Ross, E. M. and Gilman, A. G.,** Biochemical properties of hormone-sensitive adenylate cyclase, *Annu. Rev. Biochem.,* 49, 533, 1980.
287. **Hildebrandt, J. D., Codina, J., Tash, J. S., Kirchick, H. J., Lipschultz, L., Sekura, R. D., and Birnbaumer, L.,** The membrane-bound spermatozoal adenylyl cyclase system does not share coupling characteristics with somatic cell adenylyl cyclases, *Endocrinology,* 116, 1357, 1985.
288. **Cheng, C. Y. and Boettcher, B.,** Effects of cholera toxin and 5′-guanylylimidodiphosphate on human spermatozoal adenylate cyclase activity, *Biochem. Biophys Res. Commun.,* 91, 1, 1979.
289. **Casillas, E. R., Elder, C. M., and Hoskins, D. D.,** Adenylate cyclase activity of bovine spermatozoa during maturation in the epididymis and the activation of sperm particulate adenylate cyclase by GTP and polyamines, *J. Reprod. Fertil.,* 59, 297, 1980.
290. **Hyne, R. V. and Lopata, A.,** Calcium and adenosine affect human sperm adenylate cyclase activity, *Gamete Res.,* 6, 81, 1982.
291. **Stein, D. M., Fraser, L. R., and Monks, N. J.,** Adenosine and Gpp(NH)p modulate mouse sperm adenylate cyclase, *Gamete Res.,* 13, 151, 1986.
292. **Kopf, G. S. and Garbers, D. L.,** Calcium and a fucose sulfate-rich polymer regulate sperm cyclic nucleotide metabolism and the acrosome reaction, *Biol. Reprod.,* 22, 1118, 1980.
293. **Kopf, G. S. and Vacquier, V. D.,** Characterization of a calmodulin-stimulated adenylate cyclase from abalone spermatozoa, *J. Biol. Chem.,* 259, 7590, 1984.
294. **Kopf, G. S. and Vacquier, V. D.,** Characterization of a calcium-modulated adenylate cyclase from abalone spermatozoa, *Biol. Reprod.,* 33, 1094, 1985.
295. **Gross, M. K., Toscano, D. G., and Toscano, W. A.,** Calmodulin-mediated adenylate cyclase from mammalian sperm, *J. Biol. Chem.,* 262, 8672, 1987.
296. **Bentley, J. K., Garbers, D. L., Domino, S. E., Noland, T. D., and VanDop, C.,** Spermatozoa contain a guanine nucleotide binding protein ADP-ribosylated by pertussis toxin, *Biochem. Biophys. Res. Commun.,* 138, 728, 1986.

297. **Jones, J., Kopf, G. S., and Schultz, R. M.,** Variability in electrophoretic mobility of G_i-like proteins; effect of SDS, *FEBS Lett.*, 243, 409, 1989.
298. **Gilman, A. G.,** G proteins and dual control of adenylate cyclase, *Cell*, 36, 577, 1984.
299. **Florman, H. M., Tombes, R. M., First, N. L., and Babcock, D. F.,** An adhesion-associated agonist from the zona pellucida activates G protein-promoted elevations of internal Ca^{2+} and pH that mediate mammalian sperm acrosomal exocytosis, *Dev. Biol.*, 135, 133, 1989.
300. **Garty, N. B. and Salomon, Y.,** Stimulation of partially purified adenylate cyclase from bull sperm by bicarbonate, *FEBS Lett.*, 218, 148, 1987.
301. **Lee, H. C.,** Internal GTP stimulates the speract receptor mediated voltage changes in sea urchin spermatozoa membrane vesicles, *Dev. Biol.*, 126, 91, 1988.
302. **Yanagimachi, R. and Usui, N.,** Calcium dependence of the acrosome reaction and activation of guinea pig spermatozoa, *Exp. Cell Res.*, 89, 161, 1974.
303. **Yanagimachi, R.,** Acceleration of the acrosome reaction and activation of guinea pig spermatozoa by detergents and other agents, *Biol. Reprod.*, 13, 519, 1975.
304. **Green, D. P. L.,** The induction of the acrosome reaction in guinea pig sperm by the divalent cation ionophore A23187, *J. Cell Sci.*, 32, 137, 1978.
305. **Triana, L. R., Babcock, D. F., Lorton, S. P., and Lardy, H. A.,** Release of acrosomal hyaluronidase follows increased membrane permeability to calcium in the presumptive capacitation sequence for spermatozoa of the bovine and other mammalian species, *Biol. Reprod.*, 23, 47, 1980.
306. **Hyne, R. V. and Garbers, D. L.,** Calcium-dependent increase in adenosine, 3′,5′-monophosphate and induction of the acrosome reaction in guinea pig spermatozoa, *Proc. Natl. Acad. Sci. U.S.A.*, 76, 5699, 1979.
307. **Singh, J. P., Babcock, D. F., and Lardy, H. A.,** Induction of accelerated acrosome reaction in guinea pig sperm, *Biol. Reprod.*, 22, 566, 1980.
308. **Kazazoglou, T., Schackmann, R. W., Fosset, M., and Shapiro, B. M.,** Calcium channel antagonists inhibit the acrosome reaction and bind to plasma membranes of sea urchin sperm, *Proc. Natl. Acad. Sci. U.S.A.*, 82, 1462, 1985.
309. **Thomas, P. and Meizel, S.,** An influx of extracellular calcium is required for initiation of the human sperm acrosome reaction induced by human follicular fluid, *Gamete Res.*, 20, 397, 1988.
310. **Hyne, R. V.,** Bicarbonate and calcium-dependent induction of rapid guinea pig sperm acrosome reactions by monovalent ionophores, *Biol. Reprod.*, 31, 312, 1984.
311. **Roldan, E. R. S., Shibata, S., and Yanagimachi, R.,** Effect of Ca^{2+} channel antagonists on the acrosome reaction of guinea pig and golden hamster spermatozoa, *Gamete Res.*, 13, 281, 1986.
312. **Working, P. K. and Meizel, S.,** Correlation of increased intraacrosomal pH with the hamster sperm acrosome reaction, *J. Exp. Zool.*, 227, 97, 1983.
313. **Murphy, S. J. and Yanagimachi, R.,** The pH dependence of motility and the acrosome reaction of guinea pig spermatozoa, *Gamete Res.*, 10, 1, 1984.
314. **Hyne, R. V. and Garbers, D. L.,** Requirement of serum factors for capacitation and the acrosome reaction of guinea pig spermatozoa in buffered medium below pH 7.8, *Biol. Reprod.*, 24, 257, 1981.
315. **Babcock, D. F. and Pfeiffer, D. F.,** Independent elevation of cytosolic [Ca^{2+}] and pH of mammalian sperm by voltage-dependent and pH sensitive mechanisms, *J. Biol. Chem.*, 262, 15041, 1987.
316. **Hyne, R. V., Higginson, R. E., Kohlman, D., and Lopata, A.,** Sodium requirement for capacitation and membrane fusion during the guinea pig sperm acrosome reaction, *J. Reprod. Fertil.*, 70, 83, 1984.
317. **Ruknudin, A.,** Cytochemical study of intracellular calcium in hamster spermatozoa during the acrosome reaction, *Gamete Res.*, 22, 375, 1989.
318. **Rufo, G. A., Schoff, P. K., and Lardy, H. A.,** Regulation of calcium content in bovine spermatozoa, *J. Biol. Chem.*, 259, 2547, 1984.
319. **Mrsny, R. J. and Meizel, S.,** Potassium ion influx and Na^+, K^+-ATPase activity are required for the hamster sperm acrosome reaction, *J. Cell Biol.*, 91, 77, 1981.
320. **Boatman, D. E. and Bavister, B. D.,** Regulation of hamster sperm capacitation by bicarbonate ion-carbon dioxide, *J. Cell Biol.*, 97, 12a, 1983.
321. **Lee, M. A. and Storey, B. T.,** Bicarbonate is essential for fertilization of mouse eggs: mouse sperm require it to undergo the acrosome reaction, *Biol. Reprod.*, 34, 349, 1986.
322. **Madshus, I. H. and Olsnes, S.,** Selective inhibition of sodium-linked and sodium-independent bicarbonate/chloride antiport in Vero cells, *J. Biol. Chem.*, 262, 7486, 1987.
323. **Jentsch, T. J., Stahlknecht, T. R., Hollwede, H., Fischer, D. G., Keller, S. K., and Wiederholt, M.,** A bicarbonate-dependent process inhibitable by disulfonic stilbenes and a Na^+/H^+ exchange mediate $^{22}Na^+$ uptake into cultured bovine corneal endothelium, *J. Biol. Chem.*, 260, 795, 1985.

324a. **Garbers, D. L., Tubb, D. J., and Hyne, R. V.,** A requirement of bicarbonate for Ca^{2+}-induced elevations of cyclic AMP in guinea pig spermatozoa, *J. Biol. Chem.*, 257, 8980, 1982.

324b. **Garbers, D. L., Tubb, D. J., and Kopf, G. S.,** Regulation of sea urchin sperm cyclic AMP dependent protein kinases by an egg-associated factor, *Biol. Reprod.*, 22, 526, 1980.

325. **Stein, D. M. and Fraser, L. R.,** Cyclic nucleotide metabolism in mouse epididymal spermatozoa during capacitation in vitro, *Gamete Res.*, 10, 283, 1984.
326. **Kopf, G. S., Garbers, D. L., and Noland, T. D.,** unpublished data.
327. **Noland, T. D., Garbers, D. L., and Kopf, G. S.,** An elevation in cyclic AMP concentration precedes the *zona pellucida*-induced acrosome reaction of mouse spermatozoa, *Biol. Reprod.*, 38 (Suppl.), 9, 1988.
328. **Watkins, H. D., Kopf, G. S., and Garbers, D. L.,** Activation of sperm adenylate cyclase by factors associated with eggs, *Biol. Reprod.*, 19, 890, 1978.
329. **Hyne, R. V. and Garbers, D. L.,** Regulation of guinea pig sperm adenylate cyclase by calcium, *Biol. Reprod.*, 21, 1135, 1979.
330. **Wasco, W. M. and Orr, G. A.,** Function of calmodulin in mammalian sperm: presence of calmodulin-dependent cyclic nucleotide phosphodiesterase associated with demembranated rat cauda epididymal sperm, *Biochem. Biophys. Res. Commun.*, 118, 636, 1984.
331. **Jones, H. P., Bradford, M. M., McRorie, R. A., and Cormier, M. J.,** High levels of a calcium-dependent modulator protein in spermatozoa and its similarity to brain modulator protein, *Biochem. Biophys. Res. Commun.*, 82, 1264, 1978.
332. **Garbers, D. L., Hansbrough, J. R., Radany, E. W., Hyne, R. V., and Kopf, G. S.,** Purification and characterization of calmodulin from sea urchin spermatozoa, *J. Reprod. Fertil.*, 59, 377, 1980.
333. **Feinberg, J., Weinman, J., Weinman, S., Walsh, M. P., Harrican, M. C., Gabrion, J., and Demaille, J. G.,** Immunocytochemical and biochemical evidence for the presence of calmodulin in bull sperm flagellum: isolation and characterization of sperm calmodulin, *Biochim. Biophys. Acta*, 673, 303, 1981.
334. **Okamura, N. and Sugita, Y.,** Activation of spermatozoan adenylate cyclase by a low molecular weight factor in porcine seminal plasma, *J. Biol. Chem.*, 258, 13056, 1983.
335. **Okamura, N., Tajima, Y., Soejima, A., Masuda, H., and Sugita, Y.,** Sodium bicarbonate in seminal plasma stimulates the motility of mammalian spermatozoa through direct activation of adenylate cyclase, *J. Biol. Chem.*, 260, 9699, 1985.
336. **Wolf, D. E., Hagopian, S. S., and Ishijima, S.,** Changes in sperm plasma membrane lipid diffusibility after hyperaction during in vitro capacitation in the mouse, *J. Cell Biol.*, 102, 1372, 1986.
337. **Papahadjopoulos, D., Hui, S., Vail, W. J., and Poste, G.,** Studies on membrane fusion. I. Interactions of pure phospholipid membranes and the effect of myristic acid, lysolecithin, proteins and DMSO, *Biochem. Biophys. Acta*, 448, 245, 1976.
338. **Fleming, A. D. and Yanagimachi, R.,** Effects of various lipids on the acrosome reaction and fertilizing capacity of guinea pig spermatozoa, with special reference to the possible involvement of lysophospholipid in the acrosome reaction, *Gamete Res.*, 4, 253, 1981.
339. **Llanos, M. N., Lui, C. W., and Meizel, S.,** Studies of phospholipase A_2 related to the hamster sperm acrosome reaction, *J. Exp. Zool.*, 221, 107, 1982.
340. **Ohzu, E. and Yanagimachi, R.,** Acceleration of the acrosome reaction in hamster spermatozoa by lysolecithin, *J. Exp. Zool.*, 224, 259, 1982.
341. **Berridge, M. J.,** Inositol trisphosphate and diacylglycerol: two interacting second messengers, *Annu. Rev. Biochem.*, 56, 159, 1987.
342. **Amy, C. M. and Kirshner, N.,** Phosphorylation of adrenal medulla cell proteins in conjunction with stimulation of catecholamine secretion, *J. Neurochem.*, 36, 847, 1981.
343. **Creutz, C. E., Dowling, L. G., Kyger, E. M., and Franson, R. C.,** Phosphatidylinositol-specific phospholipase C activity of chromaffin granule-binding proteins, *J. Biol. Chem.*, 260, 7171, 1985.
344. **Summers, T. A. and Creutz, C. E.,** Phosphorylation of a chromaffin granule-binding protein by protein kinase C, *J. Biol. Chem.*, 260, 2437, 1985.
345. **Domino, S. E. and Garbers, D. L.,** The fucose-sulfate glycoconjugate that induces an acrosome reaction in spermatozoa stimulates inositol 1,4,5-trisphosphate accumulation, *J. Biol. Chem.*, 263, 690, 1988.
346. **Roomans, G. M.,** Calcium binding to the acrosomal membrane of human spermatozoa, *Exp. Cell Res.*, 96, 23, 1975.
347. **Berruti, G. and Franchi, E.,** Calcium and phosphoinositides: their distribution in relation to the membrane changes occurring in the head of boar spermatozoa, *Eur. J. Cell Biol.*, 41, 238, 1986.
348. **Irvine, R. F. and Moor, R. M.,** Micro-injection of inositol 1,3,4,5-tetrakisphosphate activates sea urchin eggs by a mechanism dependent on external Ca^{2+}, *Biochem. J.*, 240, 917, 1986.
349. **Kuno, M. and Gardner, P.,** Ion channels activated by inositol 1,4,5-trisphosphate in plasma membrane of human T-lymphocytes, *Nature*, 326, 301, 1986.
350. **von Tscharner, V., Prod'hom, B., Baggiolini, M., and Reuter, H.,** Ion channels in human neutrophils activated by a rise in free cytosolic calcium concentration, *Nature*, 324, 369, 1986.
351. **Kopf, G. S. and Girard, P.,** unpublished data, 1989.
352. **Endo, Y., Komatsu, S., Hirai, M., Suzuki, S., Iizuka, R., and Shimizu, N.,** Protein kinase C activity and protein phosphorylation in mouse sperm, submitted.
353. **Ribbes, H., Plantavid, M., Bennet, P. J., Chap, H., and Douste-Blazy, L.,** Phospholipase C from human sperm specific for phosphoinositides, *Biochim. Biophys. Acta*, 912, 245, 1987.

354. **Bennet, P. J., Moatti, J. P., Mansat, A., Ribbes, H., Cayrac, J. C., Pontonnier, F., Chap, H., and Douste-Blazy, L.,** Evidence for the activation of phospholipases during acrosome reaction of human sperm elicited by calcium ionophore A23187, *Biochim. Biophys. Acta,* 919, 255, 1987.
355. **Barritt, G. L., Dalton, K. A., and Whiting, J. A.,** Evidence that phosphatidic acid stimulates the uptake of calcium by liver cells but not calcium release from mitochondria, *FEBS Lett.,* 125, 137, 1981.
356. **Pelech, S. L. and Vance, D. E.,** Signal transduction via phosphatidylcholine cycles, *Trends Biochem. Sci.,* 14, 28, 1989.
357. **Roldan, E. R. S. and Harrison, R. A. P.,** Polyphosphoinositide breakdown and subsequent exocytosis in the Ca^{2+}/ionophore-induced acrosome reaction of mammalian spermatozoa, *Biochem. J.,* 259, 397, 1989.
358. **Domino, S. E., Bocckino, S. B., and Garbers, D. L.,** Activation of phospholipase D by the fucose-sulfate glycoconjugate that induces an acrosome reaction in spermatozoa, *J. Biol. Chem.,* 264, 9412, 1989.
359. **Wilde, M. W., Carlson, K. E., Manning, D. R., and Zigmond, S. H.,** Chemoattractant-stimulated GTPase activity is decreased on membranes from polymorphonuclear leukocytes incubated in chemoattractant, *J. Biol. Chem.,* 264, 190, 1989.

Chapter 8

MAMMALIAN EGG CORTICAL GRANULES AND THE CORTICAL REACTION

Thomas Ducibella

TABLE OF CONTENTS

I. INTRODUCTION

The blocks of polyspermic fertilization are necessary to prevent the incorporation of two sperm nuclei into the zygote's genome which results in abnormal development. In mammals, triploid development is usually prevented by a combination of the fertilized egg's block to polyspermy and low sperm number in the fallopian tube. However, despite these natural deterrents, greater than 1% of all recognized natural human conceptions and 5% of those from *in vitro* fertilization result in triploid embryonic abortions, the majority from dispermic fertilization.[1-5] With the increasing use of reproductive technologies (e.g., *in vitro* fertilization and gamete intrafallopian transfer) in which the sperm concentration is about 100 to 1000 times that in the oviduct during natural conception, the egg must rely exclusively on its blocks to polyspermy. Because the prevention of polyspermy is crucial to ensure reproductive success in other organisms, similar sperm exclusion mechanisms appear to have evolved in many invertebrates and vertebrates (recently reviewed in nonmammalian species[6]).

Although evidence[7,8] was presented for a mammalian block to polyspermy over 30 years ago, little is known about the origin, number, and biochemistry of the different mechanisms the fertilized mammalian egg employs. There are four potential blocks which have been identified in studies of other species (Figure 1). These include (1) changes in the extracellular investment (analogous to the zona pellucida) of the egg, (2) inhibitory substances to sperm in the milieu between this outer investment and the egg (perivitelline space), (3) the egg plasma membrane block, and (4) within the egg's cytoplasm, prevention of incorporation of more than one sperm nucleus into the embryonic genome when two or more sperm enter the egg.

Studies to date provide evidence that mammalian eggs utilize both an extracellular zona block to polyspermy as well as a plasma membrane block. The relative contributions of these blocks in different species appear to vary profoundly. Although little is known about the plasma membrane block,[6] recently our knowledge has increased about how the zona becomes modified and prevents polyspermy. Immediately after fertilization, the egg undergoes the cortical reaction which releases substances which appear to modify the zona.[9] The cortical reaction involves the release of cortical granule contents and perhaps other substances from the egg's surface. Since the biochemistry of the cortical reaction is understood in far greater detail in sea urchin and amphibian eggs, a brief review of this work is necessary to put the mammalian studies in perspective.

II. THE CORTICAL REACTION: IMPORTANT PATHWAYS ILLUSTRATED BY NONMAMMALIAN EGGS

From studies of sea urchins and amphibians, it is known that the oocyte cortex houses most, if not all, of the necessary components required for the cortical reaction. Many studies are consistent with the following pathway for cortical activation (see reviews[10-18]). After fertilization, an increase in egg inositol 1,4,5-trisphosphate (IP_3), activated by sperm-egg interaction, appears to stimulate the release of intracellular calcium sequestered in a cortical network of endoplasmic reticulum. This elevated concentration of cytoplasmic free calcium ion has been demonstrated to activate the secretion of mature egg cortical granules or their disruption in preparations of isolated cortex. The rise of cortical calcium occurs transiently as a wave which spreads radially from the point of sperm penetration throughout the cortex. Cortical granule exocytosis appears to follow immediately behind the calcium wave. Exocytosis appears to occur in a manner typical of other secretory cells with the granule membrane fusing with the overlying plasma membrane. Thus the surface area of the oocyte is significantly increased by the added cortical granule membrane, but compensation is made by increased endocytosis of membrane and changes in microvilli which may be necessary for

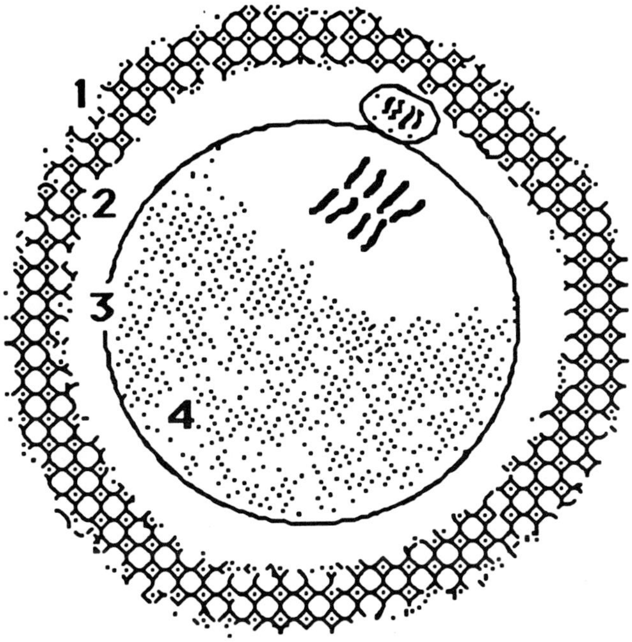

FIGURE 1. Diagram of a metaphase II mammalian egg with the four potential sites which could constitute blocks to abnormal polyspermic development: (1) zona pellucida, (2) perivitelline space, (3) egg plasma membrane, and (4) egg cytoplasm. See text for a discussion. The first polar body and cortical granules are schematically illustrated.

normal development.[19,20] Other cortical vesicles, distinct from cortical granules, become modified after egg activation and may play a role in intracellular pH regulation.[21] The increase in cytoplasmic pH after fertilization is necessary for the stimulation of important synthetic pathways and pronuclear movements in the zygote (see reviews[12,15,22]).

The dispersal of the contents of the cortical granules and the release of other substances from the oocyte initiate development of the extracellular blocks to polyspermy in sea urchins and amphibians.[6,17,23] In the sea urchin, an extracellular coat closely apposed to the oocyte plasma membrane begins to elevate about 30 s after fertilization in response to cortical granule release. Cortical granule proteases cause delamination of the vitelline envelope from the plasma membrane and release of bound supernumery sperm by digesting sperm receptors on this investment. Another enzyme, ovoperoxidase, catalyzes dityrosine cross-linking and hardens what is now called the fertilization envelope. However, all participants in fertilization envelope assembly are not derived from cortical granules. For example, H_2O_2 is not released from cortical granules, but other regions of the oocyte; and transglutaminase is activated on the egg cell surface.[23,24] In the sea urchin, within about a minute of sperm attachment, the fertilization envelope block to polyspermy is effective. While hardening and thickening may play a protective role, the events involved in envelope elevation appear to be sufficient to prevent polyspermy. Cortical granules also release hyalin which forms a new extracellular coat important for maintaining cell contact during cleavage.

In the sea urchin, these so called "slow" blocks resulting from the cortical reaction may not develop sufficiently rapidly after sperm-egg fusion to prevent polyspermy due to the relatively high concentration of sperm during insemination. During this time lag prior to fertilization envelope elevation, two other blocks have been shown to be effective. One of these is the "fast" plasma membrane block to polyspermy which occurs within 3 s of fertilization and may be related to the depolarization of the oocyte plasma membrane from the influx of sodium. This depolarization lasts about 10 min. The molecular mechanism of this blockade remains to be established. Another protective mechanism may involve the spermicidal agent H_2O_2, released by the oocyte within 15 s of insemination. The oocyte protects itself from this dangerous oxidant by producing high concentrations of cytoplasmic ovothiol.[25]

In anuran amphibians (frogs), available evidence is consistent with a similar cortical G-protein, IP_3, calcium-mediated activation of the cortical reaction.[18,26] In addition, the other G-protein pathway leading to activation of protein kinase C may be involved.[27] Instead of separating a thin envelope attached to the plasma membrane, the oocyte releases its cortical granules into a preexisting perivitelline space. Soon thereafter, the overlying vitelline envelope is elevated farther from the plasma membrane and its outer surface becomes impermeable to sperm. This fertilization envelope's block appears to be due to the localized, peripheral binding of a cortical granule-derived lectin (reviewed by Schmell et al.[13]). Another enzyme cleaves two envelope proteins. The resulting fertilization envelope acts as a mechanical barrier to sperm penetration, fails to induce the acrosomal reaction (normally occurring in the vitelline envelope), appears to lack sperm binding sites, and is resistant to sperm lysins. Frogs also have a fast electrical membrane block to polyspermy which is followed by the extracellular block described above. Urodele amphibians (salamanders) are unusual in that they normally become polyspermic, but exclude more than one sperm nucleus from the embryonic genome. An analogous rare event has been reported in mammals.[5,28]

The greater depth of research on sea urchin and anuran fertilization has provided us with working models of the cortical reaction including similar mechanisms of activation of the cortical reaction and release of cortical granule constituents. Although the biochemistry of the extracellular envelope block to polyspermy is different in these two organisms, it is mediated by the action of cortical granule contents. These paradigms have shaped our relatively meager present knowledge about the cortical reaction and block to polyspermy in mammals.

III. THE CORTICAL REACTION IN MAMMALS

A. THE ORGANIZATION OF THE CORTEX

With the exception of the region overlying the metaphase II spindle, the mature mammalian egg cortex has important structural similarities with that of sea urchins and amphibians.[29] Most of the cortex (region ≤ 2 μm from the plasma membrane) includes cortical granules, a variety of other membrane systems (discussed later), and localizations of cytoskeletal proteins.

In both the mouse and hamster, the most well-studied species, a smaller domain of cortex associated with the spindle chromosomes is free of cortical granules and other organelles (Figure 2).[30-38] Its overlying plasma membrane has different properties than the remaining egg membrane. It is free of microvilli,[39] has reduced Concanavalin A fluorescent staining perhaps due to the lack of microvillous surface amplification,[40] and membrane protein lateral diffusibility is also different.[41] These membrane characteristics are associated with a change in the organization and distribution of actin.[35] In the spindle-associated cortex, microvillous core filaments are not observed and there appears to be higher density of actin than in the surrounding cortical granule-occupied cortex from both electron microscopic[31-33] and fluorescence studies.[35] By virtue of this difference in cortical structure including the lack of cortical granules and vesicles, this region in the mouse and hamster may not participate in a complete cortical reaction at fertilization. Whether additional mosaicism of the egg cortex and plasma membrane is present remains to be investigated.

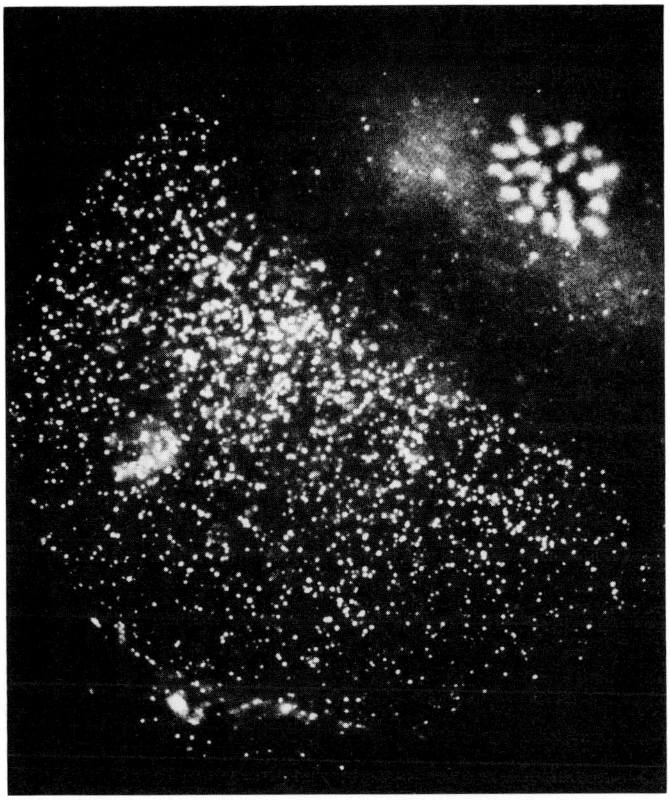

FIGURE 2. Mouse egg demonstrating lectin-stained cortical granules and metaphase II chromosomes stained with a fluorescent lectin probe.[36,38] The cortical granules (about 0.3 μm in diameter) are punctate spots which have a similar localization, distribution, and size as reported in electron microscopic studies. A cortical granule-free domain of cortex is always associated with the metaphase chromosomes.

B. FERTILIZATION: THE ROLE OF THE SPERM IN INITIATING CORTICAL GRANULE RELEASE

After penetration of the cumulus and zona in the rat, mouse, and hamster, the fertilizing sperm's head almost immediately makes contact with the egg surface, adheres to it, and initiates the block to polyspermy.[8] The interaction between the sperm and egg membranes appears to require at least some specificity in order to initiate cortical granule release. In the hamster, uncapacitated spermatozoa fail to elicit a cortical reaction in zona-free oocytes.[42] The necessary changes in the sperm postacrosomal membrane, which fuses with the egg upon fertilization, may occur during capacitation or at the time of the acrosomal reaction after binding to the zona. Cell surface interactions appear to be sufficient to cause the cortical reaction and sperm penetration into the cortical cytoplasm may not be required.[43] If this is the case, occupation of a sperm receptor on the egg surface could lead to a transmembrane G-protein/phosphoinositide cascade-mediated cortical reaction for which there is experimental evidence in amphibians.[26] Alternatively, sperm-egg fusion may result in the passage of a sperm cytoplasmic constituent which initiates the events leading to the cortical reaction. The molecular interactions between the acrosome-reacted sperm and egg membranes remain unelucidated.

In hamster eggs, experiments show that heterologous sperm[43,44] and a variety of treatments (including lectins, neuriminidase, the monovalent K ionophore boromycin, and a guanidine compound) can cause cortical granule loss in the hamster egg.[43] Although it is not clear that these chemical agents cause granule loss by the normal pathway utilized at fertilization, it was proposed that they cause activation by depolarizing the egg plasma membrane. Studies employing means other than sperm to activate cortical granule release have been reviewed.[45]

C. BIOCHEMICAL REGULATION OF THE CORTICAL REACTION

Our meager knowledge of the biochemistry of the mammalian cortical reaction also implicates an inositol phosphate-induced calcium-dependent mechanism of activation after sperm binding as described previously for sea urchins and amphibians. Calcium increases during the activation of the mammalian egg[46,47] and also causes activation after microinjection.[48] Calcium ionophore, A23187, also results in cortical granule loss in mouse and hamster eggs (Figure 3).[37,38,49,50] Cortical granule loss with A23187 in divalent cation-free medium indicates that the egg's endogenous calcium stores are sufficient to cause exocytosis.[37,38,49] There is also evidence that extracellular calcium is involved in calcium transients, perhaps to reload intracellular stores.[51]

Inositol 1,4,5-trisphosphate microinjection causes G-protein-mediated calcium transients in hamster eggs,[51,52] cortical granule secretion in hamster and sheep eggs,[53] and modifications of the zona pellucida normally observed after fertilization in mouse eggs.[54] These results indicate that exogenous IP_3 can induce the cortical reaction. Microinjection of GTP-γ-S, a stimulator of G-protein-mediated IP_3 production, also results in calcium transients as well as cortical granule loss, and suggests that IP_3 may be an endogenous regulator as well.[53] Direct evidence for fertilization-induced IP_3 production with the appropriate kinetics and localization to account for the cortical reaction awaits future investigation.

Recent evidence suggests that both pathways (diacyl glycerol and IP_3, Figure 4) arising from GTP-mediated phosphatidylinositol bisphosphate (PIP_2) hydrolysis may be involved in activation of cortical granule release.[53-56] Diacyl glycerol (DAG), a product of PIP_2 hydrolysis and stimulator of protein kinase C, inhibits the fertilization of mouse eggs by some of the egg-induced modifications of the zona which normally occur at fertilization.[55,56] The partial modification of the zona suggests that this prong of the pathway alone does not result in a normal (complete) cortical reaction which may also require the participation of IP_3 and calcium. Each pathway could activate different hypothetical components of the cortical reaction or act in concert. For example, it has been suggested that calcium released in response to IP_3 could activate the calcium, phospholipid-dependent form of protein kinase C or indirectly activate it by stimulating PIP_2 hydrolysis (Figure 4).[54]

D. CORTICAL GRANULE EXOCYTOSIS

Although the cortical reaction in mammals undoubtedly results in many biochemical and structural changes in the egg and zona, the only routine microscopic "assay" in mammals for the cortical reaction is the loss of cortical granules. Although elevation of the vitelline envelope occurs in sea urchins and amphibians, an easily demonstrable microscopic change in the zona has not been reported. In amphibian eggs, other changes are frequently reported.[26] In mammalian eggs, due to the difficulties in routinely measuring changes in membrane conductance, depolarization, and lipid lateral diffusion rates,[57] the cortical reaction has become synonymous with the exocytosis of cortical granules. It is unfortunate that experiments designed to study the cortical reaction frequently do not result in information about other relevant parameters.

In mammalian eggs, the exocytosis of cortical granules appears to occur by the fusion of the granule membrane with the overlying plasma membrane.[31,33] Less frequently, granules fuse laterally with themselves, but also appear to release their contents into the perivitelline

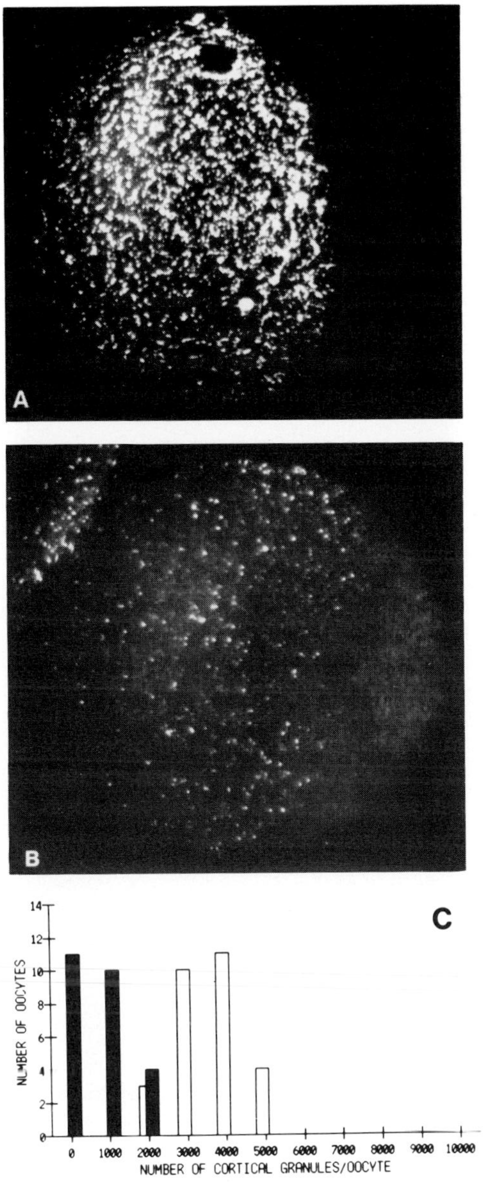

FIGURE 3. The ionophore, A23187, which mobilizes intracellular calcium, causes the loss of cortical granules from metaphase II mouse egg. Cortical granules were localized with fluorescent lectin staining.[36,38] (A) An unstimulated egg with a typical cortical granule domain (left) and cortical granule-free domain (right). (B) an ionophore-treated egg with a significant reduction in cortical granules, (C) quantitation of cortical granules from fluorescent micrographs demonstrates the extent of granule loss after ionophore treatment (black bars) compared to untreated controls (white bars).

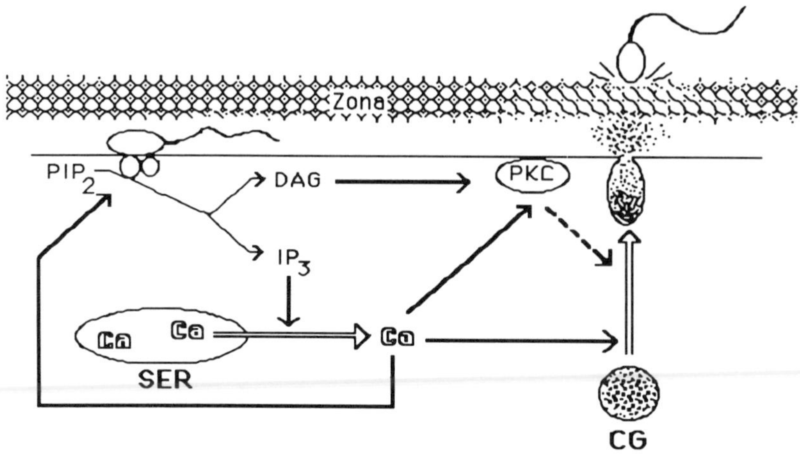

FIGURE 4. Hypothesized activation sequence by sperm of the cortical reaction, cortical granule release, and the zona block to polyspermy. This model incorporates studies of many organisms (see text for an explanation). G-protein and phospholipase C, which activate the conversion of PIP_2 to DAG and IP_3, are depicted directly under the fertilizing sperm. Heavy arrows indicate stimulation of a particular enzyme or process. Recent indirect evidence for PKC stimulation of exocytosis is indicated by the dashed line. Outlined arrows stand for Ca and CG release. CG—cortical granule, SER—smooth endoplasmic reticulum, PKC—protein kinase C, DAG—diacyl glycerol, PIP_2—phosphatidylinositol biphosphate, IP_3—inositol trisphosphate, Ca—calcium.

space.[31] Upon release at fertilization cortical granule contents appear to increase in size,[37] presumably undergoing hydration prior to diffusion across the perivitelline space and into the zona to biochemically establish the zona block to polyspermy. In the hamster, soybean trypsin inhibitor labels granules undergoing exocytosis to a greater extent than intact granules, suggesting the activation of granule enzymes upon release.[36] It is possible that granule enzymes are stored in a proenzyme state. In addition, recent reports are consistent with two other possible fates for released granule glycoproteins, including binding to the egg membrane[36] and aggregation into a large mass (in the perivitelline space) which disappears by the two-cell stage.[37]

The kinetics of cortical granule loss have been reported in phase microscopic studies of egg *in vitro*.[58] In zona-free hamster eggs, there is significant granule loss 2.5 to 5 min after insemination with homologous sperm,[43,59,60] while this process takes longer and causes release in fewer oocytes with mouse sperm.[60] In zona-enclosed, but cumulus-free mouse and rabbit eggs, significant granule loss occurred within 30 and 45 to 60 min of insemination, respectively.[60,61]

Due to the addition of cortical granule membrane at fertilization, there is a transient increase in the area of the oocyte plasma membrane. In some nonmammalian eggs, this may result in a large increase followed by plasma membrane retrieval from postfertilization endocytosis (reviewed by Longo[29]). When compared to sea urchin oocyte cortical granules, those of the mammal are smaller in size, fewer in number, and probably result in less membrane addition. For example, the release of approximately 4000 mouse cortical granules about 0.25 μm in diameter[33,38,62] from a metaphase II mouse egg would increase its surface less than 5%. After fertilization in both sea urchin and mouse eggs, large bulk membrane viscosity increases have not been observed, but changes in the lipid domain organization or composition may occur.[41,57,63]

IV. THE RELATIONSHIP OF THE CORTICAL REACTION AND THE BLOCKS TO POLYSPERMY IN MAMMALS

A. HISTORICAL ORIGINS

The pioneering work of Austin and colleagues[7,8] established the concepts of the zona and egg plasma membrane blocks to polyspermy in mammals. They recorded the number of sperm in the perivitelline space and within the oocyte after natural matings.[7] After studying a number of different mammals, they concluded that the two blocks to polyspermy occur to varying degrees depending on the species. At one extreme, they observed fertilized dog, hamster, and sheep oocytes which usually had no sperm in the perivitelline space and are consistent with a zona block to polyspermy. At the other extreme, the rabbit and gopher have significant numbers of perivitelline sperm while maintaining monospermic fertilization. This group appeared to have an oocyte plasma membrane block, while a third group (cat, ferret, guinea pig, mouse, rat) was classified as intermediate because sperm were sometimes found in smaller numbers in the perivitelline space. Much to the tribute of these early investigators, the zona block was inferred from experimental results, despite the absence of dramatic morphological changes (e.g., egg envelope elevation) which have been observed in other organisms upon fertilization. More recent *in vitro* studies of fertilized human eggs are consistent with a block primarily based at the zona.[64,65] No sperm were observed in the perivitelline space of 14 monospermic zygotes and one to three sperm were observed in an equal number of unfertilized oocytes, some of which may have been unable to undergo normal sperm binding.[65]

The next important historical step was to correlate the cortical and zona reactions (also taken by Austin and Braden[8]). They suggested that the zona block was due to the loss of cortical granules which they observed at fertilization in the hamster egg (they also had the benefit of earlier analogous sea urchin studies). With the light microscope, they observed that cortical granule disappearance occurs "as soon as the spermatozoon head makes contact with the vitellus" (egg plasma membrane) and that all 725 eggs (1.6% polyspermic) had no sperm in the perivitelline space. They concluded that hamster oocytes did not develop a plasma membrane block, but rely on the "efficiency of the zona reaction". In contrast to hamster eggs, about 20% of mouse and rat eggs were observed to have supplementary sperm in the perivitelline space.[7] Since the *in vivo* rate of polyspermy was about ten-fold lower, they proposed that these species utilize both zona and membrane blocks even though the cortical reaction had not been documented yet in these species.[8] A small number of degenerating eggs from these two and other species were highly polyspermic and was attributed to the degenerating egg's inability to initiate the blocks to polyspermy.[8]

These observations and conclusions resulted in the concept that the blocks to polyspermy were egg-induced and that it was the role of the sperm to elicit this cortical response from the egg upon fertilization. Even more prophetic was their proposal that the "tanning agent" from the egg would originate and spread from the point of sperm contact, resulting in a progressive, wavelike zona reaction. Recently a wave of calcium and cortical granule release[14] has been demonstrated from the point of sperm-egg fusion in the hamster.[47]

Since the studies of Austin and co-workers, it has been appreciated that the zona block to polyspermy is correlated with a loss of sperm binding and penetration in some species. The loss of binding has been demonstrated in the zona of fertilized eggs of the hamster[59] and mouse.[66-71] In dog and sheep eggs fertilized *in vivo*, numerous sperm remain bound to the surface of the zona.[7] These published differences between these two groups could be due to egg/zona differences or dissimilar assays and insemination conditions. Once sperm penetration of the zona has begun, further penetration is inhibited in some species. Inhibition appears to be accompanied by continued zona association in the dog, sheep,[7] and human,[65] while detachment has been observed in the mouse.[67] Sperm-zona association studies conducted *in vitro* in the absence of sufficient serum may require reinterpretation due to limited cortical granule release prior to fertilization (discussed later).

B. CORTICAL REACTION PRODUCTS AND THE ZONA REACTION

Barros and Yanagimachi[72] provided the first direct evidence that the cortical reaction released substances which were responsible for the zona reaction and block to polyspermy. After zona-free hamster eggs were activated by insemination with sperm, the surrounding medium was centrifuged and shown to induce the zona reaction in zona-intact eggs. Sperm binding and penetration were not prevented in control medium incubated with sperm alone. Evidence that the zona reaction was induced solely by egg-derived substances was provided by Gwatkin et al.[73] in studies with electrically activated hamster eggs. Released soluble substance(s) induced the zona reaction in both hamster and mouse eggs, was heat labile, and was inactivated by trypsin inhibitors. Studies with activated mouse eggs also have revealed a trypsin-inhibitor-sensitive substance which reduces sperm-zona binding.[74] A cortical granule-derived, trypsin-like protease was proposed to biochemically modify the zona and prevent sperm binding. Recently, exogenous soybean trypsin inhibitor has been localized in a cortical granule-like distribution in the hamster oocyte.[36]

By analogy with the envelope hardening reaction in sea urchin eggs,[75] an ovoperoxidase-catalyzed zona hardening by dityrosine cross-linking has been proposed in mouse eggs.[76,77] In unfertilized mouse eggs, peroxidase activity was localized to cortical granules, while in artificially activated eggs, it was associated with the egg surface, zona, and material in the perivitelline space. Several peroxidase inhibitors and tyrosine analogues prevented zona hardening.

These important experiments demonstrated that the zona block was due to cortical reaction products which were likely to have a cortical granule origin. Other studies support this hypothesis. In electrically stimulated hamster eggs, cortical granule loss appears to precede the zona block.[73] Calcium ionophore (A23187) activation also causes cortical granule release with a loss of sperm-zona binding in hamster eggs.[49] As reviewed by Wolf,[58] several studies on inseminated mouse eggs indicate that cortical granule release precedes the development of the zona block.[37,60,68] Although a zona block was not observed in mouse oocytes treated with A23187,[50] the extent of cortical granule release in this study may not have been sufficient since divalent cation-containing medium was used (see Steinhardt et al.[49]). Another interesting correlation between the cortical and zona reactions comes from studies on the KE mouse strain in which reduced egg fertilizability appears to develop in response to a premature release of a limited number of cortical granules (discussed later).

During mouse oocyte meiotic maturation *in vitro* without serum, a precocious zona block develops in the mouse and, as discussed later, appears to be due to limited cortical granule release. When compared to culture in 10% serum, serum-free maturation results in an 85% reduction in zona pellucida penetration by sperm,[78] a greater than 90% reduction in fertilization,[79] and an increased resistance of zona digestion by chymotrypsin which is indicative of the zona "hardening" reaction.[79-81] During normal meiotic maturation *in vivo* and *in vitro* with 10% serum in hamster and mouse oocytes, a limited number of cortical granules are lost,[34,38,62] but major zona modifications are not detected[62] and appear to be prevented by follicular fluid and serum.[81] In contrast, the zona modifications which occur in response to the cortical reaction after fertilization in the oviduct appear not to be inhibited probably due to the low concentration of serum proteins in oviduct fluids.[82] It is of interest to note that experiments designed to achieve both meiotic maturation and subsequent fertilization in the oviduct from transferred primate immature oocytes (after ovarian retrieval) result in failed fertilization,[83] perhaps due to the uninhibited activity of cortical granule contents released during maturation in that location.

Fortunately, the mouse zona glycoproteins, ZP2 and ZP3, which are responsible for sperm penetration and binding, respectively, have been intensively characterized[9] (see Chapter 6). Undoubtedly, changes in these zona glycoproteins will be used as bioassays in purifying and identifying important cortical reaction constituents. In this regard, A23187-activated mouse egg exudates have been found to contain a 21- to 34-kDa proteinase which appears within 5 min of treatment, consistent with a cortical granule origin.[84] This proteinase cleaves ZP2 and proteolysis correlates with zona hardening as determined by dissolution time at pH 3.6. Because ZP2 appears to be responsible for binding acrosome-reacted sperm and is proteolytically cleaved during the zona reaction after fertilization, it may have important sites which are modified by cortical granule constituents. This proteinase activity is not sensitive to trypsin inhibitors and appears to represent a different enzyme activity than described in previous exudate studies.[73,74,85]

Substances released from the mammalian egg surface, cytosol, or other cortical vesicles which do not fit the classical description of cortical granules have not been ruled out as accomplices in the zona reaction. As in the sea urchin, a collaboration of cortical granule and other egg factors may be required. Biochemical proof that cortical granule constituents cause discrete modifications of individual zona glycoproteins which are responsible for the zona block to polyspermy awaits further investigation.

C. THE CORTICAL REACTION AND OTHER BLOCKS TO POLYSPERMY

The cortical reaction has been implicated in non-zona blocks to polyspermy. Except in those species with a strong zona block, sperm have been observed in the perivitelline space after monospermic fertilization.[7] Their ability to penetrate the zona, but not the fertilized egg, has been taken as evidence for a plasma membrane block to polyspermy. In nonmammalian species an electrical membrane block has been described, but these electrical changes have not been observed in mammalian oocytes and the basis for their membrane block is not known.[6] In addition, after fertilization, substances may exist in the perivitelline space which prevent a second sperm from binding to or fusing with the oocyte.

The rabbit egg, whose cortical granules undergo release after fertilization, is cited as the classical example of a mammalian oocyte with a strong membrane block and the weakest zona block.[7] The fertilizability of artificially activated eggs[86] and the observation that rare polyspermic eggs fail to release their cortical granules suggest that their contents could be involved in establishing the membrane block.[87] The contents of the granules appear to change the zona since it undergoes hardening[88,89] without preventing sperm penetration.

Both a membrane and zona block have been reported in the mouse oocyte,[58] but the role of cortical granules in the membrane block is not yet clear. Evidence against cortical granule involvement came from an A23187 study of zona-free oocytes in which 50% granule loss occurred. These eggs could be fertilized and then exhibited a membrane block.[50] A role for cortical granules was suggested by a reduction in sperm penetration after zona-free oocytes were treated with cortical reaction exudates.[74] Also, reduced fertilization was observed in zona-free KE oocytes which were shown to have undergone premature release of some cortical granules during maturation.[90]

With an extremely effective zona block, the hamster oocyte membrane may rarely encounter supernumary sperm *in vivo* and its membrane block has been considered very weak. The zona-free hamster egg becomes highly polyspermic to its own and heterologous sperm.[44,59] However, two recent *in vitro* studies suggest a membrane block can develop after the cortical reaction is stimulated by fertilization[91] or electrical activation.[36] In the latter study, the plasma membrane block was temporally related to cortical granule exocytosis, and lectin binding, originally observed within cortical granules, was found on the plasma membrane after exocytosis.

Finally, an interesting but untested hypothesis in mammals is that the contents of the cortical granules may provide a transient hostile environment to sperm in the perivitelline space until the zona block is established. In both fish and sea urchins, spermicidal substances are released which can cause sperm to undergo agglutination and peroxide damage, respectively (reviewed by Jaffe and Gould[6]). An intermediate zone of polyspermy prevention also may exist between the zona and plasma membrane in mammals, especially since peroxidases appear to be released.[76,77]

V. MAMMALIAN OOCYTE CORTICAL GRANULES

A. ELECTRON MICROSCOPIC DEFINITION

Unlike sea urchin cortical granules, which have a characteristic ultrastructure distinct from any other organelle,[92] those in many mammalian eggs do not possess a unique morphology and have been described as a family of similar organelles in electron microscopic studies.[45,93] The significance of this heterogeneity with respect to exocytosis *in vivo* and the blocks to polyspermy is not known.

In electron micrographs, mammalian cortical granules are described as granules. 0.1 to 1.0 μm in diameter, bounded by a membrane, and characterized by a "light" to "dark" density matrix which is usually homogeneous and lacking an elaborate internal ultrastructure. Studies of their morphology, production, and cortical localization in different species have been reviewed.[45,93] The term "cortical" refers only to their final cytoplasmic localization in the cortex (defined previously). Although this is a convenient definition, cortical granules have been described in deeper cytoplasmic regions during their development or oocyte aging. Cortical granules appear to be produced by Golgi complexes below the cortex and migrate to their final location.

Although granules within the cortex have a relatively uniform and similar electron density in some species (e.g., rabbit and primate), cortical granules in others (e.g., rodents) may be light and dark within the same region of the cortex. It is not known if these differences are always due to different stages of maturation (e.g., matrix condensation) or a true structural or even functional heterogeneity. In the pig egg, the ratio of dark to light granules increases over ten-fold (around the time of ovulation after human chorionic gonadotropin [hCG] administration), suggesting maturation-associated conversion of light to dark granules. In a report of premature sperm-induced release of mouse cortical granules in mature eggs *in vitro,* primarily light granules were lost.[33] The observation that developing cortical granules in close proximity to the Golgi apparatus have a light matrix does not definitively resolve this question. We are left with the possibility that granule heterogeneity may exist, which has been recently reported at the biochemical level in the sea urchin.[94] Resolution of this question in mammalian oocytes will be possible when the contents of their granules are biochemically characterized and probes become available to localize these constituents in one or more granule populations.

It is not known if those structures which are classically defined as cortical granules by electron microscopy constitute all of the granules and vesicles which undergo secretion into the perivitelline space upon fertilization. For example, the mature mouse and rat egg cortex contains a dense array of diverse membrane systems, organelles, vesicles, as well as light and dark cortical granules.[31,33,38,95] Although some of these structures resemble cortical granules, they have neither been classified as such (due to their smaller size or less dense matrix) nor quantitated before and after fertilization. It is possible that some of them participate in the cortical reaction within the cortex, some may undergo release, and others may represent transitional stages. In conclusion, it has not been proven that secretion into the perivitelline space at fertilization and the establishment of the blocks of polyspermy are due solely to cortical granules as defined by electron microscopy.

B. LIGHT MICROSCOPIC STUDIES

The need for quantitative studies of cortical granule density as well as the assessment of their distribution throughout the entire oocyte cortex has resurrected the need for light microscopic analysis. In fact, the earliest phase microscopic studies took advantage of the highly refractile properties of hamster oocyte cortical granules[96] and continues to be used to study changes in cortical granule distribution.[34] Cortical granules were operationally defined by their size, appropriate location, and loss upon fertilization. The smaller size or lesser density of mouse than hamster cortical granules[60] makes them less desirable for phase microscopic studies.

Taking advantage of many studies demonstrating a high glycoprotein concentration in cortical granules,[30,45,97] recent lectin localization studies[36-38] have provided methods to quantitate cortical granule density and number/oocyte.[38,62] The lectin, *Lens culinaris* agglutinin (LCA), appears to recognize the overwhelming majority of CGs because studies with lectin localization and electron microscopy are in good agreement. First, the number of LCA-positive staining granules determined by the two methods is in within 10%.[33,38] Second, the number of granules decreases following either fertilization or ionophore, A23187, activation.[33,36,38,50] Third, by both methods, the granules have a similar cortical localization, distribution with respect to the cortical granule-free domain, and size. Fourth, LCA- and other lectin-staining material are released into the perivitelline space upon activation and fertilization of hamster and mouse eggs.[36,37] As in the case of the electron microscopic identification, it is not known if all lectin-staining granules undergo secretion at fertilization (vs. modification) or if a small population of granules is not included. In addition, in the mouse, the precise relationship between light and dark granules observed by electron microscopy and those visualized by lectin fluorescence is not known, due to the limitations of electron microscopic lectin localization.[38]

Cortical granule quantitation has become necessary to analyze granule production, cortical density, and release, as well as to evaluate changes in their distribution during meiotic maturation (discussed below). Lectin studies have reported similar numbers of cortical granules (4100 to 4500/oocyte)[38,62] as electron microscopic estimates (4500/oocytes)[33] in the metaphase II mouse oocyte. Electron microscopic studies have reported the number of granules in the cortex below 100 μm of plasma membrane: 32 (mouse[33]); 95 (rat[98]); and about 60 (rabbit[86]). However, the calculation of granules/egg from analyses of relatively small areas of cortex and numbers of eggs is not optimal due to different zones of granule density and granule-free areas.[38,62] Fluorescence analysis can quantitatively account for these parameters, include a significantly large number of eggs for analysis, and provide rapid through-focus observation for determination of egg orientation with respect to the meiotic spindle as well as for an assessment of stained granules in the egg interior.[36-38]

C. CHANGES IN CORTICAL GRANULE DISTRIBUTION

During oogenesis and meiotic maturation of mammalian oocytes, cortical granules undergo changes in distribution, cortical density, and number. Electron microscopic studies report that developing cortical granules are present in the vicinity of the Golgi and migrate to the subcortical cytoplasm or the cortex (reviewed by Gulyas[45] and Guraya[93]). In the fully grown (maximum diameter) germinal vesicle stage mouse oocyte, lectin-staining cortical granules are located throughout the entire cortex (Figure 5).[38] At this stage, some granules are found in subcortical domains, especially around the germinal vesicle, and after germinal vesicle breakdown, these perinuclear granules are no longer observed.[99] Because the number of granules in the cortex also increases after germinal vesicle breakdown, some granules may migrate from a perinuclear to a cortical location upon meiotic resumption.[62]

It has also been suggested from electron microscopic studies that cortical granule production and migration take place during meiotic maturation in many mammals,[31,100-103]

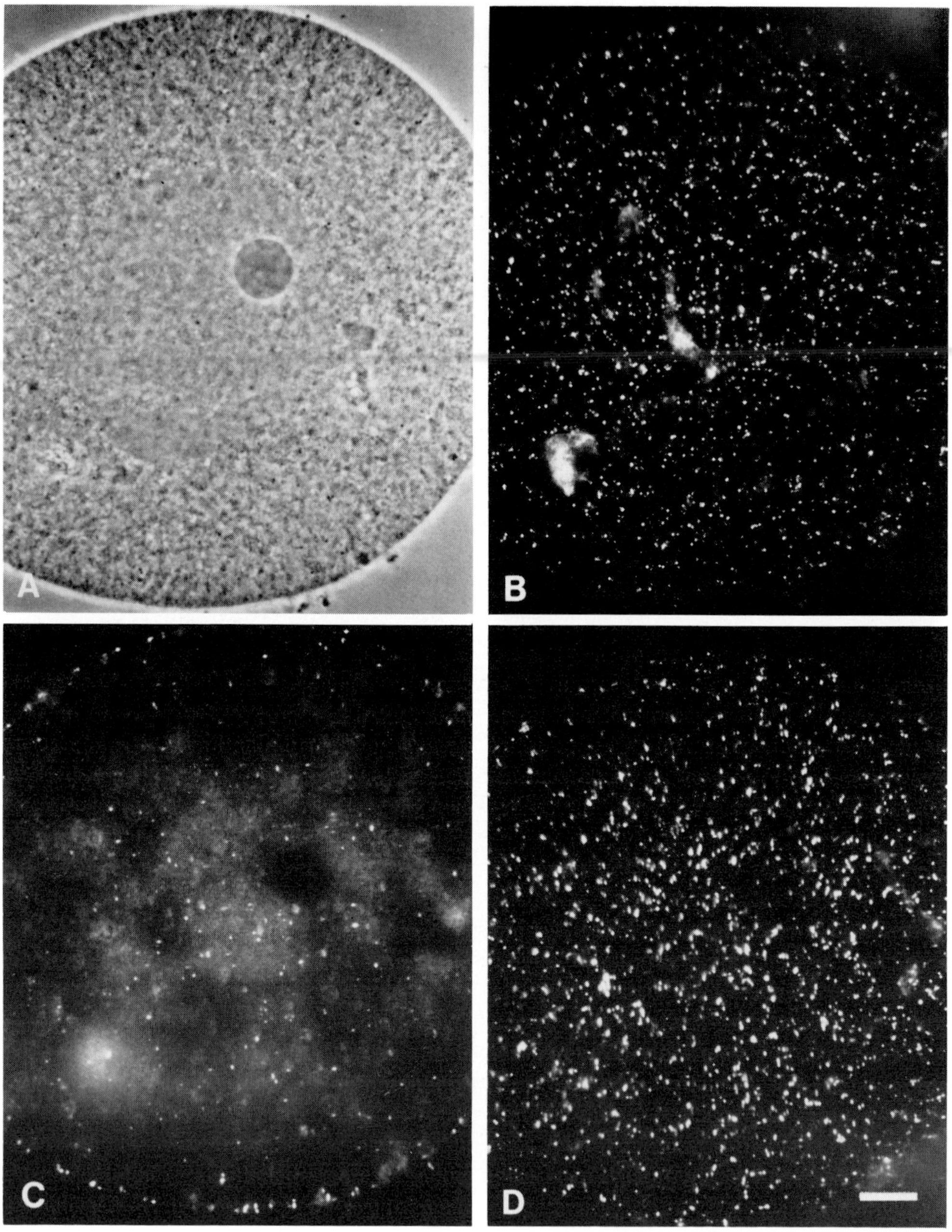

FIGURE 5. An ovarian mouse germinal vesicle stage oocyte demonstrating cortical granules throughout the entire cortex. (A) Phase microscopy after fixation and extraction (the nucleolus is still apparent); (B to D) fluorescence microscopy after lectin staining; (B) cortex; (C) equator; (D) cortex (opposite from the pole in B). Fluorescent granules are primarily confined to the cortex with some punctate staining in the interior. Bar, 10 μm. (From Ducibella, T., Anderson, E., Albertini, D. F., Aalberg, J., and Rangarajan, S., *Dev. Biol.*, 130, 184, 1988. With permission.)

including the human.[104,105] In pig eggs, morphometric studies indicate that cortical granules may undergo two phases of migration toward the cortex, one after production and the other near the time of ovulation after which most cortical granules lie within 4 nm of the plasma membrane.[103] However, in the mouse egg, cortical granules do not finally reside in such close proximity to the plasma membrane even after ovulation.[95] The biochemistry of cortical granule production and mechanisms involved in cortical granule migration as well as cortical stabilization await future study.

After the resumption of meiotic maturation in both hamster[34] and mouse[62] oocytes, a cortical granule-free domain (CGFD) develops over the metaphase I spindle chromosomes. Because immature mouse oocytes lack a microvillous-free region, the loss of microvilli and cortical granules may occur in the same region of cortex at the same time during maturation.[104] The peripheral migration of the meiotic spindle has been proposed to induce the development of this area of cortex which also appears to be the site of first polar body formation.[35] A similar domain may be present in the human egg because cortical granules are usually absent over the developing meiotic spindle.[104]

By metaphase II, the area of the CGFD has enlarged (Figures 2 and 6). A phase microscopic study of hamster eggs reported mean areas of about 700 and 1300 μm^2 at 14 and 17 h after hCG administration, respectively.[34] A small CGFD of similar size was also observed in a fluorescent lectin study of cortical granules.[36] In the mouse egg, the area of the CGFD was reported to be about 20% of the cortex from electron microscopic sections[33] at 12 to 14 h post hCG and to be double that size (about 6800 μm^2) by lectin staining of cortical granules at 16 to 18 h.[38]

The enlargement of the CGFD between metaphase I and II may include both exocytosis and redistribution of granules. In the hamster oocyte at metaphase I, granules identified by phase microscopy had a higher density at the edge of the CGFD. This was interpreted as evidence for granule redistribution while exocytosis was also reported in this region.[34] In the mouse oocyte, a denser peripheral rim of cortical granules was not observed by lectin staining at the very onset of CGFD development at metaphase I, but a higher peripheral density was present after meiosis I and formed a zone in some eggs by meiosis II.[62]

During meiotic maturation, a loss of cortical granules from the egg cortex appears to take place by exocytosis based on several different lines of evidence in mouse and hamster eggs. In the latter, electron micrographs revealed cortical granules undergoing release into the perivitelline space over the meiotic spindle.[34] In a quantitative study of granules in the mouse egg, significantly fewer granules in the cortex were found in meiosis II eggs than in germinal vesicle stage oocytes or in those immediately after germinal vesicle breakdown.[38,62] It is likely that this reduction was due to exocytosis because maturation in serum-free medium was associated with the disappearance of cortical granules over the meiotic spindle as well as the modification of the zona protein ZP2. Because modification of ZP2 normally takes place at fertilization[107] and after exposure to cortical reaction exudates from ionophore A23187 activation,[84] maturation-associated cortical granule loss is likely to occur at least in part by exocytosis. Exocytosis is also a plausible hypothesis to explain this loss of granules in the cortex, because calcium increases locally and transiently in the vicinity of the metaphase spindle in eggs[108] and other cell types[109,110] as well as stimulates cortical granule release.[49] Although a reduction of cortical granules in the cortex was not reported in an electron microscopic study,[95] further analysis of those data indicates that they appear to be consistent with the lectin staining results reflecting cortical granule loss during maturation.[62]

An abnormal prefertilization release of cortical granules has been proposed to account for the reduced fertility of the KE mouse strain.[111] Ovulated KE eggs had fewer cortical granules in the cortex by electron microscopy than ovarian KE oocytes.[112] Since this cortical granule reduction is associated with a form of premature zona hardening as assessed by increased chymotrypsin digestion time,[113,114] precocious maturation-associated exocytosis

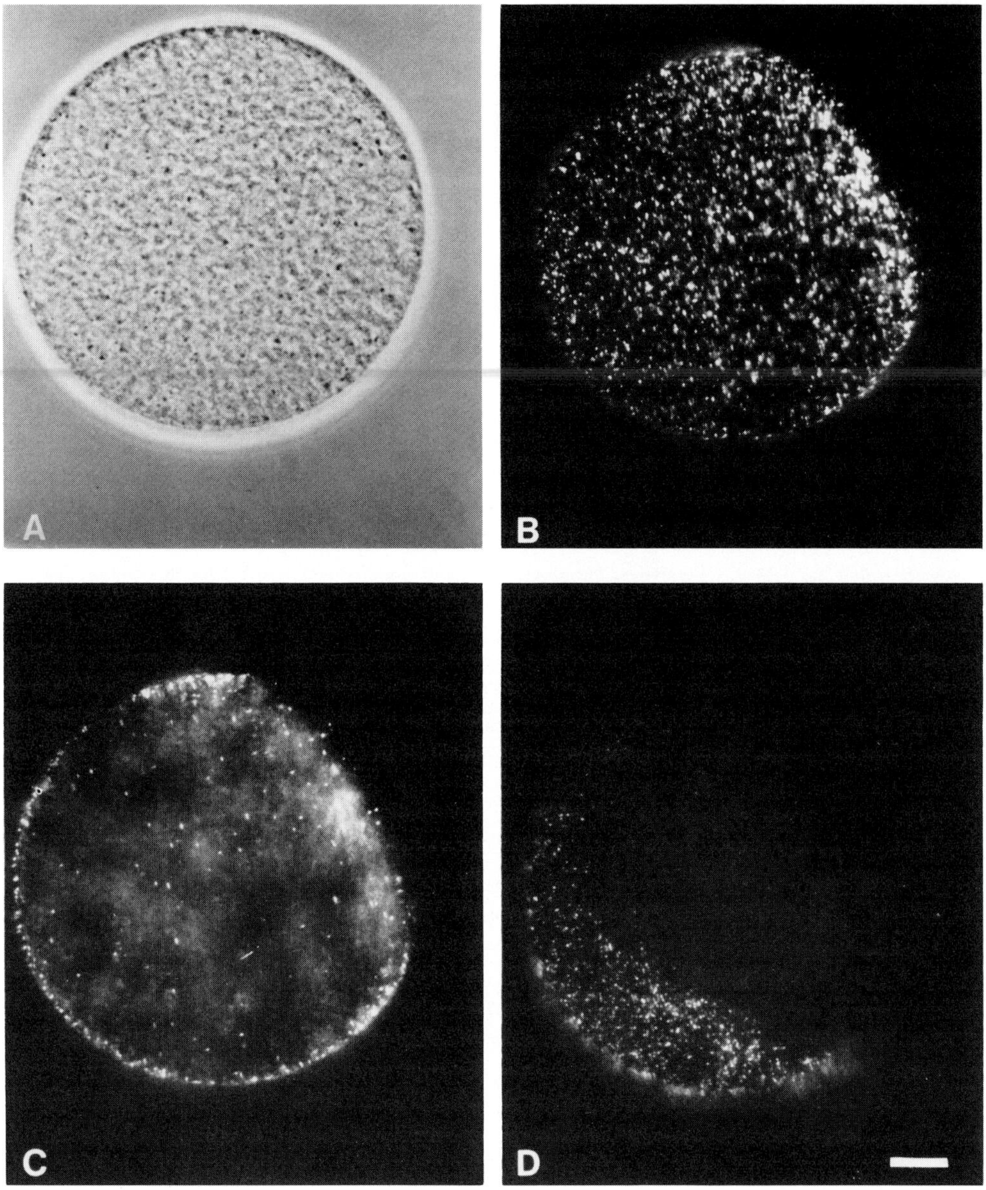

FIGURE 6. An ovulated mouse egg after meiotic maturation (the staining and optical sections in A to D are described in Figure 5). The cortical granule-free domain is best observed in (D) and present to a much lesser extent in (B) and (C). The metaphase II chromosomes are not visible (see Figure 2). Bar, 10 μm. (From Ducibella, T., Anderson, E., Albertini, D. F., Aalberg, J., and Rangarajan, S., *Dev. Biol.*, 130, 184, 1988. With permission.)

was proposed. It is not known if these events in the KE oocyte are related to the cortical granule loss and zona modification observed by others.[34,62]

The formation of a small CGFD at meiosis I can account for the scarcity and polarized distribution of cortical granules in the first polar body.[33,34] Since these cortical granules are observed primarily in the polar body cortex facing the oocyte, the cleavage furrow may arise just outside the edge of the CGFD and thereby include some cortical granules. Cortical granule redistribution may also occur during spindle rotation. A dense ring of cortical granules is present in the vicinity of the midbody between the oocyte and first polar body in the

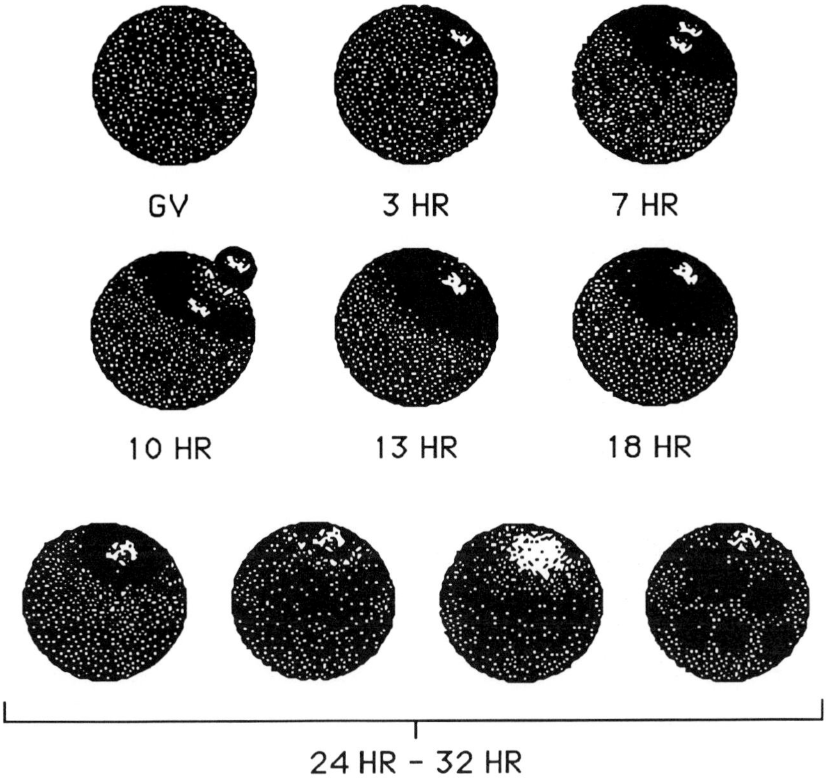

FIGURE 7. Diagrams of cortical granule distributions in the cortex of oocytes during meiotic maturation and tubal aging as assessed with fluorescent lectin staining (see Figures 5 and 6). The chromosomes are also depicted, first within the cortical granule-free domain. After 24 to 32 h from the resumption of meiotic maturation, aged eggs demonstrate smaller domains free of granules, multiple domains, or aggregated granules. (Diagram courtesy of Paul Duffy.)

human[104] and mouse[99] (Figure 7). The enlargement of the CGFD by metaphase II is likely to account for the absence or rare appearance of cortical granules in the second polar body.[31,115]

Aging of the egg beyond the normal time of fertilization results in further changes in cortical granule distribution (reviewed by Gulyas[45]) and, perhaps, in competence to undergo the cortical reaction. In aged mouse,[116] hamster,[117] and rabbit[118] eggs, organelles, including some cortical granules, appear to migrate centripedally. In the rat egg, cortical granule number in the cortex is reduced in half.[98] In the mouse, major changes in the CGFD occur within 24 h of ovulation. Many eggs have one or more CGFDs of a smaller area than that of a mature egg, while others have a dense ring or cap of granules (Figure 7).[99] The loss of granules from the egg cortex may also occur by extrusion of cytoplasmic blebs into the perivitelline space.[117,118] Changes in cortical actin may take place during aging and result in cortical granule-containing blebs and inwardly displaced granules.

It is clear that cortical granules undergo significant changes in distribution and perhaps limited release during meiotic maturation. Cortical granule production and loss during maturation are not incompatible and both may occur. Certain cortical domains may lose granules while others may maintain a relatively high density, especially at the periphery of the CGFD.[62] The relative contributions of redistribution and exocytosis await further investigation. Redistribution carries with it the interesting implication of cortical movements in the mammalian egg which are so important in relation to polarity and localizations in amphibian and other

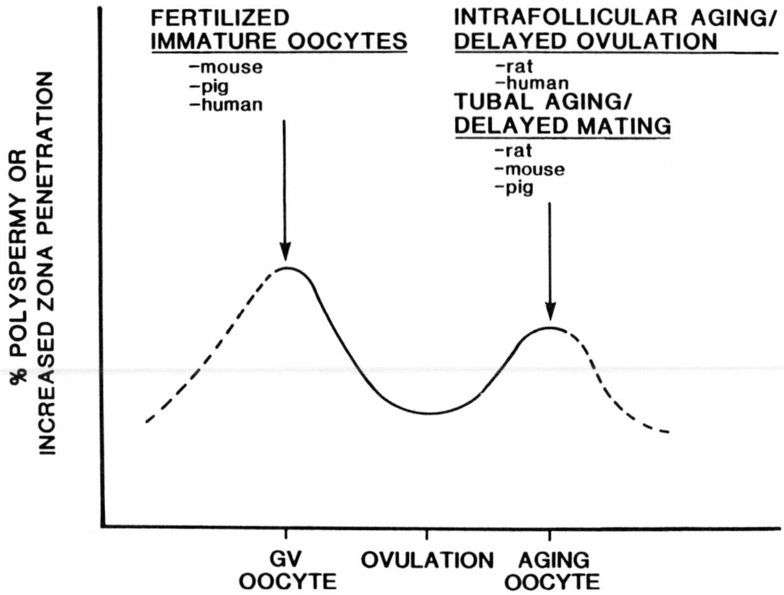

FIGURE 8. Hypothetical graph of % polyspermy or increased zona penetration vs. stage of the oocyte based on many studies. The egg's ability to mount an optimal block(s) to polyspermy is likely to be present during a narrow window of time (see text). References: fertilized immature oocytes,[120,122,124] intrafollicular aging/delayed ovulation,[139,140] tubal aging/ delayed mating.[142-144]

eggs. Exocytosis prior to fertilization appears paradoxical, but its effects appear to be mitigated by the follicular microenvironment. It is interesting that fertilization has not been observed in the CGFD of the mouse egg.[33,40] This may be due to maturation-associated granule release resulting in zona changes above this domain which have not yet been detected.[62] Of additional interest is whether there are different subsets of mammalian cortical granules (as proposed in the sea urchin[94]) which undergo release at different times.

VI. COMPETENCE TO UNDERGO THE CORTICAL REACTION

There is growing evidence that the egg must be at an appropriate stage of maturation when fertilized in order for the cortical reaction and block(s) to polyspermy to take place normally. In mammals, failure of the cortical reaction and the establishment of the block(s) to polyspermy can be lethal.[1]

Many studies of different mammalian species indicate that immature eggs (i.e., germinal vesicle stage through meiosis I) exhibit a higher incidence of polyspermic fertilization than their counterparts having completed maturation *in vivo* or *in vitro* (Figure 8). The low rate of polyspermic fertilization *in vivo,* about 1%, in fertilized mature mammalian eggs[1,7,8,87,119] attests to their competence in developing an effective block to polyspermy. The incidence of polyspermy in human *in vitro* fertilization is approximately 5%.[1] Greater polyspermy rates have been reported in cases of fertilization of immature pig[120,121] and human[122,123] oocytes. About 90% polyspermic fertilization occurs in immature pig oocytes after ovulation due to precocious administration of hCG.[120]

Greater zona penetration by sperm is also observed after insemination of immature eggs. Fertilized meiosis I mouse eggs were characterized by an unusually large number of sperm in the perivitelline space.[124] The fertilized immature egg of the mouse, unlike that of the pig, is not characterized by a very high incidence of polyspermy which may be due to an

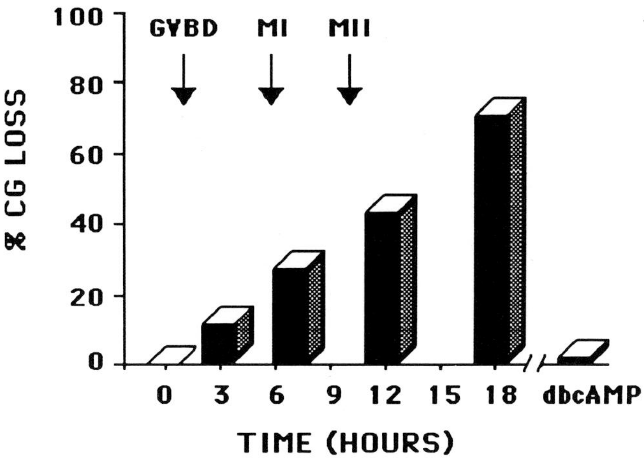

FIGURE 9. The development of competence to undergo an ionophore A23187-induced cortical reaction as indicated by % cortical granule (CG) loss is depicted during oocyte maturation *in vitro*.[62] Competence gradually develops, but does not attain normal levels (about 70% with A23187) until metaphase II (MII). The presence of dibutyryl cAMP during 18 h of culture of germinal vesicle stage oocytes inhibits not only germinal vesicle breakdown (GVBD), but also the development of competence to undergo the cortical reaction. Metaphase II eggs treated with dibutyryl cAMP are not inhibited (not shown). In oocytes matured *in vivo,* a normal level of competence develops just prior to ovulation.[127]

effective membrane block to polyspermy. Such a membrane block[8] and/or the released contents of at least some cortical granules may be responsible for these perivitelline sperm in fertilized mouse and rat eggs.

The increase in sperm penetration of the immature egg and its zona may be due to the incompetence of these oocytes to undergo a normal cortical reaction including the exocytosis of a sufficient number of cortical granules. In both the human[125] and sea urchin[126] cortical granule exocytosis was not detected in immature eggs after sperm penetration. In amphibians, competence to propagate a global cortical reaction develops after hormone-induced meiotic maturation and is correlated with the appearance of a cortical endoplasmic reticulum which is the proposed site of calcium sequestration to activate the cortical reaction (these studies have been reviewed by Elinson[17]).

In mammals, meiotic maturation has been studied with respect to the development of the oocyte's competence to undergo the cortical reaction which has been assessed by 5 μM ionophore A23187-induced cortical granule loss (Figure 9).[62,127] Competence develops progressively and a mature level of cortical granule loss is present at 10 h post hCG or about 2 h prior to ovulation. This finding emphasizes the importance of the proper timing of ovulation and oocyte maturation because ovulation of an oocyte incompetent to undergo the cortical reaction would probably result in higher incidence of polyspermy as discussed earlier. In addition, for *in vitro* fertilization in which oocytes and eggs are retrieved from follicles prior to ovulation, incompetent oocytes may be harvested. These oocytes may require *in vitro* culture to undergo nuclear and cytoplasmic maturation, including the development of cortical reaction competence. This hypothesis is supported by studies of the effects of delayed insemination in human *in vitro* fertilization in which immediate insemination resulted in increased polyspermy[122] and insemination prior to 3 h of culture resulted in a decreased pregnancy rate.[128]

The development of cortical reaction competence *in vitro* has been studied in mouse germinal vesicle stage oocytes cultured to metaphase II.[62] Ionophore-inducible competence was not detected at the germinal vesicle stage, was apparent at metaphase I, and reached a normal level by metaphase II. The development of competence *in vitro* was inhibited by dibutyryl cAMP which prevents germinal vesicle breakdown and the resumption of meiotic maturation.[129,130] It is not known if cAMP directly affects the cortex or acts through the inhibition of nuclear maturation. Thus, the development of competence could be due to either cytoplasmic maturation or mixing of the nucleoplasm and cytoplasm after germinal vesicle breakdown, or both. After the resumption of meiotic maturation in the mouse and hamster oocyte, other significant cytoplasmic events take place including alterations in the pattern of protein synthesis[131] and phosphorylation[132] as well as the development of competence to form a male pronucleus[124,133] and sperm chromosomes.[134] The mixing of cytoplasm and nucleoplasm is necessary for the formation of the male pronucleus within the egg cytoplasm.[135]

Morphological changes may be involved in cortical maturation and the development of competence. After meiotic maturation in pig[103] and sea urchin[126] eggs, cortical granules are observed closer to the plasma membrane and in this location they may become associated with factors involved in the functional maturation of the cortex. However, this change in location has not been observed in the mouse in which both light and dark cortical granules are a mean distance 0.4 to 0.5 μm from the plasma membrane in both germinal vesicle stage oocytes and metaphase II eggs.[95] The significance of other changes is not known, such as a change in the ratio of dark to light cortical granules during maturation[95,103] and the differential release of light granules after *in vitro* insemination.[33]

The development of cortical reaction competence in mammalian eggs may be due to the arrival of an important regulator(s) in the cortex after germinal vesicle breakdown. The hypothesis is based on sea urchin and amphibian studies. These indicate that one candidate for a cortically localized regulatory substance is calcium housed in the endoplasmic reticulum which develops in association with cortical granules and during the onset of cortical reaction competence in amphibians[136] (see reviews[16,17,29]). By analogy with the storage and sudden release of calcium from the sarcoplasmic reticulum of the muscle cell, specific calcium-binding proteins[137] in the egg's endoplasmic reticulum could house calcium which, at fertilization, is released to propagate cortical granule exocytosis. In mammals, similar profiles of endoplasmic reticulum have been reported in mature mouse eggs.[138] These membranes and other cortical vesicles (Figure 10) which are not classical cortical granules are three times more abundant in the cortex of the mature mouse egg than the germinal vesicle stage oocytes.[95] Their scarcity in immature eggs may represent an insufficient storage of cortical calcium and account for the incompetence of the immature mammalian oocyte to undergo the cortical reaction.

The egg's ability to mount a normal cortical reaction arises rapidly before ovulation and may last only a short period of time (Figure 8). As described previously, aging of mammalian eggs soon after ovulation results in degenerative changes in the cortex and position of cortical granules. These changes in the cortex from intrafollicular aging (delayed ovulation) or tubal aging (delayed mating) are associated with a higher incidence of polyspermy.[139-144]

With the increasing use of ovulation induction methods (to promote oocyte maturation), *in vitro* fertilization, gamete intrafallopian transfer (GIFT), and artificial insemination, the normal mechanisms which control oocyte maturation and/or timing of insemination are being altered and manipulated externally. This may result in the sperm interactions with immature or aged eggs, especially if the normal fertilizable life span of the egg is relatively short. Current best estimates for this life span in mammalian eggs are 6 to 24 h.[145] Therefore, normal maturation and optimal timing of insemination are necessary in order to insure that the egg is competent to undergo the cortical reaction when fertilized and to avoid polyspermic fertilization.

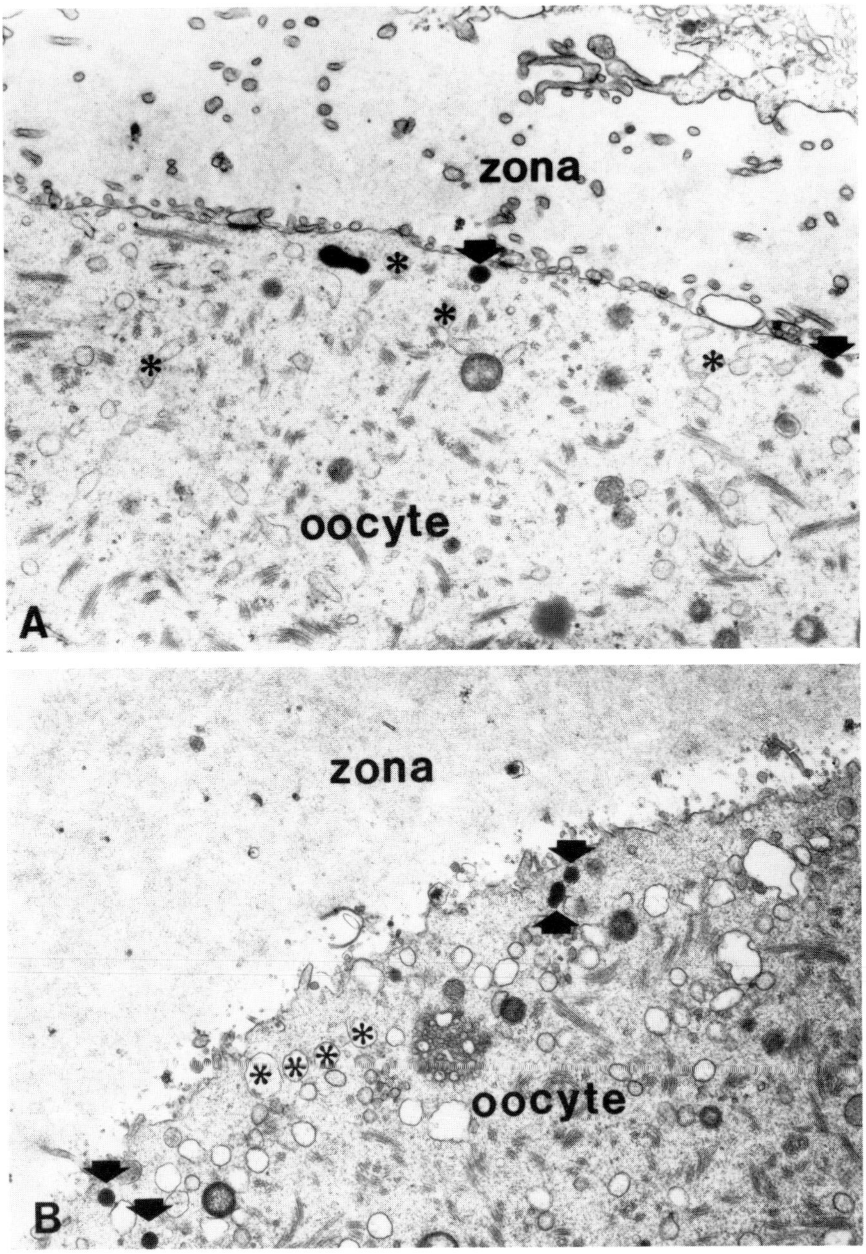

FIGURE 10. Electron micrographs of the cortex from (A) a germinal vesicle stage oocyte and (B) an ovulated egg. A larger number of cortical vesicles and membrane profiles (*) are observed in the mature egg. Arrows indicate cortical granules. (From Ducibella, T., Rangarajan, S., and Anderson, E., *Dev. Biol.*, 130, 789, 1988, with permission and unpublished data.)

VII. CONCLUDING REMARKS

The biosynthesis of cortical granules, the regulation of their release, and biochemical activities of their contents are of central importance to understanding the zona reaction and block(s) to polyspermy in mammalian eggs. Studies using cell biological, physiological, and molecular approaches will be required to unravel the story of cortical granule migration and release as well as the substrate specificity of their contents.

The major biochemical questions ahead include the characterization of the granular contents and their effects on the zona, the oocyte plasma membrane, and possibly, the sperm. In addition, it will be of interest to note which of these effects constitute the block(s) to polyspermy and know why the zona and membrane blocks appear to vary so significantly and even inversely in some eggs. Additional studies will be needed to address how egg cortical activation occurs, including the roles of the fertilizing sperm, the egg plasma membrane and endoplasmic reticulum, the IP_3 and diacyl glycerol pathways, and the fluxes of calcium and other ions. We are also left with cell biological questions including why these granules migrate and become transiently stabilized in such close proximity to the plasma membrane, why limited cortical granule loss occurs during meiotic maturation, what is the fate of the membrane of released granules, and why sperm penetration appears to preferentially take place away from the cortical granule-free domain. Answers to many of these questions should help us address the larger question of why, after the resumption of meiotic maturation, is the egg's useful, fertile life span so strikingly short relative to the time it has already spent in the ovary.

REFERENCES

1. **Kola, I. And Trounson, A.,** Dispermic human fertilization: violation of expected cell behavior, in *The Cell Biology of Fertilization,* Schatten, H. and Schatten, G., Eds., Academic Press, New York, 1989, 277.
2. **Jacobs, P. A., Angell, R. R., Buchanan, J. M., Hassold, T. J., Matusyama, A. M., and Manuel, B.,** The origin of human triploids, *Ann. Hum. Genet.,* 42, 49, 1978.
3. **Beatty, R. A.,** The origin of human triploidy: an integration of qualitative and quantitative evidence, *Ann. Hum. Genet.,* 41, 299, 1978.
4. **Uchida, I. A. and Freeman, V. C. P.,** Triploidy and chromosomes, *Am. J. Obstet. Gynecol.,* 151, 65, 1985.
5. **Kola, I., Trounson, A., Dawson, G., and Rogers, P.,** Tripronuclear human oocytes: altered cleavage patterns and subsequent karyotypic analysis of embryos, *Biol. Reprod.,* 37, 395, 1987.
6. **Jaffe, L. A. and Gould, M.,** Polyspermy-preventing mechanisms, in *Biology of Fertilization,* Vol. 3, Metz, C. and Monroy, A., Eds., Academic Press, New York, 1985, 223.
7. **Braden, A. W. H., Austin, C. R., and David, H. A.,** The reaction of the zona pellucida to sperm penetration, *Aust. J. Biol. Sci.,* 7, 391, 1954.
8. **Austin, C. R. and Braden, A. W. H.,** Early reactions of the rodent egg to spermatozoon penetration, *J. Exp. Biol.,* 33, 358, 1956.
9. **Wassarman, P. M.,** The biology and chemistry of fertilization, *Science,* 235, 553, 1987.
10. **Schuel, H.,** Secretory functions of egg cortical granules in fertilization and development: a critical review, *Gamete Res.,* 1, 299, 1978.
11. **Schuel, H.,** Functions of egg cortical granules, in *Biology of Fertilization,* Vol. 3, Metz, C. B. and Monroy, A., Eds., Academic Press, New York, 1985, 1.
12. **Epel, D.,** Mechanisms of activation of sperm and egg during fertilization of sea urchin gametes, *Curr. Top. Dev. Biol.,* 12, 185, 1978.
13. **Schmell, E. D., Gulyas, B. J., and Hedrick, J. L.,** Egg surface changes during fertilization and the molecular mechanism of the block to polyspermy, in *Mechanism and Control of Animal Fertilization,* Hartmann, J. F., Ed., Academic Press, New York, 1983, 365.
14. **Jaffe, L. F.,** The role of calcium explosions, waves, and pulses in activating eggs, in *Biology of Fertilization,* Vol. 3, Metz, C. B. and Monroy, A., Eds., Academic Press, New York, 1985, 127.
15. **Whitaker, M. J. and Steinhardt, R. A.,** Ionic signalling in the sea urchin egg at fertilization, in *Biology of Fertilization,* Vol. 3, Metz, C. B. and Monroy, A., Eds., Academic Press, Orlando, FL, 1985, 168.
16. **Trimmer, J. S. and Vacquier, V. D.,** Activation of sea urchin gametes, *Annu. Rev. Cell Biol.,* 2, 1, 1986.
17. **Elinson, R. P.,** Fertilization in amphibians: the ancestry of the block to polyspermy, *Int. Rev. Cytol.,* 101, 59, 1986.
18. **Turner, P. R. and Jaffe, L. A.,** G-proteins and the regulation of oocyte maturation and fertilization, in *The Cell Biology of Fertilization,* Schatten, H. and Schatten, G., Eds., Academic Press, New York, 1989, 297.

19. **Rebhun, L. I. and Fisher, G. W.,** Initiation of two-way cortical traffic after fertilization in sea urchin eggs, *J. Cell Biol.*, 99, 129, 1984.
20. **Fisher, G. W., Summers, R. G., and Rebhun, L. I.,** Analysis of sea urchin egg cortical transformation in the absence of cortical granule exocytosis, *Dev. Biol.*, 109, 489, 1985.
21. **Lee, H. C. and Epel, D.,** Changes in intracellular acidic compartments in sea urchin eggs after activation, *Dev. Biol.*, 98, 446, 1983.
22. **Epel, D.,** Arousal of activity in sea urchin eggs at fertilization, in *The Cell Biology of Fertilization,* Schatten, H. and Schatten, G., Eds., Academic Press, New York, 1989, 361.
23. **Kay, E. S. and Shapiro, B. M.,** The formation of the fertilization membrane of the sea urchin egg, in *Biology of Fertilization,* Vol. 3, Metz, C. B. and Monroy, A., Eds., Academic Press, New York, 1985, 45.
24. **Battaglia, D. E. and Shapiro, B. M.,** Hierarchies of protein cross-linking in the extracellular matrix: involvement of an egg surface transglutaminase in early stages of fertilization envelope assembly, *J. Cell Biol.*, 107, 2447, 1988.
25. **Turner, E., Hager, L. J., and Shapiro, B. M.,** Ovothiol replaces glutathione peroxidase as a hydrogen peroxide scavenger in sea urchin eggs, *Science,* 242, 939, 1988.
26. **Kline, D., Simoncini, L., Mandel, G., Maue, R. A., Kado, R. T., and Jaffe, L. A.,** Fertilization events induced by neurotransmitters after injection of mRNA in Xenopus eggs, *Science,* 241, 464, 1988.
27. **Bement, W. M. and Capco, D. G.,** Activators of protein kinase C trigger cortical granule exocytosis, cortical contraction, and cleavage furrow formation in *Xenopus laevis* oocytes and eggs, *J. Cell Biol.*, 108, 885, 1989.
28. **Yu, S. and Wolf, D. P.,** Polyspermic mouse eggs can dispose of supernumerary sperm, *Dev. Biol.*, 82, 203, 1981.
29. **Longo, F. J.,** Egg cortical architecture, in *The Cell Biology of Fertilization,* Schatten, H. and Schatten, G., Eds., Academic Press, New York, 1989, 105.
30. **Yanagimachi, R. and Chang, M. C.,** Fertilizable life of golden hamster ova and their morphological changes at the time of losing fertilizability, *J. Exp. Zool.*, 148, 185, 1961.
31. **Szollosi, D.,** Development of cortical granules and the cortical reaction in rat and hamster eggs, *Anat. Rec.*, 159, 431, 1967.
32. **Zamboni, L.,** Ultrastructure of mammalian oocytes and ova, *Biol. Reprod.*, 2(Suppl.), 44 1970.
33. **Nicosia, S., Wolf, D. P., and Inoue, M.,** Cortical granule distribution and cell surface characteristics in mouse eggs, *Dev. Biol.*, 57, 56, 1977.
34. **Okada, A., Yanigimachi, R., and Yanigimachi, H.,** Development of a cortical granule-free area of cortex and the perivitelline space in the hamster oocyte during maturation and following ovulation, *J. Submicrosc. Cytol.*, 18, 233, 1986.
35. **Longo, F. J. and Chen, D. Y.,** Development of cortical polarity in mouse eggs: involvement of the meiotic apparatus, *Dev. Biol.*, 107, 382, 1985.
36. **Cherr, G. N., Drobnis, E. Z., and Katz, D. F.,** Localization of cortical granule constituents before and after exocytosis in the hamster egg, *J. Exp. Zool.*, 246, 81, 1988.
37. **Lee, S. H., Ahuja, K. K., Gilburt, D. J., and Whittingham, D. J.,** The appearance of glycoconjugates associated with cortical granule release during mouse fertilization, *Development,* 102, 595, 1988.
38. **Ducibella, T., Anderson, E., Albertini, D. F., Aalberg, J., and Rangarajan, S.,** Quantitative studies of changes in cortical granule number and distribution in the mouse oocyte during meiotic maturation, *Dev. Biol.*, 130, 184, 1988.
39. **Eager, D. D., Johnson, M. H., and Thurley, K. W.,** Ultrastructural studies on the surface membrane of the mouse egg, *J. Cell Sci.*, 22, 345, 1976.
40. **Johnson, M. H., Eager, D., Muggleton-Harris, A., and Grave, M. H.,** Mosaicism in organization of concanavalin A receptors on surface membrane of mouse egg, *Nature,* 257, 321, 1975.
41. **Wolf, D. E. and Ziomek, C. A.,** Regionalization and lateral diffusion of membrane proteins in unfertilized and fertilized mouse eggs, *J. Cell Biol.*, 96, 1786, 1983.
42. **Yanagimachi, R. and Noda, Y. D.,** Physiological changes in the postnuclear cap region of the mammalian sperm: a necessary preliminary to the membrane fusion between sperm cells, *J. Ultrastruct. Res.*, 31, 48, 1970.
43. **Gwatkin, R. B. L., Rasmusson, G. H., and Williams, D. T.,** Induction of the cortical reaction in hamster eggs by membrane-active agents, *J. Reprod. Fertil.*, 47, 299, 1976.
44. **Yanagimachi, R., Yanagimachi, H., and Rogers, B. J.,** The use of zona-free animal ova as a test-system for the assessment of the fertilizing capacity of human spermatozoa, *Biol. Reprod.*, 15, 471, 1976.
45. **Gulyas, B. J.,** Cortical granules of mammalian eggs, *Int. Rev. Cytol.*, 63, 357, 1980.
46. **Cuthbertson, K. S. R., Whittingham, D. G., and Cobbold, P. H.,** Free Ca^{2+} increases in exponential phases during mouse oocyte activation, *Nature,* 294, 754, 1981.
47. **Miyazaki, S., Hashimoto, N., Yoshimoto, Y., Kishimoto, T., Igusa, Y., and Hiramoto, Y.,** Temporal and spatial dynamics of the periodic increase in intracellular free calcium at fertilization of golden hamster eggs, *Dev. Biol.*, 118, 259, 1986.

48. **Fulton, B. P. and Whittingham, D. G.,** Activation of mammalian oocytes by intracellular injection of calcium, *Nature,* 273, 149, 1978.
49. **Steinhardt, R. A., Epel, D., Carroll, E. J., and Yanagimachi, R.,** Is calcium ionophore a universal activator for unfertilized eggs?, *Nature,* 252, 41, 1974.
50. **Wolf, D. P., Nicosia, S. V., and Hamada, M.,** Premature cortical granule loss does not prevent sperm penetration of mouse eggs, *Dev. Biol.,* 71, 22, 1979.
51. **Miyazaki, S.,** *Mechanisms of Egg Activation,* Nuccitelli, R., Cherr, G. N., and Clark, W. H., Jr., Eds., Plenum Press, New York, 1989, 231.
52. **Miyazaki, S.,** Inositol 1,4,5-trisphosphate-induced calcium release and guanine nucleotide-binding protein-mediated periodic calcium rises in golden hamster eggs, *J. Cell Biol.,* 106, 345, 1988.
53. **Cran, D. B., Moor, R. M., and Irvine, R. F.,** Initiation of the cortical reaction in hamster and sheep oocytes in response to inositol trisphosphate, *J. Cell Sci.,* 91, 139, 1988.
54. **Kurasawa, S., Schultz, R. M., and Kopf, G. S.,** Egg-induced modifications of the zona pellucida of mouse eggs: effects of microinjected inositol 1,4,5-trisphosphate, *Dev. Biol.,* 133, 295, 1989.
55. **Endo, Y., Schultz, R. M., and Kopf, G. S.,** Effects of phorbol esters and a diacylglycerol on mouse eggs: inhibition of fertilization and modification of the zona pellucida, *Dev. Biol.,* 119, 199, 1987.
56. **Endo, Y., Mattei, P., Kopf, G. S., and Schultz, R. M.,** Effects of a phorbol ester on mouse eggs: dissociation of sperm receptor activity from acrosome reaction-inducing activity of the mouse zona pellucida protein ZP3, *Dev. Biol.,* 123, 574, 1987.
57. **Wolf, D. E., Edidin, M., and Handyside, A. H.,** Changes in the organization of the mouse egg plasma membrane upon fertilization and first cleavage: indications from the lateral diffusion rates of fluorescent lipid analogs, *Dev. Biol.,* 85, 195, 1981.
58. **Wolf, D. P.,** The mammalian egg's block to polyspermy, in *Fertilization and Embryonic Development In Vitro,* Mastroianni, L., Jr. and Biggers, J. D., Eds., Plenum Press, New York, 1981, 183.
59. **Barros, C. and Yanagimachi, R.,** Polyspermy-preventing mechanisms in the golden hamster egg, *J. Exp. Zool.,* 180, 251, 1972.
60. **Fukuda, Y. and Chang, M. C.,** The time of cortical granule breakdown and sperm penetration in mouse and hamster eggs inseminated in vitro, *Biol. Reprod.,* 19, 261, 1978.
61. **Fraser, L. R., Dandekar, P. V., and Gordon, M. K.,** Loss of cortical granules in rabbit eggs exposed to spermatozoa in vitro, *J. Reprod. Fertil.,* 29, 295, 1972.
62. **Ducibella, T., Kurasawa, S., Rangerajan, S., Kopf, G., and Schultz, M.,** Precocious loss of cortical granules during mouse oocyte meiotic maturation and correlation with an egg-induced modification of the zona pellucida, *Dev. Biol.,* 137, 46, 1990.
63. **Wolf, D. E., Kinsey, W., Lennarz, W., and Edidin, M.,** Changes in the organization of the sea urchin egg plasma membrane upon fertilization: indications from the lateral diffusion rates of lipid-soluble fluorescent dyes, *Dev. Biol.,* 81, 133, 1981a.
64. **Gordon, J. W., Grunfeld, L., Garrisi, G. J., Talansky, B. E., Richards, C., and Laufer, N.,** Fertilization of human oocytes by sperm from infertile males after zona pellucida drilling, *Fertil. Steril.,* 50, 68, 1988.
65. **Soupart, P. and Strong, P. A.,** Ultrastructural observations on human oocytes fertilized in vitro, *Fertil. Steril.,* 25, 11, 1974.
66. **Gwatkin, R. B. L. and Williams, D. T.,** Heat sensitivity of the cortical granule protease from hamster eggs, *J. Reprod. Fertil.,* 49, 55, 1977.
67. **Sato, K.,** Polyspermy-preventing mechanisms in mouse eggs fertilized in vitro, *J. Exp. Zool.,* 210, 353, 1979.
68. **Inoue, M. and Wolf, D. P.,** Fertilization-associated changes in the murine zona pellucida: a time sequence study, *Biol. Reprod.,* 13, 546, 1975.
69. **Sato, K. and Blandau, R. J.,** Time and process of sperm penetration into cumulus-free mouse eggs fertilized in vitro, *Gamete Res.,* 2, 295, 1979.
70. **Bleil, J. D. and Wassarman, P. M.,** Mammalian sperm-egg interaction: identification of a glycoprotein in mouse egg zonae pellucidae possessing receptor activity for sperm, *Cell,* 20, 873, 1980.
71. **Bleil, J. D. and Wassarman, P. M.,** Sperm-egg interactions in the mouse: sequence of events and induction of the acrosome reaction by a zona pellucida glycoprotein, *Dev. Biol.,* 95, 317, 1983.
72. **Barros, C. and Yanagimachi, R.,** Induction of zona reaction in golden hamster eggs by cortical granule material, *Nature,* 233, 268, 1971.
73. **Gwatkin, R. B. L., Williams, D. T., Hartmann, J. F., and Kniazuk, M.,** The zona reaction of hamster and mouse eggs: production in vitro by a trypsin-like protease from cortical granules, *J. Reprod. Fertil.,* 32, 259, 1973.
74. **Wolf, D. P. and Hamada, M.,** Induction of zonal and oolemmal blocks to sperm penetration in mouse eggs with cortical granule exudate, *Biol. Reprod.,* 17, 350, 1977.
75. **Foerder, C. A. and Shapiro, B. M.,** Release of ovoperoxidase from sea urchin eggs hardens the fertilization membrane with tyrosine crosslinks, *Proc. Natl. Acad. Sci. U.S.A.,* 79, 4214, 1977.

76. **Gulyas, B. J. and Schmell, E. D.,** Ovoperoxidase activity in ionophore treated mouse eggs. I. Electron microscopic localization, *Gamete Res.*, 3, 267, 1980.
77. **Schmell, E. D. and Gulyas, B. J.,** Ovoperoxidase activity in ionophore treated mouse eggs. II. Evidence for the enzyme's role in hardening the zona pellucida, *Gamete Res.*, 3, 279, 1980.
78. **Choi, T. S., Mori, M., Kohmoto, K., and Shoda, Y.,** Beneficial effect of serum on the fertilizability of mouse oocytes matured in vitro, *J. Reprod. Fertil.*, 79, 565, 1987.
79. **Downs, S. M., Schroeder, A. C., and Eppig, J. J.,** Developmental capacity of mouse oocytes following maintenance of meiotic arrest in vitro, *Gamete Res.*, 15, 115, 1986.
80. **DeFelici, M. and Siracusa, G.,** Spontaneous hardening of the zona pellucida or mouse oocytes during in vitro culture, *Gamete Res.*, 6, 107, 1982.
81. **DeFelici, M., Salustri, A., and Siracusa, G.,** "Spontaneous" hardening of the zona pellucida of mouse oocytes during in vitro culture. II. The effect of follicular fluid and glycosaminoglycans, *Gamete Res.*, 12, 227, 1985.
82. **Leese, H. J.,** The formation and function of oviduct fluid, *J. Reprod. Fertil.*, 82, 843, 1988.
83. **Balmaceda, J. P., Heitman, T. O., and Borrero, R. H.,** *Am. Fertil. Soc. Abstr.*, 5(Abstr. 015), 1986.
84. **Moller, C. C. and Wassarman, P. M.,** Characterization of proteinase that cleaves zona pellucida glycoprotein ZP2 following activation of mouse eggs, *Dev. Biol.*, 132, 103, 1989.
85. **Gwatkin, R. B. L. and Williams, D. T.,** Heat sensitivity of the cortical granule protease from hamster eggs, *J. Reprod. Fertil.*, 39, 153, 1974.
86. **Flechon, J. E., Haneau, D., Solari, A., and Thibault, C.,** Reaction cortical et blocage de la polyspermie dans l'oeuf de lapine, *Ann. Biol. Anim. Biochim. Biophys.*, 15, 9, 1975.
87. **Gulyas, B.,** Electron microscope observations on advanced stages of spontaneous polyspermy in rabbit zygotes, *Anat. Rec.*, 179, 285, 1974.
88. **Conrad, K., Buckley, J., and Stambaugh, R.,** Studies on the nature of the block to polyspermy in rabbit ova, *J. Reprod. Fertil.*, 27, 133, 1971.
89. **Overstreet, J. W. and Bedford, J. M.,** Comparison of the penetrability of the egg vestments in follicular oocytes, unfertilized and fertilized ova of the rabbit, *Dev. Biol.*, 41, 185, 1974.
90. **Kaleta, E.,** The influence of genetic factors on fertilization of mouse ova in vitro, *J. Reprod. Fertil.*, 51, 375, 1977.
91. **Stewart-Savage, J. and Bavister, B. D.,** A cell surface block to polyspermy occurs in golden hamster eggs, *Dev. Biol.*, 128, 150, 1988.
92. **Anderson, E.,** Oocyte differentiation in the sea urchin, *Arbacia punctulata* with particular reference to the origin of cortical granules and their participation in the cortical reaction, *J. Cell Biol.*, 37, 514, 1968.
93. **Guraya, S. S.,** Recent progress in the structure, origin, composition, and function of cortical granules in animal egg, *Int. Rev. Cytol.*, 78, 257, 1982.
94. **Anstrom, J. A., Chin, J. E., Leaf, D. S., Parks, A. L., and Raff, R. A.,** Immunocytochemical evidence suggesting heterogeneity in the population of sea urchin egg cortical granules, *Dev. Biol.*, 125, 1, 1988.
95. **Ducibella, T., Rangarajan, S., and Anderson, E.,** The development of mouse oocyte cortical reaction competence is accompanied by major changes in cortical vesicles and not cortical granule depth, *Dev. Biol.*, 130, 789, 1988.
96. **Austin, C. R.,** Cortical granules in hamster eggs, *Exp. Cell Res.*, 10, 533, 1956.
97. **Selman, K. and Anderson, E.,** The formation and cytochemical characterization of cortical granules in ovarian oocytes of the golden hamster (*Mesocricetus auratus*), *J. Morphol.*, 147, 251, 1975.
98. **Peluso, J. J. and Butcher, R. L.,** The effect of follicular aging on the ultrastructure of the rat oocyte, *Fertil. Steril.*, 25, 494, 1974.
99. **Ducibella, T., Duffy, P., Reindollar, R., and Su, B.,** Changes in the distribution of mouse oocyte cortical granules and ability to undergo the cortical reaction during gonadotropin-stimulated meiotic maturation and aging in vivo, *Biol. Reprod.*, in press.
100. **Szollosi, D.,** Oocyte maturation and paternal contribution to the embryo in mammals, in *Current Topics in Pathology*, Vol. 62, Grundmann, E. and Kirstern, N. H., Eds., Springer-Verlag, Berlin, 1976, 9.
101. **Baca, M. and Zamboni, L.,** The fine structure of human follicular oocytes, *J. Ultrastruct. Res.*, 19, 354, 1967.
102. **Kruip, T. A. M., Cran, D. G., van Benedan, T. H., and Dieleman, S. J.,** Structural changes in bovine oocytes during final maturation in vivo, *Gamete Res.*, 8, 29, 1983.
103. **Cran, D. and Cheng, W.,** Changes in cortical granules during porcine oocyte maturation, *Gamete Res.*, 11, 311, 1985.
104. **Sathananthan, A. H. and Trounson, A. O.,** Ultrastructural observations on cortical granules in human follicular oocytes cultured in vitro, *Gamete Res.*, 5, 191, 1982.
105. **Sathanathan, A. H. and Trounson, A. O.,** Ultrastructure of cortical granule release and zona interaction in monospermic and polyspermic human ova fertilized in vitro, *Gamete Res.*, 6, 225, 1982.
106. **Longo, F. J. and Chen, D. Y.,** Development of surface polarity in mouse eggs, *Scanning Electron Microsc.*, p. 703, 1984.

107. **Bleil, J. D., Beall, C. F., and Wassarman, P. M.,** Mammalian sperm-egg interaction: fertilization of mouse eggs triggers modification of the major zona pellucida glycoprotein, ZP2, *Dev. Biol.,* 86, 189, 1981.
108. **Poenie, M., Alderton, J., Tsien, R. Y., and Steinhardt, R. A.,** Changes in free calcium levels with stages of the cell division cycle, *Nature,* 315, 147, 1985.
109. **Ratan, R. R., Shelanski, M. L., and Maxfield, R. R.,** The transition from metaphase to anaphase is accompanied by local changes in cytoplasmic free calcium in PtK2 cells, *Proc. Natl. Acad. Sci. U.S.A.,* 83, 5136, 1986.
110. **Ratan, R. R., Maxfield, F. R., and Shelanski, M. L.,** Long-lasting and rapid calcium changes during mitosis, *J. Cell Biol.,* 107, 993, 1988.
111. **Krzanowska, H.,** Relation between fertilization rate and penetration of eggs by supplementary spermatozoa in different mouse strains and crosses, *J. Reprod. Fertil.,* 22, 199, 1970.
112. **Wabik-Sliz, B.,** Number of cortical granules in mouse oocytes from inbred strains differing in efficiency of fertilization, *Biol. Reprod.,* 21, 89, 1979.
113. **Krzanowska, H.,** Rapidity of removal in vitro of the cumulus oophorus and the zona pellucida in different strains of mice, *J. Reprod. Fertil.,* 31, 7, 1972.
114. **Kaleta, E. and Polak, Z.,** Solubility properties of the zona pellucida of mouse oocytes from inbred strains differing in the number of cortical granules, *Zwiertzeta Lab.,* 15, 53, 1978.
115. **Lopata, A., Sathananthan, A. H., McBain, J. C., Johnston, W. I. H., and Speirs, A. L.,** The ultrastructure of the preovulatory human egg fertilized in vitro, *Fertil. Steril.,* 33, 12, 1980.
116. **Szollosi, D.,** Morphological changes in mouse eggs due to aging in the fallopian tubes, *Am. J. Anat.,* 130, 209, 1971.
117. **Longo, F. J.,** Spontaneous activation of the hamster egg in vitro, in *Ultrastructure Endocrine and Reproductive Organs,* Vol. 179, Hess, M., Ed., John Wiley & Sons, New York, 1974, 35.
118. **Longo, F. J.,** Ultrastructural changes in rabbit eggs aged in vitro, *Biol. Reprod.,* 11, 22, 1974.
119. **Takagi, N. and Sasaki, M.,** Digynic tripoloidy after superovulation in mice, *Nature,* 264, 278, 1976.
120. **Hunter, R. H. F., Cook, B., and Baker, T. G.,** Dissociation of response to injected gonadotropin between the graafian follicle and oocyte in pigs, *Nature,* 260, 156, 1976.
121. **Polge, D. and Dziuk, P.,** Recovery of immature eggs pentrated by spermatozoa following induced ovulation in the pig, *J. Reprod. Fertil.,* 9, 357, 1965.
122. **Trounson, A. O., Mohr, L. R., Wood, C., and Leeton, J. F.,** Effect of delayed insemination on in-vitro fertilization, culture and transfer of human embryos, *J. Reprod. Fertil.,* 64, 285, 1982.
123. **Marrs, R. P., Saito, H., Yee, B., Sato, F., and Brown, J.,** Effect of variation of in vitro culture techniques upon oocyte fertilization and embryo development in human in vitro fertilization procedures, *Fertil. Steril.,* 41, 519, 1984.
124. **Iwamatsu, T. and Chang, M. C.,** Sperm penetration in vitro of mouse oocytes at various times during maturation, *J. Reprod. Fertil.,* 31, 127, 1972.
125. **Lopata, A.,** personal communication.
126. **Longo, F. J.,** Insemination of immature sea urchin (*Arbacia punctulata*) eggs, *Dev. Biol.,* 62, 271, 1978.
127. **Ducibella, T., Su, B., and Duffy, P.,** Determination of the onset of competence to undergo the cortical reaction in the mouse oocyte after human chorionic gonadotropin (HCG) administration, *Soc. Gyn. Invest.,* Abstr., 276, 1989.
128. **Harrison, K. L., Wilson, L. M., Breen, T. M., Pope, A. K., Cummins, J. M., and Hennessey, J. F.,** Fertilization of human oocytes in relation to varying delay before insemination, *Fertil. Steril.,* 50, 294, 1988.
129. **Cho, W. K., Stern, S., and Biggers, J. D.,** Inhibitory effect of dibutyryl cAMP on mouse oocyte maturation in vitro, *J. Exp. Zool.,* 187, 383, 1974.
130. **Wassarman, P. M., Joseforwicz, W. J., and Letourneau, G. A.,** Meiotic maturation of mouse oocytes in vitro: inhibition of maturation at specific stages of nuclear progression, *J. Cell Sci.,* 22, 531, 1976.
131. **Schultz, R. M. and Wassarman, P. M.,** Biochemical studies of mammalian oogenesis: protein synthesis during oocyte growth and meiotic maturation in the mouse, *J. Cell Sci.,* 24, 167, 1977.
132. **Endo, Y., Kopf, G. S., and Schultz, R. M.,** Stage specific changes in protein phosphorylation accompanying oocytes and fertilization of mouse eggs, *J. Exp. Zool.,* 239, 401, 1986.
133. **Perrault, S. D., Barbee, R. R., and Slott, V. L.,** Importance of glutathione in the aquisition and maintenance of sperm nuclear decondensing activity in maturing hamster oocytes, *Dev. Biol.,* 125, 158, 1988.
134. **Clarke, H. J. and Masui, Y.,** Dose-dependent relationship between oocyte cytoplasmic volume and transformation of sperm nuclei to metaphase chromosomes, *J. Cell Biol.,* 104, 831, 1987.
135. **Balakier, H. and Tarkowski, A. K.,** The role of germinal vesicle karyoplasm in the development of the male pronucleus in the mouse, *Exp. Cell Res.,* 128, 79, 1980.
136. **Gardiner, D. M. and Grey, R. D.,** Membrane junctions in xenopus eggs: their distribution suggests a role in calcium regulation, *J. Cell Biol.,* 96, 1159, 1983.

137. **Henson, J. H., Beaulieu, S. M., Begg, D. A., Fishkind, D. J., Bonder, E. M., Lebeche, D., and Kaminer, B.,** Localization of a calsequestrin-like protein in the ER of sea urchin eggs, *J. Cell Biol.,* 107(Abstr. 987), 172a, 1988.
138. **Luttmer, S. and Longo, F. J.,** Ultrastructural and morphometric observations of cortical endoplasmic reticulum in Arbacia, Spisula and mouse eggs, *Dev. Growth Differ.,* 27, 349, 1985.
139. **Fugo, N. W. and Butcher, R. L.,** Overripeness and the mammalian ova. I. Overripeness and early embryonic development, *Fertil. Steril.,* 17, 804, 1966.
140. **Austin, C. R. and Braden, A. W. H.,** An investigation of polyspermy in the rat and rabbit, *Aust. J. Biol. Sci.,* 6, 674, 1953.
141. **Odor, D. L. and Blandau, R. J.,** Incidence of polyspermy in normal and delayed mating in rats of the Wistar strain, *Fertil. Steril.,* 7, 456, 1956.
142. **Marston, J. H. and Chang, M. C.,** The fertilizable life of ova and their morphology following delayed insemination in mature and immature mice, *J. Exp. Zool.,* 155, 237, 1964.
143. **Hunter, R. H. F.,** The effects of delayed insemination on fertilization and early cleavage in the pig, *J. Reprod. Fertil.,* 13, 133, 1967.
144. **Ben-Rafael, Z., Meloni, F., Strauss, J. F., Blasco, L., Mastroianni, L., and Flickinger, G. L.,** Relationships between polypronuclear fertilization and follicular fluid hormones in gonadotropin-treated women, *Fertil. Steril.,* 47, 284, 1987.
145. **Austin, C. R.,** Fertilization, in *Concepts of Development,* Lash, J. and Whittaker, J. R., Eds., Sinauer Associates, Inc., Stamford, 1974, 48.

Chapter 9

BIOCHEMISTRY AND FUNCTION OF ACROSIN

Umbert A. Urch

TABLE OF CONTENTS

I. INTRODUCTION

Sperm hydrolases and their function in fertilization have been the subject of much research and investigation (for reviews see References 1 to 5). Perhaps the first definitive observation of a sperm hydrolase having an effect on egg envelopes was that of Yamane in 1935.[6] He demonstrated that a toluene extract of rabbit sperm dispersed cumulus cells and solubilized the zona pellucida (ZP) of the rabbit egg and he proposed the presence of proteases by analogy with the similar action of a crude pancreatin extract. Later, Tyler found that an extract of keyhole limpet sperm dissolved the egg membranes without any other effect on the egg.[7] This observation and others led him in 1948 to hypothesize that sperm hydrolases assisted sperm in egg membrane penetration. The first direct evidence that a proteolytic enzyme from acrosomal preparations dissolved the zona of rabbit eggs was put forward by Srivastava et al. in 1965.[8] These observations became the foundation for the hypothesis that a sperm-associated protease aided in sperm penetration of the mammalian ZP by hydrolyzing this glycoproteinaceous coat.[1,3,9]

This protease was named acrosin because of its localization in the sperm acrosome, a modified lysosome on the sperm head, and by analogy to trypsin as a digestive enzyme. Contemporary studies on acrosin began with the observations that sperm extracts hydrolyzed synthetic trypsin substrates[9] and that pH 3 effectively dissociated enzyme-inhibitor complexes.[10] However, the definition of acrosin as trypsin-like is counterproductive to understanding its function in fertilization. Acrosin is immunologically unique,[11] and larger[12] than trypsin from the homologous species. These structural differences reflect different functional domains of acrosin and are manifested in its unique role in fertilization.[13] The intent of this review is to highlight information regarding acrosin sequence and structure, including hydrophobicity, to examine acrosin localization studies, and to relate these structural and localization features to the proposed functions of acrosin in fertilization.

II. STRUCTURAL CHARACTERIZATION OF ACROSIN

A. GENERAL STRUCTURAL FEATURES

Acrosin from several species has been described as a serine active-site endoprotease with trypsin-like specificity for Arg/Lys peptide bonds in proteins, ester, and amide synthetic substrates.[12,14-16] Most of the structural evidence collected thus far has concerned the most stable form of boar acrosin. This acrosin (E.C. 3.4.21.10) is a glycosylated enzyme composed of 8.5% carbohydrate and 328 amino acid residues for a combined M_r or 38K.[17] Boar acrosin is 50% larger than porcine trypsin, with higher amounts of proline (10.9%) and arginine 7.6%), and with a higher polarity index of 41%.[18] This latter property may reflect portions of the macromolecule that are membrane-associating domains,[19] and will be the subject of a later section.

The purification of the proenzyme form of acrosin necessitates the use of low pH and the presence of trypsin inhibitors such as benzamidine. Proacrosins have been described in the following species: boar (53 to 55K),[20] bull,[21] guinea pig (62K—testicular, 55K and 43K—epididymal),[22] hamster (70K),[23] human[24] (53 to 55K),[25] mouse,[26] rabbit (68K),[27] and ram (51.3K).[28] The different M_rs for the guinea pig proacrosins led the investigators to hypothesize that the enzymes were processed during transfer through the male tract. Another finding was that proacrosin is capable of autoactivation, resulting in the sequential formation of active acrosins.[29]

Active acrosins have been purified from a number of species: toad,[30] ascidian,[31] boar,[17,32,33] cock,[34] guinea pig,[35] goat,[36] human,[37] and mouse.[26] The activation cascade of proenzyme to active acrosins forms has again been best characterized in the boar, and the cascade appears similar wherever a proacrosin has been characterized. The first isolated active form in boar is designated m_α-acrosin (49K M_r by SDS-PAGE), which is unstable and quickly degraded to a more stable form called m_β-acrosin (38K M_r by SDS-PAGE).[33] An acrosin at 25K M_r, designated m_γ-acrosin, is reportedly formed only after prolonged incubation at basic pH.[20] When boar acrosin is purified in the absence of trypsin inhibitors, the main form obtained has a molecular weight at 38K M_r, with several low molecular weight presumptive hydrolysis products of proacrosin and acrosin.[17,32] Acrosin prepared in the presence and absence of detergents cross-reacts immunologically, indicating antigenic identity in the preparations.[38]

In two systems where carbohydrate analysis was undertaken, rabbit proacrosin and boar acrosin were found to be glycosylated.[17,27] Carbohydrate analysis of boar acrosin demonstrated the presence of fucose, *N*-acetyl glucosamine, mannose, and galactose.[17] The presence of *N*-acetyl glucosamine supports the data from the N-terminal amino acid sequence data that there is a carbohydrate side chain at Asn3.[39] Protein sequence data for goat acrosin are consistent with the results for the boar enzyme.[26] Sequence analysis of the human acrosin cDNA shows the presence of two consensus sequences for N-linkage (Asn-X-Ser/Thr), one at the N terminal, and one at position 191.[40] The characterization of the oligosaccharide side chain(s) of acrosin has not yet been accomplished. The function of these side chains in fertilization is also not known at present, and remains an intriguing question for future work.

B. LIPOPHILICITY OF ACROSIN

Much contradictory and confusing information exists on the lipophilicity of acrosin. Acrosin is an unusual protease in that it aggregates with itself and with other proteins in the absence of detegents,[41,42] has high affinity for lipids and detergents, has a polarity index of 41%,[17] and is enzymatically stimulated by nonionic detergents.[19] It was these properties of boar acrosin that led to its description as an extrinsic membrane protein, or one that is loosely associated with membrane lipids or proteins. Boar acrosin purified in the presence of detergents, and then treated to remove detergent, did not bind a micelle of Triton X-100 based on gel filtration studies.[19] The enzyme was significantly activated by Triton X-100 and deoxycholate, an effect typical on intrinsic membrane proteins, e.g., cytochrome b_5. On the other hand, intrinsic membrane proteins span the lipid bilayer and bind detergent micelles. Acrosin purified in the absence of added detergent not only bound Triton X-100 micelles[43] but behaved as an amphiphilic protein in the coupled detergent electrophoretic assay.[44] In this assay, amphiphilic proteins bind charged detergent-couples resulting in a mobility shift. In contrast, hydrophilic proteins are unaffected by such treatment.[43] Adding further to these conflicting data, the sequence of human acrosin does not have any long stretches that are hydrophobic, or presumptive hydrophobic tails with which to penetrate the membrane.[40]

Boar m_β-acrosin has been reported to be inhibited by phospholipid vesicles,[19] whereas anionic phospholipid dispersions stimulated the autolysis of proacrosin.[45] In addition, when sperm membrane vesicles were prepared by nitrogen cavitation,[46,47] acrosin was found associated with these preparations.[43] Acrosin amidase activity was maximally liberated from these vesicles with the sequential addition of 1 m*M* HCl and Triton X-100. Conditions that should have resulted in the maximal extraction of an extrinsic membrane protein (salt extraction, or salt and acid) did not release significant levels of acrosin protein or activity. The interaction of acrosin with hydrophobic agents in general, and sperm acrosomal membranes in particular, is of paramount importance in understanding the roles acrosin plays in fertilization.

C. SEQUENCE ANALYSIS

The amino acid sequences of human and boar proacrosins have been deduced from cDNA clones obtained from testis cDNA libraries,[40,48,49] and the deduced boar proacrosin sequence is identical to the initial 75 residues of boar acrosin that have been sequenced.[39] When proacrosin and acrosin sequences are compared to trypsin, chymotrypsin, and hepsin, according to the sequence comparison of Leytus et al.,[50] there is sequence identity of 71% between goat and human and 69% between boar and human (Figure 1). While acrosins are only approximately 32 to 36% identical to the three serine esterases listed, 77% of the highly conserved positions for serine esterases, and all the invariant amino acids at positions for the catalytic triad, have been retained. The amino acid sequence divergence rate for goat and boar acrosin has been estimated by Hardy et al. to be 1.8 million years per 1% change.[36] This divergence rate is similar to proteins that lack a species-specific function, and the authors reason that if acrosin does have a species-specific interaction with the homologous ZP, it must reside in the ZP binding domains of acrosin, since the portions of the acrosin sequence compared for this calculation were homologous with the other serine proteases.

Boar proacrosin and boar m_α-acrosin have the same 12 N-terminal amino acid sequences,[51] while the m_α- and m_β-forms of acrosin have at least the same amino terminal dipeptide, Val-Val.[32] Initial attempts to determine the N-terminal amino acid sequence of boar acrosin were hampered by the presence of a secondary sequence cross-linked to it by sulfhydryl bridges.[52] Removal of this minor sequence required reducing agents in the presence of 6 *M* guanidine-HCl. A similar two-chain structure was also observed for goat acrosin[36] and for guinea pig acrosins.[22] The molecular weight of the light and heavy chains, respectively, for the various species examined thus far are 4200 and 37K M_r for the boar,[39] 3700 and 38K M_r for the goat,[36] and a 5000 $\pm$ 1000-M_r light chain for the guinea pig proacrosins.[22] The light chain in the boar had a 40% sequence homology with the activation peptide of chymotrypsin[39] leading Fock-Nuzel et al. to propose that the mechanism of proacrosin activation is limited hydrolysis of the single proacrosin polypeptide chain to produce the two-chain active enzyme.[52] When the N terminal of boar proacrosin was sequenced further, the sequence of the acrosin heavy chain was encountered in the sequence immediately after the C-terminal arginine of the light chain,[53] further indicating that proacrosin activation was the cleavage of a single arginine bond at position 23.

The amino acid composition of various acrosins indicates a relatively high proline content.[39] A recent amino acid composition comparison of SDS-PAGE separated and electroeluted acrosins showed that in the conversion from proacrosin to active acrosins of the 85 residues removed, 31 of them are proline.[54] The 18K polypeptide formed in the conversion of boar m_α- to m_β-acrosin has been purified and the amino acid composition corresponds to the difference between m_α- to m_β-acrosin.[55] Taking these data together, it is apparent that acrosin activation occurs at both the N- and C-terminal ends of the molecule. The C-terminal end of the deduced proacrosin sequence contains repeats of three or more prolines, with a stretch of 23 prolines in the boar. There are also arginine and lysine residues interspersed with the prolines, possibly autolytic sites for conversion and maturation of the acrosin molecule. It is not yet known whether this unusual fragment is removed intact or in pieces, but there appear to be several sites for acrosin autolysis.

Since the C-terminal portion of both boar and human acrosin, and therefore proacrosin necessarily, has an unusually high proline content, it seems unlikely that this region of the acrosin sequence is homologous to the other serine proteases. However, there is less identity between acrosins at the C-terminal end (71/124, 57%) than overall (265/380, 69%). While the C-terminal portion of acrosin presents a very unusual structure, further structural analysis and comparison will be required to determine the functions of the region of the molecule.

III. LOCALIZATION OF PROACROSIN AND ACROSIN

While acrosin has been localized to the sperm acrosome by cytochemical, histochemical, and immunological localization studies,[57-60] there are conflicting reports as to exactly where

```
                          10         20         30         40         50         60         70         80
Human Proacrosin1   IVGGKAAQHG AWPWMVSLQI FTY-NSHRYH TCGGSLLNSR WVLTAAHCFV GKNNVHDWRL VFGAKEITYG NNKPVKAPVQ
Porcine Acrosin2    VVGGMSAEPG AWPWMVSLQI FMYHNNRRYH TCGGILLNSH WVLTAAHCFK NKKKVTDWRL IFGANEVVWG SNKPVKPPLQ
Caprine Acrosin3    IIGGQDAAHG SWPWMVSL-- FTYHNNRRYH VCGGSLL
Bovine Chymotryp4   IVNGEEAVPG SWPWCVSLQD -----KTGFH FCGGSLINEN WVVTAAHCG- ----VTTSFD VVAG----EF DQGSSSEKIQ
Bovine Trypsin4     IVGGYTCGAN TVPYCVSL-- -----NSGYH FCGGSLINSQ WVVSAAHCY- -----KSGIQ VRLG----QD NINVVEGNQQ
Hepsin4             IVGGRDTSLG RWPWCVSL-- ----RYDGAH LCGGSLLSGD WVLTAAHCFP ERNRVLSRWR VFAGAVAQAS PHGLQLGVQA

                          90        100        110        120        130        140        150        160
Human Proacrosin    ERYVEKIIIH EKYNSATEGN DIALVEITPP ISCGRFIG-P GCLPHLKAGL PRGSQSCWVA GWGYIEEKAP RPSSILMEAR
Porcine Acrosin     ERFVEEIIIH EKYVSGLEIN DIALIKITPP VPCGPFIG-P GCLPQFKAGP PRAPQTCWVT GWGYLKEKGP RTSPTLQEAR
Bovine Chymtryp.    KLKIAKVFKN SKYNSLTINN DITLLKLSTA ASFSQTVSA- VCLP-SASDD FAAGTTCVTT GWGLTRYTNA NTPDRLQQAS
Bovine Trypsin      FISASKSIVH PSYNSETLNN DIMLIKLKSA ASLNSRVAS- ISLP---TSC ASAGTQCLIS GWGNTKSSGT SYPDVLKCLK
Hepsin              VVYHGGYLPF RDPNSEENSN DIALVHLSSP LPLTEYIQ-P VCLP-AAGQA LVDGKICTVT GWGNTQ YYG QQAGVLQEAR

                         170        180        190        200        210        220        230        240
Human Proacrosin    VDLIDLDLCN STQWTNGRVQ PTNVCAGYPV GKIDTCQGDS GGPLMCKDSK ESAYVVV--G ITSWG-VGCA RAKRPGIYTA
Porcine Acrosin     VALIDLELCN STRWTNGRIR STNVCAGYPR GKIDTCQGDS GGPLMCRDRA ENTFVVV--G ITSWG-VGCA RAKRPGVYTS
Bovine Chymotryp.   LPLLSNTNCK --KYWGTKIK DAMICAG--A SGVSSCMGDS GGPLVCKKN- ---GAWTLVG IVSWGSSTCS TST PGVYAR
Bovine Trypsin      APILSNSSCK --SAYPGQIT SNMFCAGYLE GGKDSCQGDS GGPVVCSGK- -------LQG IVSWG-SGCA QKNKPGVYTK
Hepsin              VPIISNDVCN GADFYGNQIK PKMFCAGYPE GGIDACQGDS GGPFVCEDSI SRTPRWRLCG IVSWG-TGCA LAQKPGVYTK

                         250        260        270        280        290        300        310        320
Human Proacrosin    TWPYLNWIAS KIGSNALRMI QSATPPPPTT RPPPIRPPFS HPISAHLPWY FQPPPRPLPP RPPAAQPRPP PSPPPPPPPP
Porcine Acrosin     TWPYLNWIAS KIGSNALQMV QLGTPPRPST PAPPVRPP-S VQTPVRPPWY FQRP--PGPS QQPGSRPRPP APPPPPPPPP
Bovine Chymotryp.   VTALVNWVQQ TLAAN
Bovine Trypsin      VCNYVSWIKQ TIASN
Hepsin              VSDFREWIFQ AIKTHSEASG MVTQL

                         330        340        350        360        370        380        Positional Identity

Human Proacrosin    ASPLPPPPPP PPPTPSSTTK LPQGLSFAKRL QQLIEVLKG KTYSDGKNHY DMETTELPEL TSTS
Porcine Acrosin     PPPPPPPPPP PPPPQQVSAK PPQALSFAKRL QQLIEALKG TAFSSGRSYY ETETTDLQEL -PAS     265/380    69%
Caprine Acrosin                                                                                  25/35    71%
Bovine Chymotryp.                                                                                74/230    32%
Bovine Trypsin                                                                                   80/223    36%
Hepsin                                                                                           90/265    34%
```


1) Baba, T., Watanabe, K., Kashiwabara, S., and Arai, Y., Primary structure of human proacrosin deduced from its cDNA sequence, *FEBS Lett.*, 244, 296, 1989.
2) Baba, T., Kashiwabara, S., Watanabe, K., Michikawa, Y., Kimura, K., Takada, M., Fukamizu, A., and Arai, Y., Activation and maturation mechanisms of boar acrosin zymogen based on the deduced primary sequence, J. Biol. Chem., 264, 11920, 1989.
3) Hedrick, J.L., Urch, U.A., and Hardy, D.M., Structure-function properties of the sperm enzyme acrosin, in *Biocatalysis in Agricultural Biotechnology*, Whitaker, J.R. and Sonnet, P.E., Eds., ACS Symposium Series No. 389, ACS, Chicago, 1989, chap. 15.
4) Leytus, S.P., Loeb, K.R., Hagen, F.S., Kurachi, K., and Davie, E., A novel serine protease (Hepsin) with a putative transmembrane domain expressed by human liver and hepatoma cells, *Biochemistry*, 27, 1067, 1988.

FIGURE 1. The amino acid sequence of the active portion human proacrosin compared to the partial sequence analyses of caprine and porcine acrosins. Included are the sequences of three serine proteases: trypsin, chymotrypsin, and hepsin. The sequence homology and best fit are reproduced after the analysis of Leytus et al., *Biochemistry*, 27, 1067, 1988.

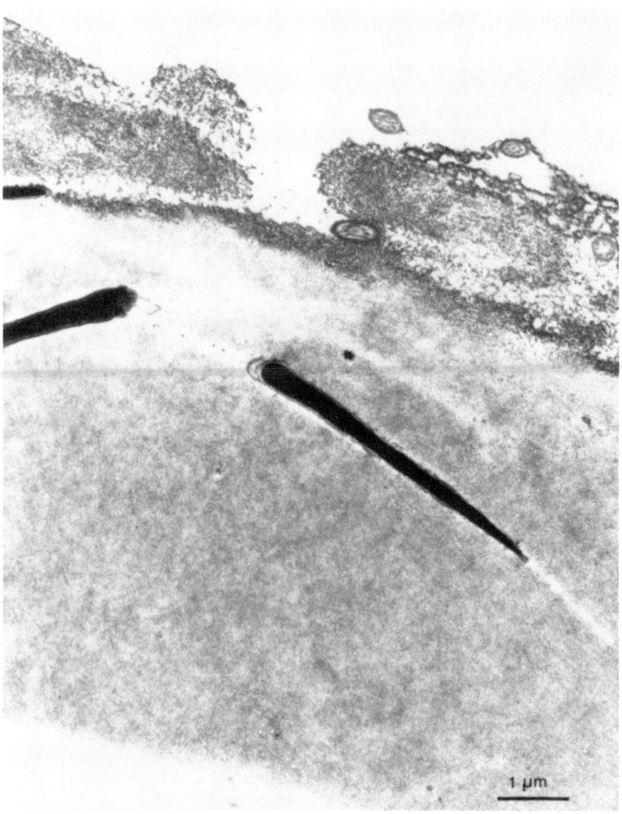

FIGURE 2. The porcine zona pellucida (ZP) containing penetrating boar sperm. In the ZP, the plasma membrane of the sperm is lost over the head region, as are the outer acrosomal membrane and acrosomal contents, as the sperm progresses into the ZP. The inner acrosomal membrane remains at this stage. (Reprinted from Russe, I. and Wendt, V., *Med. Welt,* 28, 746, 1977. With permission.)

acrosin is located in the acrosomal compartment before the acrosome reaction, and how the enzyme is released in the fertilization process. Acrosin has been localized to both the inner acrosomal[59] and outer acrosomal[60] membrane, and to the acrosomal granule matrix.[61] Part of the problem may be that acrosin is ''sticky'' and interacts with membranes and membrane proteins as mentioned above. The localization studies are thus limited by the possible leakage of acrosin from the acrosome, or perhaps by the presence of acrosin from other sperm. Another potential problem may have to do with the acrosome reaction paradigm. The acrosome reaction has been described to proceed with acrosomal membrane vesiculation, followed by acrosomal matrix dissolution, leading to the speculation that acrosin disperses before the sperm penetrates the ZP.[62] However, in the case of boar sperm, the acrosomal contents are carried along with the sperm, and they gradually erode away as the sperm penetrates the ZP (Figure 2).[63] Recently, several studies have indicated that acrosin, or a portion of it, is on the surface of the intact sperm.[64-66] These data may reflect two pools of acrosin, one in the acrosome and one on the sperm, or they may be a reflection of the inherent stickiness of the molecule. How acrosin gets to the sperm surface and whether it is the same molecular form of acrosin as found in the acrosome are important questions that will further define how acrosin functions in fertilization.

In many mammalian species, the acrosome consists of a semisolid matrix composed of acrosin and other molecules, and an enzyme inhibited by trypsin inhibitors, perhaps acrosin itself, is required to degranulate it.[61,67-70] This organelle has enough integrity to be fractionated away from the rest of the sperm contents, and the acrosomal granule minus the limiting membranes has been isolated in the guinea pig.[71] The outer acrosomal membrane and a portion of the matrix have been isolated in both the boar and guinea pig.[72-74] In these preparations high levels of acrosin were found associated with the acrosomal matrix as proacrosin. Soluble acrosin was liberated from guinea pig sperm even in the presence of trypsin inhibitors that inhibited the acrosome reaction from occurring, and this constituted as much as 74% of the total sperm acrosin.[61] A slightly different conclusion was obtained in hamster, where Olson et al. suggested that acrosin is not a major part of the detergent-resistant acrosomal matrix.[75] However, acrosin inhibitors are a necessary requirement in acrosomal extraction and electrophoresis, and in the absence of inhibitors these authors obtained many smaller products on SDS-PAGE, presumably from acrosin hydrolysis. Biochemical fractionation procedures have also indicated enzymatically active acrosin bound to the inner acrosomal membrane.[76-78]

In the ram, boar, and guinea pig, proacrosin is found tightly associated with binding protein(s) which in the case of the boar behaved as membrane protein(s).[71,79,80] Since the binding interaction required either 8 *M* urea or 6 *M* guanidine HCl to disrupt it,[80] this protein-protein interaction may act to anchor proacrosin to the acrosomal membranes or to other proacrosin molecules. At present not much is known about the structure or other functions of this binding protein, and whether or not it interacts with other forms of acrosin. Also, it is not clear at present how this binding interaction is disrupted naturally to allow acrosin to function in fertilization.

IV. FUNCTIONS OF ACROSIN

A. ZONA PELLUCIDA LYSIN

From the early observations of Yamane and Tyler, sperm-associated hydrolases have been postulated to function in fertilization by facilitating sperm passage through the egg envelopes. The description of these enzymes as sperm lysins was based largely on circumstantial evidence, such as egg envelope dissolution by sperm extracts. Hoshi[2] suggested three criteria for sperm enzymes to qualify as sperm lysins. These criteria were recently modified, with the addition of a fourth criterion.[13] These are paraphrased as follows: (1) the enzyme must be in the right place at the right time to function in gamete interaction; (2) the enzyme must selectively hydrolyze peptide bonds in the natural substrate, the egg envelope glycoproteins; (3) inhibition of enzyme activity must prevent egg envelope penetration by the sperm; and (4) chemical or immunological modification of the egg envelope should impart resistance to enzyme hydrolysis and result in the failure of sperm penetration. Each of these items will be addressed for acrosin.

Acrosin has been localized to both the inner and outer acrosomal membranes by histochemical and immunochemical procedures,[60,64] and is therefore at the correct place to function in fertilization. As mentioned above, acrosomal contents are carried with the boar sperm as it penetrates the porcine ZP.[63] The localization of acrosin to the surface of the intact sperm[64-66] adds a new dimension to spatial and temporal considerations for a lysin. It appears that acrosin is both in the acrosome and on the sperm surface, and thus more than satisfies this lysin criterion.

Hydrolysis of the natural substrate, the egg membranes, by a lysin was originally equated with complete and visual dissolution of the egg membranes. This was also one of the criteria used to determine whether acrosin had a function in fertilization.[1,9] However, complete visual dissolution of the ZP by homologous acrosin often required other enzymes.[81] With boar acrosin, dissolution of the porcine ZP was never obtained, but boar acrosin did cause

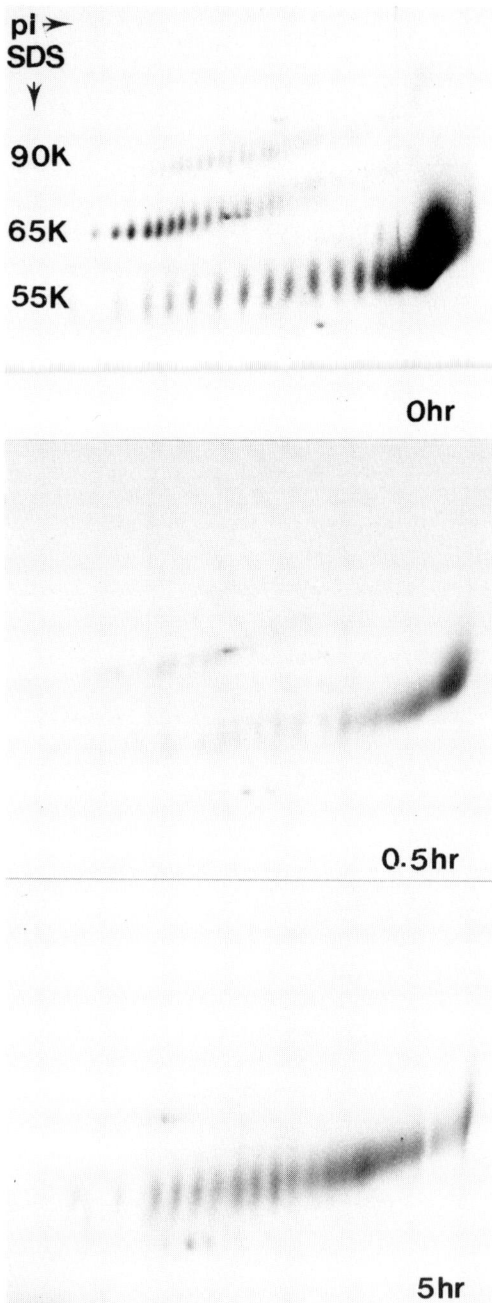

FIGURE 3. Hydrolysis action patterns of acrosin on particulate zona pellucida. (Reprinted from Urch, U. A., *The Molecular and Cellular Biology of Fertilization, Adv. Exp. Med. Biol.*, Vol. 207, Hedrick, J. L., Ed., Plenum Press, New York, 113. With permission.)

limited hydrolysis of certain families of the ZP.[82-85] The porcine ZP is composed of three families of glycoproteins, denoted 90K, $55K_{\alpha}$, and $55K_{\beta}$, and two 90K-derived components, 65K and 25K.[86] Only the 90K family and the 90K and 65K components were specifically hydrolyzed by acrosin to small polypeptides not retained in SDS-PAGE (Figure 3).[84] The ZP continued to remain visually intact even after extended acrosin hydrolysis reduced the

90K family to small peptides,[85] and the interacting hydrolyzed pieces could only be liberated with mercaptoethanol and denaturants. Although these hydrolysis observations do not by themselves prove that acrosin is a lysin of the ZP, they are consistent with early cytochemical observations of reaction products in the ZP penetration slit,[87] and with the hypothesis that acrosin aids the sperm in penetration of the ZP.

The third lysin criterion is partially applicable to acrosin. When relatively high concentrations of trypsin and acrosin inhibitors were added to capacitated sperm, fertilization was largely suppressed both *in vitro*[88] and *in vivo*.[89] Fertilization was also poor in the presence of the active-site inhibitor, tosyllysylchloromethyl ketone.[90] However, inhibition approached but was never 100%, and once the sperm had tightly bound the ZP, trypsin active-site inhibitors did not prevent further penetration or hydrolysis of the ZP.[91] This may be due to the inability of the inhibitors to cross the acrosomal membrane,[92] or it may reflect differences in inhibitor sensitivity between bound and free acrosin.[77] While information gained from the use of inhibitors is limited in nature, more specific inhibitors for acrosin may be used to more fully determine how well acrosin satisfies this criterion.

With regard to the last criterion of envelope modification, Bedford and Cross demonstrated resistance to proteolytic digestion of rabbit zonae by pretreatment with wheat germ agglutinin, and that such pretreatment did not prevent normal fertilization.[62] In discussing their findings, the authors mention that bound acrosin exhibits different kinetics than free,[77] and may be responsible for part of the result obtained with their study. With the advent of monoclonal and polyclonal antibodies to the 90K ZP glycoproteins, this criterion may be more applicable in the future.

Therefore, while acrosin does not strictly follow all of the criteria for a ZP lysin, it does cause limited hydrolysis of the ZP and it is in the correct place to act in penetration of the ZP.

B. BINDING TO THE ZONA PELLUCIDA

Several molecular mechanisms have been proposed for sperm-egg binding. Sperm binding to the ZP has been suggested to involve protein-protein interaction, carbohydrate-protein interaction, or both.[3,93] With regard to protein-protein interaction, acrosin, or a molecule similar to acrosin, has been implicated in binding to the ZP via the proteolytic domain of the molecule. Catalytically active boar acrosin bound porcine ZP preferentially over a control protein, as expected from a normal enzyme-substrate relationship.[5] In addition, mouse sperm binding to mouse eggs was inhibited with trypsin inhibitors,[67,68] and rabbit sperm binding to rabbit eggs was inhibited with antiacrosin antibodies.[94] In the mouse system, only acrosome intact sperm bind to the ZP,[95] and the acrosome reaction is induced by the ZP.[96] From the localization studies described above, acrosin was located on both inner and outer acrosomal membrane and on the sperm surface. The studies of a mouse sperm surface protein that bound seminal plasma trypsin inhibitor by Poirier's group,[97,98] and studies of a mouse sperm surface molecule that turned over active-site titrant inhibitors of trypsin, but was not proteolytically active,[99] indicate that a surface molecule with trypsin-like proteolytic domain may be involved in gamete interaction. It may be that catalytically inactive acrosin, or a portion of the proteolytic substrate binding domain of acrosin, resides on the sperm surface and is involved in binding ZP.

Recently Bleil et al. suggested that a mouse ZP glycoprotein, ZP2, acts as a ligand for sperm, albeit through a much weaker interaction.[100] Once the sperm has undergone the acrosome reaction and the outer membranes (plasma membrane and outer acrosomal membrane) have vesiculated, the inner acrosomal membrane and the acrosomal matrix itself become the limiting surface of the sperm and are thus important in maintaining gamete interaction. The authors suggest that this weaker ZP interaction may be with acrosin, and that ZP2 may be the endogenous substrate for acrosin. In the hamster, only acrosome-reacted sperm binding to the ZP, and the fertilizing sperm appears to remain bound to the ZP by a remnant of the vesiculated acrosomal membrane or acrosomal ghost.[101] Therefore, with an acrosome-reacting or -reacted sperm, acrosin would be in the right place in fertilization for both binding and hydrolysis as long as the enzyme was held in place for its action, and did not diffuse away.

With regard to a protein-carbohydrate interaction between gametes, the addition of carbohydrate inhibits the binding of sperm to the surface of the egg envelope in a number of species: fucose and fucoidin in the guinea pig, hamster, and human;[102] mannose, fucose, and fucoidin in the rat;[103] dextran sulfate and fucoidin in the pig;[104] dextran sulfate, fucoidin, and heparin in *Xenopus laevis;*[105] sulfated fucans in the sea urchin.[106] When all acrosins (including small autolytic forms of 12 to 18K M_r) were electroblotted onto nitrocellulose, all of the forms bound ^{125}I-ZP even when the enzyme was inhibited with the active-site inhibitor, diisopropyl fluorophosphate.[107,108] These observations implied a second binding site on acrosin for ZP. Other workers found similar binding of ^{125}I-ZP to nitrocellulose-blotted acrosins.[109] Recently a fucose binding protein with a molecular weight corresponding to one (53K) of the two reported for boar proacrosin[20] was described on the surface of acrosome-intact boar sperm (Figure 4).[110] This lectin has also been shown to have the same amino-terminal residues as acrosin.[111] In addition, the low molecular weight molecules involved in the species-specific binding of rabbit sperm to rabbit ZP have very recently been described as fucose-binding lectins.[112] These data are consistent with acrosin binding a fucose-containing ligand on the ZP, and also suggest that the binding domain resides in the small, autolytically produced forms of acrosin. Recently, Peterson and Hunt reported that the porcine ZP adhesion molecule in boar sperm is not proacrosin.[113] The authors observed that antibodies to proacrosin did not inhibit sperm binding, and that sera preabsorbed with proacrosin still contained sperm-binding inhibitor ability. Since there appears to be no immunological relationship between this adhesion molecule to the fucose lectin described as proacrosin, it remains to be seen if these two molecules share amino acid sequence domains or similar oligosaccharide side chains.

Carbohydrates have also been shown to inhibit acrosin's enzymatic activity. Several monosaccharides were reported to inhibit the esterase activity of human acrosin.[114,115] Simple carbohydrates did not inhibit boar acrosin amidase activity,[116] but sulfated polysaccharides that inhibited sperm binding significantly inhibited both amidase and protease activities.[117] Nonsulfated polysaccharides had no effect on acrosin's activities. Trypsin's amidase and ZP protease activities were unaffected by sulfated polysaccharides. Although initial inhibition kinetics were competitive in nature, inhibition kinetics at high inhibitor concentration were consistent with the presence of more than one enzyme, or multiple forms of the same enzyme, having similar substrate specificity but different inhibitor sensitivity.[118] m_α-Acrosin and hydrolytic products of m_β-acrosin were not as inhibitable as m_β-acrosin. It is clear from these studies that the sulfated carbohydrates can serve as important models for the natural sulfated macromolecules involved in gamete interaction, as they appear to specifically modulate acrosin's activity.

V. CONCLUSIONS

Acrosin possesses several structural features that set it apart from typical serine active-site proteases. Although acrosin retains all of the invariant amino acid residues and the catalytic triad for serine proteases, there is only approximately 33% positional identity with trypsin. The activation of the acrosin proteolytic domain results from the cleavage of a single peptide bond, Arg22—Val23, and the activation peptide remains covalently linked through disulfide bonds. In addition to the presence of a proteolytic domain, acrosin has at least two other domains: a carbohydrate binding domain and a hydrophic binding domain.

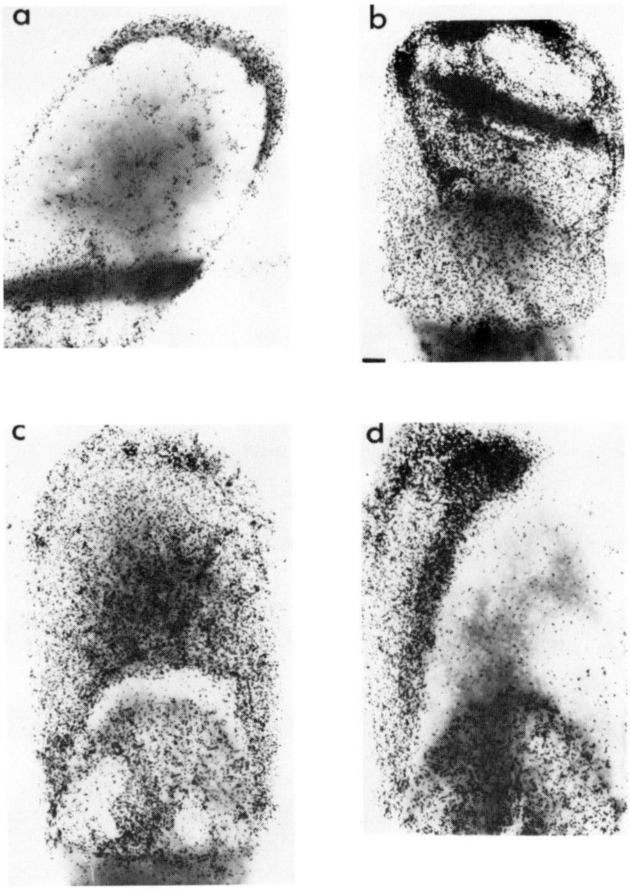

FIGURE 4. Topography of the fucose-binding protein during the acrosome reaction demonstrated by transmission electron microscopy using the gly-HRP-colloidal gold techniques. (a) Total specimen preparation of intact spermatozoon, when only the plasma membrane overlying the apical part of the acrosome is labeled with colloidal gold; (b and c) exposure of fucose-binding activity after induction of acrosome reaction, (d) during the acrosome reaction, the acrosomal cap together with fucose-binding sites loses its adherence to the sperm head. The inner acrosomal membrane shows only slight labeling, whereas the equitorial segment is heavily labeled. (Reprinted from Topfer-Petersen, E., Friess, A. E., and Schill, W.-B., *Hum. Reprod.*, 3, 319, 1988. With permission.)

Acrosin interacts with carbohydrate through a binding domain that has been inferred from fucose binding studies in the pig to be similar to the domain that interacts with the ZP. This domain is physically distant and distinct from the proteolytic domain, since active site inhibitors do not interfere with ZP binding. Polysulfated polysaccharide inhibits the amidase and protease activities of acrosin indicating that the two domains interact. This interaction of domains may function in ZP binding, by conferring specificity to the interaction. Acrosin interaction with polyanions suggests a highly complementary region in the acrosin molecule. The likely candidates for such an interaction are arginine residues, and the C-terminal end of acrosin has several arginine-interspersed-with-proline sequences. m_β-Acrosin is inhibited by these polyanions, and as it ages autolytically it loses the ability to be inhibited, suggesting that minor proteolytic trimming of the carbohydrate binding domain results in the loss of inhibition.

The hydrophobic domain of acrosin interacts with itself and with membranes, even though the enzyme does not appear to have any membrane spanning regions from sequence analysis. These interactions appear involved in the packaging of proacrosin into the acrosome with binding molecules, the activation of proacrosin, and the release of active enzyme from the acrosome. Membrane-bound acrosin has been shown to have different kinetics than the soluble enzyme, indicating that interaction with membranes may play an important part in fertilization.

At least two functions have been proposed for acrosin in fertilization—as a ZP lysin and as a ZP binding agent. Although acrosin does not fully obey all the criteria listed above for a lysin, it does cause limited and specific hydrolysis of the ZP. The examination of membrane-bound acrosin may satisfy all these criteria. Acrosin has been characterized as a fucose binding lectin of the ZP, the major ZP binding agent in the pig, although there are alternative hypotheses. Gamete interaction has been described as a species-specific event, and it would be difficult to obtain such specificity with acrosin since there is substantial positional identity (−70%) between acrosins (goat, boar, human). There may be unique regions of the acrosin sequence, or species differences could be accommodated by post-translational modifications, such as the addition of carbohydrate. Oligosaccharide side chains could possibly confer the necessary specificity.

The following model may demonstrate how the two functions of acrosin interact. There is an initial ZP binding event by catalytically inactive acrosin (proacrosin), or a portion of acrosin that contains either the carbohydrate or proteolytic binding domain. Binding induces the acrosome reaction (in those species where only intact acrosome sperm bind ZP), and this is followed by the activation of proacrosin to m_α-acrosin with subsequent hydrolysis of the ZP. Further autolysis of m_α-acrosin to m_β-acrosome results in continued binding, but the newly formed m_β-acrosin is inhibited when bound to the ZP. This dead end binding requires freshly activated acrosin to continue both autolysis of acrosin and hydrolysis of the ZP, and for the continued progress of the sperm through the ZP. Figure 2 shows just such acrosomal erosion as boar sperm penetrates the porcine ZP. This model provides for the penetration of ZP by sperm by a sequential binding and hydrolysis that would be necessary for sperm penetration of the ZP.

Future work on acrosin will concentrate on understanding the structure of acrosin domains, how they interact, and a better understanding of their physiological function in fertilization. Also, studies will show how these domains function in proacrosin activation and the acrosome reaction. Two very intriguing areas for further work are the sequence determination of the acrosin oligosaccharide side chains and how they function in fertilization, and the functions of the extended proline region in the C-terminal end of acrosin.

ACKNOWLEDGMENT

This work was supported in part by USDA 85-CRCR-1-1855.

REFERENCES

1. **McRorie, R. A. and Williams, W. L.,** Biochemistry of mammalian fertilization, *Annu. Rev. Biochem.*, 42, 777, 1974.
2. **Hoshi, M.,** Lysins, in *Biology of Fertilization*, Vol. 2, Metz, C. B. and Monroy, A., Eds., Academic Press, New York, 1985, 431.
3. **Yanagimachi, R.,** Mechanisms of fertilization in mammals, in *Fertilization and Embryonic Development In Vitro*, Mastroianni, L., Jr. and Biggers, J. D., Eds., Plenum Press, New York, 1981, 812.

4. **Meizel, S.,** The importance of hydrolytic enzymes to an exocytotic event, the mammalian sperm acrosome reaction, *Biol. Rev.*, 59, 125, 1984.
5. **Urch, U. A.,** The action of acrosin on the zona pellucida, in *The Molecular and Cellular Biology of Fertilization, Adv. Exp. Med. Biol.*, Vol. 207, Hedrick, J. L., Ed., Plenum Press, New York, 1986, 113.
6. **Yamane, J.,** Kausal-analytische studien uber die befructung des kaninchenensis. II. Die isoliering der auf das eizytoplasma auflosend wirkenden substanzen aus den spermatozoen, *Cytologia*, 6, 233, 1935.
7. **Tyler, A.,** Extraction of an egg membrane lysin from sperm of the giant keyhole limpet (*Megathura crenulata*), *Proc. Natl. Acad. Sci. U.S.A.*, 25, 317, 1939.
8. **Srivastava, P. N., Adams, C. E., and Hartree, E. F.,** Enzymatic action of acrosomal preparations on the rabbit ovum in vitro, *J. Reprod. Fertil.*, 10, 61, 1965.
9. **Stambaugh, R. and Buckley, J.,** Zona pellucida dissolution enzymes of the rabbit sperm head, *Science*, 161, 585, 1968.
10. **Polakoski, K. L., Zaneveld, L. J. D., and Williams, W. L.,** An acrosin-acrosin inhibitor complex in ejaculated boar sperm, *Biochem. Biophys. Res. Commun.*, 45, 381, 1971.
11. **Dudkiewicz, A. B., Lehnhardt, W. F., Johnson, D. A., Travis, J., and Williams, W. L.,** Rabbit acrosin: immunological dissimilarity to rabbit trypsin, *J. Exp. Zool.*, 203, 307, 1978.
12. **Schleuning, W.-D., Hell, R., and Fritz, H.,** Boar acrosin. I. Isolation procedure, active-site titration and evidence for the presence of sialic acid, *Hoppe Seylers Z. Physiol. Chem.*, 356, 1915, 1975.
13. **Hedrick, J. L., Urch, U. A., and Hardy, D. M.,** Structure-function properties of the sperm enzyme acrosin, in *Biocatalysis in Agricultural Biotechnology*, ACS Symp. Ser. No. 389, Whitaker, J. R. and Sonnet, P. E., Eds., American Chemical Society, Washington, D.C., 1989, chap. 15.
14. **Schiesler, H., Schleuning, W.-D., and Fritz, H.,** Cleavage specificity of boar acrosin on polypeptide substrates, ribonuclease and insulin β-chain, *Hoppe Seylers Z. Physiol. Chem.*, 356, 1933, 1975.
15. **Paulson, J. D., Parrish, R. F., and Polakoski, K. L.,** Comparative synthetic substrate kinetics of porcine acrosin and porcine trypsin, *J. Biochem.*, 10, 247, 1979.
16. **Parrish, R. F., Straus, J. W., Paulson, J. D., and Polakoski, K. L.,** Structure-activity relationships for the inhibition of acrosin by benzamidine derivatives, *J. Med. Chem.*, 21, 1132, 1978.
17. **Muller-Esterl, W., Kuffer, S., and Fritz, H.,** Purification and properties of boar acrosin, *Hoppe Seylers Z. Physiol. Chem.*, 361, 1811, 1980.
18. **Fock-Nuzel, R., Lottspeich, F., Henschen, A., Muller-Esterl, W., and Fritz, H.,** N-terminal amino acid sequence of boar acrosin, *Hoppe Seylers Z. Physiol. Chem.*, 361, 1823, 1980.
19. **Muller-Esterl, W. and Fritz, H.,** Interaction of boar acrosin with detergents, *Hoppe Seylers Z. Physiol. Chem.*, 361, 1673, 1980.
20. **Polakoski, K. L. and Parrish, R. F.,** Boar acrosin. Purification and preliminary activation studies of proacrosin isolated from ejaculated boar sperm, *J. Biol. Chem.*, 252, 1888, 1977.
21. **Shams Borhan, G., Schleuning, W.-D., Tschesche, H., and Fritz, H.,** Occurrence of multiple forms of bull and ram acrosin during proenzyme activation and inhibition of activation by *p*-nitro-*p*′-quanidinium benzoate, *Hoppe Seylers Z. Physiol. Chem.*, 357, 667, 1976.
22. **Hardy, D. M., Wild, G. C., and Tung, K. S. K.,** Purification and initial characterization of proacrosins from guinea pig testes and epididymal spermatozoa, *Biol. Reprod.*, 37, 189, 1987.
23. **Meizel, S. and Mukerji, S. K.,** Biochemical studies of proacrosin and acrosin from hamster cauda epididymal spermatozoa, *Biol. Reprod.*, 14, 444, 1976.
24. **Tobias, P. S. and Schumacher, G. F. B.,** Observation of two proacrosins in extracts of human spermatozoa, *Biochem. Biophys. Res. Commun.*, 74, 434, 1977.
25. **Siegel, M. S., Bechtold, D. S., Kopta, C. I., and Polakoski, K. L.,** The rapid purification and partial characterization of human sperm proacrosin using automated fast protein liquid chromatography (FPLC) system, *Biochim. Biophys. Acta*, 883, 567, 1986.
26. **Brown, C. R.,** Purification of mouse sperm acrosin, its activation from proacrosin and effect on homologous egg investments, *J. Reprod. Fertil.*, 69, 289, 1983.
27. **Mukerji, S. K. and Meizel, S.,** Rabbit testis proacrosin. Purification, molecular weight estimation, and amino acid and carbohydrate composition of the molecule, *J. Biol. Chem.*, 254, 11721, 1979.
28. **Harrison, R. A. P. and Brown, C. R.,** The zymogen form of acrosin in testicular, epididymal, and ejaculated spermatozoa from ram, *Gamete Res.*, 2, 75, 1979.
29. **Brown, C. R. and Harrison, R. A. P.,** The activation of proacrosin from ram, bull, and boar, *Biochim. Biophys. Acta*, 526, 202, 1978.
30. **Iwao, Y. and Katagiri, C.,** Properties of the vitelline coat lysin from toad sperm, *J. Exp. Zool.*, 219, 87, 1982.
31. **Sawada, H., Yokosawa, H., and Ishii, S.,** Purification and characterization of two types of trypsin-like enzymes from sperm of the ascdian (prochordata), *Halocynthia roretzi*, *J. Biol. Chem.*, 259, 2900, 1984.
32. **Zelezna, B. and Cechova, D.,** Boar acrosin. Isolation of two active forms from boar ejaculated sperm, *Hoppe Seylers Z. Physiol. Chem.*, 363, 757, 1982.

33. **Parrish, R. F. and Polakoski, K. L.,** Boar M_α-acrosin. Purification and characterization of the initial active enzyme resulting from the conversion of boar proacrosin to acrosin, *J. Biol. Chem.*, 253, 8428, 1978.
34. **Brown, C. R. and Hartree, E. F.,** Comparison of neutral proteinase activities in cock and ram spermatozoa and observations on a proacrosin in cock spermatozoa, *J. Reprod. Fertil.*, 46, 155, 1976.
35. **Adekunle, A. O., Storey, B. T., and Teuscher, C.,** Guinea pig testicular acrosin system: further characterization of the active enzyme, *Biol. Reprod.*, 40, 127, 1989.
36. **Hardy, D. M., Schoots, A. F. M., and Hedrick, J. L.,** Caprine acrosin. Purification, characterization and proteolysis of the porcine zona pellucida, *Biochem. J.*, 257, 447, 1989.
37. **Schleuning, W.-D., Hell, R., and Fritz, H.,** Multiple forms of human acrosin: isolation and characterization, *Hoppe Seylers Z. Physiol. Chem.*, 357, 855, 1976.
38. **Urch, U. A., Berger, T., and Hedrick, J. L.,** The binding of the zona pellucida by boar sperm membrane proteins and the antigenic similarity with acrosin, submitted.
39. **Fock-Nuzel, R., Lottspeich, F., Henschen, A., Muller-Esterl, W., and Fritz, H.,** N-terminal amino acid sequence of boar sperm acrosin, *Hoppe Seylers Z. Physiol. Chem.*, 361, 1823, 1980.
40. **Baba, T., Watanabe, K., Kashiwabara, S., and Arai, Y.,** Primary structure of human acrosin deduced from its cDNA, *FEBS Lett.*, 244, 296, 1989.
41. **Muller-Esterl, W., Zippel, B., and Fritz, H.,** Oligomerization of boar acrosin, *Hoppe Seylers Z. Physiol. Chem.*, 361, 1381, 1980.
42. **Parrish, R. F. and Polakoski, K. L.,** An apparent high molecular weight form of boar proacrosin resulting from the presence of a protein that binds to proacrosin, *Anal. Biochem.*, 87, 108, 1978.
43. **Urch, U. A. and Hedrick, J. L.,** The hydrophobic nature of the boar sperm lectin-protease, acrosin, *J. Cell Biol.*, 107, 167a, 1989.
44. **Helenius, A. and Simons, K.,** Charge shift electrophoresis: simple method for distinguishing between amphiphilic and hydrophilic proteins in detergent solution, *Proc. Natl. Acad. Sci. U.S.A.*, 74, 529, 1977.
45. **Parrish, R. F., Straus, J. W., Polakoski, K. L., and Dombrose, F. A.,** Phospholipid vesicle stimulation of proacrosin activation, *Proc. Natl. Acad. Sci. U.S.A.*, 75, 149, 1978.
46. **Gillis, G., Peterson, R., Russell, L., Hook, L., and Freund, M.,** Isolation and characterization of membrane vesicles from human and boar spermatozoa: methods using nitrogen cavitation and ionophore induced vesiculation, *Prep. Biochem.*, 8, 363, 1978.
47. **Peterson, R., Russell, L., Bundman, D., and Freund, M.,** Evaluation of the purity of boar sperm plasma membranes prepared by nitrogen cavitation, *Biol. Reprod.*, 23, 637, 1980.
48. **Baba, T., Kashiwabara, S., Watanabe, K., Itoh, H., Michikawa, Y., Kimura, K., Takada, M., Fukamizu, A., and Arai, Y.,** Activation and maturation mechanisms of boar acrosin zymogen based on the deduced primary structure, *J. Biol. Chem.*, 264, 11920, 1989.
49. **Adham, I. M., Klemm, U., Maier, W.-M., Hoyer-Fender, S., Tsaousidou, S., and Engel, W.,** Molecular cloning of preproacrosin and analysis of its expression pattern in spermatogenesis, *Eur. J. Biochem.*, 182, 563, 1989.
50. **Leytus, S. P., Loeb, K. R., Hagen, F. S., Kurachi, K., and Davie, E.,** A novel serine protease (Hepsin) with a putative transmembrane domain expressed by human liver and hepatoma cells, *Biochemistry*, 27, 1067, 1988.
51. **Fridberger, A., Klint, M., Sundelin, J., and Peterson, P. A.,** The amino terminal sequences of boar sperm proacrosin and active acrosin are identical, *Biochem. Biophys. Res. Commun.*, 121, 884, 1984.
52. **Fock-Nuzel, R., Lottspeich, F., Henschen, A., and Muller-Esterl, W.,** Boar acrosin is a two-chain molecule. Isolation and the structure of the light chain; homology with the pro-part of other serine proteinases, *Eur. J. Biochem.*, 141, 441, 1984.
53. **Cechova, D., Topfer-Petersen, E., and Henschen, A.,** Boar proacrosin is a single-chain molecule which has the N-terminus of the acrosin A-chain (light chain), *FEBS Lett.*, 241, 136, 1988.
54. **Baba, T., Michikawa, Y., Kawahura, K., and Arai, Y.,** Activation of boar proacrosin is effected by processing at both N-terminal and C-terminal portions of the zymogen molecule, *FEBS Lett.*, 244, 132, 1989.
55. **Zelezna, B., Cechova, D., and Henschen, A.,** Isolation of the boar sperm acrosin peptide released during the conversion of the α-form into β-form, *Biol. Chem. Hoppe Seyler*, 370, 325, 1989.
56. **Morton, D. B.,** Acrosomal enzymes: immunochemical localization of acrosin and hyaluronidase in ram spermatozoa, *J. Reprod. Fertil.*, 45, 375, 1975.
57. **Flechon, J.-E., Huneau, D., Brown, C. R., and Harrison, R. A. P.,** Immunocytochemical localization of acrosin in the anterior segment of the acrosomes of ram, boar and bull spermatozoa, *Ann. Biol. Anim. Bioch. Biophys.*, 17, 749, 1977.
58. **Garner, D. and Easton, M. P.,** Immunofluorescent localization of acrosin in mammalian spermatozoa, *J. Exp. Zool.*, 200, 157, 1977.
59. **Green, D. P. L. and Hockaday, A. R.,** The histochemical localization of acrosin in guinea pig sperm after the acrosome reactions, *J. Cell Sci.*, 32, 177, 1978.

60. **Johnson, L. A., Garner, D. L., Truitt-Gibert, A. J., and Lessley, B. A.,** Immunocytochemical localization of acrosin on the acrosomal membranes of porcine spermatozoa, *J. Androl.*, 2, 16, 1981.
61. **Huang, T. T. F., Jr., Hardy, D., Yanagimachi, H., Teuscher, C., Tung, K., Wild, G., and Yanagimachi, R.,** pH and protease control of acrosomal content stasis and release during the guinea pig sperm acrosome reaction, *Biol. Reprod.*, 32, 451, 1985.
62. **Bedford, J. M. and Cross, N. L.,** Normal penetration of the rabbit spermatozoa through a trypsin- and acrosin-resistant zona pellucida, *J. Reprod. Fertil.*, 54, 385, 1978.
63. **Russe, I. and Wendt, V.,** Ultrastrucktur kapazitierter eberspermien wahrend der befruchtung, *Med. Welt,* 28, 746, 1977.
64. **de Vries, J. W. A., Willemsen, R., and Geuze, H. J.,** Immunocytochemical localization of acrosin and hyaluronidase in epididymal and ejaculated spermatozoa, *Eur. J. Cell Biol.*, 37, 81, 1985.
65. **Topfer-Petersen, E., Friess, A. E., and Schill, W.-B.,** The acrosome reaction in boar spermatozoa, *Hum. Reprod.*, 3, 319, 1988.
66. **Urch, U. A. and Berger, T.,** Inhibition of sperm binding by acrosin antibodies, in preparation.
67. **Frazer, L. R.,** p-Aminobenzamidine, an acrosin inhibitor, inhibits mouse sperm penetration of the zona pellucida but not the arosome reaction, *J. Reprod. Fertil.*, 65, 185, 1982.
68. **Saling, P. M.,** Involvement of a trypsin-like activity in binding of mouse spermatozoa, *Proc. Natl. Acad. Sci. U.S.A.*, 78, 6231, 1981.
69. **Dravland, J. E. and Meizel, S.,** The effect of inhibitors of trypsin and phospholipase A2 on the penetration of zona pellucida-free eggs by acrosome-reacted hamster sperm, *J. Androl.*, 3, 388, 1982.
70. **Dravland, J. E., Llanos, M. N., Munn, R. J., and Meizel, S.,** Evidence for the involvement of a sperm trypsin-like enzyme in membrane events of the hamster sperm acrosome reaction, *J. Exp. Zool.*, 232, 117, 1984.
71. **Stojanoff, A., Bourne, H., Andrews, A. G., and Hyne, R. V.,** Isolation of a stable apical segment of the guinea pig sperm acrosome, *Gamete Res.*, 17, 321, 1987.
72. **Topfer-Petersen, E. and Schill, W. B.,** A new separation method of subcellular fractions of boar spermatozoa, *Andrologia,* 13, 174, 1981.
73. **Topfer-Petersen, E., Schmoekel, C., and Schill, W.-B.,** The acrosomal membrane system of boar spermatozoa—morphological and biochemical studies, *Andrologia,* 15, 62, 1983.
74. **Olson, G. E., Winfrey, V. E., Winer, M. A., and Davenport, G. R.,** Outer acrosomal membrane of guinea pig: isolation and structural characterization, *Gamete Res.*, 17, 77, 1987.
75. **Olson, G. E., Winfrey, V. P., and Davenport, G. R.,** Characterization of matrix domains of the hamster acrosome, *Biol. Reprod.*, 39, 1145, 1988.
76. **Bhattacharyya, A. and Zaneveld, L. J. D.,** Acrosin release from spermatozoa, *Fed. Proc.*, 36, 371, 1977.
77. **Brown, C. R. and Hartree, E. F.,** Studies on ram acrosin. Activation of proacrosin accompanying the isolation of acrosin from spermatozoa, and purification of the enzyme by affinity chromatography, *Biochem. J.*, 175, 227, 1978.
78. **Zahler, W. L. and Doak, G. A.,** Isolation of the outer acrosomal membrane from bull sperm, *Biochem. Biophys. Acta,* 480, 479, 1975.
79. **Parrish, R. F. and Polakoski, K. L.,** An apparent high molecular weight form of boar proacrosin resulting from the presence of a protein that binds to proacrosin, *Anal. Biochem.*, 87, 108, 1978.
80. **Harrison, R. A. P.,** The interaction of ram proacrosin and acrosin forms with anitsera raised against ram m_{β}-acrosin, *J. Reprod. Immunol.*, 4, 231, 1982.
81. **Srivastava, P. N., Akruk, S. R., and Williams, W. L.,** Dissolution of rabbit zona by sperm acrosomal extract: effect of calcium, *J. Exp. Zool.*, 207, 521, 1979.
82. **Brown, C. R. and Cheng, W. T. K.,** Limited proteolysis of the porcine zona pellucida by homologous sperm acrosin, *J. Reprod. Fertil.*, 74, 257, 1985.
83. **Dunbar, B. S., Dudkiewicz, A. B., and Bundman, D. S.,** Proteolysis of specific zona pellucida glycoproteins by boar acrosin, *Biol. Reprod.*, 32, 619, 1985.
84. **Urch, U. A., Wardrip, N. A., and Hedrick, J. L.,** Limited and specific proteolysis of the zona pellucida, *J. Exp. Zool.*, 233, 479, 1985.
85. **Urch, U. A., Wardrip, N. A., and Hedrick, J. L.,** Proteolysis of the zona pellucida by acrosin: the nature of the hydrolysis products, *J. Exp. Zool.*, 236, 239, 1985.
86. **Hedrick, J. L. and Wardrip, N. J.,** On the macromolecular composition of the zona pellucida from porcine oocytes, *Dev. Biol.*, 121, 478, 1987.
87. **Stambaugh, R.,** Sperm proteinase release during fertilization of rabbit ova, *J. Exp. Zool.*, 197, 121, 1976.
88. **Stambaugh, R., Brackett, B. G., and Mastroianni, L.,** Inhibition of *in vitro* fertilization of rabbit ova by trypsin inhibitors, *Biol. Reprod.*, 1, 223, 1969.
89. **Zaneveld, L. J. D., Robertson, R. I., Kessler, M., and Williams, W. L.,** Inhibition of fertilization *in vivo* by pancreatic and seminal plasma trypsin inhibitors, *J. Reprod. Fertil.*, 25, 387, 1971.

90. **Zaneveld, L. J. D., Robertson, R. I., and Williams, W. L.,** Synthetic enzyme inhibitors as antifertility agents, *FEBS Lett.*, 11, 345, 1970.
91. **Hartmann, J. F. and Hutchison L. F.,** Nature of the pre-penetration contact interactions between hamster gametes *in vitro*, *J. Reprod. Fertil.*, 36, 49, 1974.
92. **Schill, W. B., Feifel, M., Fritz, H., and Hammerstein, J.,** Inhibitors of acrosomal proteinase as anti-fertility agents. A problem of acrosomal membrane permeability, *Int. J. Androl.*, 4, 25, 1981.
93. **Ahuja, K. K.,** Fertilization studies in the hamster. The role of cell-surface carbohydrates, *Exp. Cell Res.*, 140, 353, 1982.
94. **Dudkiewicz, A. B.,** Inhibition of fertilization in the rabbit by anti-acrosin antibodies, *Gamete Res.*, 8, 183, 1983.
95. **Florman, H. M. and Storey, B. T.,** Mouse gamete interaction: the zona pellucida is the site of the acrosome reaction leading to fertilization in vitro, *Dev. Biol.*, 91, 121, 1982.
96. **Bleil, J. D. and Wassarman, P. M.,** Sperm-egg interactions in the mouse: sequence of events and induction of the acrosome reaction by a zona pellucida glycoprotein, *Dev. Biol.*, 95, 317, 1984.
97. **Aarons, D., Speake, J. L., and Poirier, G. R.,** Evidence for a proteinase inhibitor binding with murine spermatozoa, *Biol. Reprod.*, 31, 811, 1984.
98. **Robinson, R., Richardson, R., Hinds, K., Clayton, D., and Poirier, G. R.,** Features of a seminal proteinase inhibitor-zona pellucida-binding component on murine spermatozoa, *Gamete Res.*, 16, 217, 1987.
99. **Benau, D. A. and Storey, B. T.,** Characterization of the mouse sperm plasma membrane zona-binding site sensitive to trypsin inhibitors, *Biol. Reprod.*, 36, 282, 1987.
100. **Bliel, J. D., Greve, J. M., and Wassarman, P. M.,** Identification of a secondary sperm receptor in the mouse egg zona pellucida: role in maintenance of binding of acrosome-reacted sperm to eggs, *Dev. Biol.*, 128, 376, 1988.
101. **Yanagimachi, R. and Phillips, D. M.,** The status of acrosomal caps of hamster spermatozoa immediately before fertilization in vivo, *Gamete Res.*, 9, 1, 1984.
102. **Huang, T. T. F., Ohzu, E., and Yanagimachi, R.,** Evidence suggesting that L-fucose is part of a recognition signal for sperm-zona pellucida attachment in mammals, *Gamete Res.*, 5, 355, 1982.
103. **Shalgi, R., Matityahu, A., and Nebel, L.,** The role of carbohydrates in sperm-egg interaction in rats, *Biol. Reprod.*, 34, 446, 1986.
104. **Peterson, R. N., Russell, L. D., and Hunt, W. P.,** Evidence for specific binding of uncapacitated boar spermatozoa to porcine zonae pellucidae in vitro, *J. Exp. Zool.*, 231, 137, 1984.
105. **Lindsay, L. L. and Hedrick, J. L.,** Identification of Xenopus laevis sperm and egg envelope binding components on nitrocellulose, *J. Exp. Zool.*, 245, 286, 1988.
106. **Glabe, C. G., Grabel, L. B., Vacquier, V. D., and Rosen, S. D.,** Carbohydrate specificity of sea urchin binding: a cell surface lectin mediated sperm egg adhesion, *J. Cell Biol.*, 94, 123, 1982.
107. **Urch, U. A. and Hedrick, J. L.,** Acrosin derived peptides bind the zona pellucida, *J. Cell Biol.*, 99, 387a, 1985.
108. **Jones, R. and Brown, C.,** Identification of a zona-binding protein from boar spermatozoa as proacrosin, *Exp. Cell Res.*, 171, 503, 1987.
109. **Brown, C. R. and Jones, R.,** Binding of zona pellucida proteins to a boar sperm polypeptide of Mr 53,000 and identification of zona moieties involved, *Development*, 99, 333, 1987.
110. **Topfer-Petersen, E. and Henschen, A.,** Acrosin shows zona and fucose binding, novel properties for a serine proteinase, *FEBS Lett.*, 226, 18, 1987.
111. **Topfer-Petersen, E. and Henschen, A.,** Zona pellucida binding and fucose-binding of boar sperm acrosin is not correlated with proteolytic activity, *Biol. Chem. Hoppe Seyler*, 309, 69, 1988.
112. **Fisher, S. J., Widgren, E. E., and O'Rand, M. G.,** Characterization of the rabbit sperm membrane autoantigen, RSA, as a lectin which binds zona pellucida, *J. Cell Biol.*, 105, 255a, 1987.
113. **Peterson, R. N. and Hunt, W. P.,** Identification, isolation, and properties of a plasma membrane protein involved in the adhesion of boar sperm to the porcine zona pellucida, *Gamete Res.*, 23, 103, 1989.
114. **Anderson, R. A., Jr., Oswald, C., Leto, S., and Zaneveld, L. J. D.,** Inhibition of human acrosin by fructose and other monosaccharides, *Biol. Reprod.*, 22, 1079, 1980.
115. **Anderson, R. A., Jr., Oswald, C., and Zaneveld, L. J. D.,** Inhibition of human acrosin by monosaccharides and related compounds. Structure-activity relationships, *J. Med. Chem.*, 24, 1288, 1981.
116. **Urch, U. A. and Hedrick, J. L.,** Inhibition of boar acrosin by carbohydrates, *Fed. Proc.*, 42, 1952, 1983.
117. **Urch, U. A. and Hedrick, J. L.,** The inhibition of boar acrosin amidase activity by sulfated polysaccharides, *Biol. Chem. Hoppe Seyler*, 369, 727, 1988.
118. **Segel, I.,** *Enzyme Kinetics*, John Wiley & Sons, New York, 1975.

Chapter 10

ELECTRON MICROSCOPY OF MAMMALIAN FERTILIZATION

David M. Phillips

TABLE OF CONTENTS

I. INTRODUCTION

Biologists at the end of the last century and the early part of this century were intensely interested in gametes. Many of the best-known biologists of those times, such as E. B. Wilson, Frank Lillie, and Jack Loeb, made the morphology of gamete interaction a major focus of their research. Their early studies on interaction of mammalian gametes served as the basis for the physiological studies that followed, studies only made possible in recent years by the development of systems to capacitate, acrosome react, and fertilize mammalian gametes *in vitro*.

A resurgence of interest in the structure of gametes and the fertilization process occurred when the electron microscope was applied to biological systems. Descriptions of the ultrastructure of mammalian spermatozoa and eggs and structural changes of the spermatozoon during the acrosome reaction and during interaction with the oocyte and its investment stimulated molecular analysis of sperm-egg interactions. Today the electron microscope remains an invaluable complement to other techniques for investigating problems of mammalian fertilization.

In this chapter, I will briefly describe mammalian spermatozoon and egg structures that are involved in fertilization and the structural changes that characterize sperm-egg interactions. Emphasis will be placed on observations on the ultrastructure of fertilization that appear related to the molecular biological observations presented in other chapters of this book.

II. THE MAMMALIAN SPERMATOZOON

Although the principal organelles that compose mammalian sperm cells can also be found in spermatozoa of other animal groups, mammalian spermatozoa are structurally unique in the way that organelles are shaped and arranged within the cell. The demarcation between head and tail is particularly obvious in mammalian spermatozoa because the head is flattened. In nearly all other animal groups, the sperm heads are pencil-shaped or conical. The flattened shape of the mammalian sperm head may make more surface area available for contact between the sperm and zona pellucida or vitteline membrane during gamete interactions. Although the arrangement of organelles in the sperm tail of mammals is also distinctive, studies of fertilization have primarily concerned the sperm head, as it is the head which interacts with the egg and its vestments.

Because of burgeoning interest in cell surface receptors, the sperm cell membrane is of central interest. The cell membrane over the sperm head and the acrosomal inner and outer membranes seem to be specifically important in attachment to the oocyte and in sperm penetration. The mammalian acrosome always assumes a precise species-specific shape. In many species, regions of acrosomal contents are arranged in paracrystalline arrays.[1] In others, different portions of the acrosome display different electron densities,[1,2] and in some, acrosomal protrusions occur in specific regions.[3,4] The distinctive shapes of acrosome, arrangement of paracrystalline material, and occurrence of areas of different densities suggest that the acrosome is not simply a bag of enzymes. It seems likely that structural proteins may be involved in maintaining acrosome shape and the positioning of different functional components within the acrosomal matrix. A high degree of acrosome organization may be required to assure the release of acrosome components in a precise order as the spermatozoon traverses the egg vestments. Although acrosomes vary in size and shape among species, all mammalian acrosomes possess two distinct regions: the acrosome cap and the equatorial segment (Figure 1). The acrosome cap participates in the acrosome reaction; the equatorial segment does not.

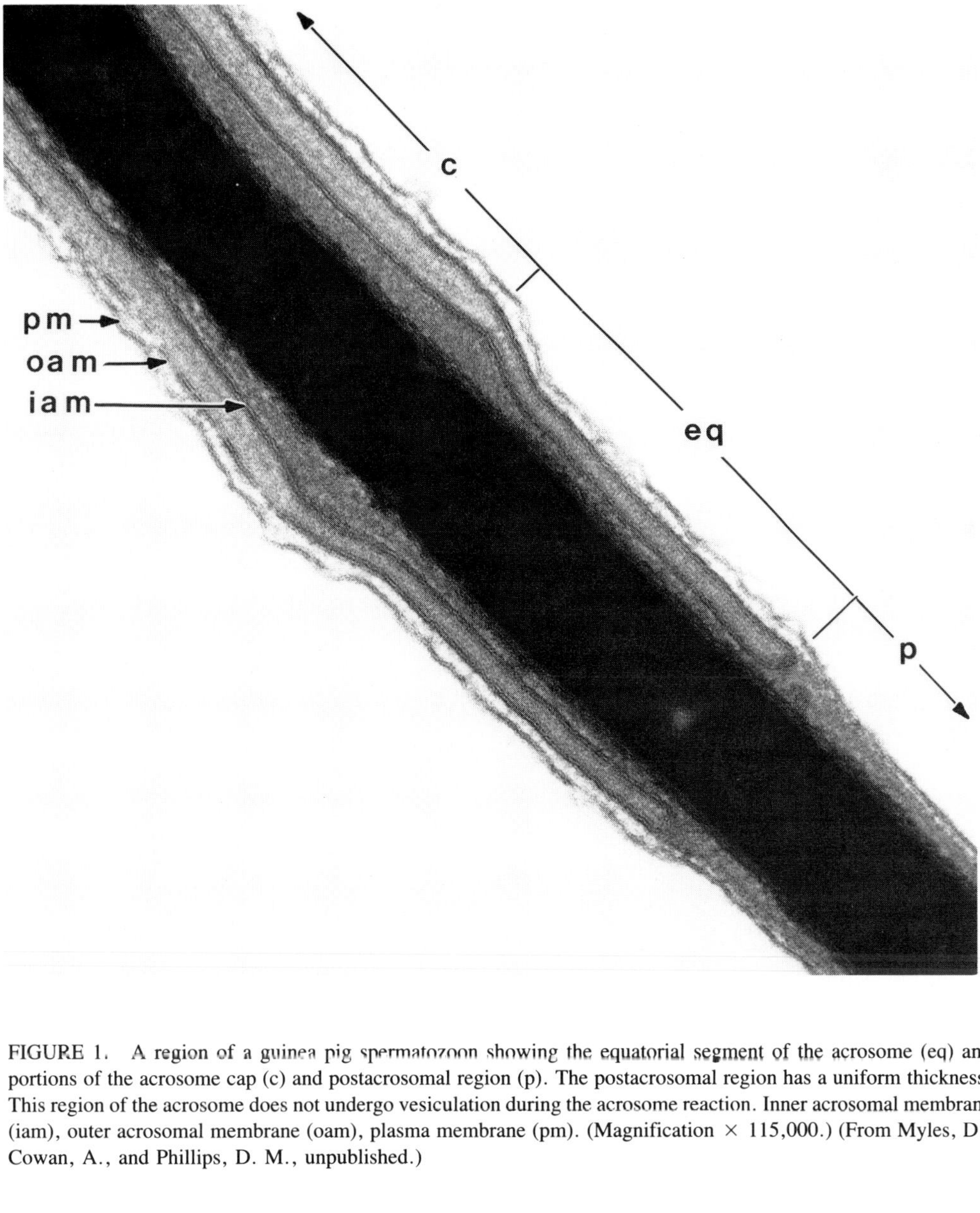

FIGURE 1. A region of a guinea pig spermatozoon showing the equatorial segment of the acrosome (eq) and portions of the acrosome cap (c) and postacrosomal region (p). The postacrosomal region has a uniform thickness. This region of the acrosome does not undergo vesiculation during the acrosome reaction. Inner acrosomal membrane (iam), outer acrosomal membrane (oam), plasma membrane (pm). (Magnification × 115,000.) (From Myles, D., Cowan, A., and Phillips, D. M., unpublished.)

III. THE MAMMALIAN OOCYTE AND ITS VESTMENTS

A. THE CUMULUS

The mammalian egg and zona pellucida are surrounded by a cumulus oophorus composed of thousands of cells embedded in mucus matrix. Synthesis of the mucus by cumulus cells is one of the maturational changes triggered by the luteinizing hormone (LH) surge.[5,6] The mucus provides a substance through which spermatozoa pass to reach the oocyte; spermatozoa will not traverse the nonexpanded cumulus of proestrus animals.[7] In addition, it has been demonstrated that the mucified cumulus has the correct consistency to be picked up and transported by the fimbrae of the oviduct, as other types of mucus are transported by the fimbrae of the oviduct but cumulus-free ova are not transported.[8,9] There is also evidence that the glycoproteins in the cumulus oophorus may protect the zona pellucida from hardening.[10,11] Zona hardening refers to changes that zonae normally undergo after fertilization. These changes leave the zonae more resistant to proteolysis. The function of cumulus mucus in the fertilization process will be discussed below.

There is controversy concerning the structure of the cumulus mucus. In the SEM, the mucus generally appears smooth.[12,13] This may not be an accurate representation of the true structure because even with careful critical point drying, very hydrated structures such as mucus shrink and become deformed. In standard preparations for the TEM, the mucus is hardly visible, but when tannic acid is employed with the fixative, the mucus is seen as small fibers. Often fibers are seen to be orientated in groups, suggesting that mucus is not randomly distributed.[14] When ruthinium red is used with glutaraldehyde, the cumulus matrix is seen as both fibers and granules.[15,16] Recently, Yudin et al.[17] examined freeze-substitution preparations of mucus. These workers observed channels, and they suggested that spermatozoa swam through these mucus channels. Channels have not been observed with standard fixation techniques. Thus, the cumulus mucus appears morphologically different with different fixation techniques. More work needs to be done to determine which method gives the most accurate picture of cumulus structure.

At the time of fertilization most of the cells of the cumulus oophorus appear healthy by morphological criteria. They have prominent, rough endoplasmic reticulum and Golgi apparatus, and mitochondria with tubular cristae and some smooth endoplasmic reticulum. The presence of rough endoplasmic reticulum and Golgi apparatus suggest that even after the cumulus-oocyte complex reaches the oviduct, it continues to synthesize and secrete mucus (Figures 2a and b). The presence of smooth ER and mitochondria with tubular cristae suggest that these could be steroid-producing cells. Cumulus cells derived from the follicle cells are undergoing luteinization and producing progesterone. The recent observations that cytochrome P-450 occurs in mitochondria of these cells[18] and that progesterone induces the acrosome reaction[19] support the idea that the cumulus is an endocrine tissue.

B. THE ZONA PELLUCIDA

In transmission electron micrographs the substance of the zona pellucida appears filamentous. The filaments appear to be more loosely arranged toward the zona surface facing the cumulus oophorus than near the surface that faces the oocyte. The groups of fibers are arranged with their long axes perpendicular to the zona surfaces.[2] These differences in the ultrastructure of the inner and outer surfaces of the zona pellucida are more obvious in the scanning electron microscope (SEM). To observe both the inner and outer surfaces of the zona pellucida, we removed the cumulus oophorus from hamster oocytes by brief treatment with hyaluronidase. Then, using a mouth pipette, we drew up oocytes in a finely drawn capillary tube. In some oocytes, this treatment peels back part of the zona pellucida, revealing the inner surface. Thus, we are able to view both the inner and outer surface of the same zona pellucida. Visualized in this way, the outer surface has an appearance reminiscent of layers of Swiss cheese, whereas the inner surface is covered with small bumps.[20] To determine if these structural differences are correlated with functional differences, we incubated *in vitro* capacitated hamster spermatozoa with fragments of the zona pellucida. Using the SEM to distinguish between the inner and outer surfaces of the zona pellucida, we observed that capacitated spermatozoa bound only to the outer surface of the zona pellucida. Thus, the inner and outer surfaces of the zona pellucida are functionally, as well as structurally, dissimilar.[20] Before sperm meets egg, it encounters the zona pellucida. Since the sperm moves from outside to inside, it is not surprising to find differences in the outer and inner surfaces of the zona. The function of the zona pellucida which surrounds the mammalian oocyte will be further discussed in other chapters of this book.

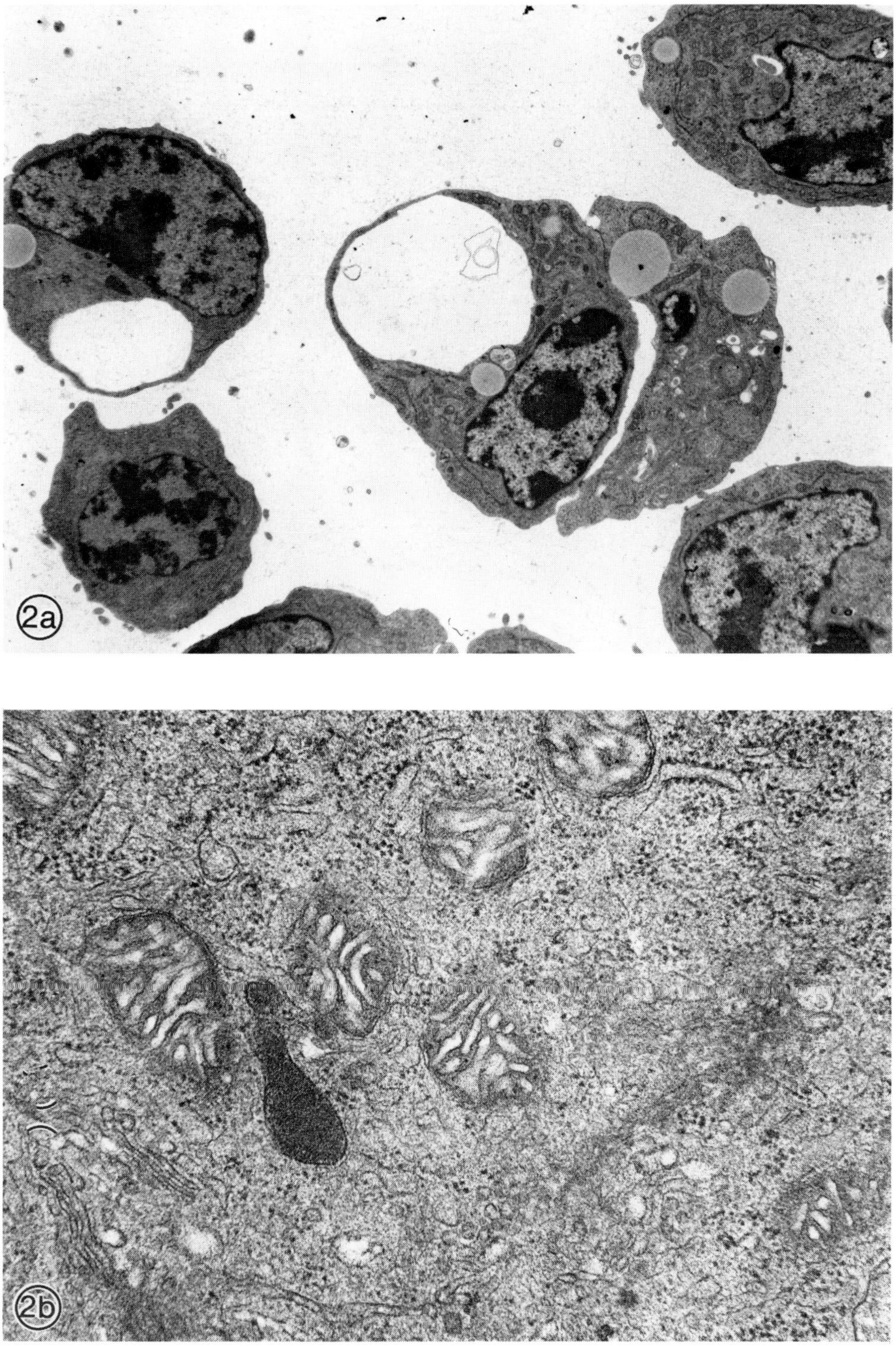

FIGURE 2. (a) At the time of fertilization, cumulus cells contain abundant mitochondria, rough ER, and lipid. They do not show signs of degradation. (Magnification × 4000.) (b) The cytoplasm of prefertilization cumulus cells contains abundant rough ER and prominent Golgi. (Magnification × 41,000.)

C. THE OOCYTE

Mammalian eggs are only 60 to 100 μm in diameter, very small compared to eggs of other groups of animals. Their surface is covered with microvilli.[21] This is not unusual; eggs of many vertebrates and invertebrates have microvillar surfaces.[22] As discussed in Chapter 4, the first meiotic division in mammals is induced by the LH surge. The mammalian oocyte remains in metaphase II until the sperm fusion initiates the completion of the second meiotic division. These meiotic events influence cell surface morphology; in some species, notably the mouse, a region of the oocyte surface devoid of microvilli bulges from the surface of the oocyte over the metaphase chromosomes.[21,23]

IV. FERTILIZATION—WHAT ELECTRON MICROSCOPY CAN AND CANNOT DO

Studying mammalian fertilization is difficult because mammalian eggs are small and very few in number. Relatively recently, techniques to isolate larger quantities of zona pellucida and methodology to carry out immunization, electrophoresis, and Western blots with minute amounts of material have facilitated work on molecular biology of fertilization. However, since only small amounts of material are required for morphological studies, biologists have been looking at ultrastructural aspects of mammalian fertilization for 30 years.

Relatively few studies of fertilization have been conducted on eggs fertilized *in vivo*. Physiological and biochemical studies generally cannot be carried out *in vivo* at the site of fertilization in the ampulla of the oviduct because these types of studies require a controlled environment. On the other hand, structural studies, though difficult, are possible. Under normal conditions, eggs are fertilized over a period of several hours. Therefore, one of the problems with studying fertilization *in vivo* is that it is not possible to time fertilization precisely. To add to the challenge, very few spermatozoa are present in the ampulla of the oviduct at the time of fertilization, and generally only one spermatozoon is associated with the egg. These problems notwithstanding, spermatozoa have been observed *in vivo* in the cumulus oophorus, in association with the zona pellucida, and bound to the egg plasma membrane. Since differences may exist between *in vitro* systems and natural fertilization, morphological analyses of *in vivo* fertilization are especially important. They are, in many cases, our source of information about natural fertilization. For this reason, this chapter will emphasize the findings of ultrastructural studies which have been carried out on fertilization *in vivo*.

V. ULTRASTRUCTURE OF THE ACROSOME REACTION

The acrosome reaction, as originally defined, referred to the multipoint fusion between the plasma membrane over the acrosome cap and the outer acrosomal membrane.[24] Acrosomal contents are released through the fenestrations caused by the fusion. The result is an empty acrosome covered with the porous, fused, membranous product of the plasma membrane and outer acrosomal membrane. Eventually this fused membrane is lost. The equatorial segment of the acrosome does not participate in the acrosome reaction. Although originally the term acrosome reaction referred to the initial fusion between the plasma membrane and the outer acrosomal membrane, now many workers refer to a spermatozoon as being acrosome reacted only when the acrosome reaction has been completed to the extent that only the inner acrosomal membrane remains over the acrosome cap region.

Recently several workers have independently observed, using ultrastructural analysis, that human spermatozoa can undergo an unorthodox spontaneous acrosome reaction.[25-27] In these studies, human spermatozoa were incubated under capacitating conditions for long periods of time, and only a minority of sperm cells underwent the acrosome reaction. Under these conditions, the outer acrosomal membrane and plasma membrane did not undergo the kind of fusion described above. Instead, vesicles formed in the acrosome matrix which subsequently became swollen and distended. Eventually the acrosome matrix disintegrated in an irregular fashion. Since human spermatozoa can also be induced to undergo a typical acrosome reaction, the unorthodox spontaneous acrosome reaction could represent changes in the acrosome that occur as cells die.

Recently Flaherty and Olson[28] described the acrosome reaction in *in vitro*-capacitated guinea pig spermatozoa. The spermatozoon of the guinea pig has a giant acrosome. Although this is not the rule among mammals, spermatozoa of certain other animals, including squirrels,[29] shrews,[30-32] the loris,[33] and the naked-tail armadillo,[34] also have giant acrosomes. Large regions of vesiculating membranes are found during the acrosome reaction in guinea pig spermatozoa. Flaherty and Olson[28] concluded that all of this vesiculating membrane was formed by the fusion of the plasmalemma and the outer acrosomal membrane. However, the mass of vesicles suggests that additional membrane material is contribued from elsewhere. It is possible that some of the acrosomal contents may somehow rearrange into membranous material during the acrosome reaction (Figures 3a and b). The above-mentioned possible exceptions to the typical acrosome reaction in human and guinea pig suggest that interspecies variation may occur in some aspects of the acrosome reaction.

VI. SPERM INTERACTION WITH THE CUMULUS OOPHORUS

A key to understanding acrosome function is to understand where the acrosome reaction occurs. If acrosomal enzymes are released in the cumulus, this suggests that they function in cumulus penetration; if they are released in the zona pellucida, it suggests a function in zona penetration. To make the distinction, a number of morphologists have attempted to determine the site of the acrosome reaction. Austin and Bishop[7] examined the status of the acrosome reaction in rabbit spermatozoa in the ampulla of the oviduct and in the cumulus oophorus. Similar studies were carried out more recently by Cummins and Yanagimachi[35] in the hamster. In these studies, live acrosome-reacting and acrosome-reacted spermatozoa were found in the cumulus and the ampulla of the oviduct. It was concluded that the acrosome reaction occurs in or near the cumulus oophorus. Cummins and Yanagimachi[36] also observed that *in vitro* only the hamster sperm that penetrated the cumulus had undergone initial morphological changes of the acrosome. In their studies, penetration was inhibited by a hyaluronidase inhibitor. These observations led to the general conclusion that the acrosome reaction was initiated in the ampulla of the oviduct shortly before or shortly after the fertilizing spermatozoon entered the cumulus oophorus. This supports the findings of earlier studies by Bedford.[37,38] He reported on the ultrastructure of spermatozoa in rabbit cumulus oocyte complexes fixed at the time of fertilization. Most spermatozoa in the cumulus-oophorus were observed to be undergoing the acrosome reaction. Those that had reached the zona pellucida had lost virtually all of their acrosomal contents; however, the vesiculated, fused cell membrane and outer acrosomal membrane remained. Spermatozoa undergoing acrosome reaction have also been observed in human cumuli taken from the oviduct.[39] Clearly, spermatozoa undergo the acrosome reaction in the cumulus oophorus. However, since in these studies spermatozoa may have been in the cumulus for many hours, it is not clear whether or not the spermatozoa did so when they were still alive. Spermatozoa can undergo a ''false acrosome reaction'' after they die, and some spermatozoa may have been in the cumulus for hours before they were fixed.

Recently Suarez et al.[40] purified from human follicular fluid a 45-kDa factor that is capable of rapidly inducing the acrosome reaction. The factor has very recently been found in the cumulus oophorus.[41] *In vitro*-capacitated human spermatozoa to which this factor was

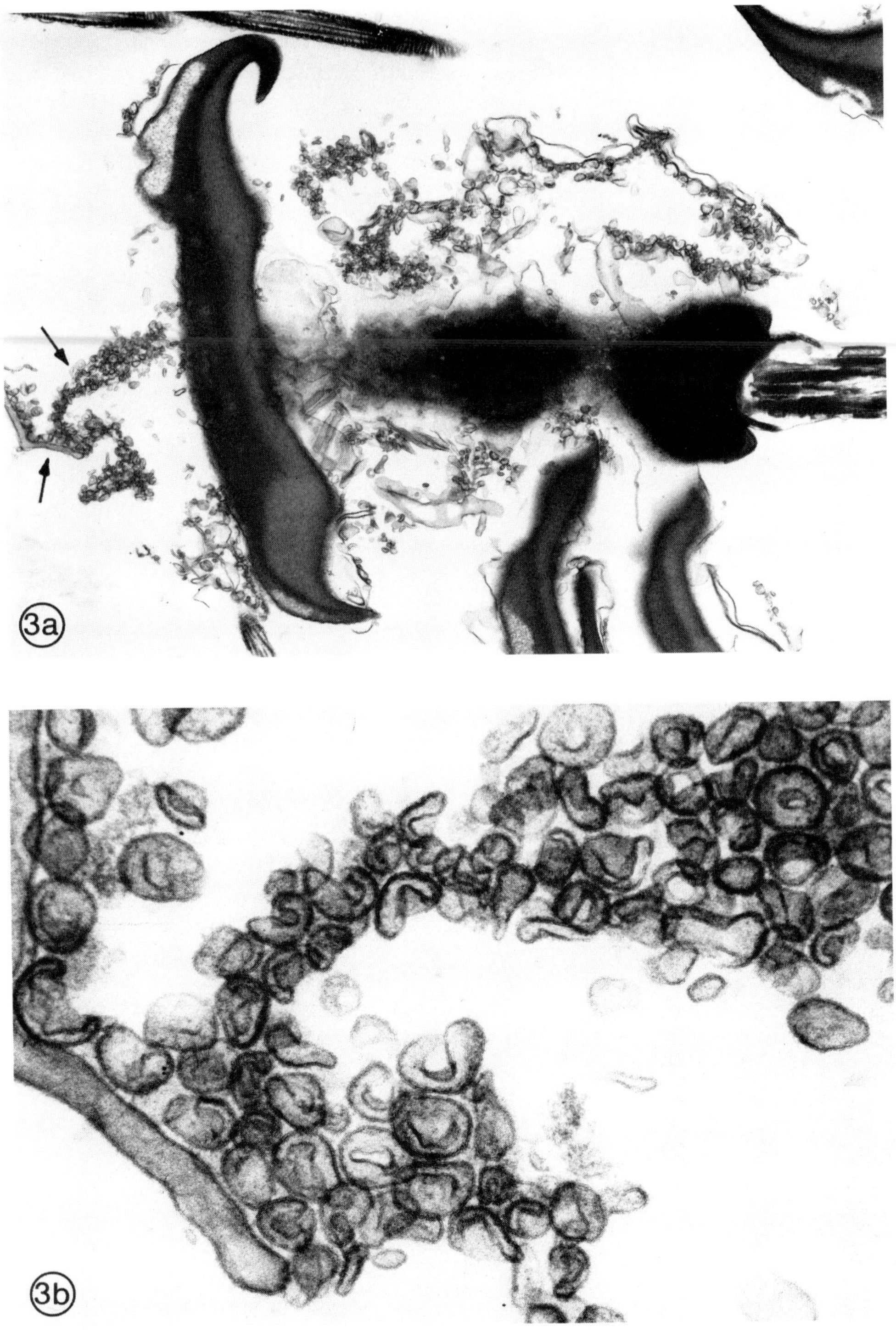

FIGURE 3. (a) Guinea pig spermatozoon *in vitro* undergoing an induced acrosome reaction. (Magnification × 9000.) (From Myles, D., Cowen, A., and Phillips, D. M., unpublished.) (b) High magnification of the area of (a) indicated by the arrows. Spherical vesicles formed during the acrosome reaction in this species appear to be involuted. (Magnification × 93,000.)

added underwent a rapid acrosome reaction. Within 1 min of adding the factor, the plasma membrane and outer acrosomal membrane had fused and vesiculated, and most of the acrosomal contents were no longer visible. The micrographs depicting the acrosome reaction are particularly convincing in the recent report by Yudin et al.,[42] which details the ultrastructure of the induced acrosome reaction. The use of thiocarbohydrazine in the fixative made it possible to distinguish the plasma membrane from the outer acrosomal membrane. Their micrographs showed that the vesicles observed during the acrosome reaction were composed partially of plasma membrane and partially of outer acrosomal membrane.

We wondered whether adding a factor to *in vitro*-capacitated spermatozoa was comparable to the situation in the intact cumulus. To test this, we recently examined human spermatozoa in the human cumulus oophorus.[76] Spermatozoa were first incubated under capacitating conditions that we used for *in vitro* fertilization. They were next incubated for 5 to 30 min with pieces of normal cumulus-oophorus from IVF patients. In thin sections of this material, all of the 48 spermatozoa we observed in the cumulus oophorus had intact acrosomes. These observations suggest that in humans the cumulus oophorus may not induce the acrosome reaction under physiological conditions.

Ten years ago, it was the predominant opinion that the acrosome reaction occurred in the cumulus oophorus. It is not generally accepted that the acrosome reaction occurs on the zona pellucida and that the cumulus oophorus has no role in the acrosome reaction. As you will read in other chapters of this book, the strongest reason for this conclusion is that ZP3 binds to and induces the acrosome reaction in *in vitro*-''capacitated'' mouse spermatozoa. If one considers all of the evidence, however, it is more likely that the acrosome undergoes initial changes in the cumulus-oophorus and subsequently undergoes further changes on the zona pellucida. There is also evidence that there are species differences in the site and function of the acrosome reaction. Thus the situation in the mouse could be different than in other species.

VII. SPERM INTERACTION WITH THE ZONA PELLUCIDA *IN VIVO*

Spermatozoa in contact with the zona pellucida have been observed in oocyte-cumulus complexes removed from the oviducts of the rabbit,[37] sheep,[43] hamster,[44,45] cow,[46] rat,[47] and goat.[48] Observations in all these mammalian species are similar. The TEM reveals a vesiculated acrosome with most of the contents missing. Although the acrosome is clearly depleted at this point, the time at which it reacted is in dispute. Bedford[37,49] interpreted these observations to mean that fusion of the plasma membrane and outer acrosomal membrane takes place in the cumulus, and acrosomal contents are released there. However, Crozet[46] favors the hypothesis that the acrosome is still intact when the spermatozoon reaches the zona pellucida and reacts when it makes contact with the zona surface. Although other authors of this book may disagree with Bedford's hypothesis, it does offer an explanation of why intact spermatozoa have not been observed on the zona pellucida of oocytes fertilized *in vivo*. Whether or not the initial stages of the acrosome reaction occur in the cumulus or on the zona pellucida, it is clear that it is the sperm plasma membrane, vesiculated or not, which initially interacts with the zona pellucida.

Yanagimachi and I have employed phase microscopy in conjunction with scanning electron microscopy to examine *in vivo* sperm-zona interactions in the hamster.[45] In this species the acrosome is large enough that phase-contrast microscopy can distinguish between spermatozoa with normal intact acrosomes, crenulated acrosomes (Figure 4), or totally reacted (missing) acrosomes. We observed that some spermatozoa that reached the zona pellucida had acrosomes, whereas others did not. When we examined *in vivo* fertilized eggs with the SEM, all spermatozoa bound to zona pellucida had acrosomes. The SEM doesn't

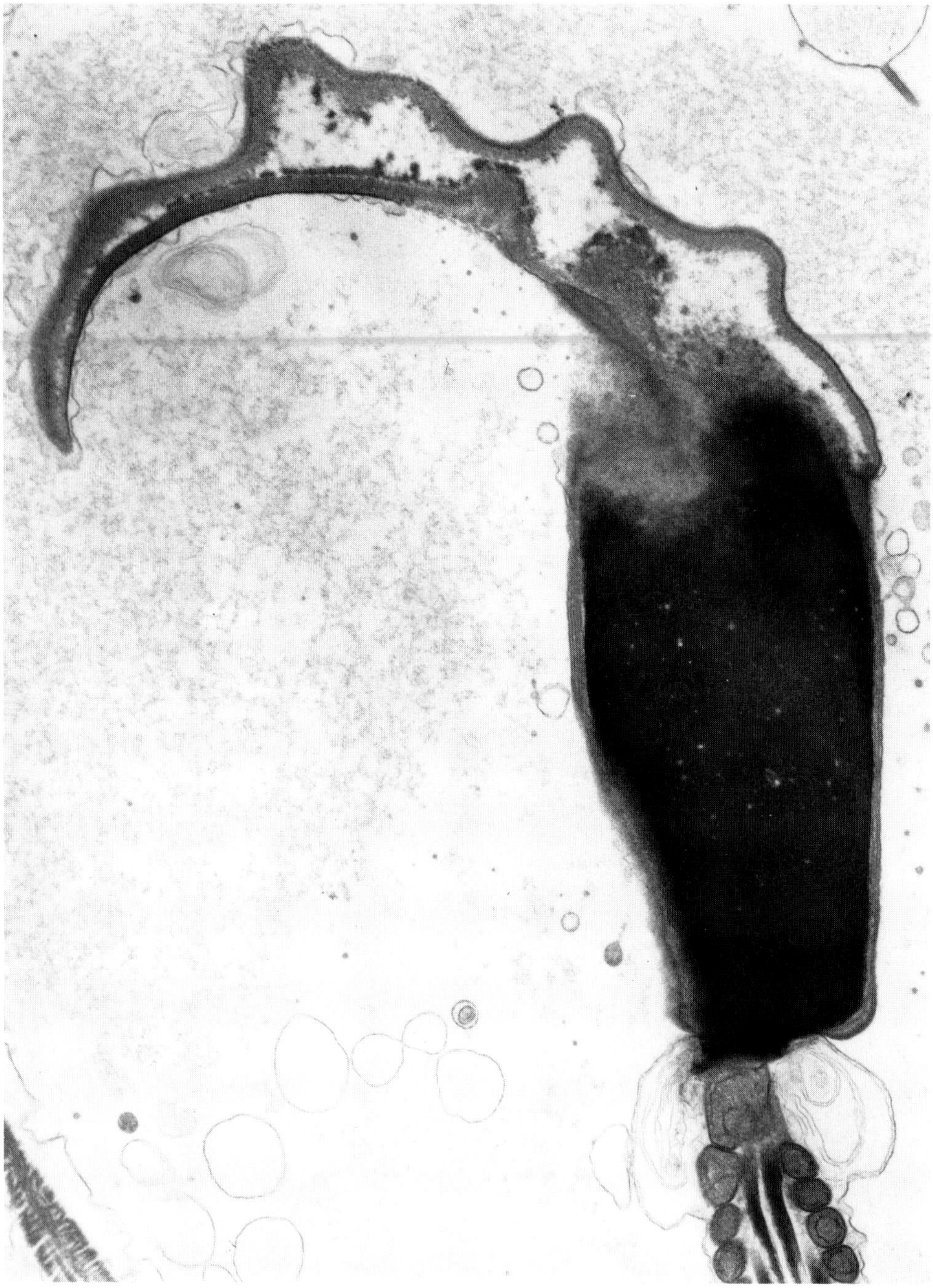

FIGURE 4. The changes in the hamster acrosome, seen here as a bumpy surface and internal structure with distinct electron-dense and electron-lucid regions, are also visible by phase microscopy, making it possible to distinguish those sperm which have undergone acrosome modification. Live spermatozoa with acrosomes modified in this way have been observed in the ampulla of the oviduct and in the cumulus oophorus. Cummins and Yanagimachi[36] found that only spermatozoa with modified acrosomes penetrate the cumulus oophorus. Thus, hamster spermatozoa undergo acrosomal modifications before they reach the zona pellucida. (Magnification × 22,000.)

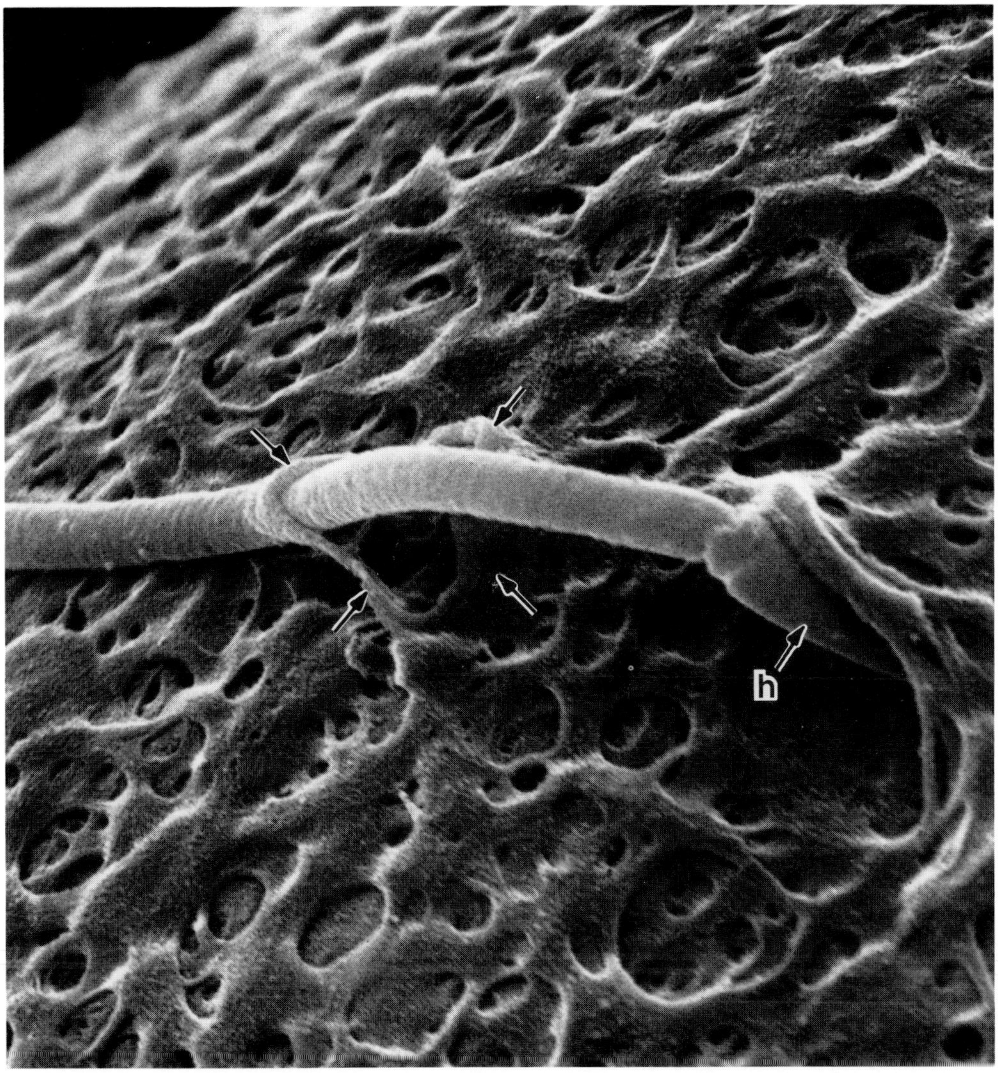

FIGURE 5. In the hamster, the acrosomal ghost forms a ring which tethers the spermatozoon to the zona pellucida. Here an acrosomal ghost (arrows) surrounds the midpiece of a hamster spermatozoon as the sperm head (h) penetrates the zona pellucida. The egg was removed from the ampulla of the oviduct and fixed after stripping off the cumulus oophorus.[45] (Magnification × 8000.)

distinguish between spermatozoa with intact acrosomes and those that have undergone membrane fusion and lost contents. However, based on phase microscopy, these cells probably had undergone membrane fusion. We referred to these acrosomes as "acrosome ghosts".[45] SEM observations of subsequent events suggest that these acrosome ghosts are firmly bound to the zona pellucida. The open acrosome ghost subsequently acts as a collar to tether the fertilizing spermatozoon as it penetrates the zona pellucida (Figure 5); it splits open, and the spermatozoon passes through the hole. This may solve the problem which faces the spermatozoon which has secured itself on the zona pellucida. It must detach from the zona pellucida in order to penetrate, but if it detaches, it could fall off. Using the acrosome ghost as a collar may be the hamster sperm's way of hanging on.

VIII. SPERM INTERACTION WITH THE ZONA PELLUCIDA *IN VITRO*

Before discussing ultrastructural studies of sperm-zona interactions, I would like to emphasize that, *in vitro,* one can obtain attachment of spermatozoa to the zona pellucida under a variety of nonphysiological conditions. Noncapacitated spermatozoa, as well as capacitated sperm, will attach to the zona pellucida. Spermatozoa from one species will attach to zonae pellucidae of another.[50,51] It is not known whether these nonspecific associations are due to nonspecific factors, such as surface charge, or whether they are mediated through specific receptor binding, i.e., to ZP3.

In contrast to the *in vivo* findings discussed above, hamster spermatozoa *in vitro* have been found to associate with the zona pellucida with intact or partially intact acrosomes.[52] An early ultrastructural study demonstrated that *in vitro*-capacitated mouse spermatozoa have an intact acrosome when they bind to the zona pellucida.[53] The acrosome in the guinea pig is so large that one can easily distinguish it without the electron microscope. Results of *in vitro* experiments are conflicting. Huang et al.[54] observed that only motile spermatozoa with missing acrosomes bound to the zona pellucida. However, in a more recent study, Myles et al.[55] observed spermatozoa with both intact and missing acrosomes bound to the zona pellucida. The reason for the differences is not clear, although their techniques for *in vitro* capacitation and reaction differed somewhat.

Recently Cross et al.[56] used the TEM to examine the interaction between *in vitro*-capacitated human spermatozoa and previously frozen human zonae pellucida. They found that spermatozoa bound to the zona pellucida with their acrosomes intact and subsequently underwent a rapid typical acrosome reaction involving fusion of the plasma membrane and the outer acrosomal membrane. The authors concluded that the acrosome reaction in human spermatozoa is morphologically similar to that in other species and that it is induced by contact with the zona pellucida.

IX. FUSION WITH THE VITELLUS *IN VIVO*

It is difficult to observe the initial fusion between the sperm plasma membrane and the oolemma in oocytes fertilized *in vivo*. In nature, fertilization occurs over a period of several hours.[35,57] Thus the probability of observing a fertilizing spermatozoon fusing with the vitellus in a thin section of an egg fixed at about the expected time of fertilization is very small. Even if a spermatozoon was observed in association with the oolemma there would be no way to determine how long it had been since it first made contact with the oolemma. The SEM makes it much easier to observe associations between sperm and egg. Examining thin sections taken through an entire egg might take days, but the entire egg surface can be viewed with the SEM in minutes.

In the SEM, spermatozoa attached to oocytes always occur with their broad surface parallel to the surface of the oocyte. We have observed in hamster and rat spermatozoa that as the sperm head disappears under the oocyte surface, egg mircovilli usually appear over the equatorial segment but not over the acrosome cap or postacrosomal region.[47,58] This suggests that initial fusion between the egg plasma membrane and the sperm plasma membrane occurs over the equatorial segment. When Moore and Bedford[59] examined hamster oocytes fertilized *in vivo* with the TEM, they used immature oocytes which lack a mechanism to block polyspermy. This greatly increased their chances of finding sperm-egg associations. Although this was not, strictly speaking, a natural situation, they also observed that initial fusion took place at the plasma membrane overlying the equatorial segment of the acrosome.

Some of the most intriguing aspects of mammalian fertilization are the surface alterations that accompany nuclear events in the ooplasm. The surface region over the meiotic spindle

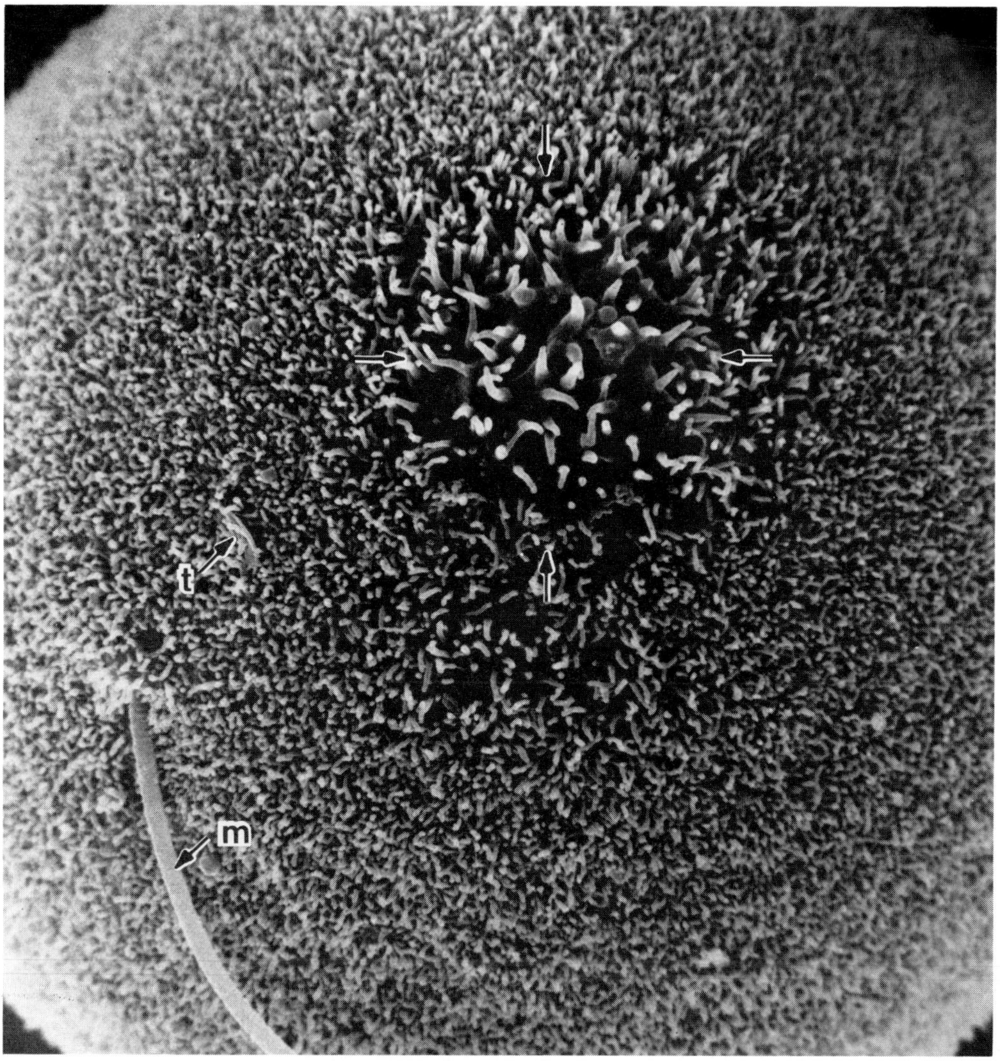

FIGURE 6. Fusion of the sperm head with the vitellus triggers the second meiotic division. The anterior-most tip of the sperm head (t) and midpiece (m) are seen in this SEM. Long, broad microvilli are observed in the region over the meiotic chromosomes where the second polar body will form. *In vivo* fertilized hamster egg. (Magnification × 3000.) (From Shalgi, R. and Phillips, D. M., unpublished.)

is devoid of microvilli.[21] Just preceding the formation of the polar bodies, however, long microvilli extend from this region. This occurs at both the first and second meiotic divisions[60,61] (Figure 6). Concurrently, the region over the meiotic spindle becomes free of cortical granules. Okada et al.[60] have shown that during meiosis II, the disappearance of cortical granules is caused at least in part by exocytosis. The formation of the polar bodies themselves is unusual in that the surfaces of the polar bodies are free of microvilli. Therefore, the contrast between the surface of the polar body and the microvillar surface of the oocyte is striking (Figure 7). An elevated microvillus-free cone forms at the site of the decondensing sperm nucleus.[62-64] It is obvious from these phenomena that there are intimate interactions between nuclear events and the egg surface or structures underlying the egg surface.

Decondensation of the sperm nucleus has been observed in the TEM in several species. In most species, chromatin begins to condense from the end of the nucleus nearest the tail

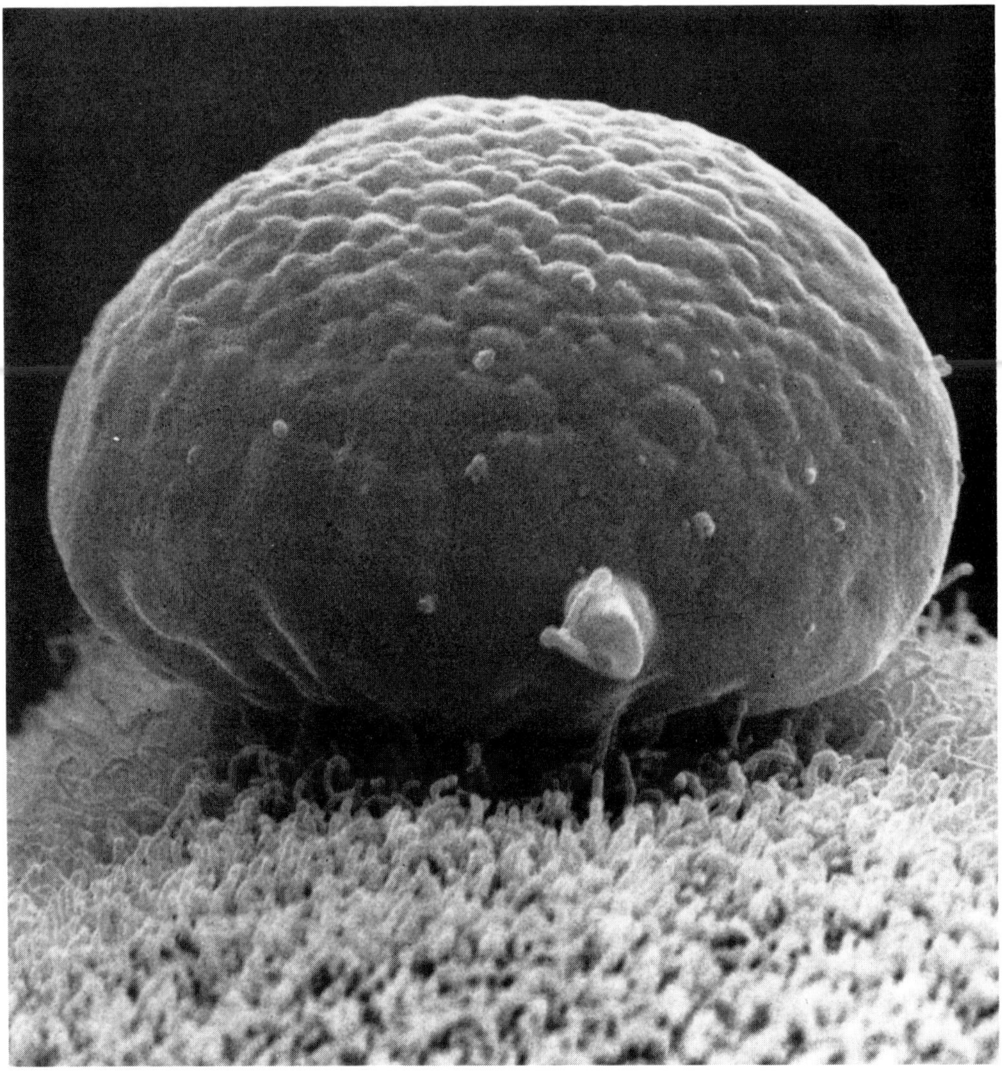

FIGURE 7. The second polar body in an *in vivo* fertilized hamster egg. The surface of the polar body is characterized by small bumps, but the microvilli have disappeared. (Magnification × 7000.) (From Shalgi, R. and Phillips, D. M., unpublished.)

and progresses forward.[65,66] Human spermatozoa are an exception. In the human sperm, chromatin decondensation appears to occur fairly uniformly, but the last chromatin to decondense is at the posterior end of the nucleus.[67] As chromatin decondenses, the nucleus takes on the spherical shape of the male pronucleus. In mammals, the entire sperm tail is incorporated into the egg. As it does so, it is thrown into serpentine curves which protrude from the egg surface[68] (Figure 8).

A. SPERM FUSION WITH THE OOLEMMA *IN VITRO*

In vitro, especially with cumulus- and zona-free eggs, one can precisely time gamete interactions. By adding many sperm one can visualize the interaction of more spermatozoa with the egg surface. These types of *in vitro* manipulations have facilitated a number of structural studies of mammalian gamete interactions. It is clear, however, that the *in vitro*

FIGURE 8. Ten hours after fusion, serpentine protusions (arrows) are observed where the sperm tail underlies the oolemma.[58] (Magnification × 5500.)

situation is not a perfect simulation of the natural fertilization process. There appears to be a difference in the angle at which spermatozoa pass through the zona pellucida and contact the egg. *In vivo,* sperm pass through a zona pellucida and contact the oolemma at a very acute angle.[44,48,49,69] However, sperm *in vitro* pass through the zona pellucida and enter the oocyte at a right angle.[70,71] The difference in angles may have to do with the lack of an intact cumulus *in vitro*. The way that spermatozoa push against the cumulus may cause them to traverse the zona pellucida at an acute angle. In any event, fusion of spermatozoa with the oolemma *in vitro* appears to occur over the equatorial segment of the acrosome on the sperm plasma membrane just as it does *in vivo*. This is true of sperm-egg fusion in human[67] (Figure 9), guinea pig,[66] and hamster.[72]

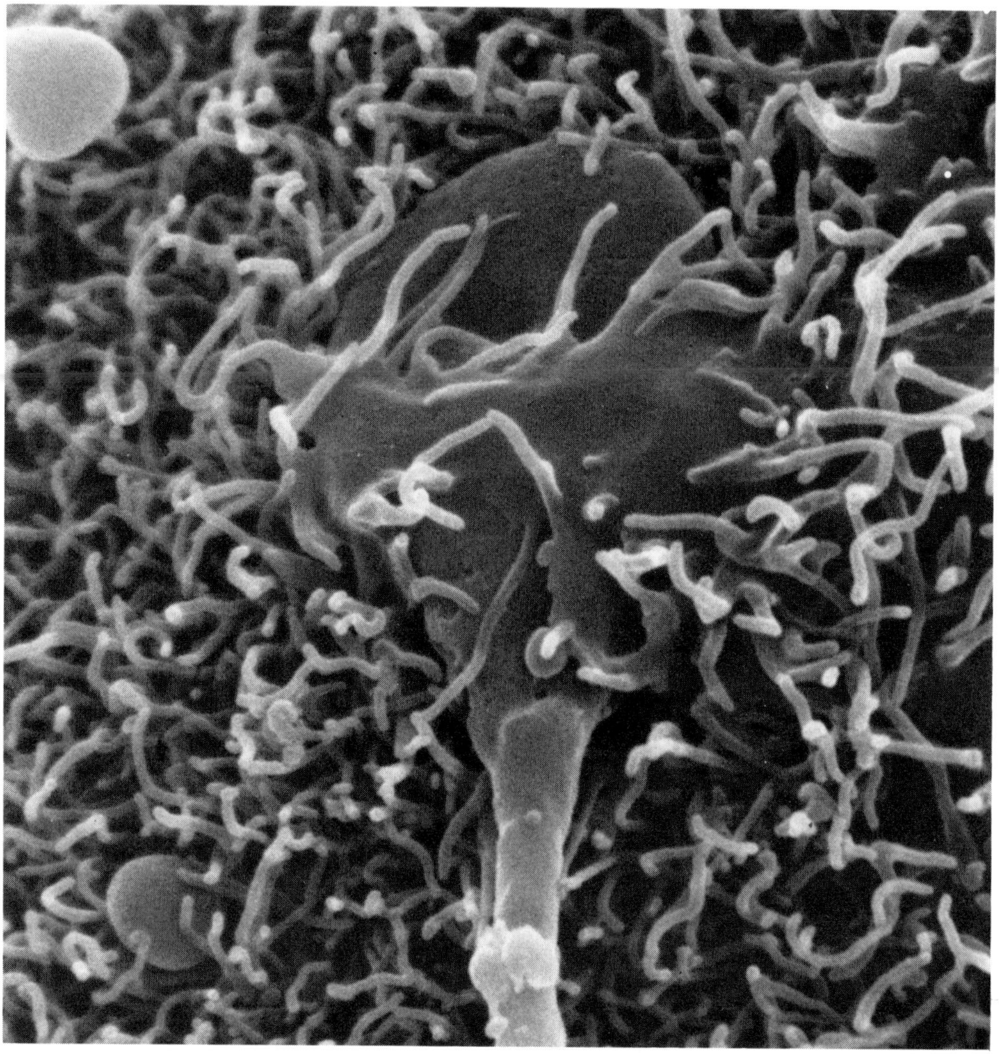

FIGURE 9. Human sperm fusing with a hamster oocyte. What appears to be oocyte microvilli and membrane are observed over the equatorial segment of the acrosome. (Magnification × 7000.) (From Phillips, D. M., Bronson, R., and Cooper, G., unpublished.)

X. SUMMARY

I have discussed the ultrastructure of mammalian gametes and ultrastructural aspects of gamete interactions. Although there are many similarities among gametes of different mammalian species, there are also differences. Hopefully, in our search for universal mechanisms of fertilization in mammals, we will not ignore the differences. The exceptions may prove to be just as enlightening as the general rule.

REFERENCES

1. **Phillips, D. M.,** Substructure of the mammalian acrosome, *J. Ultrastruct. Res.,* 38, 591, 1972.
2. **Phillips, D. M. and Dryden, G.,** A comparative morphology of mammalian gametes, in *Comparative Overview of Mammalian Gametes,* Dunbar, B. S. and O'Rand, M. G., Eds., in preparation.
3. **Berrios, M., Flechon, J. E., and Barros, C.,** Ultrastructure of *Octodon degus* spermatozoon with special reference to the acrosome, *Am. J. Anat.,* 151, 39, 1978.
4. **Breed, W. G. and Leigh, C.,** Sperm head morphology of Australian Molossid bats with special reference to the acrosomal structure, *Mammalia,* 49, 403, 1985.
5. **Dekel, N. and Phillips, D. M.,** c-AMP, prostaglandin E_2 and steroids—possible mediators in the rat cumulus oophorus mucification, *Biol. Reprod.,* 22, 289, 1980.
6. **Phillips, D. M. and Dekel, N.,** Effects of gonadotropins and prostaglandin on cumulus mucification in cultures of intact follicles, *J. Exp. Zool.,* 221, 275, 1982.
7. **Austin, C. R. and Bishop, M. W. H.,** Role of the rodent acrosome and perforatorium in fertilization, *Proc. R. Soc. London Ser. B,* 149, 241, 1958.
8. **Blandau, R. J.,** Gamete transport—comparative aspects, in *The Mammalian Oviduct,* Hafez, E. S. E. and Blandau, R. J., Eds., University of Chicago Press, Chicago, 1969, 129.
9. **Mahi-Brown, C. A. and Yanagimachi, R.,** Parameters influencing ovum pick up by oviductal fimbria in the Golden Hamster, *Gamete Res.,* 8, 1, 1983.
10. **De Felici, M. and Siracusa, G.,** "Spontaneous" hardening of the zona pellucida of mouse oocytes during vitro culture, *Gamete Res.,* 6, 107, 1982.
11. **De Felici, M., Salustri, A., and Siracusa, G.,** "Spontaneous" hardening of the zona pellucida of mouse oocytes during in vitro culture. II. The effect of follicular fluid and glycosaminoglycans, *Gamete Res.,* 12, 227, 1985.
12. **Phillips, D. M., Shalgi, R., and Dekel, N.,** Mammalian fertilization as seen with the scanning electron microscope, *Am. J. Anat.,* 174, 357, 1985.
13. **Dekel, N., Kraicer, P. F., Phillips, D. M., Sanchez, R. S., and Segal, S.,** Cellular associations in the rat oocyte cumulus cell complex, *Gamete Res.,* 1, 47, 1978.
14. **Phillips, D. M.,** Ultrastructure of mammalian fertilization, *11th Int. Congr. Electron Microsc.,* p. 1893, 1986.
15. **Talbot, P. and DiCarlantonio, G.,** The architecture of the hamster oocyte-cumulus complex, *Gamete Res.,* 9, 261, 1984.
16. **Talbot, P. and DiCarlantonio, G.,** The oocyte-cumulus complex: ultrastructure of the extracellular matrix in hamsters and mice, *Gamete Res.,* 10, 127, 1984.
17. **Yudin, A. I., Cherr, G. N., and Katz, D. F.,** Structure of the cumulus matrix and zona pellucida in the golden hamster. A new view of sperm interactions with oocyte-associated extracellular matrices, *Cell Tissue Res.,* 251, 555, 1988.
18. **Goldschmit, D., Kraicer, P., and Orly, J.,** Periovulatory expression of cholesterol side-chain cleavage cytochrome P-450 in cumulus cells, *Endocrinology,* 124, 369, 1989.
19. **Osman, R. A., Andria, M. L., Jones, A. D., and Meizel, S.,** Steroid induced exocytosis: the human acrosome reaction, *Biochem. Biophys. Res. Commun.,* 160, 828, 1989.
20. **Phillips, D. M. and Shalgi, R.,** Surface properties of the zona pellucida, *J. Exp. Zool.,* 213, 1, 1980.
21. **Phillips, D. M. and Shalgi, R.,** Surface properties of the mouse and hamster zona pellucida and oocyte, *J. Ultrastruct. Res.,* 72, 172, 1980.
22. **Phillips, D. M. and Shalgi, R.,** Comparative ultrastructure of fertilization, in *Development and Function of the Reproductive Organs,* Parviven, M., Ed., Raven Press, New York, 1988, 11.
23. **Longo, L. J. and Chen, D. Y.,** Development of cortical polarity in mouse eggs: involvement of the meiotic apparatus, *Dev. Biol.,* 107, 382, 1984.
24. **Austin, C. R. and Short, R. V.,** *Reproduction in Mammals. 1. Germ Cells and Fertilization,* Cambridge University Press, Cambridge, 1982.
25. **Stock, C. E. and Fraser, L. R.,** The acrosome reaction in human sperm from men of proven fertility, *Hum. Reprod.,* 2, 109, 1987.
26. **Bronson, R. A., Cooper, G. W., and Phillips, D. M.,** Effects of sperm-reactive antibodies and complement on the ultrastructure of human spermatozoa, *Ann. N. Y. Acad. Sci.,* 513, 574, 1987.
27. **Nagae, T., Yanagimachi, R., Srivastava, P. N., and Yanagimachi, H.,** Acrosome reaction in human spermatozoa, *Fertil. Steril.,* 45, 701, 1986.
28. **Flaherty, S. A. and Olson, G. E.,** Membrane domains in guinea pig sperm and their role in the membrane fusion events of the acrosome reaction, *Anat. Rec.,* 220, 267, 1988.
29. **Martan, J. and Hruban, Z.,** Unusual spermatozoan formations in the epididymis of the flying squirrel, *Glaucomys volans, J. Reprod. Fertil.,* 21, 167, 1970.
30. **Koehler, J. K.,** Fine structure of spermatozoa of the Asiatic musk shrew, *Suncus murinus, Am. J. Anat.,* 149, 135, 1977.

31. **Ploen, L., Ekwall, H., and Afzelius, B. A.,** Spermiogenesis and the spermatozoa of the European common shrew, *Sorex aruneusl, J. Ultrastruct. Res.,* 68, 149, 1979.
32. **Phillips, D. M. and Bedford, J. M.,** Ultrastructure of spermatozoa of the musk shrew *Suncus murinus, J. Exp. Zool.,* 235, 119, 1985.
33. **Phillips, D. M. and Bedford, J. M.,** Sperm-sperm associations in the loris, *Gamete Res.,* 18, 17, 1987.
34. **Heath, E., Schaffer, N., Merit, D. A., and Jeyendran, R. S.,** Rouleaux formation by spermatozoa in the naked-tail armadillo, *Cabassous unicinctus, J. Reprod. Fertil.,* 79, 153, 1987.
35. **Cummins, J. M. and Yanagimachi, R.,** Sperm-egg ratios and the site of the acrosome reaction during *in vivo* fertilization in the hamster, *Gamete Res.,* 5, 239, 1982.
36. **Cummins, J. M. and Yanagimachi, R.,** Development of ability to penetrate the cumulus oophorus by hamster spermatozoa capacitated *in vitro* in relation to the timing of the acrosome reaction, *Gamete Res.,* 15, 187, 1986.
37. **Bedford, J. M.,** An electron microscope study of sperm penetration into the rabbit egg after natural mating, *Am. J. Anat.,* 133, 213, 1972.
38. **Bedford, J. M.,** Mechanisms involved in penetration of spermatozoa through the vestments of the mammalian egg, in *Physiology and Genetics of Reproduction Part B,* Coutinho, E. M. and Fuchs, F., Eds., Plenum Press, New York, 1974, 55.
39. **Pereda, J. and Coppo, M.,** An electron microscopic study of sperm penetration into human egg investments, *Anat. Embryol. (Berlin),* 173, 247, 1985.
40. **Suarez, S. S., Wolf, D. P., and Meizel, S.,** Induction of the acrosome reaction in human spermatozoa by a fraction of human follicular fluid, *Gamete Res.,* 14, 107, 1986.
41. **Siiteri, J. E., Dandekar, P., and Meizel, S.,** Human sperm acrosome reaction-initiating activity associated with the human cumulus oophorus and mural granulosa cells, *J. Exp. Zool.,* 246, 71, 1988.
42. **Yudin, A. I., Gottlieb, W., and Meizel, S.,** Ultrastructural studies of the early events of the human sperm acrosome reaction as initiated by human follicular fluid, *Gamete Res.,* 20, 11, 1988.
43. **Crozet, N. and Dumont, M.,** The site of the acrosome reaction during *in vivo* penetration of the sheep oocyte, *Gamete Res.,* 10, 97, 1984.
44. **Yanagimachi, R. and Noda, Y. D.,** Ultrastructural changes in the hamster sperm head during fertilization, *J. Ultrastruct. Res.,* 31, 465, 1970.
45. **Yanagimachi, R. and Phillips, D. M.,** The status of acrosomal caps of hamster spermatozoa in egg investments prior to fertilization *in vivo, Gamete Res.,* 9, 1, 1984.
46. **Crozet, N.,** Ultrastructural aspects of *in vivo* fertilization in the cow, *Gamete Res.,* 10, 241, 1984.
47. **Phillips, D. M. and Shalgi, R.,** Sperm penetration into rat ova fertilized *in vivo, J. Exp. Zool.,* 221, 373, 1982.
48. **Crozet, N., Theron, M. C., and Chemineau, P.,** Ultrastructure of *in vivo* fertilization in the goat, *Gamete Res.,* 18, 191, 1987.
49. **Bedford, J. M.,** Ultrastructural changes in the sperm head during fertilization in the rabbit, *Am. J. Anat.,* 123, 329, 1968.
50. **Bedford, J. M.,** Sperm-egg interactions: the specificity of human spermatozoa, *Anat. Res.,* 166, 477, 1977.
51. **Swenson, C. E. and Dunbar, B. S.,** Specificity of sperm-zona interactions, *J. Exp. Zool.,* 219, 97, 1982.
52. **Yanagimachi, R.,** Mechanisms of fertilization in mammals, in *Fertilization and Embryo Development In Vitro,* Mastroianni, L. and Biggers, J. D., Eds., Plenum Press, New York, 1981.
53. **Saling, P., Sowinski, J., and Storey, B.,** An ultrastructural study of epididymal mouse sperm binding to the zona pellucida *in vitro*: sequential relationship to the acrosome reaction, *J. Exp. Zool.,* 209, 229, 1979.
54. **Huang, T. T. F., Fleming, A. D., and Yanagimachi, R.,** Only acrosome reacted spermatozoa can bind to and penetrate the zona pellucida: a study using the guinea pig, *J. Exp. Zool.,* 217, 287, 1981.
55. **Myles, D. G., Hyatt, H., and Primakoff, P.,** Binding of both acrosome-intact and acrosome-reacted guinea pig sperm to the zona pellucida during *in vitro* fertilization, *Dev. Biol.,* 121, 559, 1987.
56. **Cross, N. L., Morales, P., Overstreet, J. W., and Hanson, F. W.,** Induction of acrosome reactions by the human zona pellucida, *Biol. Reprod.,* 38, 235, 1988.
57. **Shalgi, R. and Phillips, D. M.,** Motility of rat spermatozoa at the site of fertilization, *Biol. Reprod.,* 39, 1207, 1988.
58. **Shalgi, R. and Phillips, D. M.,** Mechanics of sperm entry in cycling hamsters, *J. Ultrastruct. Res.,* 71, 156, 1980.
59. **Moore, H. D. M. and Bedford, J. M.,** Ultrastructure of the equatorial segment of hamster spermatozoa during penetration of oocytes, *J. Ultrastruct. Res.,* 62, 110, 1978.
60. **Okada, A., Yanagimachi, R., and Yanagimachi, H.,** Development of a cortical granule-free area of cortex and paravitelline space in the hamster oocyte during maturation and following ovulation, *J. Submicrosc. Cytol.,* 18, 233, 1986.
61. **Ebensperger, E. and Bairos, C.,** Changes at the hamster oocyte surface from germinal vesicle stage to ovulation, *Gamete Res.,* 9, 387, 1984.

62. **Stefanini, M., Chikayoshi, O., and Zamboni, L.,** Ultrastructure of fertilization in the mouse. II. Penetration of sperm into the ovum, *J. Submicrosc. Cytol.*, 1, 1, 1969.
63. **Shalgi, R., Phillips, D. M., and Kraicer, P. F.,** Observations on the incorporation cone in the rat, *Gamete Res.*, 1, 27, 1978.
64. **Lopata, A., Sathananthan, A. H., McBain, J. C., Johnston, W. I., and Speirs, A. L.,** The ultrastructure of preovulatory human eggs fertilized *in vivo, Fertil. Steril.*, 33, 12, 1980.
65. **Crozet, N.,** Fine structure of sheep fertilization *in vitro, Gamete Res.*, 19, 291, 1988.
66. **Noda, Y. D. and Yanagimachi, R.,** Electron microscopic observations of guinea pig spermatozoa penetrating eggs *in vivo, Dev. Growth Differ.*, 18, 15, 1976.
67. **Sathananthan, A. H., Ng, S. C., Edirisinghe, R., Ratnam, S. S., and Wong, P. C.,** Human sperm-egg interaction *in vitro, Gamete Res.*, 15, 317, 1986.
68. **Hirao, Y. H. and Hiraoka, J. I.,** Surface architecture of sperm tail entry into the hamster oocyte, *Dev. Growth Differ.*, 29, 123, 1987.
69. **Mori, T. and Uchida, T. A.,** Ultrastructural observations of fertilization in the Japanese long-fingered bat *Miniopterus schreibersii fuliginosus, J. Reprod. Fertil.*, 63, 231, 1981.
70. **Shalgi, R. and Phillips, D. M.,** Mechanics of *in vitro* fertilization in the hamster, *Biol. Reprod.*, 23, 433, 1980.
71. **Shalgi, R. and Phillips, D. M.,** Sperm penetration into rat fertilized *in vitro, J. Androl.*, 3, 382, 1982.
72. **Yanagimachi, R. and Noda, Y. D.,** Physiological changes in the post-nuclear cap region of mammalian spermatozoa: a necessary preliminary to the membrane fusion between sperm and egg cells, *J. Ultrastruct. Res.*, 31, 486, 1970.
73. **Myles, D., Cowen, A., and Phillips, D. M.,** unpublished.
74. **Shalgi, R., and Phillips, D. M.,** unpublished.
75. **Phillips, D. M., Bronson, R., and Cooper, G.,** unpublished.
76. **White, D., Bedford, J. M., and Phillips, D. M.,** unpublished.

Chapter 11

VIDEOMICROSCOPY OF MAMMALIAN FERTILIZATION

Erma Z. Drobnis and David F. Katz

TABLE OF CONTENTS

I. WHAT QUESTIONS CAN BE ANSWERED USING VIDEOMICROSCOPY OF MAMMALIAN FERTILIZATION?

In mammals, fertilization occurs in a complex microenvironment within the female reproductive tract. This microenvironment is not accessible for direct observation of detailed events at the cellular level. Information about fertilization has been obtained by removing the gametes from their natural environment, and either chemically fixing them for examination or observing their interaction directly under *in vitro* conditions.

Ultrastructural studies on gametes collected from mated animals have provided important information about many aspects of mammalian fertilization. However, since immotile and other presumably nonfertilizing sperm may be present at the site of fertilization,[1-10] evaluation of fixed sperm which are associated with the oocyte cannot resolve questions regarding the fertilizing spermatozoon. Direct observations of living gametes have been used to investigate questions relating, for example, to: the characteristics of the fertilizing sperm; the mechanisms for completion of each of the final phases of fertilization; and the temporal sequence of events during fertilization.

Many questions of great interest to mammalian fertilization biologists relate to the complex natural history of spermatozoa within the female reproductive tract. In mammals, numerous sperm are deposited at mating, but very few actually participate in fertilization.[11] The events which accompany sperm transport to the site of fertilization[12-15] result in a large reduction of sperm numbers, producing sperm-egg ratios close to unity at the time of fertilization.[8,16-18] Accompanying this selection process are final maturational changes, collectively termed capacitation, which render sperm capable of fertilization.[11,12,19] During transport, sperm accumulate in various anatomical regions within the female reproductive tract, some of which probably function as sperm reservoirs.[13-15] The sperm within these various sites have different physiological characteristics.[12,14]

Most studies of sperm capacitation and fertilization have utilized sperm which are collected from the epididymis or ejaculate, and which are then capacitated by hours of incubation in culture medium. The results of such *in vitro* studies must be tested for their relevance to the complex environment in which fertilization occurs *in vivo*. Direct observations of the final stages of sperm transport and of fertilization events have been used to study the characteristics (e.g., acrosomal status and motility) of "the fertilizing sperm", as opposed to characteristics seen in the heterogeneous population of sperm cells present in the female reproductive tract or in semen. Observations of fertilization have also been used to compare sperm collected from different sites in the female reproductive tract with those collected from the male and capacitated *in vitro*, and to compare ovarian and ovulated oocytes.

The recent review of Gaddum-Rosse[21] gives an excellent overview and historical perspective of experiments in which the interactions between living mammalian gametes were observed directly. These experiments include cinemicrography of the later events of fertilization which follow fusion of the gametes and precede the first cleavage of the zygote. In the present review, we will focus instead on the methods used to make direct observations of mammalian fertilization, some of the technical problems associated with these methods, what species have been used in these studies, and how videorecording techniques have been used, primarily over the last 10 years, to *document* observations of the early events in mammalian fertilization. These early events include the complex interactions of the fertilizing sperm with the extracellular matrices, the cumulus cell matrix, and the zona pellucida, which invest the mammalian oocyte. We have included a discussion of changes in sperm motility during capacitation, since hyperactivated motility, which has been studied extensively *in vitro*, occurs at the site of fertilization, and is thought to be important in penetration of the oocyte investments. Unless specifically mentioned, we will use the terms mammal and mammalian to refer only to eutherian mammals.

II. METHODOLOGY AND CONTROL FOR ARTIFACTS

A. INTRODUCTION

Because videomicroscopy of mammalian fertilization requires *in vitro* conditions, the experimenter must consider the possibility of artifacts; that is, the behavior of the gametes observed *in vitro* may not be identical to their behavior under natural conditions *in vivo.* In fact, fertilization events observed under the conditions required for videomicroscopy may even differ from those occurring under more optimal conditions of culture *in vitro.* For example, a number of investigators have reported that sperm motility decreases and oocytes or embryos lose their natural appearance during microscopic observation of fertilization events *in vitro,* while, concomitantly, fertilization continues normally in control drops under more appropriate culture conditions.[22-24] Important sources of artifact include: use of culture media which differ significantly from ampullar fluid (e.g., in pH, Ca^{2+} concentration, protein composition); failure to provide optimal culture conditions (e.g., humidified environment, CO_2 tension); removal of oocyte investments to improve visibility; use of epididymal, seminal, or uterine sperm instead of ampullar sperm; use of ovarian oocytes; damage to gametes during preparation; and exposure of gametes to light. Over the years, investigators have used improved microscopic and videorecording techniques, and have developed specialized chambers designed to produce conditions comparable to those present in an incubator. Although the possibility of artifacts cannot be avoided, identification of the potential sources of artifact is important for designing appropriate control experiments. In many cases, the experimental question is not what actually occurs *in vivo,* but what conditions are required for a given event or process to occur.

B. PHYSIOLOGICAL STATE OF GAMETES

To date, the most directly relevant method used to observe fertilization has been to visualize the gametes within the oviduct.[25-27] However, in order to make more detailed observations, it is necessary to remove the gametes from the female reproductive tract. Ideally, sperm and oocytes are collected from the oviductal ampulla just prior to fertilization; unfortunately, there are extremely few sperm present at this site at this time. In mammals, sperm are stored in a quiescent state in the oviductal isthmus.[14,15,28] By a mechanism which has not been well characterized, very small numbers of sperm are released into the ampulla once the products of ovulation are present in the oviduct. A final sperm-to-oocyte ratio close to unity is produced.[8,17-19] Following ovulation, penetration of the cumulus cell matrix and zona pellucida by the fertilizing spermatozoon is very rapid, making it extremely difficult to collect the gametes from the ampulla before sperm have completed vestment penetration and fused with the oocytes.[1,8,21,29-31] Moreover, if sperm are penetrating the cumulus matrix or the zona pellucida when the gametes are collected from the ampulla, motility may cease or be inhibited when the cumulus matrix is compressed for observation.[8,30,31] This could be due to the rearrangement of the microstructure of the cumulus cell matrix around the sperm flagellum. In a number of studies, investigators have collected oocytes and eggs from large numbers of mated or artificially inseminated females, and have observed fertilization events.[8-10,29-31] However, because of difficulties in observing the interaction of gametes collected from the natural site of fertilization, and the inability to observe details (e.g., flagellar kinematics) in these experiments, other studies have used sperm collected from the epididymis or from the uteri of mated females. Sperm derived from these sources are in a different physiological state than ampullar sperm, and incubation of sperm *in vitro* is required to complete the changes necessary for sperm participation in fertilization. Even when sperm are collected from the oviductal isthmus, instead of the oviductal ampulla, incubation is required to initiate the motility seen in ampullar sperm.[5,6] It is important to recognize that even though sperm which have been cultured *in vitro* are able to fertilize oocytes, the

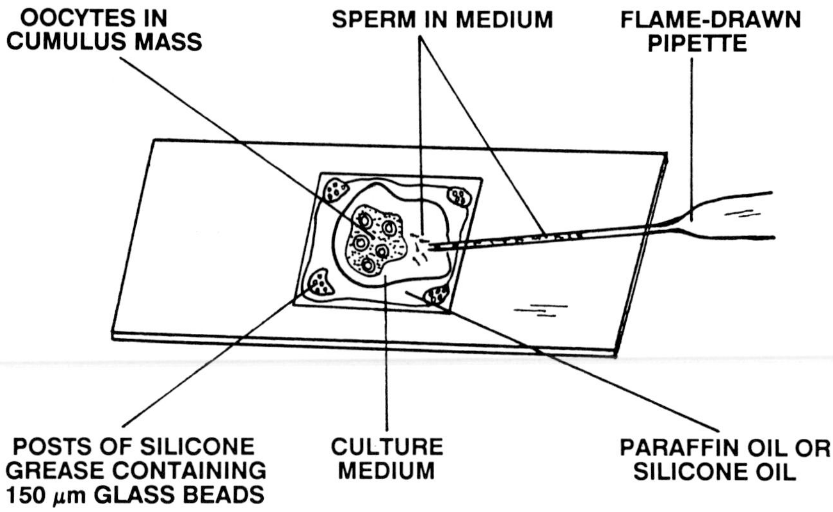

FIGURE 1. A simple slide preparation for videomicroscopy of fertilization. Four posts composed of silicone grease, which contains acid-washed glass beads, are positioned on a siliconized slide. A cumulus mass is transferred to the slide in a small drop of culture medium. After covering the crop with a siliconized coverglass, the preparation is compressed to the diameter of the glass beads (150 μm). The culture medium drop is surrounded with washed, incubator-equilibrated paraffin oil, then a small number of sperm are added at one side of the drop. Individual sperm can be observed and videorecorded as they interact with the oocyte and its investing layers.

resulting fertilization events may not be identical to those which occur during fertilization within the ampulla.

The number of sperm used to inseminate oocytes in the observation chamber is also a critical factor. Ideally, only one sperm per oocyte would be used in an attempt to reproduce the conditions present at the time of fertilization *in vivo.* However, even under ideal culture conditions *in vitro,* fertilization rates are usually low unless more sperm are present.[19,32] If large numbers of sperm are inseminated, there may be contamination by enzymes and metabolites released from the sperm, and these might affect fertilization events. An example of this occurs in the hamster. In this species, if too many sperm are applied to cumulus masses, the sperm appear to tear up the cumulus cell matrix, apparently by both enzymatic and mechanical means.[23,33,34] Several methods have been used to reduce this artifactual degradation. Oocytes with their associated sperm have been transferred to slides and covered with a coverglass; then fresh medium has been drawn through this preparation to rinse out extra sperm.[33,35] Another strategy is to allow brief (1 to 3 min) interaction of sperm with oocytes in a culture drop, then draw the oocytes and associated sperm into a flat capillary tube for observation.[23,36] Finally, small numbers of sperm can be added to one corner of the observation chamber[24,34,37] (Figure 1); individual sperm cells can then be observed as they approach the oocyte and interact with the cumulus cell matrix and the zona pellucida.

The physiological age of the oocytes may not be as critical as that of the sperm. Hamster oocytes collected before ovulation have been used to observe penetration of the cumulus matrix by sperm.[23,36] These individual oocytes, with their associated cumuli, are convenient for studying penetration of the cumulus material surrounding individual oocytes; and, if oocytes are to be recovered from living animals, it is easier to collect them from the follicle than by flushing the oviduct. However, there is evidence that the composition of the zona pellucida changes following ovulation,[38-41] and that the cumulus cell matrix expands within the oviduct as the cumulus cells continue to secrete their matrix material.[42] These changes may be important to fertilization events *in vivo.*

Another oocyte factor which may influence gamete interaction is "spontaneous hardening'' of the zona pellucida.[43-47] This process results in decreased *in vitro* fertilization rates in mouse oocytes cultured for a few hours *in vitro.* Other spontaneous changes in the zona pellucida occur during incubation of oocytes *in vitro,*[47] and the possibility of changes in properties of the zona pellucida during culture should be considered in studies using immature oocytes which have been matured *in vitro.*

The physiology of oocytes may also be affected by the use of hormone treatments to superovulate and/or induce ovulation. Although these treatments increase the number of gametes available for study, there is evidence that superovulation with FSH affects the properties of the zona pellucida, perhaps by premature recruitment of follicles.[48,49] The cumulus cell matrix may also be affected by superovulation treatments, since secretion of this extracellular matrix is under gonadotropin control.[50,51] We have noted that the larger cumulus masses produced by superovulation in hamsters have a reduced distance between neighboring cumulus cells, probably due to inability of the cumulus mass to expand completely within the oviductal lumen.[184]

In many observations of mammalian fertilization, in order to improve the visualization of sperm, the cumulus cell matrix was removed from the oocytes, or oocytes with dispersed cumulus matrices were chosen for observation. This has been justified by the demonstration that fertilization occurs when oocytes are transferred to the oviduct after removal of their cumulus cell matrix, albeit at reduced rates.[52] However, as discussed below (Section III.C), the cumulus cell matrix may modulate events in a variety of ways during fertilization *in vivo,* although these might be ancillary roles of the cumulus cell matrix which become important under suboptimal conditions (e.g., when mating and ovulation are not well timed). Removal of this material undoubtedly alters subsequent fertilization events. On the other hand, a disadvantage of leaving the matrix intact derives from the pressure of having so many cumulus cells present in the small observation drop. During incubation in a chamber *in vitro,* these cells produce metabolites and release putatively damaging substances which would normally be removed *in vivo.* It has been proposed that this may be responsible for loss of sperm motility which occurs within a few of hours during observations using cumulus-enclosed oocytes.[33] Removal of the cumulus cell matrix permitted the first observation of large numbers of sperm during penetration of the zona pellucida,[57,58] and it was in this system that the most complete observations of fertilization were made, including oocyte activation, pronuclear events, and first cleavage. These later events require several hours to occur, and thus could not have been observed in the presence of the cumulus cells.

In studies of sperm penetration of the cumulus cell matrix, considerable care must be taken not to damage this fragile material. Tearing or shearing the matrix during collection may produce ''holes'' through which even cells of low motility can pass. Such damage can be minimized by flushing the cumulus mass from the oviduct, instead of the more common method of teasing the cumulus mass out of the oviduct through a hole which has been pierced in the oviductal wall. Once collected, all transfers of the cumulus mass should utilize a needle or pipette with an inner diameter greater than the diameter of the expanded cumulus mass (e.g., a normal Pasteur pipette). Compression of the cumulus cell matrix is usually required for clear visualization of the sperm, and this may also alter the properties of the microenvironment experienced by them.

C. CULTURE CONDITIONS AND OBSERVATION CHAMBERS

In order to collect gametes in native oviductal fluid, the oviduct can be transferred to a slide, covered with silicone or paraffin oil to prevent evaporation, and then carefully dissected to release the contents.[8,30] Alternatively, the contents of the oviduct can be collected by flushing the oviduct with oil.[5,6,9] However, even these maneuvers will not provide the normal milieux for fertilization; in the absence of the oviductal epithelium, culture conditions will change (e.g., metabolites will build up from the living cells in the chamber). It is also difficult during collection of gametes in oviductal fluid to prevent contamination with blood and tissue fluids from the damaged area where the oviduct has been opened. Moreover, the oviductal fluid is largely uncharacterized, and fertilization in a completely defined culture environment may be desirable for investigations of basic mechanisms.

The use of a culture medium introduces a number of factors which may influence fertilization events. *In vitro* fertilization is sensitive to such factors as pH, Ca^{2+} concentration, and protein supplementation of the medium.[59,60] Also, critical molecules present in the oviduct or interstices of the cumulus matrix may be lost when the cumulus mass is rinsed with a culture medium.

Although *in vitro* fertilization is possible in many species with high success, appropriate culture conditions are difficult to reproduce on a microscopic stage. Technical difficulties include maintaining a constant temperature, preventing evaporation of the culture medium, maintaining the appropriate gas tension, and preventing damage from light. These problems can be mitigated by choice of the appropriate observation technique. As discussed below, the methods which have been used for observation of mammalian fertilization are transluminal illumination, use of a culture dish on an inverted microscope, and use of a flat capillary tube or a compressed preparation on a microscope slide.

Excellent culture conditions can be maintained during gamete interaction if the gametes are observed, by means of an inverted microscope, as they float free in a drop of culture medium.[61,62] With this approach, the gametes can be observed in a culture dish, and nearly optimal culture conditions can be obtained by controlling the temperature and gas environment surrounding the microscopic stage. The gametes are transferred to a small drop on a slide chamber or in a culture dish, and then covered with silicone or paraffin oil to prevent evaporation from the drop. The oil used for this purpose can be sterilized by autoclaving and is usually equilibrated in an incubator to produce the correct gas tension (generally 5% CO_2 in water-saturated air). Appropriate gas tension is required to maintain the pH in the bicarbonate-buffered media normally used for fertilization *in vitro*.[19,59,60] In some cases, the microscope has been surrounded by a chamber having the appropriate temperature[22,57,58] and atmosphere[31] for culture. If not, the appropriate temperature can be maintained with a heated microscope stage, with a thermostatically controlled air curtain,[61,62] or by working in an environmental room which is warmed to body temperature.[6,9,24,34,37,63,64] In the hamster and mouse, normal hyperactivated motility is produced in sperm when 25 m*M* Hepes is included in the buffer.[26,65] This strategy may eliminate or reduce the necessity for carbon dioxide in the atmosphere, and should be evaluated for its ability to support *in vitro* fertilization under air. As sperm undergo capacitation, at least under *in vitro* conditions, they begin to stick to glass surfaces. In all methods of observing fertilization which involve glass chambers, the surfaces should be coated with a siliconizing agent.[24,34,62] We have found that sperm do not stick to plastic, and the problem of sticking is minimized if plastic culture dishes are used on an inverted microscope (unpublished observations). While the inverted microscope method can, in principle, provide optimal culture conditions, it is only possible to visualize the sperm at relatively low magnifications using this technique. The preparation is too deep to maintain individual sperm in focus at high magnification. Also, if the cumulus matrix has been removed, the flagellar motions of spermatozoa attached to the surface of the zona pellucida cause the oocytes to move about within the drop, making it difficult to focus on individual sperm. Movement of oocytes can be reduced by collecting them on a coverglass which has been coated with either the lectin DBA or WGA,[66] and positioning the coverglass in the bottom of the culture dish. These lectins bind to glycoproteins of the zona pellucida and help immobilize the oocytes.

The optics can be improved by compressing the fertilization drop or by drawing it into a flat capillary tube. Flat capillary tube preparations (100 μm deep) can be viewed on a conventional microscope.[5,6,23,36] Recently, a microdiffusion chamber has been described in which gametes are observed in a flat capillary tube, and the medium can be changed continuously.[67,68] This technique is potentially important since it allows the chemical composition of the microenvironment to be controlled during observation of the events of fertilization. A disadvantage of the flat capillary tube is that the cumulus material is sheared as it is drawn into the tube. This is a particular problem with the very thin tubes (50 and 100 μm) which provide the best optical resolution.

Fertilization events have also been observed on microscopic slide preparations.[8,9,24,27,30,33-35,37,57,58,61,63,69] The gametes are transferred to the slide in a small amount of medium, covered with a coverglass, and surrounded with oil. The coverglass is supported with "posts" to prevent excessive compression of the observation chamber. Investigators have used various materials for the supporting posts, including strips of parafilm (American Can Co., Greenwich, CT),[25,63] strips cut from No. 1 coverglass material (250 to 300 μm depth),[37,63] transfer letter material (32 μm depth),[70] various mixtures of Vaseline, beeswax, paraffin, and silicone grease,[8,30,33,57,58] and silicone grease containing acid-washed glass beads.[24,34] This latter method is simple and convenient since the mixture of silicone grease (Dow Corning, Midland, MI) and glass beads can be packaged in a syringe equipped with a blunt, large-gauge needle. Glass beads are available in a variety of sizes (Polysciences, Warrington, PN; Sigma Chemical Co., St. Louis, MO) to provide chambers with various depths. Hyperactivated spermatozoa have been videorecorded in a hemocytometer chamber (100 μm depth),[71,72] and such chambers could also be used for observing gamete interactions. Compressed preparations allow excellent visualization of details such as flagellar beat shape or acrosomal status of the sperm; and, if the cumulus cell matrix has been removed, the coverglass can be used to compress the zona pellucida very slightly, thus preventing motion of the oocyte during observation.[22,57,58] Simple perfusion is possible in these slide preparations if the chamber is only sealed along two sides of the coverglass. Fresh medium can be added at one side, while medium is removed from the other side with tissue.

The major disadvantage of viewing gametes in shallow slide preparations is that the cumulus cell matrix is compressed, and this has an unknown effect on the events of fertilization. Also, the small volume of the fertilization drop in these compressed chambers exacerbates the problem of accumulation of metabolic byproducts.

Transluminal illumination is possible for the oviducts of rodents since the oviductal tissues in these species are quite transparent. The oviducts are excised from the female at the appropriate time following mating, and are mounted in a slide preparation. Few details can be observed using this method, but the general motility of the sperm and their association with the oviductal epithelium can be documented with minimal manipulation[25-27] (Figure 2).

D. MICROSCOPE OPTICS

Unless they are stained, mammalian spermatozoa are nearly transparent and are difficult to visualize with brightfield microscopic optics. With darkfield optics, excellent resolution of the sperm is possible, and such optics have been used to observe hyperactivation of sperm motility[56,73,74] and the events of fertilization.[8,10,33] However, in darkfield illumination, sperm may be obscured by cumulus cells and the zona pellucida, since all cells and solid structures appear bright white.

Phase contrast optics are excellent for visualizing sperm, and it was the advent of these optics which permitted the first detailed observations of mammalian fertilization.[22,75,76] Since different cells and structures have a different appearance with phase contrast optics, the sperm cell can be visualized as it associates with other structures such as the zona pellucida. Also, it is possible to visualize the acrosome in a number of species with phase contrast optics, and the occurrence of the acrosome reaction can be documented. Unfortunately, phase contrast optics are not suitable for observation of the details of sperm behavior in the presence of the cumulus cell matrix. This is because the cumulus cells are quite refractile

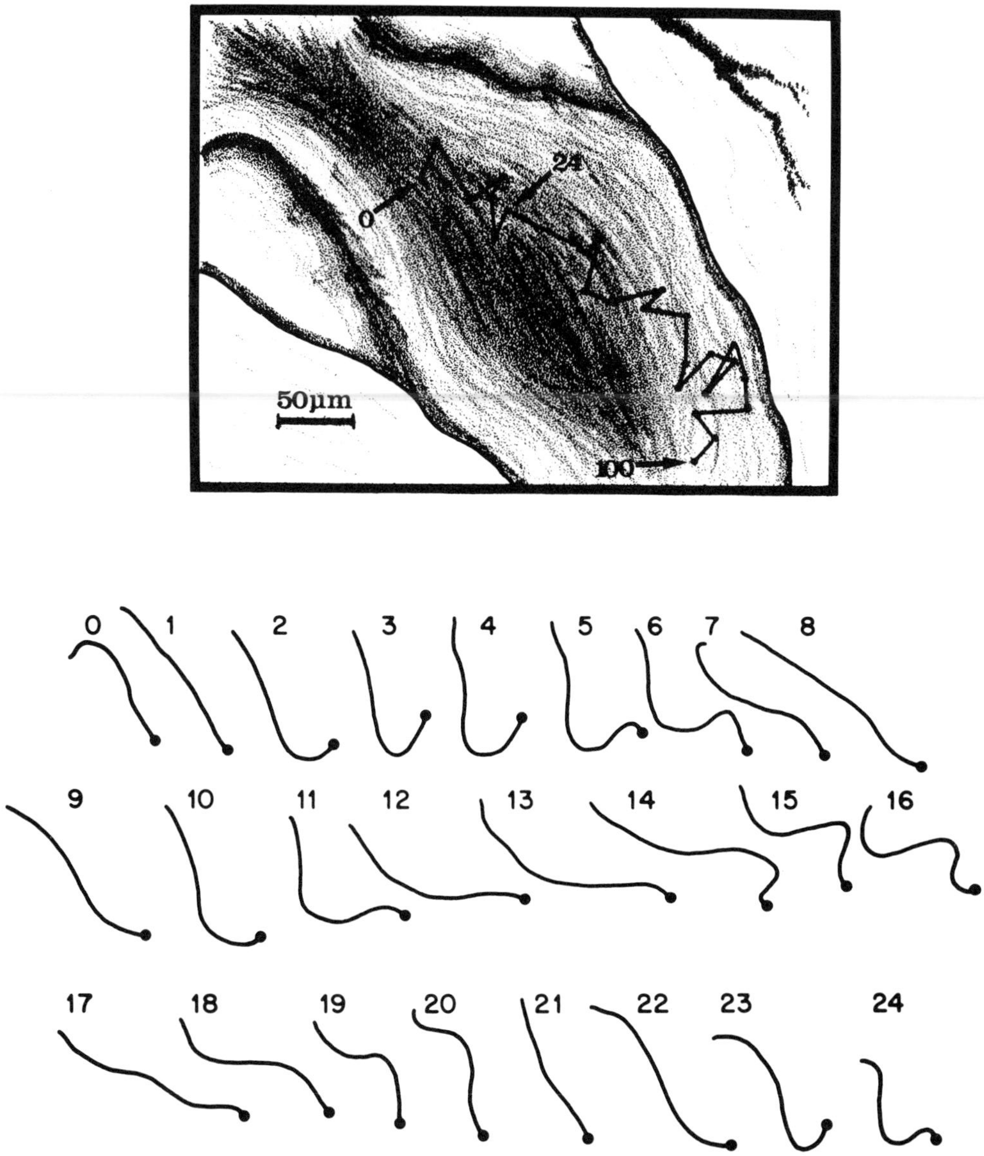

FIGURE 2. Sketches made from the monitor after cinemicroscopy of a hamster sperm within the ampulla of the oviduct. The upper panel shows the nonprogressive path taken by the hyperactivated sperm, and the lower panel shows the complex flagellar waveform during the first 24 frames. (From Katz, D. F. and Yanagimachi, R., *Biol. Reprod.*, 22, 759, 1980. With permission.)

and scatter considerable light. In order to utilize phase contrast optics, the cumulus cell matrix must be removed, or the preparation must be compressed to a very low depth to spread out the cumulus cells. This extreme compression has been used *after* the events of interest have been observed, in order to confirm the acrosomal status of a spermatozoon.[10,33]

Interference-contrast optics such as Nomarski, or Hoffman modulation, allow observations of sperm on the zona pellucida, and within the cumulus cell matrix.[9,24,34,37,63,64] The acrosomal status of rabbit[9] and hamster[37] sperm can be determined using interference-contrast microscopy. Because of the relatively small depth of field with these optics, the sperm can be brought into focus and clearly visualized, even when several layers of cells lie above and below the sperm in the optical path. While these optics are useful for observing the details of flagellar motion within the cumulus cell matrix, they present images with low contrast. The human eye can readily interpret such images, but they are not currently amenable to accurate computer digitization and image analysis.

Another concern with microscopic observations of gametes is damage by the illuminating light. In one study, light produced by a zirconium arc was tested, and it had no effect on fertilization events.[57,58] Other studies have used a red filter to reduce ultraviolet exposure of the gametes.[24,34,37]

E. VIDEORECORDING EQUIPMENT

The videorecording equipment includes a camera, recorder, and monitor. Each of these components contributes to the spatial resolution of the final image. At present, the recorder tends to be the weak link in the production of resolute images.

1. Camera

Video cameras suited for observing fertilization must have sufficient spatial resolution to provide clear images, and must require low enough light levels that the biological processes studied are not altered. The original video cameras used to observe fertilization were tube cameras. Among these cameras, those with a 1-in. Newvicon tube have proven to combine spatial resolution, light sensitivity, and durability the most expediently. More recently, the evolution of video technology has produced new generations of cameras whose sensors are solid-stage devices containing a grid of individual pixel elements. These solid-state or ''chip'' cameras are beginning to offer resolution and light sensitivity comparable to tube cameras, and are substantially more durable. Moreover, their images are inherently free of dimensional distortion and are better suited to machine vision analysis. Color cameras provide the human observer with images easier to interpret, but not necessarily better suited to machine vision. Color cameras are much more expensive than black and white cameras of comparable resolution and light sensitivity. Incorporation of a shutter into the camera reduces the exposure time per video field, although at a cost of lower light transmission to the camera sensor. Reduced light transmission can become an important consideration at high magnifications and when interference contrast optics are used. The presence of a camera shutter can reduce the blur produced by rapidly moving sperm during the exposure interval, and is quite useful for studying hyperactivated sperm. The normal video framing rate is 30 frames per second. Each video ''frame'' is comprised of the overlay (''interlacing'') of two video ''fields'', each containing half as many vertical pixels as the composite frame (which is why the image appears to ''jitter'' when a single video frame is viewed). Thus, ''breaking the interlace'' of the camera produces images that have twice the temporal resolution (now 1/60 s) but half the vertical spatial resolution. In general, analysis of sperm motion at increased rates of image capture gives increased curvilinear velocities, because more of the true texture of the sperm path is revealed.[26,77,78] If the playback on the video recorder can break the interlace of each video frame into its component video fields, the 1/60-s time resolution is often a significant improvement. However, only certain recorder technologies provide this capability (see below). True ''high-speed'' camera-recorder systems (e.g., 200 frames per second) overcome this problem, but are currently quite expensive.

Because ultraviolet radiation is damaging to cells, it is a goal during videomicroscopy to decrease exposure of the specimen to light. This can be achieved by increased light sensitivity of the camera. Silicone-intensified or SIT cameras achieve low-light sensitivity by collecting light over time. The adaptability of these sensitive cameras to computer-assisted sperm analysis (CASA) techniques requires the reconciliation of the need for extended time for light capture (to achieve higher sensitivity) with the fact that an instantaneous picture is required for motion analysis. This is an area of current interest and development.

2. Recorder

Video cassette recorders used for videomicroscopy generally utilize 1/2- or 3/4-in. tape widths. All other things being equal, the greater tape width offers greater spatial resolution. For manually driven tape playback and analysis, it is desirable to have ability on the tape recorder to vary the speed of playback, in both the forward and backward directions. A frame-by-frame or field-by-field advance button is also useful. Many of the current 3/4-in. U-matic recorders break the video field interlace during manual playback so that 1/60-s time resolution is possible. The current 1/2-in. recorders do not generally break the interlace for manual playback, and therefore provide only 1/30-s resolution. Machine vision CASA systems can in principle break the interlace on either 1/2- or 3/4-in. tapes. The most common recorder format in current use is VHS, the alternative being beta format. Recently, the resolutions of both the VHS and beta formats have been improved and are called "super VHS" and "extended definition beta". These new systems produce high clarity images that rival if not exceed 3/4-in. tape image quality.

3. Monitor

The spatial resolution and contrast control on the monitor can contribute to reduced visual stress during manual tape analysis. Such manual analysis can be extremely time consuming, but is necessary for many studies of fertilization because there is not sufficient contrast in the images for current CASA technology to be accurate (especially with the use of Nomarski or Hoffman optics), and/or because details of flagellar motion are desired. CASA technology is just beginning to address the latter.[79] Dimensional linearity on the monitor is highly desirable, but can increase the cost by a factor of three to five. It is possible to compensate for a lack in linearity by calibrating the screen of the monitor using the image of a microscopic slide micrometer.

F. MODEL SPECIES

Most observations of fertilization have been made in rodents, especially mice and golden hamsters, with a few studies in rats and guinea pigs. There has also been some work in rabbits. In order to observe fertilization in a given species, *in vitro* fertilization techniques must be available for that species. It is also an advantage if *in vitro* capacitation of spermatozoa is possible. These techniques have now been developed for a large number of mammalian species. Another consideration in choosing a model species is that not all species respond well to superovulation and induction of ovulation: species differ in the number of oocytes which are available for study. It is also desirable that the timing of ovulation be accurately determined or controlled. Preferably, estrus is detected easily, and estrus behavior or vaginal cytology are tightly coupled to ovulation. A final consideration is the size, shape, and optical characteristics of the spermatozoon. The advantages and disadvantages of various model species are briefly reviewed below.

1. Hamsters

The golden hamster (*Mesocricetus auratus*) is the mammal used most extensively for observations of vestment penetration by sperm, especially penetration of the cumulus cell matrix. This is primarily because the acrosome is visible, so the investigator can obtain, simultaneously, information about acrosomal status and interaction of sperm with the cumulus cell matrix and with the zona pellucida.[23,33,37,63] In addition, the hamster sperm cell is very large, and the flagellum is comparatively easy to visualize.

All mammalian sperm have extremely flattened heads, perhaps to enhance their ability to cross the zona pellucida more easily (this oocyte investment is unique to mammals).[80] The flagellar beat is primarily within the plane of the flattened head,[2,81-84] the B X L plane in the nomenclature of sperm axes;[85] it consists of waves which are initiated as bends in the midpiece, and are propagated along the flagellum. Hamster sperm, as well as those of mice and rats, have a hook-shaped head, with the hook in the plane of the flattened head. The orientation of the axonemal elements is constant with respect to the head hook.[81,83,86] Consequently, it is possible to delineate the direction of flagellar bending (i.e., principal bends vs. reverse bends) with respect to the head hook.

Female hamsters produce multiple oocytes and respond to superovulatory stimulus producing 30 to 80 oocytes per female. Like the mouse and rat, the hamster has a 4-day estrous cycle, and ovulation is well coupled with the onset of the dark period on the day of estrus. Estrus can be detected easily by vaginal smears or by lordosis behavior in the presence of a male.

Sperm can be collected from the caudal epididymides, capacitated *in vitro,* and the acrosome reaction induced in simple media.[87,88] The acrosome reaction can also be synchronized, and populations with a high proportion of acrosome reacted sperm produced.[89] Large numbers of ejaculated sperm can be collected from the uteri of mated females and can be capacitated *in vitro*.[24,34] *In vitro* fertilization techniques yield high fertilization rates at low sperm-to-oocyte ratios.[31,59]

Until recently, hamster embryos produced by *in vitro* fertilization would not develop past a single cleavage division (the so called "two-cell block"), but this problem has recently been overcome.[90]

The main disadvantage of the hamster as an experimental model for studies of fertilization events is that the molecular mechanisms of fertilization have not been characterized as well for this species as for the mouse. This is largely because hamsters are more expensive than mice.

Recently, fertilization events have been characterized for a second hamster species, the Chinese hamster.[91]

2. Mice

The out-bred mouse is the most inexpensive and hence the most widely used mammal for studies of fertilization. The complex interaction of the sperm with the zona pellucida has been best characterized at the molecular level in this species.[19] The mouse also shares with the hamster advantages of multiple oocytes, a response to superovulation, a 4-day estrus cycle, detection of estrus by vaginal smear, physically large sperm, and well-developed *in vitro* fertilization and *in vitro* capacitation techniques. However, since the acrosome is not visible at the light level in this species, the mouse has not been used as extensively as has the hamster for direct observations of fertilization. Nonetheless, as discussed below, considerable attention has been devoted to characterization of hyperactivated motility in mouse sperm. Also, there have been more direct observations of fertilization in mice than in any other species,[57,58] although these studies employed oocytes from which the cumulus cell matrix had been removed.

The availability of inbred strains of mice is an advantage which has not been exploited extensively by fertilization biologists. Strain differences can be used to demonstrate causal relationships between the structure and behavior of gametes and their function.

3. Rats

The rat shares many advantages with the hamster and mouse but it does not have the visible acrosome of the hamster, and it is much more expensive than the mouse. For these reasons, it has not been used as extensively for studies of fertilization and it is a relatively poorly characterized model system. The mature rat does not respond well to superovulatory stimulation, but immature animals can be superovulated.[62] *In vitro* fertilization techniques have been developed for this species,[92] and hyperactivated motility characterized.[62] *In vitro* capacitation of rat sperm collected from the epididymis, requires at least 6 h, a longer period than in mice and hamsters.[62] Fertilization events have been observed in this species.[22]

In contrast to its use in studies of fertilization, the rat has been utilized extensively in toxicology studies. There is now renewed interest in this species for studies of the effects of toxicants on gametes and on fertilization events.

4. Guinea Pigs

The guinea pig is also more expensive than the mouse, but it has the largest, most visible acrosome of any model species. Except for sperm binding to the zona pellucida,[61] the details of fertilization in this species have not been observed. As discussed below (Section III.B), the acrosomal condition of the fertilizing guinea pig sperm when it binds to the zona pellucida may differ from that of mouse sperm.[61,93,94] In this respect, the guinea pig model may resemble the human more closely than does the well-characterized mouse model.[95]

Guinea pig oocytes can be obtained from the ovaries and matured by culturing overnight.[96] Hyperactivation of motility has been well characterized for guinea pig sperm,[55] and in this species, like the hamster, sperm can be capacitated *in vitro,* then stimulated to acrosome react simultaneously, producing a population with a very high proportion of acrosome reacted sperm.[97] Sperm can be capacitated by overnight incubation, or by incubation for only 1 h in medium containing lysophosphatidyl choline (LPC).[93,98] Although not as well characterized as the mouse model, the molecular biology of fertilization is currently being studied in this species.[61]

5. Rabbits

Like the sperm of guinea pigs and humans, rabbit sperm are capable of binding to the zona pellucida both before[9] and after [99] completion of the acrosome reaction. This phenomenon warrants further study. Although much smaller than that of the hamster and guinea pig, the rabbit acrosome can be visualized at the light level using interference-contrast optics.[9]

The rabbit is the species best studied with respect to the characterization of sperm transport in the female reproductive tract, including the location, numbers, and behavior of sperm over time after mating.[14] This species is a reflex ovulator, which simplifies experimental timing of ovulation relative to sperm transport events.

One disadvantage of studying fertilization in rabbits has been the difficulty in capacitating rabbit sperm *in vitro.*[100] This may now be overcome by treatment with LPC.[101] Other disadvantages of this model species include the expense of purchasing animals, the inability to detect axonemal orientation in the paddle-shaped sperm, and poor response to stimulation of superovulation. In addition, the rabbit is unique among mammals in having no block to polyspermy at the level of the zona pellucida.[102] While this could facilitate visualization of zona pellucida penetration (since multiple sperm regularly penetrate each zona pellucida), the mechanisms of interaction of sperm with the zona pellucida in this species may not be typical of most mammals.

6. Primates

For ethical reasons, videomicroscopy of human fertilization may never be possible. However, binding of sperm to the zona pellucida and penetration of the zona pellucida have been studied and observed, using nonliving oocytes collected from ovaries which have been removed at surgery.[95,103,104] Penetration of the cumulus cell matrix by human sperm has been observed using hamster cumulus masses,[105,106] and these observations could be repeated with human cumulus material, which is often available from *in vitro* fertilization procedures.

In vitro capacitation of sperm and *in vitro* fertilization techniques have only recently been developed for nonhuman primates.[107,108] Since there is considerable variation among species in fertilization mechanisms and events, nonhuman primate species may be used to provide information about fertilization which is highly relevant to humans.

III. VIDEOMICROSCOPIC DESCRIPTION OF MAMMALIAN FERTILIZATION

A. MOTILITY OF SPERMATOZOA AT THE SITE OF FERTILIZATION

Videomicroscopy is a valuable research tool, primarily because it provides temporal data which cannot be obtained from fixed samples. This includes information about the kinetics of the spermatozoon during its interaction with the investments of the oocyte and with the oolemma. As mammalian sperm become capacitated and gain the capability of fertilization, they undergo motility changes, termed hyperactivation,[56] which are believed to be required for fertilization. Videomicroscopy has been used to characterize the motility of spermatozoa at the site of fertilization and the changes in motility which occur during incubation under capacitating conditions. These data have enabled estimations of the forces that sperm are capable of exerting against the macromolecular matrices which invest the mammalian oocyte, as well as the resistive forces exerted upon sperm by these matrices.[109]

Although the term hyperactivation had yet to be coined, this distinct motility pattern was first described for hamster spermatozoa which had been capacitated *in vitro*.[110-112] Initially, this vigorous motility was called "activated". However, since the term motility "activation" refers to the initiation of active sperm motility which occurs at spawning or ejaculation, the term "hyperactivation" was suggested[56] to describe the extremely active motility of capacitated mammalian sperm. In the initial descriptions of hyperactivation, the altered motility was described simply as an increase in the lateral motions of the sperm head. The resulting "bobbing" of the head[110] is due to increased curvature of the flagellar bends as they form behind the sperm head in the flagellar midpiece, before being propagated along the flagellum. Hyperactivation has been studied extensively since these initial reports in a large number of species including hamsters,[27,63-65,113] mice,[2,26,73,74,114-120] rats,[62] guinea pigs,[3,69,94-96,113,121,122] rabbits,[4,6,9,100] dogs,[123] dolphins,[124] bats,[125] sheep,[126,127] cattle,[128] pigs,[129] nonhuman primates,[107,130-132] and humans.[67,133-137] It is likely that hyperactivation is a universal feature of mammalian fertilization. In some of the studies cited above, the motility of sperm collected from the female reproductive tract was described; in others, the motility of sperm was observed during incubation under capacitating conditions *in vitro*.

As the term implies, hyperactivated sperm move very rapidly, often in complex, jerky paths. Early descriptions were subjective and included such terms as "bobbing",[110] "whiplash",[114] "serpentine",[96] and "dancing".[119] The ability to use videomicrography (or, in some cases, cinemicrography) enabled later investigators to define measures which more objectively characterize the changes in sperm motility which accompany capacitation.[5,6,9,37,63,64,74,113,120] The aspects of sperm motion which can be measured relate to (1) the curvature and direction of flagellar *bends,* (2) the shape (e.g., amplitude and symmetry) and kinetics of the flagellar *beat* or wave, and (3) the shape and the kinetics of the *sperm track* or the path traversed by the sperm head. In general, hyperactivation involves not only an increase in the curvature of flagellar bends; the beat shape becomes more asymmetrical because the flagellar bends in one direction have greater curvature than those in the other direction (Figure 3). Sperm normally have somewhat asymmetrical beat shapes,[2] but they continue to move progressively by rolling. During hyperactivation, the flagellar beats become more planar,[62] and this, together with increasing asymmetry of the flagellar beat, causes the sperm to swim in a nonlinear trajectory. The asymmetry may become so extreme that the sperm is completely nonprogressive and moves in circles, or even figure-of-eight trajectories.[4,7,56,63,114] Typically, these increases in bend curvature and beat asymmetry are accompanied by a significant increase in the curvilinear velocity of the head and a decrease in linearity of the sperm track, while the beat frequency may actually decrease.[26] Such changes in sperm motion are the hallmarks of hyperactivation.

Hyperactivated motility is believed to be important in penetration of the oocyte vestments,

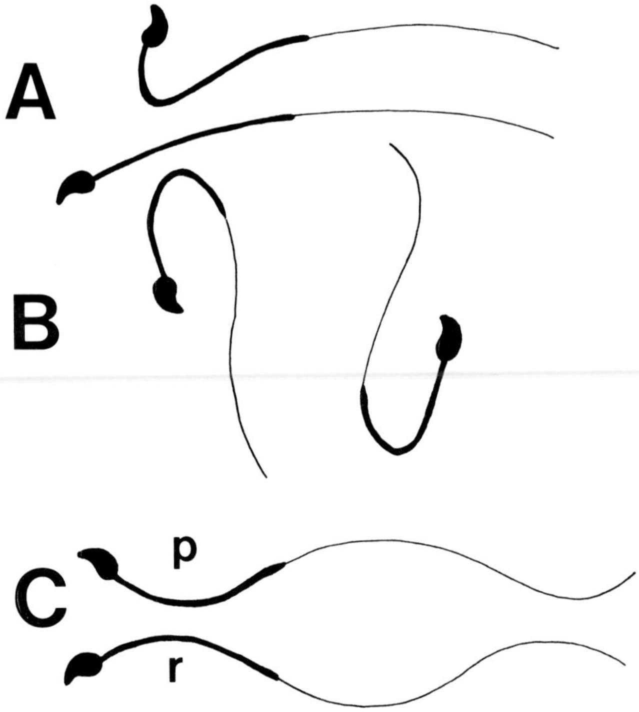

FIGURE 3. Cartoon of the types of flagellar bends and flagellar waves seen in hyperactivated hamster sperm. (A) The two sperm show the two extreme positions of the asymmetric waveform produced by an extreme principal bend in the proximal midpiece. In low viscosity medium or in the cumulus matrix, these bends form slowly and are released rapidly in a ''hatchet'' beat during which forward progression is minimal. During penetration of the zona pellucida, the sperm head is constrained within the zona matrix and the flagellum remains relatively straight to produce ''lever'' beats. (B) Extreme reverse bends initiated in the distal midpiece are propagated along the flagellum followed by principal bends of very low curvature. The resulting asymmetrical waveform, displayed by some hyperactivated sperm in low viscosity medium, may produce star-shaped or figure-of-eight trajectories. (C) Sinusoidal waves are symmetrical because principal bends (p) are propagated along the flagellum followed by reverse bends (r) of equal curvature.

and identification of the characteristics of this form of motility is required for studies of the mechanism and control of motility hyperactivation. A potential problem in characterizing those attributes of hyperactivated motility which are important in fertilization is the heterogeneity of mammalian sperm. In mammals, heterogeneity in the physiology within sperm populations in the female reproductive tract is probably important to insure that an adequate number of sperm in the appropriate physiological state are available after ovulation occurs.[11] Whether mammalian sperm are analyzed after collection from the epididymis or semen, after capacitation *in vitro,* or after collection from the female reproductive tract, a variety of subpopulations are present, based on motility or other physiological features. Even when sperm are collected from the site of fertilization, the acrosomal status and motility are not uniform, and some sperm are dead or dying.[1-10,62] Since the attributes of motility which are of primary interest are those of sperm capable of fertilization, and since sperm incapable of fertilization are present, sperm heterogeneity should be considered in studies of motility hyperactivation.

While it has been suggested that the high curvature bends generated during hyperactivation may enable the sperm to generate greater forces against the oocyte vestments,[16,56,80,110,138,139] it has not been demonstrated that such motility is strictly required. In general, three criteria have been used to distinguish the motility characteristics which correspond to "hyperactivation", and to draw the conclusion that such motility is required for fertilization. The first is that this motility is present in some of the sperm found at the site of fertilization in mated or artificially inseminated females.[2-4,9,62,111,112] The second is that the changes in motility occur at the same time, or just prior to the time during culture *in vitro,* when sperm in a suspension gain the ability to fertilize homologous oocytes, fuse with zona-free hamster oocytes, or undergo the acrosome reaction.[26,33,34,37,63,67,73,74,94,96,97,107,113-119,121,125,128-130,134-137,140,141] The third criterion is that the hyperactivated motility characteristics are lost under conditions for which the sperm are not capable of fertilization.[94,118,121,142] It is of concern that, in the first two cases, senescent sperm may be present along with functionally competent sperm in the population under observation; the motility characteristics which accompany the initial onset of senescence are not known. It is possible that some of the changes observed are present only in sperm which are in the initial stages of senescence. In the case of the third criterion, some investigators have reported that sperm lose their ability to fertilize oocytes concomitantly with loss of hyperactivated motility during prolonged incubation,[96,121] or when the chemical milieux is altered.[97,121,114-118,142,143] Such experiments have failed to establish a causal relationship between the motility observed and fertilization, since the treatment (long-term incubation or biochemical manipulation) may inhibit motility and inhibit fertilization by different mechanisms.

A direct approach to characterizing the motility of the fertilizing sperm is to observe the cells as they interact with the oocyte. As discussed below (Section III.E) we have found that as hamster sperm cross the zona pellucida, they produce very high curvature flagellar bends.[24] This motility pattern, which we have called "lever strokes", allows the sperm to generate very large forces against the zona pellucida. However, these bends have the opposite curvature with respect to the axoneme as do the extreme bends seen in the small subpopulation of the sperm in suspension which swim in figure-of-eight trajectories.[34] These observations have raised the question of the direction of flagellar bending in hyperactivated sperm. In the hook-shaped sperm of rats and hamsters, the direction of bending is important since it indicates the axonemal orientation.[81,86] The direction of flagellar bending can be controlled by altering the environment of the flagellar axoneme;[144,145] thus, this direction contributes to our understanding of the mechanism and control of sperm motility at the cellular and molecular levels. In the future, such experiments should be extended to additional species, so that more general information may be obtained regarding the kinematics of those "successful" sperm which actually penetrate the zona pellucida. Whenever possible, observations of the events of fertilization should be accompanied by a description of the direction of flagellar bending using the standard nomenclature for sperm axes.[85] With regard to the latter, there has been some confusion in the literature regarding the nomenclature of flagellar bends.[65,81,83,86,144] Most authors currently use the nomenclature which is clearly described and illustrated by Woolley[81] (Figures 1 and 2) and by Mohri and Yano[86] (Figure 3).

Measurements of flagellar parameters are usually made using sketches traced from the video monitor, a process requiring considerable time. Although it is not currently possible to analyze the flagellar motions of hyperactivated sperm automatically using CASA, it is possible to track the sperm head automatically. Hyperactivated motility is characterized by several changes which can be measured from such tracks including increased curvilinear velocity of the head (VCL) and decreased straight-line velocity (VSL). Together these changes constitute a decrease in the linearity (LIN) and an increase in the amplitude of the lateral head displacement (ALH).[9,26,71,74] The VCL, VSL, LIN, and ALH can now be measured automatically on commercially available instruments, and attempts have been made to relate such measures to hyperactivated motility. The most concerted efforts thus far have been in the mouse[146] and the human.[141,147] In humans, attempts have been made to standardize these techniques and make them available for clinical application.[145] Focus upon subpopulations of sperm in a suspension becomes particularly important here. Strictly speaking, it is not yet known in any species how those sperm which are capable of fertilization move when they are free-swimming in low viscosity medium. Additional observations of *individual* sperm as they approach the oocyte and then complete fertilization events will be important in characterization of requisite sperm motility parameters.

B. SITE OF THE ACROSOME REACTION

It is generally agreed that sperm undergo the acrosome reaction prior to penetration of the zona pellucida. However, the site at which the acrosome reaction occurs and the exact mechanism for induction of the acrosome reaction in the fertilizing sperm have not been established.[11,18,54-56] Observations of fertilization events have been used to investigate these questions, particularly for hamster sperm, in which the acrosome reaction can be visualized at the light microscopic level.

Since sperm enzymes, including hyaluronidase and acrosin, are capable of lytic action upon the cumulus cell matrix and the zona pellucida, respectively, and since these enzyme activities are released from sperm during the acrosome reaction, it is widely believed that an important function of the acrosome reaction is to release or expose these enzymes which then aid sperm in penetration of the oocyte vestments.[18,55,56] However, in early observations of oocytes collected from mated rodents, all highly motile sperm in the cumulus cell matrix were acrosome intact,[1] suggesting that sperm might initiate penetration of the cumulus matrix prior to the acrosome reaction. It has been proposed that the acrosome reaction in the hamster is initiated at the surface of the cumulus mass, that the acrosome is actually ''modified'' in hamster sperm which arrive at the surface of the zona pellucida, and that this modification exposes enzyme activity which facilitates penetration of the matrix.[18,55,56,80] These modified acrosomes are present in the majority of hamster sperm found in the oviductal ampulla or penetrating the cumulus matrix in mated females.[1,8,10] However, both hamster sperm with modified acrosomes and those with intact acrosomes have been observed to penetrate the cumulus cell matrix *in vitro,*[23,37] indicating that the acrosome need not be modified to enable the sperm to progress through the cumulus matrix. Ultrastructural observations of these sperm confirm that they have intact acrosomes and that there is no clearing of the surrounding cumulus matrix material suggestive of enzymatic dissolution of the matrix.[150] Direct observations have confirmed that, although the acrosome may be modified, the acrosome reaction is not complete in those hamster sperm which initiate binding to the zona pellucida and later complete penetration through the zona pellucida.[24,37,63] As discussed below (Section III.E) hamster sperm which have not completed the acrosome reaction attach to the zona pellucida where, following a short period of apparently loose association, they complete the acrosome reaction, establish a stronger attachment to the surface of the zona pellucida, and then initiate penetration into the zona material.

In the mouse, data have accumulated which support the hypothesis that only sperm in which the acrosome is intact are able to bind to the zona pellucida.[151-153] The specific glycoproteins of the zona pellucida, which act as species-specific binding proteins and inducers of the acrosome reaction, have been well characterized in this species,[154,155] and some investigators have concluded that the acrosome reaction in mammalian sperm occurs at the surface of the zona pellucida. However, except for one study in which no direct observations were made, and in which the large numbers of sperm inseminated make the presence of an intact cumulus matrix doubtful,[156] these studies have used oocytes from which the cumulus cell matrix was removed. Therefore, it is not possible to conclude that only mouse sperm having intact acrosomes can bind to the zona pellucida during fertilization *in vivo;* cumulus components may be required to enable binding of acrosome-reacted sperm. Because of the small acrosome of the mouse sperm, this cannot be resolved by direct observations. However, acrosomal staining can be used on living sperm to determine acrosomal status,[157] and these methods could be used in conjunction with videomicroscopy techniques to answer these questions.

In contrast to hamsters and mice, there is evidence that the sperm of guinea pigs, rabbits, and humans are capable of binding to the zona pellucida both before and after completion of the acrosome reaction. Initial experimental evidence suggested that the acrosome reaction was obligatory for binding of guinea pig sperm to the zona pellucida;[93,94] however, more recent experiments indicate that in this species, sperm can bind to the zona pellucida both before and after completion of the acrosome reaction.[60] In rabbit sperm, the acrosome is relatively small and difficult to visualize. However as in many other ''paddle-shaped'' sperm, e.g., those of ruminants, dogs, and pigs,[156] they have an apical ridge, visible with interference-contrast optics, which is lost following the acrosome reaction. Rabbit sperm with intact acrosomes have been observed crossing the cumulus cell matrix and binding to the zonae pellucidae[9] of oocytes collected from mated females. Nonetheless, sperm (presumably acrosome reacted) which were collected from the perivitelline space of fertilized eggs were able to fertilize fresh oocytes,[99] indicating that an intact acrosome may not be necessary for interaction with the zona pellucida. Experiments with oocytes collected from human ovaries have suggested that human sperm are also able to bind to the zona pellucida both before and after completing the acrosome reaction.[95] In all three of these species, it is important to determine if the same events occur when the zona pellucida is surrounded by an intact cumulus cell matrix. This is an important consideration because, as discussed below (Section III.C) hamster sperm which have completed the acrosome reaction may not be capable of penetrating the cumulus matrix. Thus, important differences among mammalian species may exist in the early events of fertilization. The experiments necessary to determine the site of the acrosome reaction during fertilization of cumulus-intact oocytes may not be possible in humans for ethical reasons and because the acrosome is so difficult to visualize (although this latter difficulty may be overcome by using vital stains). However, direct observations of acrosomal condition in the course of these events could be made in the guinea pig and rabbit, and would help resolve the controversy regarding interspecies variation in these events. Both rabbits and guinea pigs not only have sperm with visible acrosomes, but also permit methods for producing populations in which all of the sperm are acrosome reacted.

C. PENETRATION OF THE CUMULUS CELL MATRIX

The presence of numerous cumulus cells within the cumulus cell matrix makes it difficult to observe spermatozoa as they interact with the zona pellucida and with the oolemma. Largely for this reason, the cumulus cell matrix has often been removed before observation of fertilization in mammals,[22,57,58] or the oocytes chosen for observation have been those denuded of the cumulus matrix.[8,30-33] It has been rationalized that the cumulus cell matrix is relatively unimportant to normal fertilization, because when oocytes from which this matrix has been removed are transferred into the oviducts of mated hamsters, fertilization occurs, albeit at reduced rates.[52,158] However, although the cumulus cell matrix may not be *required* for fertilization in all species, it is normally present at the time of fertilization in eutherian mammals, except perhaps in ruminants,[159-161] and it is possible that it influences or modulates the events of fertilization. When conditions are not optimal (e.g., when the time between mating and ovulation is not ideal), such modulation may become important. It has been suggested that the cumulus cell matrix enhances the ability of sperm to fertilize oocytes,[53,162] and that it may have specific functions in restricting access of sperm to the zona pellucida[33,37,52,163] (i.e., as a block to polyspermy); in stimulating sperm motility;[163] in producing and/or maintaining an optimal microenvironment for fertilization;[53,164] in inducing the acrosome reaction;[18,52-55,165] in orienting sperm for penetration of the zona pellucida;[24] and in providing a viscoelastic substrate for the sperm flagellum, resulting in generation of increased sperm forces against the material of the zona pellucida.[24,56]

The first descriptions of the details of sperm penetration of the cumulus cell matrix were made using gametes flushed from the oviducts of inseminated hamsters[8] and mated rabbits.[9] Considerable further observations have been made, particularly following insemination of cumulus masses *in vitro*.[18,23,34,36,37,63] In addition to investigations of the site of the acrosome reaction, discussed above, observations of sperm penetration of the cumulus cell matrix are of biological importance in studying the access of sperm to the zona pellucida, and their behavior during interaction with this structure.

Investigators generally agree that sperm move slowly in the cumulus cell matrix relative to their motility in low viscosity medium,[9,23,34,36,37,63] reaching the zona pellucida in approximately 3 to 20 min.[8,23,37] Their paths are randomly oriented, rather than being radially directed toward the surface of the zona pellucida,[23,37] and their motility is relatively uniform throughout the cumulus mass.[37] For the most part, these sperm move through the matrix material between the cumulus cells, although occasionally sperm appear to glide along the surfaces of the cells, perhaps through matrix material of diminished resistance.[37]

Although, as discussed above, it has been proposed that hyperactivated motility might aide in penetration of the cumulus cell matrix, as well as the zona pellucida, by increasing the forces sperm can exert against these materials,[18,56,113,139] it has been reported that cells with sluggish motility, such as frog sperm, are capable of crossing the cumulus.[168] These experiments did not involve direct observations of this process, and it is possible that the fragile cumulus matrix had been damaged during handling, producing artifactual holes or channels. Ultrastructural observations of the cumulus matrix show an even matrix, devoid of holes or channels, which appears to be ''stretched'' by motion of the sperm flagellum.[150] The motility of sperm which are hyperactivated in low viscosity medium is slowed considerably in the viscoelastic material of the cumulus matrix,[8,9,23,36,37,56,63] and sperm appear to require considerable effort to penetrate this structure.[34]

Sperm videorecorded while penetrating the surface of the matrix actually enter the matrix by means of extremely high frequency, low amplitude, sinusoidal beats,[34] rather than by means of the high amplitude, high curvature beats characteristic of hyperactivation. Characterization of these sinusoidal beats within the low viscosity medium outside the cumulus mass has enabled us to calculate an effective viscosity of the cumulus matrix of 1 to 5 P.[34]

Before the kinematic details of cumulus penetration by sperm were measured, it appeared that the motions of hyperactivated sperm were altered significantly following penetration into the cumulus cell matrix.[8,9,23,36,37,63] However, we have observed that the motions of hyperactivated sperm are damped but not altered by the cumulus matrix.[34] Once within the cumulus matrix, sperm swim with flagellar beat shapes analogous to those of the inseminated sperm swimming in low viscosity medium, although beat frequency and sperm velocity are generally diminished (Figure 4). As in low viscosity medium, hyperactivated sperm in the cumulus cell matrix swim with complex beat shapes, varying between low curvature, propagated bends which produce sinusoidal beats of relatively low amplitude, and extreme bends which result in asymmetric beat shapes. During the intermittent sinusoidal beating, sperm trajectories are transiently quite progressive, and this likely corresponds to the intermittent progressive motility reported for hyperactivated sperm[4-6,34] and the ''darting''[36] of these cells. From our observations, we have concluded that hyperactivated motility per se is not required for penetration of the cumulus matrix, since sperm do progress through the matrix by virtue of low amplitude, sinusoidal beats; however, vigorous flagellation was necessary for progression through the matrix and high curvature flagellar bending did appear to facilitate penetration of the matrix.[34]

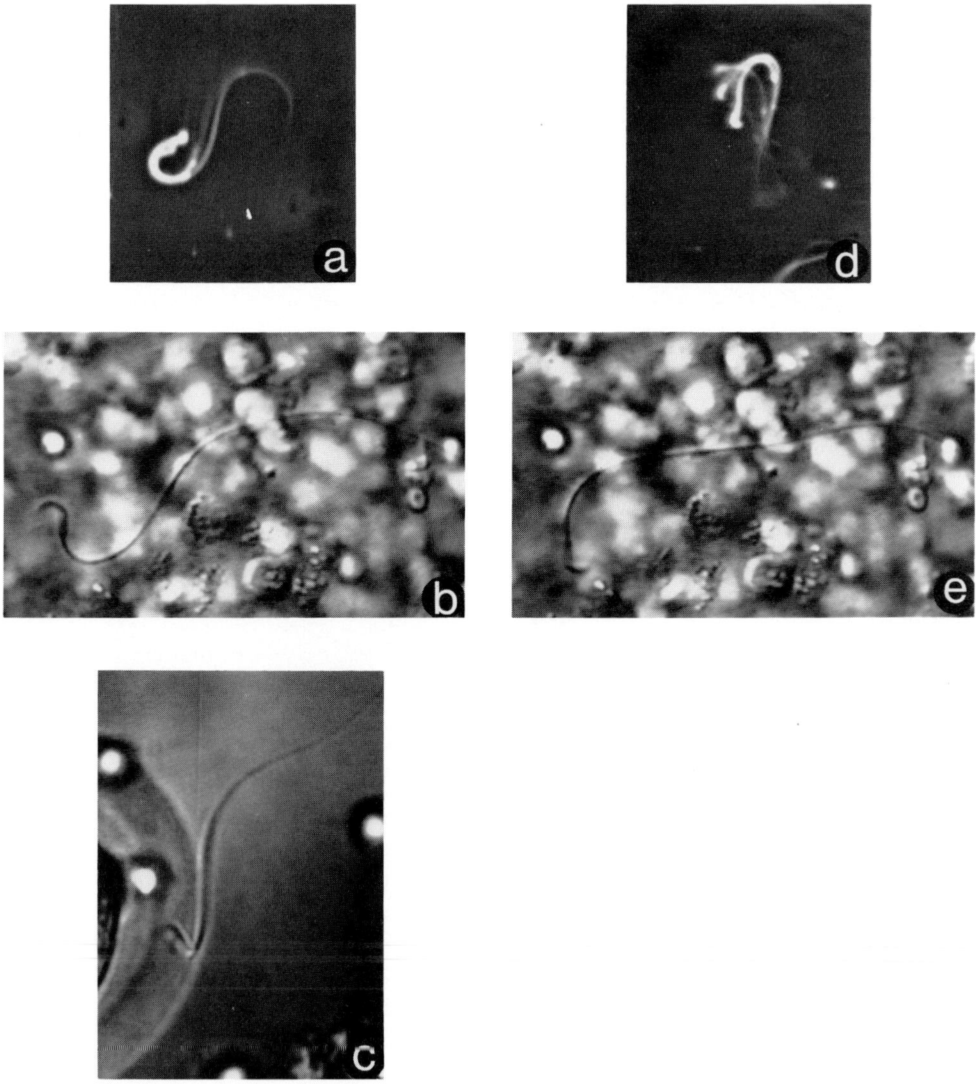

FIGURE 4. Photographs taken from the video monitor of asymmetrical flagellar waveforms seen in hyperactivated hamster sperm. (a, b, c) These three photographs show sperm which have formed extreme principal bends in the proximal midpiece. (a, b) Sperm are producing "hatchet" beats in culture medium and the cumulus matrix, respectively. (c) A sperm is penetrating the zona pellucida by means of "lever" beats during which the flagellum remains relatively straight. (d, e) These photographs show sperm which have formed extreme reverse bends in the distal midpiece. (d) The sperm is in culture medium and is swimming in a star-shaped trajectory. (e) The sperm is swimming in the cumulus matrix. Such extreme reverse bends were never observed in hamster sperm during penetration of the zona pellucida. (Adapted from figures in References 24 and 34.)

Most studies of the interactions of mammalian sperm with the zona pellucida have used oocytes from which the cumulus cell matrix had been removed, or oocytes inseminated with large numbers of sperm in which the intactness of the cumulus matrix can be questioned. The relatively few observations of sperm penetration of the matrix have demonstrated that the cumulus matrix restricts access to the zona pellucida, allowing only sperm having certain physiological characteristics to reach the zona surface.[33,36] In order to penetrate the cumulus matrix, hamster sperm must have initiated capacitation but must retain intact acrosomes. Uncapacitated sperm cannot penetrate the cumulus cell matrix until they have been incubated under capacitating conditions.[8,18,33,35,36,168] Freshly collected hamster sperm are prevented from entering the cumulus cell matrix because they adhere strongly to cumulus cells. Indeed, capacitation has been defined with respect to the time when hamster sperm lose the capacity to bind to cumulus cells in suspension.[164] This difference in the surface properties of uncapacitated sperm, which causes them to stick to other cells, may be involved in retaining the sperm that accumulate in the oviductal isthmus.[25,33] A strict requirement for capacitation is probably not needed for cumulus penetration, since sperm which have been incubated under capacitating conditions for an intermediate length of time, not sufficient to permit fertilization, are able to penetrate the cumulus cell matrix.[33,36] Also, sperm which do not yet display hyperactivated motility can cross the cumulus cell matrix and attach to the zona pellucida, where they remian for some time before acrosome reacting and penetrating the surface of the zona pellucida.[34]

Acrosome reacted hamster sperm do not appear able to penetrate the surface of the cumulus mass, or they penetrate poorly.[18,23,33,37,63] Motile sperm which have completely or partially undergone the acrosome reaction become tightly associated with cumulus cells and are prevented from entering the matrix.[23,37,63] However, in some instances, acrosome reacted hamster sperm have been observed swimming easily through the cumulus matrix.[1,8,34,63,91,149] We suspect that it is the acrosomal shroud, a structure which remains tightly associated with motile, acrosome reacted hamster sperm,[1,23] which may be responsible for the adherence of these sperm to cumulus cells and their subsequent failure to penetrate to the zona pellucida. As discussed below (Section III.D) this remnant of the acrosomal cap may be important in anchoring hamster sperm to the zona pellucida following the acrosome reaction, and it appears to be this structure which also anchors acrosome reacted sperm to glass and to cumulus cells.[1,8,23,37,63] Acrosome reacted sperm which have been observed penetrating freely through the cumulus matrix may have broken free from their acrosomal shroud, and thus are no longer excluded from the cumulus matrix. Hamster sperm have been observed to complete the acrosome reaction on the surface of the zona pellucida, then work their way through their acrosomal remnant, which remains tightly bound to the zona material, and eventually swim away into the cumulus matrix.[10,34] In conclusion, there appears to be a critical window in the capacitation process during which sperm are capable of penetrating the cumulus cell matrix and arriving at the surface of the zona pellucida.

Direct observations of cumulus penetration, made primarily with hamster gametes, have given us considerable information regarding the interactions of mammalian sperm with the cumulus cell matrix. It is important that these experiments be extended to other species in order to improve our understanding of the site of the acrosome reaction, and of the relative role of the cumulus cell matrix and the zona pellucida in modulating the behavior of the fertilizing sperm. Direct observations will be especially important in species such as the mouse, rabbit, and guinea pig, which appear to vary in the sequence of fertilization events. In these species, experiments with cumulus-intact oocytes need to be performed, to insure that the cumulus cell matrix remains undisturbed during the events under investigation.[93,98,156]

D. BINDING OF THE SPERM TO THE ZONA PELLUCIDA

Sperm binding to the zona pellucida is a complex, receptor-mediated event which may progress from a relatively loose association to the tight binding which precedes penetration of the sperm into the material of the zona pellucida.[12,56,169-172] This event is among those under the most intensive investigation by biologists studying mammalian fertilization. Observations of hamster sperm binding have been used to determine the temporal sequence of sperm association with the zona pellucida, the orientation and kinematics of sperm during this association, and, as discussed above, the acrosomal status of sperm as they approach and interact with the surface of the zona pellucida. Videomicroscopy has also been used recently to estimate the forces required to tether the bound sperm to the material of the zona pellucida.[173]

On the hamster zona pellucida, sperm bind in their equatorial region so that their heads are lying flat on the surface of the zona pellucida. They remain attached in this way for some time, while continued flagellar beats cause the sperm to rotate 360° around their point of attachment.[8,10,34,37,56,63,64] Such sperm are loosely attached, and have been observed to detach from the zona surface and swim away into the surrounding cumulus matrix.[34,37] Eventually, while maintaining this tangential orientation, the sperm undergo the acrosome reaction, after which a stronger attachment to the surface of the zona pellucida inhibits further rotation.[34,37,63] As mentioned above, on rare occasions these acrosome reacted sperm have been observed escaping from the zona surface,[10,34] but usually sperm appear to be strongly tethered by the acrosomal remnant.[10,18,37] It is quite likely that "loose binding",[174] during which sperm can be removed from the zona pellucida by gentle pipetting, corresponds to the attachment of acrosome intact sperm which are able to rotate freely around their point of attachment, and which are able to detach from the surface of the zona pellucida. "Tight binding" probably involves the association of the surface of the acrosome reacted sperm with the material of the zona pellucida.[33,37,63] These two types of sperm binding to the zona pellucida may involve different mechanisms, and these mechanisms could be studied by direct observation in the hamster system.

Sperm which bind to the zona pellucida in the equatorial region of their flat surface (e.g., hamster, guinea pig, rabbit) must move from a tangential to a nearly radial orientation before penetration into the zona material commences. The mechanism for this is complicated, since the flagellar motions are planar and restricted to the B X L plane which contains the flattened sperm head.[2,83,85] Sperm in this tangential orientation are capable of generating forces solely in a direction tangent to the surface of the zona pellucida, and such forces cannot aid in zona penetration. In one observation of the reorientation of a hamster sperm during penetration of the zona pellucida,[24] the sperm appeared to force its head forward against its point of attachment, rather than rotating as it did prior to the acrosome reaction. The forward motion forced the flagellum out of the tangent plane of attachment, while continued flagellar beats started to force the flattened head down into the material of the zona pellucida. Some clearing of the cumulus matrix around the sperm head may be desirable, providing a space for reorientation on the surface of the zona pellucida, and hyaluronidase released during the acrosome reaction may serve to provide this. We have suggested that the cumulus matrix, which surrounds much of the flagellum during this process, may be important in allowing the sperm to generate sufficient force to accomplish such reorientation. In our observations, sperm which were bound to regions of the zona pellucida, from which the cumulus matrix had been cleared completely, appeared to have difficulty initiating penetration into the material of the zona pellucida. Many of these sperm were either unable to penetrate the zona pellucida or they tended to penetrate the zona in long paths at more oblique angles.[24]

Direct observations of the interactions of hamster sperm with the surface of the zona pellucida have added to our understanding of this process. However, there is apparently considerable variation among species in this event. For example, while most mammalian sperm, including those of hamsters, rabbits, pigs, sheep, and cattle, initially associate with the surface of the zona pellucida by their flat, equatorial surface, mouse sperm may initially bind by the tip of their head. This means that for mouse sperm, there is no need for reorientation before penetration into the material of the zona. Moreover, in the guinea pig and the human, acrosome reacted sperm are capable of binding to the zona pellucida. Thus, the events in these species may be quite different than those in hamsters. Direct observations of sperm interactions with the zona pellucida in other species would increase our understanding of this important event.

E. PENETRATION OF THE ZONA PELLUCIDA

Penetration of the zona pellucida by sperm is very difficult to observe. In experiments with gametes flushed from the oviduct at the time of fertilization and fixed for observation, it is rare to find sperm with their heads within the zona material,[1] and it is not known if these rare sperm could have completed fertilization. With direct observations, the identity of "the fertilizing sperm" is known and, retrospectively, the video (or cine) record can be used to characterize various aspects of the penetration process. The initial experiments in which multiple zona penetration events were recorded focused on determining the time required for penetration and the orientation of the sperm during zona penetration.

Direct observations of large numbers of zona penetration events were first made by Yang et al.,[31] who studied 2249 oocytes within cumulus masses flushed from the oviducts of 232 mated hamsters. They were able to record by cinematography 15 sperm crossing the zona pellucida. The presence of the cumulus cell matrix, the use of phase contrast microscopy, and the relatively deep slide preparations in these experiments did not permit visualization of the details of zona penetration by sperm; however, these investigators were able to determine that penetration of the zona was accompanied by vigorous sperm motility with the sperm oriented at angles slightly oblique to the surface of the zona and required 4 to 11 min.

Sato and Blandau[57,58] observed 7754 oocytes from 585 mice. The cumulus was removed from the oocytes which were then inseminated with sperm collected from the epididymis and capacitated *in vitro*. Using this method, these investigators were able to observe the entire process of fertilization including oocyte activation, pronuclear events, and first cleavage. They were able to observe 141 sperm penetrating the material of the zona pellucida, 10 of which were observed from initial attachment to the surface of the zona pellucida. In some experiments, their observations were recorded with cinematography. They also reported that penetration involved vigorous motility, with the sperm oriented nearly vertical to the surface of the zona pellucida. Penetration required 15 to 26 min. This system was also used to study fertilization in the rat.[22]

More recently, we have used videomicroscopy to characterize the kinematics of hamster sperm as they penetrate the zona material. Such data are important to our understanding of the forces exerted by sperm against this extracellular matrix,[107] and the role of these forces in the penetration process. The mechanism by which the sperm crosses this barrier, which has considerable mechanical resistance,[47,175] is not known.[56,80,171] Acrosin, a protease capable of degrading the zona material,[176,177] may be present on the surface of the sperm and participate in this process, but sperm forces are probably involved as well.[21,56,80,109,172] The mammalian sperm head is flat and is hardened along its leading edge by a specialized structure, the perforitorium. These unique structural characteristics reduce the surface area of the leading edge of the sperm. The resulting increase in force per unit area, coupled with the oscillatory movements of the sperm head,[10,24,30,57,58] may enable mammalian sperm to cut "knife-like" into the material of the zona pellucida.[21,56,80] Videomicroscopic methods have provided data which have been used to estimate the forces that sperm are capable of exerting against the zona material.[109]

Sperm biophysicists have applied slender body resistive force theory to estimate the forces which mammalian sperm can exert when abutted against a solid object (i.e., the zona pellucida).[109] These methods require information about the flagellar wave such as its beat frequency, wavelength, and wave amplitude. In preliminary calculations investigators assumed a sinusoidal waveform, and measured the necessary experimental parameters from sperm which were either free-swimming in low viscosity medium[113,179,180] or which were held on micropipettes.[83] The forward thrust of the sperm, estimated by these methods, is in the range 1 to 10 μdyn. More recently, Baltz et al.[173] determined that, for the sperm cell, the torque (or moment) produced by the flagellar wave is much greater than is the forward thrust. Taking the torques into account, but still assuming a sinusoidal wave, they estimated that the human sperm is capable of generating forces in the order of 20 μdyn. In addition to these theoretical calculations, two direct, experimental approaches have been used to measure the forces generated by sperm. Baltz and co-workers,[173] in their studies of the forces required to tether bound sperm to the zona pellucida, held sperm on micropipettes and determined experimentally the force that the sperm required to pull free from the pipette. Similarly, Katz and deMestre[179] observed hamster sperm which were attached to (but not penetrating) cumulus-free oocytes and were rolling the oocytes. They estimated the force exerted by these sperm in rotating eggs at the observed rates. In both of these experimental studies, sperm forces in the range of 10 to 40 μdyn were estimated, in reasonable agreement to theoretical calculations which assumed a sinusoidal wave form. Forces of this order are insufficient to break covalent bonds; therefore, it was concluded that sperm forces are probably unimportant during zona penetration relative to enzymatic cleavage.[175,178]

Recently, we have used interference-contrast optics and high-speed videomicroscopy to document the details of the flagellar wave during penetration of the zona pellucida by the fertilizing hamster sperm.[24] Our results were quite surprising. In the course of zona penetration, the flagellar waveform is not constant. Asymmetrical, high curvature, high amplitude beats, which are the hallmarks of hyperactivation, alternate with episodes of low curvature, sinusoidal beats. During high curvature beats, the flagellum forms an extremely acute *principal* bend in the proximal midpiece. Then, with the flagellum remaining nearly straight, the bend is rapidly released. The resulting, high velocity "level stroke" of the flagellum is reminiscent of the effective stroke of a cilium. We have calculated that the *peak* forces produced by hamster sperm against the zona during these lever strokes are approximately 3000 μdyn, which are at least an order of magnitude greater than the sperm forces estimated by ourselves and others when a sinusoidal beat shape was assumed. Forces of this order are capable of breaking covalent bonds, and this reopens the possibility that sperm forces are critical during penetration of the zona pellucida. In answering the question of the forces generated during penetration of the zona pellucida, videomicroscopy was essential for determining the temporal experimental details, e.g., beat frequency, and for ensuring that the sperm observed were able to complete penetration.

One interesting observation made by us during our recordings of zona penetration relates to the direction of the flagellar wave. The acute *principal* bends involved in the generation of lever strokes have the opposite curvature to the acute *reverse* bends which produce the figure-of-eight motility pattern in free-swimming, hyperactivated hamster sperm[24,64] and mouse sperm (Figure 3).[114] Indeed, during our observations, hamster sperm penetrating the zona pellucida never generated extreme reverse bends. To determine if this represented a change in the direction of curvature of the flagellar wave once sperm began to penetrate the zona, we carefully determined the direction of extreme curvature of capacitated hamster sperm during penetration of the cumulus cell matrix, and while free-swimming in low viscosity culture medium prior to insemination.[34] We found that in both cases, few (less than 5%) of the extremely active sperm which we would subjectively classify as hyperactivated were swimming in figure-of-eight trajectories (only one was ever observed within the cumulus matrix). The possibility must be considered that these are senescent sperm which do not participate in fertilization. Instead, hyperactivated hamster sperm swam in extremely complex, jerky paths by means of a complex sequence of flagellar bends. Always included in the flagellar bends formed by hyperactivated sperm were the extreme principal bends which predominate during zona penetration (Figure 4). In the cumulus matrix, or in low viscosity medium, the flagellum remains relatively stationary during such beats, while the head is slowly thrown back, then rapidly brought forward in a "hatchet stroke". The bend does not appear to be propagated down the flagellum in a wave. The flagellum remains relatively straight, and the sperm does not move forward. Very similar beats predominate in the hyperactivated sperm collected from the site of fertilization in the Chinese hamster,[62] and perhaps in the rat;[91] hatchet strokes have been observed by us in human sperm during penetration of the hamster cumulus,[105] and they may well correspond to the "flexion" beats of human sperm described by Burkman.[136] Similar nonprogressive beats have been observed by us in the hyperactivated sperm of humans,[185] and are a prevalent beat pattern in the hyperactivated sperm of rhesus monkeys and cynomolgus monkeys.[186] We propose that these highly asymmetric bends, which contribute to the nonprogressive paths of hyperactivated mammalian sperm, are important in generating high forces during lever strokes, which help push the sperm head into the material of the zona pellucida.

Originally, we believed that the flagellum remained relatively straight during hatchet strokes because the bend was not propagated down the flagellum; it simply formed and was released.[34,181] If this is true, it suggests that these beats are produced by a different mechanism than that which is responsible for the more prevalent sinusoidal flagellar waves generated by mammalian sperm. However, it is also possible that the extreme principal bend is part of a wave with a very long wave length (i.e., longer than the length of the flagellum) such that the entire wave cannot be visualized on the flagellum at one time. More careful observations will be necessary to distinguish between these possibilities.

Videomicroscopy techniques have been important in determining the type of motility used by hamster sperm during penetration of the zona pellucida. This work should be extended to other species to determine if high-curvature lever strokes are a general feature of zona penetration by mammalian sperm. Eutherian mammals have uniquely flattened, hardened heads which may be an advantage in a mechanical mechanism for penetration of the zona.[80] Conversely, in the opossum, the sperm head is periform with a larger cross-sectional area, and the zona is thin and flimsy. In this marsupial species, the sperm probably crosses the zona more by enzymatic than by mechanical means.[80] This process should be visualized in the opossum to determine if lever strokes are involved in penetration of the softer marsupial zona pellucida. If lever strokes are necessary to produce forces sufficient for mechanical penetration of the eutherian zona pellucida, they are probably not required by the sperm of the opossum, and are probably absent.

IV. CONCLUSIONS AND FUTURE EXPERIMENTS

In model species, direct observations of gamete interaction *in vitro* have increased our understanding of the fertilization biology of mammals. The early events in fertilization have been documented by cinemicrography and videomicrography, and the data obtained from these records have allowed us to begin to characterize the behavior of *the fertilizing sperm*. As outlined below, we envision that there are five ways in which videomicroscopic observations of the early events of mammalian fertilization will be used in future experiments: (1) direct observations will be used as controls in many investigations of the molecular mechanisms of fertilization processes; (2) questions about interspecies variations in fertilization events will be resolved by observations in several key species; (3) more information on the biophysics of fertilization will be obtained, incorporating image-enhancement techniques into videomicroscopic observations; (4) visible, intracellular probes will be used to add to our information of the timing and control of fertilization events; and (5) the motility of functionally competent sperm will be elucidated, and the functional competence of subpopulations of sperm in suspensions will be analyzed.

Whenever possible, studies which use sperm binding or fertilization rates as experimental end points should seek to include some observations of the fertilization events. When such experiments are conducted without observation, the results can be difficult to interpret for a variety of reasons, including the following: (1) fertilization is accomplished by a small fraction of the total sperm present in the suspension, and the characteristics of that subpopulation may not be apparent in the average characteristics of the whole population; (2) it is difficult to insure that the cumulus matrix remains intact around the zona pellucida during gamete interaction, and the event being studied may not occur in the presence of intact cumulus matrix; and (3) fixatives require a finite time to stop biological activity.

Certain experiments, which were attempted without observations in the past, can now be accomplished during direct observation of gametes in perfusion chambers. For example, the role of sperm enzymes in penetration of the oocyte investments can be studied by introducing and removing inhibitors of enzyme activities. Sperm penetration can be observed before, during, and after alteration of the perfusate to insure that data are collected on functionally competent sperm. Similar experimental setups can be used to study those surface molecules on gametes, which can be visualized with probes, and those environmental factors which are critical during each fertilization event.

Now that the experimental techniques are available, the information which we have gained by videomicroscopic observations of fertilization in hamsters should be expanded to other species. This is particularly important for events in which interspecies variation may exist; our abilities to extrapolate from information gained in inexpensive model species are limited. An example is the site of the acrosome reaction of the fertilizing sperm. This question is central to our understanding of the mechanism and control of fertilization events, since the site of the acrosome reaction is important to the mechanism of induction of the reaction and binding of the sperm to the zona pellucida. Observations should be made in the presence of the intact cumulus cell matrix, to determine the site of the acrosome reaction in those sperm which are capable of penetrating the oocyte vestments. To date, such experiments have been conducted primarily in hamsters, in which sperm penetrate the cumulus matrix and bind to the zona pellucida before completing the acrosome reaction. These studies should be extended to the guinea pig and rabbit, which may differ from the hamster in the functional capacity of acrosome reacted sperm.

In the future, further analysis of the role of sperm forces in egg vestment penetration will be undertaken. This will require more resolute analysis of flagellar motion within the cumulus cell matrix. More advanced image enhancement and analysis techniques will be needed to provide such resolution.

In future observations of mammalian fertilization, molecular probes will be used to gain simultaneous information about the intracellular environment of the sperm and its motility in relation to fertilizaton events. For example, in species such as mice and humans, in which the sperm have small acrosomes which cannot be visualized at the level of the light microscope, permeant probes having fluorescence patterns sensitive to intracellular ion concentrations may be useful in determining acrosomal status. One such probe, chlortetracycline, has been used to detect early changes associated with initiation of the acrosome reaction in living mouse and human sperm.[157] In conjunction with videomicroscopy, this probe could help determine the site of the acrosome reaction in these species during interaction with the cumulus and the zona pellucida. Another area of investigation for which information about intracellular ion concentrations may be useful is in characterizing the intracellular changes which accompany changes in motility during penetration of the oocyte vestments. Since sperm motility is reduced or abolished during epifluorescent observation, low light illumination, coupled with image-enhancement techniques, may be required to make these observations.

One of the major goals in current gamete biology is evaluation of the functional competence of a population of sperm.[182] The evaluation of sperm motility has received considerable attention since it can be analyzed quickly, accurately, and precisely using current computer-aided techniques. Recently, it has been suggested that the ability of sperm to display hyperactivated motility following capacitation *in vitro* may be related to the fertility of human semen.[183] However, while it is believed that hyperactivated sperm motility is important in the penetration of the oocyte vestments, it is not known in any species what specific motility characteristics are required. We have made direct observations of individual hamster sperm as they entered the cumulus matrix, crossed the cumulus, and penetrated the zona pellucida. Further experiments in the hamster and other species will help us characterize the salient motility patterns, and allow us to determine which subpopulations of sperm in suspensions of hyperactivated sperm contain functionally competent cells.

ACKNOWLEDGMENT

The experiments here were supported in part by NIH HD12971.

REFERENCES

1. **Austin, C. R. and Bishop, M. W. H.,** Role of the rodent acrosome and perforatorium in fertilizaton, *Proc. R. Soc. London,* 149, 241, 1958.
2. **Phillips, D. M.,** Comparative analysis of mammalian sperm motility, *J. Cell Biol.,* 53, 561, 1972.
3. **Yanagimachi, R. and Mahi, C. A.,** The sperm acrosome reaction and fertilization in the guinea-pig. A study in vivo, *J. Reprod. Fertil.,* 46, 49, 1976.
4. **Cooper, G. W., Overstreet, J. W., and Katz, D. F.,** The motility of rabbit spermatozoa recovered from the female reproductive tract, *Gamete Res.,* 2, 35, 1979.
5. **Overstreet, J. W., Katz, D. F., and Johnson, L. L.,** Motility of spermatozoa in the secretions of the oviduct, *Biol. Reprod.,* 22, 1083, 1980.
6. **Johnson, L. L., Katz, D. F., and Overstreet, J. W.,** The movement characteristics of rabbit spermatozoa before and after activation, *Gamete Res.,* 4, 275, 1981.
7. **Cummins, J. M.,** Hyperactivated motility patterns of ram spermatozoa recovered from the oviducts of mated ewes, *Gamete Res.,* 6, 53, 1982.
8. **Cummins, J. M. and Yanagimachi, R.,** Sperm-egg ratios and the site of the acrosome reaction during in vivo fertilization in the hamster, *Gamete Res.,* 5, 239, 1982.
9. **Suarez, S. S., Katz, D. F., and Overstreet, J. W.,** Movement characteristics and acrosomal status of rabbit spermatozoa recovered at the site and time of fertilization, *Biol. Reprod.,* 29, 1277, 1983.
10. **Yanagimachi, R. and Phillips, D. M.,** The status of acrosomal caps of hamster spermatozoa immediately before fertilization in vivo, *Gamete Res.,* 9, 1, 1984.
11. **Bedford, J. M.,** Significance of the need for sperm capacitation before fertilization in eutherian mammals, *Biol. Reprod.,* 28, 108, 1983.
12. **Yanagimachi, R.,** Mammalian fertilization, in *The Physiology of Reproduction,* Knobil, E. and Neill, J., et al., Eds., Raven Press, New York, 1988, 135.
13. **Hunter, R. H. F.,** Transport, migration and survival of spermatozoa in the female genital tract: species with intrauterine deposition of semen, in *The Biology of Spermatozoa,* Hafez, E. S. E., Ed., Karger, Basel, 1975, 145.
14. **Overstreet, J. W.,** Transport of gametes in the reproductive tract of the female mammal, in *Mechanism and Control of Animal Fertilization,* Hartmann, J. F., Ed., Academic Press, New York, 1983, 499.
15. **Hawk, H. W.,** Transport and fate of spermatozoa after insemination of cattle, *J. Dairy Sci.,* 70, 1487, 1987.
16. **Katz, D. F., Drobnis, E. Z., and Overstreet, J. W.,** Factors regulating mammalian sperm migration through the female reproductive tract and oocyte vestments, *Gamete Res.,* 22, 443, 1989.
17. **Gwatkin, R. B. L.,** *Fertilization Mechanisms in Man and Mammals,* Plenum Press, New York, 1977, 91.
18. **Talbot, P.,** Sperm penetration through oocyte investments in mammals, *Am. J. Anat.,* 174, 331, 1985.
19. **Bavister, B. D.,** Animal in vitro fertilization and embryo development, in *Manipulation of Mammalian Development,* Gwatkin, R. B. L., Ed., Plenum Press, New York, 1986, 81.
20. **Clegg, E. D.,** Mechanisms of mammalian sperm capacitation, in *Mechanism and Control of Animal Fertilization,* Hartmann, J. F., Ed., Academic Press, New York, 1983, 177.
21. **Gaddum-Rosse, P.,** Mammalian gamete interactions: what can be gained from observations on living eggs?, *Am. J. Anat.,* 174, 347, 1985.
22. **Gaddum-Rosse, P., Blandau, R. J., Langlet, L. B., and Battaglia, D. E.,** In vitro fertilization in the rat: observations on living eggs, *Fertil. Steril.,* 42, 285, 1984.

23. **Corselli, J. and Talbot, P.,** In vitro penetration of hamster oocyte-cumulus complexes using physiological numbers of sperm, *Dev. Biol.,* 122, 227, 1987.
24. **Drobnis, E. Z., Yudin, A. I., Cherr, G. N., and Katz, D. F.,** Hamster sperm penetration of the zona pellucida: kinematic analysis and mechanical implications, *Dev. Biol.,* 130, 311, 1988.
25. **Suarez, S. S.,** Sperm transport and motility in the mouse oviduct: observations in situ, *Biol. Reprod.,* 36, 203, 1987.
26. **Suarez, S. S. and Osman, R. A.,** Initiation of hyperactivated flagellar bending in mouse sperm within the female reproductive tract, *Biol. Reprod.,* 36, 1191, 1987.
27. **Katz, D. F. and Yanagimachi, R.,** Movement characteristics of hamster spermatozoa within the oviduct, *Biol. Reprod.,* 22, 759, 1980.
28. **Hunter, R. H. F.,** Human fertilization in vivo, with special reference to progression, storage and release of competent spermatozoa, *Hum. Reprod.,* 2, 329, 1988.
29. **Austin, C. R.,** Observations on the penetration of the sperm into the mammalian egg, *Aust. J. Sci. Res. (B),* 4, 581, 1951.
30. **Yanagimachi, R.,** Time and process of sperm penetration into hamster ova in vivo and in vitro, *J. Reprod. Fertil.,* 11, 359, 1966.
31. **Yang, W. H., Lin, L. L., Wang, J. R., and Chang, M. C.,** Sperm penetration through zona pellucida and perivitelline space in the hamster, *J. Exp. Zool.,* 179, 191, 1972.
32. **Bavister, B. D.,** Fertilization of hamster eggs in vitro at sperm: egg ratios close to unity, *J. Exp. Zool.,* 210, 259, 1979.
33. **Cummins, J. M. and Yanagimachi, R.,** Development of ability to penetrate the cumulus oophorus by hamster spermatozoa capacitated in vitro, in relation to the timing of the acrosome reaction, *Gamete Res.,* 15, 187, 1986.
34. **Drobnis, E. Z., Yudin, A. I., Cherr, G. N., and Katz, D. F.,** Kinematics of hamster sperm during penetration of the cumulus cell matrix, *Gamete Res.,* 21, 367, 1988.
35. **Austin, C. R.,** Capacitation and the release of hyaluronidase from spermatozoa, *J. Reprod. Fertil.,* 3, 310, 1960.
36. **Corselli, J. and Talbot, P.,** An in vitro technique to study penetration of hamster oocyte-cumulus complexes by using physiological numbers of sperm, *Gamete Res.,* 13, 293, 1986.
37. **Cherr, G. N., Lambert, H., Meizel, S., and Katz, D. F.,** In vitro studies of the golden hamster sperm acrosome reaction: completion on the zona pellucida and induction by homologous soluble zonae pellucidae, *Dev. Biol.,* 114, 119, 1986.
38. **Kapur, R. P. and Johnson, L. V.,** An oviductal fluid glycoprotein associated with ovulated mouse ova and early embryos, *Dev. Biol.,* 112, 89, 1985.
39. **Brown, C. R. and Cheng, W. K. T.,** Changes in composition of the porcine zona pellucida during development of the oocyte to the 2- to 4-cell embryo, *J. Exp. Morphol.,* 92, 183, 1986.
40. **Hedrick, J. L., Wardrip, N. T., and Berger, T.,** Differences in the macromolecular composition of the zona pellucida isolated from pig oocytes, eggs and zygotes, *J. Exp. Zool.,* 241, 259, 1987.
41. **Oikawa, T., Sendai, Y., Kurata, S., and Yanagimachi, R.,** A glycoprotein of oviductal origin alters biochemical properties of the zona pellucida of hamster egg, *Gamete Res.,* 19, 113, 1988.
42. **Cherr, G. N., Yudin, A. I., and Katz, D. F.,** Organization of the hamster cumulus extracellular matrix: a hyaluronate-glycoprotein gel which modulates sperm access to the oocyte, *Dev. Growth Differ.,* in press.
43. **DeFelici, M. and Siracusa, G.,** "Spontaneous" hardening of the zona pellucida of mouse oocytes during in vitro culture, *Gamete Res.,* 6, 107, 1982.
44. **DeFelici, M., Salustri, A., and Siracusa, G.,** "Spontaneous" hardening of the zona pellucida of mouse oocytes during in vitro culture. II. The effect of follicular fluid and glycosaminoglycans, *Gamete Res.* 12, 227, 1985.
45. **Gianfortoni, J. G. and Gulyas, B. J.,** The effects of short-term incubation (aging) of mouse oocytes on in vitro fertilization, zona solubility, and embryonic development, *Gamete Res.,* 11, 59, 1985.
46. **Down, S. M., Schroeder, A. C., and Eppig, J. J.,** Serum maintains the fertilizability of mouse oocytes matured in vitro by preventing hardening of the zona pellucida, *Gamete Res.,* 15, 115, 1986.
47. **Drobnis, E. Z., Andrew, J. B., and Katz, D. F.,** Biophysical properties of the zona pellucida measured by capillary suction: is zona hardening a mechanical phenomenon?, *J. Exp. Zool.,* 245, 206, 1988.
48. **Conrad, K., Buckley, J., and Stambaugh, R.,** Studies on the nature of the block to polyspermy in rabbit ova, *J. Reprod. Fertil.,* 27, 133, 1971.
49. **Krzanowska, H.,** Rapidity of removal in vitro of the cumulus oophorus and the zona pellucida in different strains of mice, *J. Reprod. Fertil.,* 31, 7, 1972.
50. **Eppig, J. J.,** FSH stimulates hyaluronic acid synthesis by oocyte-cumulus cell complexes from mouse preovulatory follicles, *Nature,* 281, 483, 1979.
51. **Phillips, D. M. and Dekel, N.,** Effect of gonadotropins and prostaglandin on cumulus mucification in cultures of intact follicles, *J. Exp. Zool.,* 221, 275, 1982.

52. **Moore, H. D. M. and Bedford, J. M.,** An in-vivo analysis of factors influencing the fertilization of hamster eggs, *Biol. Reprod.,* 19, 879, 1978.
53. **Bavister, B. D.,** Evidence for a role of post-ovulatory cumulus components in supporting fertilizing ability of hamster spermatozoa, *J. Androl.,* 3, 365, 1982.
54. **Meizel, S.,** The mammalian sperm acrosome reaction, a biochemical approach, in *Development in Mammals,* Vol. 3, Johnson, M. H., Ed., North-Holland, Amsterdam, 1978, 1.
55. **Meizel, S.,** Molecules that initiate or help stimulate the acrosome reaction by their interaction with the mammalian sperm surface, *Am. J. Anat.,* 174, 285, 1985.
56. **Yanagimachi, R.,** Mechanisms of fertilization in mammals, in *Fertilization and Embryonic Development In Vitro,* Mastroianni, L. and Biggers, J. D., Eds., Plenum Press, New York, 1981, 81.
57. **Sato, K. and Blandau, R. J.,** Time and process of sperm penetration into cumulus-free mouse eggs fertilized in vitro, *Gamete Res.,* 2, 295, 1979.
58. **Sato, K. and Blandau, R. J.,** Second meiotic division and polar body formation in mouse eggs fertilized in vitro, *Gamete Res.,* 2, 283, 1979.
59. **Rogers, B. J.,** Mammalian sperm capacitation and fertilization in vitro: a critique of methodology, *Gamete Res.,* 1, 165, 1978.
60. **Bavister, B. D.,** Analysis of culture media for in vitro fertilization and criteria for success, in *Fertilization and Embryonic Development In Vitro,* Mastroianni, L. and Biggers, J. D., Eds., Plenum Press, New York, 1981, 41.
61. **Myles, D. G., Hyatt, H., and Primakoff, P.,** Binding of both acrosome-intact and acrosome-reacted guinea pig sperm to the zona pellucida during in vitro fertilization, *Dev. Biol.,* 121, 559, 1987.
62. **Shalgi, R. and Phillips, D. M.,** Motility of rat spermatozoa at the site of fertilization, *Biol. Reprod.,* 39, 1207, 1988.
63. **Suarez, S. S., Katz, D. F., and Meizel, S.,** Changes in motility that accompany the acrosome reaction in hyperactivated hamster spermatozoa, *Gamete Res.,* 10, 253, 1984.
64. **Katz, D. F., Cherr, G. N., and Lambert, H.,** The evolution of hamster sperm motility during capacitation and interaction with the ovum vestments in vitro, *Gamete Res.,* 14, 333, 1986.
65. **Suarez, S. S.,** Hamster sperm motility transformations during epididymal maturation and the development of hyperactivation in vitro, *Gamete Res.,* 19, 51, 1988.
66. **Cherr, G. N. and Cross, N. L.,** Immobilizaiton of mammalian eggs on solid substrates by lectins for electron microscopy, *J. Microsc.,* 10, 122, 1987.
67. **Burkman, L. J., Overstreet, J. W., and Katz, D. F.,** A possible role for potassium and pyruvate in the modulation of sperm motility in the rabbit oviductal isthmus, *J. Reprod. Fertil.,* 71, 367, 1984.
68. **Burkman, L. J.,** A microperfusion chamber for study of mammalian spermatozoa, *J. Androl.,* 9, 102, 1989.
69. **Katz, D. F. and Yanagimachi, R.,** Movement characteristics of hamster and guinea pig spermatozoa upon attachment to the zona pellucida, *Biol. Reprod.,* 25, 785, 1981.
70. **Mack, S. O., Wolf, D. P., and Tash, J. S.,** Quantitation of specific parameters of motility in large numbers of human sperm by digital image processing, *Biol. Reprod.,* 38, 270, 1988.
71. **Tessler, S. and Olds-Clark, P.,** Linear and non-linear mouse sperm motility patterns: a quantitative classification, *J. Androl.,* 6, 35, 1985.
72. **Olds-Clark, P.,** Motility characteristics of sperm from the uterus and oviducts of female mice after mating to congenic males differing in sperm transport and fertility, *Biol. Reprod.,* 34, 453, 1986.
73. **Cooper, T. G.,** Stroboscopic illumination for the assessment of hyperactivated motility of mouse spermatozoa, *J. Exp. Zool.,* 223, 291, 1982.
74. **Cooper, T. G.,** The onset and maintenance of hyperactivated motility of spermatozoa from the mouse, *Gamete Res.,* 9, 55, 1984.
75. **Austin, C. R.,** Observations on the penetration of the sperm into the mammalian egg, *Aust. J. Sci. Res. Ser. B,* 4, 581, 1951.
76. **Austin, C. R. and Walton, A.,** Fertilization, in *Marshall's Physiology of Reproduction,* Vol. 2 (Part 2), 3rd ed., Parkes, A. S., Ed., Little, Brown, Boston, 1960, 310.
77. **Katz, D. F. and Phillips, D. M.,** The response of rhesus monkey sperm movement to cervical mucus and solid surfaces, *Gamete Res.,* 13, 231, 1986.
78. **Mortimer, D., Serres, C., Mortimer, S. T., and Jouannet, P.,** Influence of image sampling frequency on the perceived movement characteristics of progressively motile human spermatozoa, *Gamete Res.,* 20, 313, 1988.
79. **Stephens, D. T., Vijayaraghavan, S., and Hoskins, S. S.,** Computerized flagellar analysis of the effects of the intracellular Ca^{2+} binding compounds fura 2, quin 2 and bapta on bovine sperm flagellar waveform, *Biol. Reprod.,* 38(Suppl. 1), 89, 1988.
80. **Moore, H. D. M. and Bedford, J. M.,** The interaction of mammalian gametes in the female, in *Mechanism and Control of Animal Fertilization,* Hartmann, J. R., Ed., Academic Press, New York, 1983, 453.

81. **Woolley, D. M.,** Evidence for "twisted plane" undulations in golden hamster sperm tails, *J. Cell Biol.*, 75, 851, 1977.
82. **Mohri, H., Ishijima, S., and Hiramoto, Y.,** Movement characteristics of hamster spermatozoa, in *The Sperm Cell,* Andre, J., Ed., Martinus Nijhoff, The Hague, 1983, 345.
83. **Ishijima, S. and Mohri, H.,** A quantitative description of flagellar movement in golden hamster spermatozoa, *J. Exp. Biol.*, 114, 463, 1985.
84. **Ishijima, S., Oshio, S., and Mohri, H.,** Flagellar movement of human spermatozoa, *Gamete Res.*, 13, 185, 1986.
85. **Cummins, J. M. and Jamieson, B. G. M.,** On axes and symmetry in spermatozoa, *Dev. Growth Differ.*, Suppl., 166, 1986.
86. **Mohri, H. and Yano, Y.,** Analysis of mechanism of flagellar movement with golden hamster spermatozoa, *Biomed. Res.*, 1, 552, 1980.
87. **Yanagimachi, R. and Chang, M. C.,** Fertilization of hamster eggs in vitro, *Nature,* 200, 281, 1963.
88. **Barros, C. and Austin, C. R.,** In vitro fertilization and the sperm acrosome reaction in the hamster, *Anat. Rec.*, 157, 208, 1967.
89. **Ohzu, E. and Yanagimachi, R.,** Accleration of acrosome reaction in hamster spermatozoa by lysolecithin, *J. Exp. Zool.*, 224, 259, 1982.
90. **Bavister, B. D.,** A consistently successful procedure for in vitro fertilization of golden hamster eggs, *Gamete Res.*, 23, 139, 1989.
91. **Yanagimachi, R., Kamiguchi, Y., Sagawara, S., and Mikamo, K.,** Gametes and fertilization in the Chinese hamster, *Gamete Res.*, 8, 97, 1983.
92. **Toyoda, Y. and Chang, M. C.,** Fertilization of rat eggs in vitro by epididymal spermatozoa and the development of eggs following transfer, *J. Reprod. Fertil.*, 36, 9, 1974.
93. **Huang, T. T. F., Flemming, A. D., and Yanagimachi, R.,** Only acrosome-reacted spermatozoa can bind and penetrate into zona pellucida: a study using guinea pig, *J. Exp. Zool.*, 217, 286, 1981.
94. **Flemming, A. D. and Yanagimachi, R.,** Fertile life of acrosome-reacted guinea pig spermatozoa, *J. Exp. Zool.*, 220, 109, 1982.
95. **Morales, P., Cross, N. L., Overstreet, J. W., and Hanson, F. W.,** Acrosome intact and acrosome-reacted human sperm can initiate binding to the zona pellucida, *Dev. Biol.*, 133, 385, 1989.
96. **Yanagimachi, R. and Usui, N.,** Calcium dependence of the acrosome reaction and activation of guinea pig spermatozoa, *Exp. Cell Res.*, 89, 161, 1974.
97. **Yanagimachi, R.,** Fertilization of guinea pig eggs in vitro, *Anat. Rec.*, 174, 9, 1972.
98. **Flemming, A. D. and Yanagimachi, R.,** Effect of various lipids on the acrosome reaction and fertilizing capacity of guinea pig spermatozoa with special reference to the possible involvement of lysophospholipids in the acrosome reaction, *Gamete Res.*, 4, 253, 1981.
99. **Kuzan, F., Flemming, A. D., and Siedel, G.,** Successful fertilization in vitro of fresh intact oocytes by perivitelline (acrosome reacted) spermatozoa in the rabbit, *Fertil. Steril.*, 41, 766, 1984.
100. **Brackett, B. G. and Oliphant, G.,** Capacitation of rabbit spermatozoa in vitro, *Biol. Reprod.*, 12, 260, 1975.
101. **Kim, C. K., Im, K. S., Zheng, X., and Foote, R. H.,** In vitro capacitation and fertilizing ability of ejaculated rabbit sperm treated with lysophosphatidylcholine, *Gamete Res.*, 22, 131, 1989.
102. **Wolf, D. P.,** The mammalian egg's block to polyspermy, in *Fertilization and Embryonic Development In Vitro,* Mastroianni, L. and Biggers, J. D., Eds., Plenum Press, New York, 1981, 183.
103. **Overstreet, J. W.,** The use of the human zona pellucida in diagnostic tests of sperm fertilizing capacity, in *In Vitro Fertilization and Embryo Transfer,* Crosignani, P. G. and Rubin, B. L., Eds., Academic Press, London, 1983, 145.
104. **Overstreet, J. W. and Cross, N. L.,** The biology of human sperm-zona pellucida interaction, *Ann. N.Y. Acad. Sci.*, 541, 337, 1988.
105. **Drobnis, E. Z., Yudin, A. I., and Katz, D. F.,** Characteristic motility of human and hamster spermatozoa during penetration of the hamster cumulus, *Fertil. Steril.*, Abstr. P-240, 99, 1987.
106. **Katz, D. F., Yudin, A. I., Drobnis, E. Z., Hanson, F. W., Cragun, J. R., and Chang, R. J.,** Human sperm penetration of cervical mucus and the hamster cumulus in relation to capacitation, *Abstr. Soc. Gynecol. Invest.*, p. 139, 1988.
107. **Boatman, D. E. and Bavister, B. D.,** Stimulation of rhesus monkey sperm capacitation by cyclic nucleotide mediators, *J. Reprod. Fertil.*, 71, 357, 1984.
108. **Wolf, D. P., VandeVoort, C. A., Meyer-Haas, G. R., Zelinski-Wooten, M. B., Hess, D. L., Baughman, W. L., and Stouffer, R. L.,** In vitro fertilization and embryo transfer in the rhesus monkey, *Biol. Reprod.*, 41, 335, 1989.
109. **Katz, D. F. and Drobnis, E. Z.,** Analysis and interpretation of the forces generated by spermatozoa, in *Proc. Serono Symp. on Fertilization in Mammals,* Bavister, B. D., Ed., Serono Symposia, USA, Norwell, MS, in press.

110. **Gwatkin, R. B. L. and Anderson, O. F.,** Capacitation of hamster spermatozoa by bovine follicular fluid, *Nature,* 224, 1111, 1969.
111. **Yanagimachi, R.,** In vitro acrosome reaction and capacitation of golden hamster spermatozoa by bovine follicular fluid and its fraction, *J. Exp. Zool.,* 170, 269, 1969.
112. **Yanagimachi, R.,** The movement of golden hamster spermatozoa before and after capacitation, *J. Reprod. Fertil.,* 23, 193, 1970.
113. **Katz,, D. F., Yanagimachi, R., and Dresdner, R. D.,** Movement characteristics and power output of guinea-pig and hamster spermatozoa in relation to activation, *J. Reprod. Fertil.,* 52, 167, 1978.
114. **Fraser, L. R.,** Motility patterns in mouse spermatozoa before and after capacitation, *J. Exp. Zool.,* 202, 429, 1977.
115. **Fraser, L. R.,** Dibutyryl cyclic AMP decreases capacitation time in vitro in mouse spermatozoa, *J. Reprod. Fertil.,* 62, 63, 1981.
116. **Fraser, L. R.,** Ca^{2+} is required for mouse sperm capacitation and fertilization in vitro, *J. Androl.,* 3, 412, 1982.
117. **Fraser, L. R.,** Potassium ions modulate expression of mouse sperm fertilizing ability, acrosome reaction and hyperactivated motility in vitro, *J. Reprod. Fertil.,* 69, 539, 1983.
118. **Aonuma, S., Okabe, M., Kawaguchi, M., and Kishi, Y.,** Studies on sperm capacitation. IX. Movement characteristics of spermatozoa in relation to capacitation, *Chem. Pharm. Bull.,* 28, 1497, 1980.
119. **Tessler, S., Carey, J. E., and Olds-Clark, P.,** Mouse sperm motility affected by factors in the T/t complex, *J. Exp. Zool.,* 217, 277, 1981.
120. **Olds-Clark, P.,** Motility characteristics of sperm from the uterus and oviducts of female mice after mating to congenic males differing in sperm transport and fertility, *Biol. Reprod.,* 34, 453, 1986.
121. **Barros, C., Berrios, M., and Herrera, E.,** Capacitation in vitro of guinea-pig spermatozoa in a saline solution, *J. Reprod. Fertil.,* 34, 547, 1973.
122. **Barros, C. and Berrios, M.,** Is the activated spermatozoa really capacitated?, *J. Exp. Zool.,* 201, 65, 1977.
123. **Mahi, C. A. and Yanagimachi, R.,** Maturation and sperm penetration of canine ovarian oocytes in vitro, *J. Exp. Zool.,* 196, 189, 1976.
124. **Flemming, A. D., Yanagimachi, R., and Yanagimachi, H.,** Spermatozoa of the Atlantic bottlenosed dolphin, *Tursiops truncatus, J. Reprod. Fertil.,* 63, 509, 1981.
125. **Lambert, H.,** Temperature dependence of capacitation in bat sperm monitored by zona-free hamster ova, *Gamete Res.,* 4, 525, 1981.
126. **Cummins, J. M.,** Hyperactivated motility patterns of ram spermatozoa recovered from the oviducts of mated ewes, *Gamete Res.,* 6, 53, 1982.
127. **Shams-Borham, G. and Harrison, R. A. P.,** Production, characterization, and use of ionophore-induced, calcium dependent acrosome reaction in ram spermatozoa, *Gamete Res.,* 4, 407, 1981.
128. **Singh, J. P., Babcock, D. F., and Lardy, H. A.,** Motility activation, respiratory stimulation, and alteration of Ca^{2+} transport in bovine sperm treated with amine local anesthetic and calcium transport antagonist, *Arch. Biochem. Biophys.,* 221, 291, 1983.
129. **Nagai, T., Niwa, K., and Iritani, A.,** Effect of sperm concentration during preincubation in a defined medium on fertilization in vitro of pig follicular oocytes, *J. Reprod. Fertil.,* 70, 271, 1984.
130. **Moore, H. D. M.,** An assessment of the fertilizing ability of spermatozoa in the epididymis of the marmoset monkey (*Callithrix jacchus*), *Int. J. Androl.,* 4, 321, 1981.
131. **Young, L. G., Gould, K. G., and Hinton, B. T.,** Changes in binding of a 27-kilodalton chimpanzee cauda epididymal protein glycoprotein component to chimpanzee sperm, *Gamete Res.,* 18, 163, 1987.
132. **Behboodi, E., Katz, D. F., Overstreet, J. W., and Hendrickx, A. G.,** Movement of Cynomolgus and Rhesus monkey spermatozoa collected from the lower female reproductive tract, *Gamete Res.,* 24, 333, 1989.
133. **Aitken, R. J., Best, F. S. M., Richardson, D. W., Djahanbakhch, O., and Lees, M. M.,** The correlates of fertilizing capacity in normal fertile men, *Fertil. Steril.,* 38, 68, 1982.
134. **Mortimer, D., Cortot, A. M., Giovangrandi, Y., and Jeulin, C.,** Do capacitated human spermatozoa show an "activated" pattern of motility?, in *The Sperm Cell,* Andre, J., Ed., Martinus Nijhoff, The Hague, 1983, 349.
135. **Mortimer, D., Coutot, A. M., Giovagrandi, Y., Jeulin, C., and David, G.,** Human sperm motility after migration into, and incubation in, synthetic media, *Gamete Res.,* 9, 131, 1984.
136. **Burkman, L. J.,** Characterization of hyperactivated motility by human spermatozoa during capacitation: comparison of fertile and oligospermic sperm populations, *Arch. Androl.,* 13, 153, 1984.
137. **Morales, P., Overstreet, J. W., and Katz, D. F.,** Changes in human sperm motion during capacitation in vitro, *J. Reprod. Fertil.,* 83, 119, 1988.
138. **Katz, D. F. and Overstreet, J. W.,** Mammalian sperm movement in the secretions of the male and female genital tracts, in *Testicular Development, Structure and Function,* Steinberger, A. and Steinberger, E., Eds., Raven Press, New York, 1980, 481.

139. **Katz, D. F., Drobnis, E. Z., Baltz, J., Cherr, G. N., Yudin, A. I., Cone, R. A., and Cheng, L. Y.,** The biophysics of sperm penetration of the cumulus and zona pellucida, in *New Horizons in Sperm Cell Research,* Mohri, H., Ed., Japan Scientific Societies Press, Tokyo, 1987, 275.
140. **Mack, S. O., Wolf, D. P., and Tash, J. S.,** Quantitation of specific parameters of motility in large numbers of human sperm by digital image processing, *Biol. Reprod.,* 38, 270, 1988.
141. **Mack, S. O., Tash, J. S., and Wolf, D. P.,** Effect of measurement conditions on quantification of hyperactivated human sperm subpopulations by digital image analysis, *Biol. Reprod.,* 40, 1162, 1989.
142. **Fraser, L. R. and Quinn, P. J.,** A glycolytic product is obligatory for initiation of sperm acrosome reaction and whiplash motility required for fertilization in the mouse, *J. Reprod. Fertil.,* 61, 25, 1981.
143. **Yanagimachi, R.,** Requirement of extracellular calcium ions for various stages of fertilization and fertilization related phenomena in the hamster, *Gamete Res.,* 5, 323, 1982.
144. **Yeung, C. H.,** Inhibition of the ATP-induced reactivation of demembranated hamster spermatozoa by the action of free ATP^{4-} and $MgATP^{2-}$, *J. Reprod. Fertil.,* 81, 195, 1987.
145. **Lindemann, C. B., Goltz, J. S., and Kanous, K. S.,** Regulation of activation state and flagellar waveform in epididymal rat sperm: evidence for the involvement of both Ca^{2+} and cAMP, *Cell Motil. Cytoskel.,* 8, 324, 1987.
146. **Neill, J. M. and Olds-Clark, P.,** A computer-assisted assay for mouse sperm hyperactivation demonstrates that bicarbonate but not bovine serum albumin is required, *Gamete Res.,* 18, 121, 1987.
147. **Robertson, L., Wolf, D. P., and Tash, J. S.,** Temporal changes in motility parameters related to acrosomal status: identification and characterization of populations of hyperactivated human sperm, *Biol. Reprod.,* 39, 797, 1988.
148. **Boyers, S. P., Davis, R. O., and Katz, D. F.,** Automated semen analysis, *Curr. Probl. Obstet. Gynecol. Fertil.,* 12, 165, 1989.
149. **Yanagimachi, R. and Noda, Y. D.,** Ultrastructural changes in the hamster sperm head during fertilization, *J. Ultrastruct. Res.,* 31, 465, 1970.
150. **Yudin, A. I., Cherr, G. N., and Katz, D. F.,** Structure of the cumulus matrix and zona pellucida in the golden hamster: a new view of sperm interaction with oocyte-associated extracellular matrices, *Cell Tissue Res.,* 251, 555, 1988.
151. **Saling, P. M., Sowinski, J., and Storey, B. T.,** An ultrastructural study of epididymal mouse spermatozoa binding to zonae pellucidae in vitro: sequential relationship to the acrosome reaction, *J. Exp. Zool.,* 209, 229, 1979.
152. **Saling, P. M. and Storey, B. T.,** Mouse gamete interactions during fertilization in vitro: chlortetracycline as a fluorescent probe for the mouse sperm acrosome reaction, *J. Cell Biol.,* 83, 554, 1979.
153. **Florman, H. M. and Storey, B. T.,** Mouse gamete interactions: the zona pellucida is the site of the acrosome reaction leading to fertilization in vitro, *Dev. Biol.,* 91, 121, 1982.
154. **Wassarman, P. M.,** Zona pellucida glycoproteins, *Annu. Rev. Biochem.,* 57, 415, 1988.
155. **Wassarman, P. M.,** Early events in mammalian fertilization, *Annu. Rev. Cell Biol.,* 3, 109, 1987.
156. **Storey, B. T., Lee, M. A., Muller, C., Ward, C. R., and Wirtshafter, K. G.,** Binding of mouse spermatozoa to the zona pellucida of mouse eggs in cumulus: evidence that the acrosomes remain substantially intact, *Biol. Reprod.,* 30, 1119, 1984.
157. **Cross, N. L. and Meizel, S.,** Methods for evaluating the acrosomal status of mammalian sperm, *Biol. Reprod.,* 41, 635, 1989.
158. **Causnicu, P. S. and Bedford, J. M.,** Sperm entry into zona-free oocytes in the hamster oviduct: implications for the mechanisms of acrosome reaction induction, *Gamete Res.,* 21, 85, 1988.
159. **Lorton, S. F. and First, N. L.,** Hyaluronidase does not disperse the cumulus oophorus surrounding bovine ova, *Biol. Reprod.,* 21, 301, 1979.
160. **Crozet, N.,** Ultrastructural aspects of in vivo fertilization in the cow, *Gamete Res.,* 10, 241, 1984.
161. **Crozet, N. and Dummont, M.,** The site of the acrosome reaction during in vivo penetration of the sheep oocyte, *Gamete Res.,* 10, 97, 1984.
162. **Tesarik, J., Kopecny, V., and Dvorak, M.,** Selective binding of human cumulus cell-secreted glycoproteins to human spermatozoa during capacitation in vitro, *Fertil. Steril.,* 41, 919, 1984.
163. **Bradley, M. P. and Garbers, D. L.,** The stimulation of bovine caudal epididymal sperm forward motility by bovine cumulus-egg complexes in vitro, *Biochem. Biophys. Res. Commun.,* 115, 777, 1983.
164. **Gwatkin, R. B. L., Andersen, O. F., and Hutchinson, C. F.,** Capacitation of hamster spermatozoa in vitro: the role of cumulus components, *J. Reprod. Fertil.,* 30, 389, 1972.
165. **Meizel, S. and Turner, K. O.,** Glycosaminoglycans stimulate the acrosome reaction of previously capacitated hamster sperm, *J. Exp. Zool.,* 237, 137, 1986.
166. **Talbot, P.,** Hyaluronidase dissolves a component of the hamster zona pellucida, *J. Exp. Zool.,* 229, 309, 1984.
167. **Talbot, P., DiCarlantonio, G., Zao, P., Penkala, J., and Haimo, L. T.,** Motile cells which lack hyaluronidase can penetrate the hamster oocyte-cumulus complex, *Dev. Biol.,* 108, 387, 1985.

168. **Bedford, J. M.,** Importance of capacitation for establishing contact between eggs and sperm in the rabbit, *J. Reprod. Fertil.,* 13, 365, 1967.
169. **Hartmann, J. F.,** Mammalian fertilization: gamete surface interactions in vitro, in *Mechanism and Control of Animal Fertilization,* Hartmann, J. F., Ed., Academic Press, New York, 1985, 325.
170. **Wassarman, P. M., Florman, H. M., and Greve, J. M.,** Receptor-mediated sperm-egg interactions in mammals, in *Fertilization,* Vol. 2, Metz, C. B. and Monroy, A., Eds., Academic Press, New York, 1985, 341.
171. **Macek, M. B. and Shur, B. D.,** Protein-carbohydrate complementarity in mammalian gamete recognition, *Gamete Res.,* 20, 93, 1988.
172. **Fraser, L. R. and Ahuja, K. K.,** Metabolic and surface events in fertilization, *Gamete Res.,* 20, 491, 1988.
173. **Baltz, J., Katz, D. F., and Cone, R.,** The mechanics of sperm-egg interaction at the zona pellucida, *Biophys. J.,* 54, 643, 1988.
174. **Hartmann, J. G., Gwatkin, R. B. L., and Hutchinson, C. F.,** Early contact interactions between mammalian gametes in vitro: evidence that the vitellus influences adherence between sperm and zona pellucida, *Proc. Natl. Acad. Sci.,* 69, 2767, 1972.
175. **Green, D. P. L.,** Mammalian sperm cannot penetrate the zona pellucida solely by force, *Exp. Cell Res.,* 169, 31, 1987.
176. **Urch, U. A., Wardrip, N. J., and Hedrick, J. L.,** Limited and specific proteolysis of the zona pellucida by acrosin, *J. Exp. Zool.,* 233, 479, 1985.
177. **Urch, U. A., Wardrip, N. J., and Hedrick, J. L.,** Proteolysis of the zona pellucida by acrosin: the nature of the hydrolysis product, *J. Exp. Zool.,* 236, 239, 1985.
178. **Green, D. P. L.,** Sperm thrusts and the problem of penetration, *Biol. Rev.,* 63, 79, 1988.
179. **Katz, D. F. and deMestre, N. J.,** Hydrodynamic thrust of an undulating slender body against a no-slip boundary at low Reynolds number: a model of sperm thrust during fertilization, Research Report CMA-R13-86, Centre for Mathematical Analysis, Australian National University, Canberra, 1985.
180. **Green, D. P. L. and Purves, R. D.,** Mechanical hypothesis of sperm penetration, *Biophys. J.,* 45, 659, 1984.
181. **Drobnis, E. Z., Yudin, A. I., Cherr, G. N., and Katz, D. F.,** Sperm penetration of cumulus and zona pellucida: role of non-propagated, high-curvature flagellar bends, *J. Androl.,* 9, 34-P, 1988.
182. **Amann, R. P.,** Can the fertility potential of a seminal sample be predicted accurately, *J. Androl.,* 10, 89, 1989.
183. **Chan, S. Y. W., Wang, C., Ng, M., Tam, G., Lo, T., Tsoi, W. L., Nie, G., and Leung, J.,** Evaluation of computerized analysis of sperm movement characteristics and differential sperm tail swelling patterns in predicting human sperm in vitro fertilizing capacity, *J. Androl.,* 10, 133, 1989.
184. **Yudin, A. I. and Drobnis, E. Z.,** unpublished observations.
185. **Drobnis, E. Z. and Davis, R. O.,** unpublished observations.
186. **Drobnis, E. Z. and VandeVoort, C. A.,** unpublished observations.

INDEX

A

B

C

G

H

I

J

L

M

Q

R

S

BOX N1+1+1+1+

B00/CS 24+26+15+14+